Problems in Quantum Mechanics and Field Theory with Mathematical Modelling

In *Problems in Quantum Mechanics and Field Theory with Mathematical Modelling*, a number of exactly solvable problems in electrodynamics and in quantum-mechanics of particles with different spins are presented.

The main topics covered include: the Cox scalar particle with intrinsic structure in presence of the magnetic field in the spaces of constant curvature, Euclid, Riemann, and Lobachevsky; Cox particle in the Coulomb field; tunneling effect through Schwarzschild barrier for a spin 1/2 particle; electromagnetic field in Schwarzschild space-time, the Majorana - Oppenheimer approach in electrodynamics; scalar particle with polarizability in the Coulomb field; Dirac particle in the Coulomb field on the background of hyperbolic Lobachevsky and spherical Riemann models; particle with spin 1 in the Coulomb field; geometrical modeling of the media in Maxwell electrodynamics; P-asymmetric equation for a spin 1/2 particle; fermion with two mass parameters in the Coulomb field; helicity operator for a spin 2 particle in presence of the magnetic field.

The book will be of interest to researchers, and is accessible enough to serve as a self-study resources for courses at undergraduate and graduate levels.

Aleksander V. Chichurin is a Professor at The John Paul II Catholic University of Lublin, Poland. His research interests currently span differential equations, mathematical modelling, computer algebra systems.

Elena M. Ovsiyuk is currently head of Department of Theoretical Physics and Applied Informatics at Mozyr State Pedagogical University in Belarus. Her current research areas include elementary particles with spin in external electromagnetic and gravitational fields.

Viktor M. Red'kov is Chief Researcher for the Institute of Physics at the National Academy of Sciences of Belarus. His primary fields of research include elementary particles with spin in external electromagnetic and gravitational fields.

Problems in Quantum Mechanics and Field Theory with Mathematical Modelling

Edited by
Aleksander V. Chichurin, Elena M. Ovsiyuk,
and Viktor M. Red'kov

CRC Press
Taylor & Francis Group
Boca Raton London New York

CRC Press is an imprint of the
Taylor & Francis Group, an **informa** business

Designed cover image: Shutterstock_386127955

First edition published 2025
by CRC Press
2385 NW Executive Center Drive, Suite 320, Boca Raton FL 33431

and by CRC Press
4 Park Square, Milton Park, Abingdon, Oxon, OX14 4RN

CRC Press is an imprint of Taylor & Francis Group, LLC

ISBN: 978-1-032-75097-2 (hbk)
ISBN: 978-1-032-75100-9 (pbk)
ISBN: 978-1-003-47237-7 (ebk)

DOI: 10.1201/9781003472377

Typeset in CMR10
by KnowledgeWorks Global Ltd.

Publisher's note: This book has been prepared from camera-ready copy provided by the authors.

Contents

Foreword

The book is devoted to investigating the particles with spins in external fields and non-Euclidean space-time backgrounds. The key problems are:

Extended Cox scalar particle in the external magnetic field in the spaces of constant curvature: Euclid, Riemann, Lobachevsky; the Cox particle in the Coulomb field; the tunnelling effect through the Schwarzschild barrier for a spin 1/2 particle; electromagnetic field in Schwarzschild space-time, matrix Duffin – Kemmer and Majorana – Oppenheimer approaches; the spin zero particle with polarisability in the Coulomb field; the Dirac particle in the Coulomb field on the background of hyperbolic Lobachevsky and spherical Riemann models; particles with spin 1 in the Coulomb field; geometrical modelling of the media in Maxwell electrodynamics; P-asymmetric equation for a spin 1/2 particle in external fields; fermion with two mass parameters in the Coulomb field; the theory of a fermion with three mass parameters and the geometric modelling of the neutrino oscillations. In the book there will be presented a number of exact solutions for new problems in electrodynamics, in quantum-mechanics of particles with different spins in presence of external electromagnetic fields and curved space-time background. The book will provide the readers with many technical tools for treating the physical problems and a number of new physical ideas.

The book may be interesting for researchers; it may well serve as a pedagogical tool for either self study or in courses at both the undergraduate and graduate levels. Bibliographies complete many chapters, and an index covers the entire book.

Preface

The book is devoted to investigating the particles with spins in external fields and non-Euclidean space-time backgrounds.

Chapter 1. An extended Cox scalar particle in the external uniform magnetic field on the background of the Lobachevsky space. Generalised Schrödinger equation for a scalar Cox particle is studied in the presence of a magnetic field on the background of a 3-dimensional Lobachevsky space. Separation of the variables is performed. An equation describing motion along the axis z turns out to be much more complicated than that when describing the Cox particle in Minkowski space. The form of the effective potential curve says that we have a quantum-mechanical problem of tunnelling type. The derived equation has six regular singular points. To the physical domains $z = \pm\infty$, there correspond the singular points 0 and 1 of the derived equation. Solution of the equation are constructed with the help of power series. Convergence of the series is examined by Poincaré–Perron method. These series are convergent in the whole physical domain z belongs $(-\infty, +\infty)$. Visualisation of constructed solutions and numerical study of the tunnelling effect are performed.

Chapter 2. An extended Cox scalar particle in the external uniform magnetic field on the background of the spherical Riemann space. Generalised Schrödinger equation for a spin zero particle with intrinsic structure by Darwin–Cox is studied in the presence of a magnetic field on the background of 3-dimensional spherical Riemann space. The separation of the variables is done. An equation describing the motion of the particle along the axis z is studied. The form of the effective potential indicates that we have a quantum-mechanical problem with the complicated box-type potential. Frobenius solutions of the equation are constructed. The convergence of the relevant series is proved by Poincaré–Perron method. These series are convergent in the all physical domain of the variable z, which belongs to $[-\pi/2, +\pi/2]$. Due to the compactness of the spherical space, the existence of discrete energy levels is assumed; however, any exact quantization rule is not known. An approximate method for producing the discrete spectrum for energy is developed, it is based on the use of polynomials instead of power series involved in exact Frobenius solutions. Numerical study and visualisation of constructed solutions are performed.

Chapter 3. Quantum-mechanical scalar particle with the Cox structure in the Coulomb field. Generalised Klein–Fock–Gordon equation for a scalar particle with the Darwin–Cox structure, which takes into account the distribution of the electric charge of the particle inside a finite spherical region is studied in the presence of the external Coulomb field. There are constructed exact Frobenius-type solutions of the derived equations, and the convergence of the relevant power series with 8-term recurrent relations is studied. As the analytical quantization rule takes so-called transcendency conditions. It provides us with a 4th-order algebraic equation with respect to energy values, which has four sets of roots. One set of roots, $0 < E_{nj} < 1$, depending on the angular momentum $j = 0, 1, 2, ...$ and the main quantum number $n = 0, 1, 2, ...$, may be interpreted as corresponding to some bound states of the particle in the Coulomb field. In the same manner, a generalised nonrelativistic Schrödinger equation for such a particle is studied, and the final results are similar.

Chapter 4. The tunnelling effect through the Schwarzschild barrier for a spin 1/2 particle. For massless Dirac particle, the general mathematical and numerical study of the tunnelling process through the effective potential barrier generated by Schwarzschild black hole geometry is done. The study will be based on the use of eight Frobenius solutions of related 2nd-order differential equations with nonregular singularities of rank 2. We construct these solutions in explicit form and prove that the power series involved in them are converged in all physical regions of the physical region of the variable r, which belongs to $(1, +\infty)$. Results for the tunnelling effect significantly differ for two situations: one when the particle falls on the barrier from within and another when the particle falls from outside. The mathematical structure of the derived asymptotic relations is exact; however, analytical expressions for involved convergent powers series are not known, and further study is based on numerical summing of the series. The calculations are implemented using the Mathematica system.

Chapter 5. On Maxwell equations in Schwarzschild space-time. It is shown that the generally covariant extended method of Riemann–Silberstein–Majorana–Oppenheimer in electrodynamics, specified in Schwarzschild metrics, after separating the variables provides us with the possibility to reduce the problem to a differential equation similar to that arising in the case of a scalar field in the Schwarzschild space-time. This differential equation is recognised as a confluent Heun equation. We have considered the electromagnetic field on the basis of the 10-dimensional Duffin–Kemmer approach, when in addition to six components of the strength tensor one uses four components of an electromagnetic potential. After the separation of the variable, we have arrived at a system of ten radial equations, which were simplified by the use of additional constraints followed by an eigenvalue equation for spatial parity operator $\hat{\Pi}\Psi = P\Psi$; the radial system has been divided into two subsystems of four and six equations, respectively. In this second approach, the problem of electromagnetic field has been reduced to the confluent Hein differential equation as well. In particular, we have shown explicitly how solutions found in complex form are embedded into matrix 10-dimensional formalism; besides, we determine radial functions that are responsible for gauge degrees of freedom.

Chapter 6. Quantum mechanical spin-zero particles with polarisability in the Coulomb field, analytical, and numerical consideration. Methods for solving the differential equation describing the wave functions of a polarisable particle in the Coulomb potential are discussed. Relations between the coefficients under which the general solutions of this equation can be found in analytical form are detailed. For the case of zero polarisability, the general solution to this equation in terms of special functions is obtained; for the first values of the parameter j, plots of the corresponding solutions are presented. For nonzero polarisability and certain specially chosen values of the parameters, solutions possessing the required physical properties are constructed with the use of numerical methods and functional objects of the type Differential Root. Instructions in Mathematica are presented which permit to apply elaborated methods in studying other problems in physics and mathematics.

Chapter 7. Dirac particle in the Coulomb field on the background of hyperbolic Lobachevsky and spherical Riemann models. The known systems of radial equations describing relativistic hydrogen atom on the basis of Dirac equation in spherical Riemann spaces are investigated. The relevant 2nd-order differential equations have six regular singular points, there solutions of Frobenius type are constructed. To produce the quantisation rule for energy values, we use the known condition separating transcendental Frobenius solutions. This provides us with energy spectra which are physically interpretable and similar to those arising from the scalar Klein–Fock–Gordon equation in these geometrical models. The spectra coincide with those previously found when studying the same radial equations within the semi-classical method. The convergence of the series involved is proved analytically and numerically. The squared integrability of solutions is demonstrated numerically. Visualisation of the results is given.

Chapter 8. Particles with spin 1 in the Coulomb field, exact wave functions, and the energy spectra. We have studied the system of six equations which describe the quantum states of a spin 1 particle with parity $P = (-1)^j$ in an external Coulomb field. It is shown that, due to the Lorentz condition, one of the radial functions must be equal to zero. Any of the five remaining functions may be taken as a primary one. For such a primary function, we derive two different 2nd-order differential equations. Their Frobenius solutions are constructed, and the convergence of the involved power series is studied. As a quantisation rule, we apply so called transcendency condition to Frobenius solutions. In this way, for both equations, we have found different reasonable, from physical point of view, energy spectra.

Chapter 9. Geometrical modelling of the media in Maxwell electrodynamics. It is known that vacuum Maxwell equations are being considered in the background of any pseudo-Riemannian space-time may be interpreted as Maxwell equations in Minkowski space but specified in some effective medium, which constitutive relations are determined by metric of curved space-time. In that context, we will consider space-time models with event horizon. All of them have a metric of one the same structure, we restrict ourselves to spherically symmetric case, and consider the de Sitter, anti de Sitter, and Schwarzschild models. Also we will study hyperbolic Lobachevsky and spherical Riemann models, parameterised coordinates with spherical and cylindric symmetry. We will construct these solutions explicitly, applying Maxwell equations in spinor form.

Chapter 10. The P-asymmetric equation for a spin 1 particle in external fields. Within the theory of relativistic wave equations with extended sets of Lorentz group representations, a new P-noninvariant 20-component wave equation for the spin 1/2 particle is proposed. The presence of an external electromagnetic field and a Riemannian space-time background have been taken into account. Due to internal structure of the particle, additional interaction terms appear, it relates to anomalous magnetic moment of the particle. Exact solutions of the equation in presence of the external Coulomb field have been constructed, radial wave functions are expressed in terms of confluent Heun functions.

Chapter 11. Fermion with two mass parameters in the Coulomb field, relativistic and non-relativistic theories. Generalised wave equation for a spin 1/2 particle with two mass parameters is studied in the presence of an external Coulomb field. After separating the variables the problem reduces the system eight differential equations of the 1st-order. Taking into account the diagonalisation of the space reflection operator, we derive two independent systems of four equations, referring to states of opposite parity. When considering these equations on a large distance from the centre, they take the form of two subsystems for two ordinary Dirac particles in an external Coulomb field, with masses of M_1 and M_2, respectively. To simplify the problem, we perform a transition to the nonrelativistic description of the system. In this way, we derive two systems of linked 2nd-order equations, referring to states with different parities. They lead to 4th-order differential equations for separate functions. Their solutions of the Frobenius type have been constructed; they involve power series with 10-term recurrent relations. Two solutions are appropriate to describe bound states. As a quantisation rule, we apply the known transcendency condition; in this way, we derive two analytical formulas for energy spectra. They are similar to nonrelativistic spectra for ordinary spin 1/2 particles, but are governed by masses M_1 and M_2. Results on constructing solutions end obtaining the energy spectra are extendable to relativistic theory as well.

Chapter 12. On modelling neutrinos oscillations by geometry methods in the frames of the theory for a fermion with three mass parameters. In this chapter, starting from the general Gel'fand–Yaglom approach a new wave equation for spin 1/2 fermion, which is characterised by three mass parameters, is derived. On the basis of the 20-component wave function, three auxiliary bispinors are introduced. In absence of an external field, these bispinors obey three separate Dirac-like equations with different masses M_1, M_2, and M_3. It is shown that in the presence of external fields, electromagnetic fields, or gravitational non-Euclidean backgrounds with non-vanishing Ricci scalar curvature, the main equation is not split into separated equations; instead, a quite definite mixing of three Dirac-like equations arises. It is shown that a generalised equation for a Majorana particle with three mass parameters exists as well; such a generalised Majorana equation is not split into three separate equations in a curved background if the Ricci scalar of the space-time model does not vanish.

Chapter 13. The eigenvalue problem for the helicity operator for a spin 2 particle in the presence of an external magnetic field. The explicit form of the helicity operator for a symmetric 2nd-rank tensor describing the spin 2 particle is specified in cylindrical coordinates. After separating the variables the system of 10 differential 1st-order equations is derived. It is split into two independent subsystems of four and six equations. The system of four equations is solved straightforwardly in terms of confluent hypergeometric functions; there are corresponding eigenvalues and eigenfunctions. A subsystem of six equations can be reduced to one ordinary differential equation of the 4th-order. The corresponding 4th-order operator is factorised into permutable 2nd-order operators, so the problem reduces to solving two differential equations of the 2nd-order. Their solutions are constructed in terms of Bessel functions. This analysis is extended to the presence of an external uniform magnetic field when solutions are constructed in terms of confluent hypergeometric functions.

Writing a handbook is a long process. It might not have been possible for us without the support and encouragement of the people whose names we want to mention here.

We are thankful to our colleges in the Institute of Mathematics, Informatics, and Landscape Architecture, The John Paul II Catholic University of Lublin, the Department "Fundamental Interactions and Astrophysics" of the Institute of Physics of the National Academy of Sciences of Belarus, and Mozyr State Pedagogical University for helpful discussions and useful comments.

We would like to thank especially our colleagues, Rev. Profesor Jacek Łapinski, and Dr. Marcin Płonkowski from the John Paul II Catholic University of Lublin.

Also we wish to thank our friends, Dr. G.G. Krylov and Dr. G.V. Grushevskaya, from Belarus State University, and Prof. V.A. Pletyukhov from Brest State University. Their help, stimulating suggestions, critics, interest, and valuable hints helped us in writing this book.

Author

Prof., Dr Sci. **Aleksander V. Chichurin**
Institute of Mathematics, Informatics, and Landscape Architecture
Faculty of Natural and Technical Sciences
The John Paul II Catholic University of Lublin
ul. Konstantynów 1 H, Lublin 20-708, Poland
E-mail: achichurin@kul.pl

Dr Sci. **Elena M. Ovsiyuk**
Department of Theoretical Physics and Applied Informatics
Mozyr State Pedagogical University named after I.P. Shamyakin
28/1 Studencheskaya Str. Mozyr 247760, Belarus
E-mail: e.ovsiyuk@mail.ru

Prof., Dr Sci. **Viktor M. Red'kov**
Department of Fundamental Physics and Astrophysics
National Academy of Sciences of Belarus
68 Nezaleznosti Avenue, Minsk 220072 Belarus
E-mail: v.redkov@ifanbel.bas-net.by

1

Cox scalar particle in the magnetic field in the Lobachevsky space

Generalised Schrödinger equation for scalar Cox particle is studied in presence of magnetic field on the background of 3-dimensional Lobachevsky space. Separation of the variables is performed. An equation describing motion along the axis z turns out to be much more complicated than that when describing the Cox particle in Minkowski space.

The form of the effective potential curve says that we have a quantum-mechanical problem of tunnelling type. The derived equation has six regular singular points. To physical domains $z = \pm\infty$, there correspond the singular points 0 and 1 of the derived equation. Solution of the equation are constructed with the help of power series. Convergence of the series is examined by Poincaré–Perron method. These series are convergent in the whole physical domain $z \in (-\infty, +\infty)$. Visualisation of constructed solutions and numerical study of the tunnelling effect are performed.

The chapter is based on [1–14].

1.1 Introduction

In the frames of the theory of generalised relativistic wave equations, a special model for scalar particles was proposed by Cox [1]. Updated treatment of this theory has been seen in a recent book [2]. Such wave equations, being constructed on the basis of extended sets of representations of the Lorentz group, in the presence of external electromagnetic fields, describe after excluding additional components particles which interact non-minimally and in various ways with the electromagnetic field through electromagnetic tensor. Such additional interaction terms are referred to intrinsic electromagnetic structure of the particle. In particular, the Cox electromagnetic structure may be associated with the Darvin interaction term in nonrelativistic Schrödinger equation, this additional interaction is related to the non-point-like distribution of the electric charge in the finite volume inside the particle (see, for instance, in the book [3]. In recent papers [4, 5], the behaviour of such a particle in external magnetic and electric fields, and in spaces with non-Euclidean geometry was studied. In particular, a generalised Schrödinger wave equation for the Cox particle was derived.

In this chapter we examine the Cox particle in external magnetic field on the background of 3-dimensional Lobachevsky space. Influence of the curved space model becomes very significant at large distances. The problem reduces to a rather complex system of differential equations in two variables. The main attention is paid to studying the equation describing the motion of the particle along the axis z. Here we are to examine the quantum tunnelling effect through an effective barrier.

DOI: 10.1201/9781003472377-1

In the special system of cylindric coordinates in the Lobachevsky space, analogue of the uniform magnetic field is determined by the relations [6] (we use dimensionless coordinates):

$$dS^2 = c^2dt^2 - \text{ch}^2 z(dr^2 + \text{sh}^2 r d\phi^2) + dz^2,$$
$$\sqrt{-g} = \rho^3 \text{ sh } r \text{ ch}^2 z, \quad A_\phi = -B\rho^2(\text{ch } r - 1), \; F_{r\phi} = -B\rho \text{ sh } r,$$
$$B_3 = -B\rho \text{ sh } r, \quad B^3 = -\frac{B}{\rho \text{ sh } r \text{ ch}^4 z}, \quad B_i B^i = B^2 \text{ch}^{-4} z. \tag{1.1}$$

We start with the known form of the generalised Schrödinger equation for a Cox scalar particle

$$D_t \Psi = \frac{1}{2M\rho^2}\Big[\overset{\circ}{D}_1 \overset{*}{D}_1 + \overset{\circ}{D}_2 \frac{1}{\text{sh}^2 r} \overset{*}{D}_2 + \overset{\circ}{D}_3 \overset{*}{D}_3\Big]\Psi,$$

where

$$D_1 = i\hbar\partial_r, \quad D_2 = i\hbar\partial_\phi + \frac{e}{c}B\rho^2(\text{ch } r - 1), \quad D_3 = i\hbar\partial_z,$$

$$\overset{\circ}{D}_1 = i\hbar(\partial_r + \frac{\text{ch } r}{\text{sh } r}), \quad \overset{\circ}{D}_2 = i\hbar\partial_\phi + \frac{e}{c}B\rho^2(\text{ch } r - 1), \quad \overset{\circ}{D}_3 = i\hbar(\partial_z + 2\frac{\text{sh } z}{\text{ch } z}),$$

$$\overset{*}{D}_1 = \frac{1}{1 + \Gamma^2 B^2 \text{ch}^{-4} z}\Big[i\hbar\partial_r - \frac{\Gamma B \text{ch}^{-2} z}{\text{sh} r}(i\hbar\partial_\phi + \frac{e}{c}B\rho^2(\text{ch } r - 1))\Big],$$

$$\overset{*}{D}_2 = \frac{1}{1 + \Gamma^2 B^2 \text{ch}^{-4} z}\Big[(i\hbar\partial_\phi + \frac{e}{c}B\rho^2(\text{ch } r - 1)) + i\hbar\Gamma B \text{ch}^{-2} z \text{ sh } r\partial_r\Big],$$

$$\overset{*}{D}_3 = \frac{(D_3 + \Gamma^2 B^3 \; B_3 D_3)}{1 + \Gamma^2 B^2 \text{ch}^{-4} z} = i\hbar\partial_z.$$

Below we use notations $B\rho^2/\hbar c = bt$, $\Gamma B \text{ch}^{-2} z = \gamma(z)$. With the use of the relations

$$\frac{1}{2M\rho^2} \overset{\circ}{D}_1 g^{11} \overset{*}{D}_1 = -\frac{\hbar^2 \text{ch}^{-2} z}{2M\rho^2(1 + \gamma^2(z))}$$
$$\times\Big(\partial_r^2 + (\frac{\text{ch } r}{\text{sh } r} + i\gamma(z)b\frac{\text{ch } r - 1}{\text{sh } r})\partial_r - \frac{\gamma(z)}{\text{sh } r}\partial_r\partial_\phi + i\gamma(z)b\Big),$$
$$\frac{1}{2M\rho^2} \overset{\circ}{D}_2 g^{22} \overset{*}{D}_2 = -\frac{\hbar^2 \text{ch}^{-2} z}{2M\rho^2(1 + \gamma^2(z))}$$
$$\times\Big[\frac{1}{\text{sh}^2 r}[\partial_\phi - ib(\text{ch } r - 1)]^2 + \gamma(z)[\partial_\phi - ib(\text{ch } r - 1)] \frac{1}{\text{sh } r}\partial_r\Big],$$
$$\frac{1}{2M\rho^2} \overset{\circ}{D}_3 g^{33} \overset{*}{D}_3 = -\frac{\hbar^2}{2M\rho^2}(\partial_z + 2\frac{\text{sh } z}{\text{ch } z})\partial_z.$$

and of the substitution for wave function

$$\Psi = e^{-iEt/\hbar} e^{im\phi} Z(z) R(r), \quad \epsilon = \frac{E}{\hbar^2/2M\rho^2}; \tag{1.2}$$

we derive the following equation (by physical reason we make the change $\gamma \Longrightarrow i\gamma$)

$$\Big[\frac{\text{ch}^{-2} z}{1 - \gamma^2(z)}\Big(\partial_r^2 + \frac{\text{ch } r}{\text{sh } r}\partial_r - \frac{[m - b(\text{ch } r - 1)]^2}{\text{sh}^2 r} + b\gamma(z)\Big)$$
$$+ \epsilon + (\partial_z + 2\frac{\text{sh } z}{\text{ch } z})\partial_z\Big] R(r) Z(z) = 0. \tag{1.3}$$

After separation of the variables we arrive at two equations (note that $\gamma = B\Gamma$)

$$\Big(\frac{d^2}{dr^2} + \frac{\text{ch}\, r}{\text{sh}\, r} \frac{d}{dr} - \frac{[m - b(\text{ch}\, r - 1)]^2}{\text{sh}^2 r} + \Lambda \Big) R = 0, \tag{1.4}$$

$$\Big(\frac{d^2}{dz^2} + 2 \frac{\text{sh}\, z}{\text{ch}\, z} \frac{d}{dz} + \epsilon + \frac{b\gamma - \Lambda\, \text{ch}^2 z}{\text{ch}^4 z - \gamma^2} \Big) Z = 0. \tag{1.5}$$

1.2 The usual scalar particle in Lobachevsky space

First, let us consider the motion of the usual scalar free particle in Lobachevsky space along the axes z:

$$\Big(\frac{d^2}{dz^2} + \epsilon - 1 - U(z) \Big) f(z) = 0 \,,$$
$$U(z) = + \frac{\Lambda}{\text{ch}^2 z} \,, \qquad U(z \to \pm\infty) = +0 \,. \tag{1.6}$$

This is a Schrödinger-type equation in the effective (generated by geometry) potential of the barrier type, so there exists the possibility of the tunnelling effect.

First, we use the most simple variable Z:

$$\text{ch}^2 z = Z \,, \quad \frac{d}{dz} = 2 \text{sh}\, z \,\text{ch}\, z \frac{d}{dZ} \,, \quad \frac{d^2}{dz^2} = \frac{d}{dz} 2\text{sh}\, z\, \text{ch}\, z \frac{d}{dZ}$$
$$= 4 \text{sh}^2 z\, \text{ch}^2 z \frac{d^2}{dZ^2} + 2 (\text{ch}^2 z + \text{sh}^2 z) \frac{d}{dZ} = 4Z(Z-1) \frac{d^2}{dZ^2} + 2(2Z - 1) \frac{d}{dZ} ;$$

so the above equation takes the form

$$\Big[Z(Z-1) \frac{d^2}{dZ^2} + (Z - \frac{1}{2}) \frac{d}{dZ} + \frac{\epsilon - 1}{4} - \frac{\Lambda/4}{Z^2} \Big] f(Z) = 0 \,. \tag{1.7}$$

This is an equation of the hypergeometric type. With the use of the substitution $f(Z) = Z^a\, (Z-1)^b\, F(Z)$, we derive

$$Z\, (Z - 1)\, \frac{d^2 F}{dZ^2} + \Big((2\,a + 2\,b + 1) Z - 2\,a - \frac{1}{2} \Big) \frac{dF}{dZ}$$
$$+ \Big(\frac{\epsilon - 1}{4} + (a + b)^2 + \frac{1}{2} \frac{b(2\,b - 1)}{Z - 1} + \frac{1}{4} \frac{-4\,a^2 + 2\,a - \Lambda}{Z} \Big) F = 0 \,.$$

At the following restrictions

$$a = \frac{1}{4} \pm \frac{i}{4} \sqrt{4\,\Lambda - 1} \,, \qquad b = 0, \; \frac{1}{2} \tag{1.8}$$

it is simplified to the canonical form

$$Z\, (1 - Z)\, \frac{d^2 F}{dZ^2} + \Big(2\,a + \frac{1}{2} - (2\,a + 2\,b + 1) Z \Big) \frac{dF}{dZ} - \Big(\frac{\epsilon - 1}{4} + (a + b)^2 \Big) F = 0$$

with parameters

$$\alpha = a + b + \frac{i}{2} \sqrt{\epsilon - 1} \,, \quad \beta = a + b - \frac{i}{2} \sqrt{\epsilon - 1} \,, \quad \gamma = 2\,a + \frac{1}{2} ; \tag{1.9}$$

below for brevity we use the notation $\sqrt{4\Lambda - 1} = \lambda$.

First, let us consider a pair of complex conjugate solutions

$$\begin{aligned} b=0, \quad & a=\frac{1+i\lambda}{4}, \quad f_1(Z)=Z^a u_1(Z) \\ & = Z^{(1+i\lambda)/4} F(\frac{1+i\lambda}{4}+\frac{i\sqrt{\epsilon-1}}{2}, \frac{1+i\lambda}{4}-\frac{i\sqrt{\epsilon-1}}{2}, 1+\frac{i\lambda}{2}; Z); \\ b=0, \quad & a=\frac{1-i\lambda}{4}, \quad f_1^*(Z)=Z^a u_5(Z) \\ & = Z^{(1-i\lambda)/4} F(\frac{1-i\lambda}{4}+\frac{i\sqrt{\epsilon-1}}{2}, \frac{1-i\lambda}{4}-\frac{i\sqrt{\epsilon-1}}{2}, 1-\frac{i\lambda}{2}; Z); \end{aligned} \tag{1.10}$$

symbol u_i represents different Kummer solutions for hypergeometric equation.

Obviously, when using the variable Z, one can construct only solutions which are symmetrical relative to replacement $z \to -z$. By evident procedure, from two conjugate solutions, we can construct two independent real-valued ones.

Solutions (1.10) have simple asymptotic behaviour near the singular point $Z=0$:

$$f_1(Z \to 0) = Z^{(1+i\lambda)/4}, \quad f_1^*(Z \to 0) = Z^{(1-i\lambda)/4};$$

however, the point $Z=0$ lies outside the physical region of the variable, $Z \in (1,+\infty)$. The limit $Z \to +\infty$ corresponds to different physical infinities $z \to \pm\infty$; besides the region $Z \in (-\infty,+1)$ does not lie in the physical domain of the coordinate Z.

To find behaviour of the solutions at $Z \to +\infty$, we apply the known Kummer formula

$$u_1 = \frac{\Gamma(\gamma)\Gamma(\beta-\alpha)}{\Gamma(\gamma-\alpha)\Gamma(\beta)} u_3 + \frac{\Gamma(\gamma)\Gamma(\alpha-\beta)}{\Gamma(\gamma-\beta)\Gamma(\alpha)} u_4 \,, \tag{1.11}$$

where the standard notations are used

$$\begin{gathered} u_1(Z) = F(\alpha,\beta,\gamma;Z)\,, \\ \alpha = \frac{1}{4}+\frac{i\lambda}{4}+\frac{i\sqrt{\epsilon-1}}{2}, \quad \beta = \frac{1}{4}+\frac{i\lambda}{4}-\frac{i\sqrt{\epsilon-1}}{2}, \quad \gamma = 1+\frac{i\lambda}{2}, \\ u_3(Z) = (-Z)^{-\alpha} F(\alpha, \alpha+1-\gamma, \alpha+1-\beta; \frac{1}{Z}), \\ u_4(Z) = (-Z)^{-b} F(\beta, \beta+1-\gamma, \beta+1-\alpha, \frac{1}{Z}). \end{gathered} \tag{1.12}$$

In the region $Z \to \infty$, formula (1.11) takes the form

$$u_1 = \frac{\Gamma(\gamma)\Gamma(\beta-\alpha)}{\Gamma(\gamma-\alpha)\Gamma(\beta)} (-Z)^{-\alpha} + \frac{\Gamma(\gamma)\Gamma(\alpha-\beta)}{\Gamma(\gamma-\beta)\Gamma(\alpha)} (-Z)^{-\beta}.$$

Multiplying the last relation by Z^a, we find asymptotic behaviour of the complete function $f_1(Z)$ at infinity

$$f_1(Z \to \infty) = \frac{\Gamma(\gamma)\Gamma(\beta-\alpha)}{\Gamma(\gamma-\alpha)\Gamma(\beta)} (-Z)^{a-\alpha} + \frac{\Gamma(\gamma)\Gamma(\alpha-\beta)}{\Gamma(\gamma-\beta)\Gamma(\alpha)} (-Z)^{a-\beta},$$

or differently

$$\begin{aligned} f_1(Z \to \infty) = & \frac{\Gamma(\gamma)\Gamma(\beta-\alpha)}{\Gamma(\gamma-\alpha)\Gamma(\beta)} (-Z)^{-i\sqrt{\epsilon-1}/2} \\ & + \frac{\Gamma(\gamma)\Gamma(\alpha-\beta)}{\Gamma(\gamma-\beta)\Gamma(\alpha)} (-Z)(-Z)^{+i\sqrt{\epsilon-1}/2}. \end{aligned} \tag{1.13}$$

The formula (1.13), translated to the original variable $z \in (-\infty, +\infty)$, looks as follows:

$$z \to +\infty\,, \quad Z \to \frac{1}{4} e^{2z}\,, \qquad \sqrt{\epsilon - 1} \equiv k\,,$$

$$\begin{aligned} f_1(z \to +\infty) &= \frac{\Gamma(\gamma)\Gamma(\beta-\alpha)}{\Gamma(\gamma-\alpha)\Gamma(\beta)} (\frac{e^{2z+i\pi}}{4})^{-ik/2} + \frac{\Gamma(\gamma)\Gamma(\alpha-\beta)}{\Gamma(\gamma-\beta)\Gamma(\alpha)} (\frac{e^{2z+i\pi}}{4})^{+ik/2} \\ &= \Big(\frac{\Gamma(\gamma)\Gamma(\beta-\alpha)}{\Gamma(\gamma-\alpha)\Gamma(\beta)} e^{+\pi k/2} e^{ik\ln 2} \Big) e^{-ikz} \\ &+ \Big(\frac{\Gamma(\gamma)\Gamma(\alpha-\beta)}{\Gamma(\gamma-\beta)\Gamma(\alpha)} e^{-\pi k/2} e^{-ik\ln 2} \Big) e^{+ikz}; \end{aligned} \tag{1.14}$$

$$z \to -\infty\,, \quad Z \to \frac{1}{4} e^{-2z}\,,$$

$$\begin{aligned} f_1(z \to -\infty) &= \frac{\Gamma(\gamma)\Gamma(\beta-\alpha)}{\Gamma(\gamma-\alpha)\Gamma(\beta)} (\frac{e^{-2z+i\pi}}{4})^{-ik/2} + \frac{\Gamma(\gamma)\Gamma(\alpha-\beta)}{\Gamma(\gamma-\beta)\Gamma(\alpha)} (\frac{e^{-2z+i\pi}}{4})^{+ik/2} \\ &= \Big(\frac{\Gamma(\gamma)\Gamma(\alpha-\beta)}{\Gamma(\gamma-\beta)\Gamma(\alpha)} e^{-\pi k/2} e^{-ik\ln 2} \Big) e^{-ikz} \\ &+ \Big(\frac{\Gamma(\gamma)\Gamma(\beta-\alpha)}{\Gamma(\gamma-\alpha)\Gamma(\beta)} e^{+\pi k/2} e^{ik\ln 2} \Big) e^{+ikz}\,. \end{aligned} \tag{1.15}$$

At both infinities $z \to \pm\infty$, we have superpositions of the plane waves:

$$\begin{aligned} f_1(z \to +\infty) &= A e^{+\pi k/2} e^{ik\ln 2}\, e^{-ikz} + B e^{-\pi k/2} e^{-ik\ln 2} e^{+ikz}\,, \\ f_1^*(z \to +\infty) &= B^* e^{-\pi k/2} e^{+ik\ln 2} e^{-ikz} + A^* e^{+\pi k/2} e^{-ik\ln 2}\, e^{+ikz}\,; \end{aligned} \tag{1.16}$$

$$\begin{aligned} f_1(z \to -\infty) &= B e^{-\pi k/2} e^{-ik\ln 2}\, e^{-ikz} + A e^{+\pi k/2} e^{ik\ln 2} e^{+ikz}\,, \\ f_1^*(z \to -\infty) &= A^* e^{+\pi k/2} e^{-ik\ln 2} e^{-ikz} + B^* e^{-\pi k/2} e^{+ik\ln 2}\, e^{+ikz}\,. \end{aligned} \tag{1.17}$$

Let us detail the behaviour of the functions f_1 and f_1^* at $z \to +\infty$, schematically it is

$$f_1(z \to +\infty) = M e^{-ikz} + N e^{+ikz}, \quad f_1^*(z \to +\infty) = N^* e^{-ikz} + M^* e^{+ikz}.$$

Let us introduce a linear combination of them, so that

$$F(z) = (\frac{1}{M} f_1(z) - \frac{1}{N^*} f_1^*(z))_{z\to+\infty} = (\frac{N}{M} - \frac{M^*}{N^*}) e^{+ikz}. \tag{1.18}$$

However, at negative infinity $z \to -\infty$ we have for the main terms in asymptotic only the following:

$$\begin{aligned} F(z) &= (\frac{f_1(z)}{M} - \frac{f_1^*(z)}{N^*})_{z\to-\infty} \\ &= \frac{N e^{+ikz} + M e^{-ikz}}{M} - \frac{N^* e^{-ikz} + M^* e^{+ikz}}{N^*} \\ &= (\frac{N}{M} - \frac{M^*}{N^*}) e^{+ikz} + (1 - 1 + ...) e^{-ikz}\,; \end{aligned}$$

which means that the function $F(z)$ does not lead to description of the tunnelling effect through the barrier.

Initial equation (1.6) can also be investigated with the use of the other variable, $x = \cosh z$:

$$\frac{d}{dz} = \sinh z \frac{d}{dx}, \quad \frac{d^2}{dz^2} = \sinh z \frac{d}{dx} \sinh z \frac{d}{dx}$$
$$= \sinh^2 z \frac{d^2}{dx} + \cosh z \frac{d}{dx} = (x^2 - 1)\frac{d^2}{dx^2} + x\frac{d}{dx} \, ;$$

the basic equation takes on the form[1]

$$\Big(\frac{d^2}{dx^2} + \frac{x}{(x^2-1)} \frac{d}{dx} + \frac{\epsilon - 1}{(x^2-1)} - \frac{\Lambda}{x^2(x^2-1)} \Big) f = 0 \, .$$

It can be transformed to

$$\frac{d^2 f}{dx^2} + \Big(\frac{1/2}{x+1} + \frac{1/2}{x-1} \Big) \frac{df}{dx}$$
$$+ \Big(\frac{\Lambda}{x^2} + \frac{(\Lambda - \epsilon + 1)/2}{x+1} - \frac{(\Lambda - \epsilon + 1)/2}{x-1} \Big) f = 0; \tag{1.19}$$

here we have four singular points $x = -1, \, 0, \, 1, \, \infty$; they all are regular, so we have the general Heun equation [7], [8].

First, let us find the behaviour of its solutions in the neighbourhood of the singular point $x = 0$ (this point lies outside the physical region of the variable):

$$x \to 0, \quad \Big(\frac{d^2}{dx^2} - x\frac{d}{dx} + \frac{\Lambda}{x^2} \Big) f = 0, \quad f = x^A,$$
$$A(A-1) + \Lambda = 0, \quad A = \frac{1 \pm i\sqrt{4\Lambda - 1}}{2} \, . \tag{1.20}$$

In the neighbourhood of the points $x = -1$ and $x = +1$, solutions behave as

$$x \to -1, \quad f = (x+1)^\rho, \quad \rho = 0, \frac{1}{2} \, ;$$
$$x \to +1, \quad f = (x-1)^\sigma, \quad \sigma = 0, \frac{1}{2} \, . \tag{1.21}$$

The behaviour of solutions in the neighbourhood of the point $x \to \infty$ ($z \to \pm\infty$) is of the most interest. We make the change of variable

$$\frac{1}{x} = X, \quad \frac{d}{dx} = \frac{dX}{dx}\frac{d}{dX} = -X^2 \frac{d}{dX}, \quad \frac{d^2}{dx^2} = X^4 \frac{d^2}{dX^2} + 2X^3 \frac{d}{dX} \, ;$$

$$\frac{d^2}{dX^2} + \frac{2}{X}\frac{d}{dX} - \frac{1}{2}\Big(\frac{1}{X(1+X)} + \frac{1}{X(1-X)} \Big) \frac{d}{dX}$$
$$+ \Big(\frac{\Lambda}{X^2} - \frac{(\Lambda - \epsilon + 1)}{X^2} \frac{1}{X^2 - 1} \Big) f = 0 \, . \tag{1.22}$$

Near the point $X = 0$, the equation and its solutions have the form

$$X \to 0, \; \Big(\frac{d^2}{dX^2} + \frac{1}{X}\frac{d}{dX} + \frac{\epsilon - 1}{X^2} \Big) f = 0 \, , \; f \approx X^{\pm i\sqrt{\epsilon - 1}} = x^{\mp i\sqrt{\epsilon - 1}} \, . \tag{1.23}$$

---

[1]The structure of this equation implies the possibility to use the variable $x^2 = u$; so the power series below will have 2-term recurrent formulas.

With the use of notation

$$\frac{\Lambda - \epsilon + 1}{2} = L,$$

eq. (1.22) can be rewritten shorter as

$$\Big(\frac{d^2}{dX^2} + (\frac{1}{X} + \frac{1/2}{X-1} + \frac{1/2}{1+X})\frac{d}{dX} + \frac{\epsilon - 1}{X^2} + \frac{L}{X-1} - \frac{L}{X+1}\Big) f = 0 \; ; \qquad (1.24)$$

recall that

$$z \in (-\infty, +\infty), \quad x \in (1, +\infty), \quad X \in (0,\ 1) .$$

Let us construct Frobenius solutions in the neighbourhood of $x = \infty$ ($X = 0$):

$$f(X) = X^A \; F(X) \; ;$$

allowing for

$$\frac{d}{dX} f = AX^{A-1}F + X^A F' \; , \quad \frac{d^2}{dX^2} f = A(A-1)X^{A-2}F + 2AX^{A-1}F' + X^A F'' \; ,$$

we derive an equation for $F(X)$:

$$F'' + \Big(\frac{2A}{X} + \frac{1}{X} + \frac{1/2}{X-1} + \frac{1/2}{X+1}\Big) F'$$

$$+\Big(\frac{A(A-1)}{X^2} + \frac{A}{X^2} + \frac{A/2}{X(X-1)} + \frac{A/2}{X(X+1)} + \frac{\epsilon - 1}{X^2} + \frac{L}{X-1} - \frac{L}{X+1}\Big) F = 0 \; .$$

The condition

$$A^2 + \epsilon - 1 = 0, \qquad A = \pm i\sqrt{\epsilon - 1}$$

removes terms with X^{-2} and we get

$$F'' + \Big(\frac{2A+1}{X} + \frac{1/2}{X-1} + \frac{1/2}{X+1}\Big) F' + \Big(\frac{L + A/2}{X-1} - \frac{L + A/2}{X+1}\Big) F = 0 \; . \qquad (1.25)$$

For brevity below we use notation $K = L + A/2$; so eq. (1.25) is written as

$$X(X^2-1)F'' + (X^2-1)(2A+1)F' + \frac{1}{2}X(X+1)F' + \frac{1}{2}X(X-1)F'$$

$$+KX(X+1)F - KX(X-1)F = 0 \; ;$$

its solutions may be constructed in the form of a power series:

$$F = \sum_{n=0}^{\infty} c_n X^n, \quad F' = \sum_{n=1}^{\infty} n c_n X^{n-1}, \quad F'' = \sum_{n=2}^{\infty} n(n-1) c_n X^{n-2} .$$

In usual way, we derive

$$\sum_{n=3}^{\infty} (n-1)(n-2) c_{n-1} x^n - \sum_{n=1}^{\infty} (n+1) n c_{n+1} x^n$$

$$+(2A+1) \sum_{n=2}^{\infty} (n-1) c_{n-1} x^n - (2A+1) \sum_{n=0}^{\infty} (n+1) c_{n+1} x^n \sum_{n=2}^{\infty} (n-1) c_{n-1} x^n$$

$$+2K\sum_{n=1}^{\infty} c_{n-1}x^n = 0;$$

so we get recurrent relations for coefficients:

$$n=0, \quad -(2A+1)c_1 = 0, \qquad c_1 = 0\,;$$

$$n=1, \quad -2\cdot c_2 + 2Kc_0 - 2(2A+1)c_2 = 0\,;$$

$$n=2, \quad -6c_3 + (2A+1)c_1 - 3(2A+1)c_3 + c_1 + 2Kc_1 = 0 \quad \Longrightarrow \quad c_3 = 0\,;$$

$$n=3, \qquad 2c_2 - 12c_4 + 2(2A+1)c_2 - 4(2A+1)c_4 + 2c_2 + 2Kc_2 = 0\,.$$

Therefore, the main 2-term recurrent formula is

$$[(n-1)(n-2) + (2A+1)(n-1) + (n-1) + 2K]c_{n-1}$$
$$+[-n(n+1) - (2A+1)(n+1)]c_{n+1} = 0\,. \tag{1.26}$$

We investigate the convergence of this series by using the Poincaré–Perron method. To do this, we divide relation (1.26) by c_{n-1}:

$$[(n-1)(n-2) + (2A+1)(n-1) + (n-1) + 2K]$$
$$+[-n(n+1) - (2A+1)(n+1)]\frac{c_{n+1}}{c_n}\frac{c_n}{c_{n-1}} = 0\,.$$

The radius of convergence R of the power series is determined by the formula

$$R = \lim_{n\to\infty}\frac{c_{k+1}}{c_k}\,, \qquad R_{conv} = \frac{1}{|\,R\,|}\,.$$

To find an algebraic equation for R, the resulting equation is multiplied by n^{-2} and let n tend to infinity, $n\to\infty$. This results in a simple algebraic equation

$$1 - R^2 = 0 \quad \Longrightarrow \quad R = \pm 1 \quad \Longrightarrow \quad R_{conv} = 1. \tag{1.27}$$

Therefore, the series converges in the entire physical area of the variable:

$$z \in (-\infty, +\infty), \quad x = \cosh z \in (1, +\infty), \qquad X = \frac{1}{x} \in (0,1).$$

Obviously, the power series contains only even degrees

$$X^2, \quad X^4, \quad \ldots\,.$$

Let us try to use yet another variable

$$x = \sinh z, \qquad \Big(\frac{d^2}{dz^2} + \epsilon - 1 - \frac{\Lambda}{\cosh^2 z}\Big)f(z) = 0\,; \tag{1.28}$$

equation is transformed into

$$\Big((1+x^2)\frac{d^2}{dx^2} + x\frac{d}{dx} + \epsilon - 1 - \frac{\Lambda}{1+x^2}\Big)f = 0\,.$$

We introduce a new variable

$$y = \frac{1-ix}{2}, \quad x = \frac{1-2y}{i}, \quad 1+x^2 = 4y(1-y), \quad \frac{d}{dx} = -\frac{i}{2}\frac{d}{dy}\,;$$
$$z\to+\infty,\ x\to+i\infty\,; \qquad z\to-\infty,\ x\to-i\infty\,; \tag{1.29}$$

then above equation takes the form

$$\Big[\frac{d^2}{dy^2}+\Big(\frac{1/2}{y-1}+\frac{1/2}{y}\Big)\frac{d}{dy}+\frac{\epsilon-1}{y-1}-\frac{\epsilon-1}{y}$$
$$+\frac{\Lambda/4}{(y-1)^2}+\frac{\Lambda/4}{y^2}-\frac{\Lambda/2}{y-1}+\frac{\Lambda/2}{y}\Big]f=0\,. \tag{1.30}$$

The singular points $y=0$ and $y=1$ are regular; they lie outside the physical domain of the variable. Near the point $y=0$, the equation and its solutions have the form:

$$y\to 0,\qquad \Big(\frac{d^2}{dy^2}+\frac{1}{2y}\frac{d}{dy}+\frac{\Lambda}{4y^2}\Big)f=0\,,\quad f=y^A,\quad A=\frac{1\pm i\sqrt{4\Lambda-1}}{4}\,.$$

Near the point $y=1$, the equation and its solutions have the form

$$y\to 1,\quad \Big(\frac{d^2}{dy^2}-\frac{1}{2(1-y)}\frac{d}{dy}+\frac{\Lambda}{4(1-y)^2}\Big)f=0\,,\ f=(1-y)^B,\ B=\frac{1\pm i\sqrt{4\Lambda-1}}{4}.$$

Let us investigate the behaviour of solutions in the vicinity of the point $y=\infty$. In the variable $Y=y^{-1}$, the equation takes the form

$$\frac{d^2f}{dY^2}+\Big(\frac{1}{Y}+\frac{1/2}{Y-1}\Big)\frac{df}{dY}+\Big(-\frac{\epsilon-1}{Y-1}+\frac{\epsilon-1}{Y^2}+\frac{1}{4}\frac{\Lambda}{(Y-1)^2}+\frac{\epsilon-1}{Y}\Big)f=0\,. \tag{1.31}$$

In the neighbourhood of $Y=0$, the equation takes the form

$$Y\to 0,\qquad \Big(\frac{d^2}{dY^2}+\frac{1}{Y}\frac{d}{dY}+\frac{\epsilon-1}{Y^2}\Big)f=0\,,\quad f(Y)\sim Y^{\pm i\sqrt{\epsilon-1}};$$

the point $y=\infty$ is regular.

Evidently, the equation with three regular singular points must be reduced to the hypergeometric functions. With the use of the substitution

$$f(y)=y^A(1-y)^BF(y) \tag{1.32}$$

we get

$$F''+\Big(\frac{2A}{y}-\frac{2B}{1-y}+\big(\frac{1/2}{y-1}+\frac{1/2}{y}\big)\Big)F'$$
$$+\Big[\frac{A(A-1)}{y^2}-\frac{2AB}{y(1-y)}+\frac{B^2}{(1-y)^2}+\Big(\frac{1/2}{y-1}+\frac{1/2}{y}\Big)\Big(\frac{A}{y}-\frac{B}{1-y}\Big)$$
$$+\frac{\epsilon-1}{y-1}-\frac{\epsilon-1}{y}+\frac{\Lambda/4}{(y-1)^2}+\frac{\Lambda/4}{y^2}-\frac{\Lambda/2}{y-1}+\frac{\Lambda/2}{y}--\frac{B}{(1-y)^2}\Big]F=0\,.$$

We choose the parameters A and B as follows:

$$A=\frac{1}{4}\pm\frac{i}{4}\sqrt{4\,\Lambda-1}\,,\quad B=\frac{1}{4}\pm\frac{i}{4}\sqrt{4\,\Lambda-1}$$

and thereby eliminate the quadratic terms:

$$F''+\Big(\frac{2A+1/2}{y}+\frac{2B+1/2}{y-1}\Big)F'+\Big(-\frac{2AB+A/2+B/2}{y(1-y)}+\frac{\epsilon-1-\Lambda/2}{y(y-1)}\Big)F=0\,.$$

This equation can be identified with the equation of hypergeometric type

$$y(1-y)F''+\left[(2A+1/2)-(2A+2B+1)y\right]F'$$
$$-\left[2AB+A/2+B/2+\epsilon-1-\Lambda/2\right]F=0$$

with parameters

$$a = A + B + \frac{1}{2}\sqrt{4A^2 + 4B^2 - 2A - 2B + 2\Lambda - 4\epsilon + 4}\,,$$

$$b = A + B - \frac{1}{2}\sqrt{4A^2 + 4B^2 - 2A - 2B + 2\Lambda - 4\epsilon + 4}\,, \quad c = 2A + \frac{1}{2}\,.$$

Expressions for parameters a, b can be written differently

$$a = A + B + \sqrt{A(A - \frac{1}{2}) + B(B - \frac{1}{2}) - (\epsilon - 1) + \frac{\Lambda}{2}}\,,$$

$$a = A + B - \sqrt{A(A - \frac{1}{2}) + B(B - \frac{1}{2}) - (\epsilon - 1) + \frac{\Lambda}{2}}\,.$$

Here exist four possibilities, and for each we find the expressions for the parameters a, b, c (using the notation $\lambda = +\sqrt{4\Lambda - 1}, \lambda > 0$):

$$\begin{gathered} A = \frac{1}{4} + \frac{i\lambda}{4}, \quad B = \frac{1}{4} - \frac{i\lambda}{4}, \\ a = \frac{1}{2} + i\sqrt{\epsilon - 1}\,, \; b = \frac{1}{2} - i\sqrt{\epsilon - 1}\,, \quad c = 1 + \frac{i\lambda}{2}; \end{gathered} \tag{1.33}$$

$$\begin{gathered} A = \frac{1}{4} - \frac{i\lambda}{4}, \quad B = \frac{1}{4} + \frac{i\lambda}{4}, \\ a = \frac{1}{2} + i\sqrt{\epsilon - 1}\,, \; b = \frac{1}{2} - i\sqrt{\epsilon - 1}\,, \; c = 1 - \frac{i\lambda}{2}; \end{gathered} \tag{1.34}$$

$$\begin{gathered} A = \frac{1}{4} + \frac{i\lambda}{4}, \quad B = \frac{1}{4} + \frac{i\lambda}{4}, \\ a = \frac{1}{2} + \frac{i\lambda}{2} + i\sqrt{\epsilon - 1}\,, \; b = \frac{1}{2} + \frac{i\lambda}{2} - i\sqrt{\epsilon - 1}\,, \; c = 1 + \frac{i\lambda}{2}; \end{gathered} \tag{1.35}$$

$$\begin{gathered} A = \frac{1}{4} - \frac{i\lambda}{4}, \quad B = \frac{1}{4} - \frac{i\lambda}{4}, \\ a = \frac{1}{2} - \frac{i\lambda}{2} + \sqrt{1 - \epsilon}\,, \quad b = \frac{1}{2} - \frac{i\lambda}{2} - \sqrt{1 - \epsilon}\,, \quad c = 1 - \frac{i\lambda}{2}\,. \end{gathered} \tag{1.36}$$

Since only two linearly independent solutions can exist, we should choose two of them. Let us take the first two. They correspond to complex conjugate solutions, so it is enough to follow only one of them:

$$\begin{gathered} F(y) = y^{(1+i\lambda)/4}(1 - y)^{(1-i\lambda)/4} F(a, b, c; y), \quad y = \frac{1 - i\sinh z}{2}\,, \\ a = \frac{1}{2} + i\sqrt{\epsilon - 1}\,, \quad b = \frac{1}{2} - i\sqrt{\epsilon - 1}\,, \quad c = 1 + \frac{i\lambda}{2}\,. \end{gathered} \tag{1.37}$$

This solution has the simple behaviour near the point lying in the nonphysical region:

$$y = \frac{1 - i\sinh z}{2} = 0 \quad \Longrightarrow \quad i = \sinh z\,, \qquad F(y \to 0) \sim y^{(1+i\lambda)/4} \to 0.$$

To find solutions with simple behaviour at $y \to \infty$, we use other Kummer solutions of the hypergeometric equation, depending on the argument

$$\frac{1}{y} = \frac{2}{1 - i \sinh z}, \qquad z \to \pm\infty, \qquad \frac{1}{y} \to 0 .$$

We use solutions u_3 and u_4:

$$u_3(y) = (-y)^{-a} F(a, a+1-c, a+1-b; \frac{1}{y}) ,$$
$$u_4(y) = (-y)^{-b} F(b, b+1-c, b+1-a; \frac{1}{y}) ,$$

which generate two complete solutions of the initial equation

$$F_3(y) = y^{(1+i\lambda)/4} (1-y)^{(1-i\lambda)/4} \, (-y)^{-a} F(a, a+1-c, a+1-b; \frac{1}{y}) ,$$
$$F_4(y) = y^{(1+i\lambda)/4} (1-y)^{(1-i\lambda)/4} \, (-y)^{-b} F(b, b+1-c, b+1-a; \frac{1}{y}) . \qquad (1.38)$$

At infinity $\frac{1}{y} \to 0$, these solutions behave:

$$F_3(y) = y^{(1+i\lambda)/4} (-y)^{(1-i\lambda)/4} \, (-y)^{-\frac{1}{2} - i\sqrt{\epsilon - 1}} \sim y^{-i\sqrt{\epsilon-1}} \sim (1 - i \sinh z)^{-i\sqrt{\epsilon-1}} ,$$
$$F_4(y) = y^{(1+i\lambda)/4} (-y)^{(1-i\lambda)/4} \, (-y)^{-\frac{1}{2} + i\sqrt{\epsilon - 1}} \sim y^{+i\sqrt{\epsilon-1}} \sim (1 - i \sinh z)^{+i\sqrt{\epsilon-1}} .$$

At infinities opposite in sign, $z \to +\infty$ and $z \to -\infty$, solutions behave

$$F_3(z \to +\infty) \sim e^{-i\sqrt{\epsilon-1} z}, \quad F_3(z \to -\infty) \sim e^{+i\sqrt{\epsilon-1} z} ;$$
$$F_4(z \to +\infty) \sim e^{+i\sqrt{\epsilon-1} z}, \quad F_4(z \to -\infty) \sim e^{-i\sqrt{\epsilon-1} z} .$$

From the last asymptotic equations, we conclude that this variable is also hardly suitable for analysing the tunnelling effect.

Finally, we apply the variable that is used as the standard one for the analysis of the Schrödinger equation with the potential $U \sim \cosh^{-2} z$. This is variable $y = \tanh z$; note that this variable distinguishes between the points $z = +\infty$ and $z = -\infty$. Allowing for identities

$$y = \tanh z, \qquad \frac{d}{dz} = \frac{dy}{dz}\frac{d}{dy} = \frac{1}{\cosh^2 z}\frac{d}{dy} = (1 - y^2)\frac{d}{dy},$$

we transform the above equation to

$$(1-y)(1+y)\frac{d^2 f}{dy^2} - 2y\frac{df}{dy} + \Big(- \Lambda + \frac{1}{2}\frac{\epsilon - 1}{y+1} + \frac{1}{2}\frac{-\epsilon + 1}{y-1} \Big) f = 0;$$

whence in the variable

$$x = \frac{1-y}{2} = \frac{1 - \tanh z}{2}$$

we get

$$\Big[x(1-x)\frac{d^2}{dx^2} + (1-2x)\frac{d}{dx} \; - \; \Lambda + \frac{(\epsilon - 1)/4}{1-x} + \frac{(\epsilon-1)/4}{x} \Big] f = 0 . \qquad (1.39)$$

Using the substitution $f(x) = x^A (1-x)^B F(x)$, we arrive at

$$x(1-x)\frac{d^2 F}{dx^2} + [2A + 1 - (2A + 2B + 2)x]\,\frac{dF}{dx}$$

$$+\Big[-(A+B)(A+B+1)-\Lambda+\frac{1}{4}\,\frac{4A^2+\epsilon-1}{x}+\frac{1}{4}\,\frac{4B^2+\epsilon-1}{1-x}\Big]F=0\,.$$

At the following restrictions

$$A=\pm\frac{i}{2}\sqrt{\epsilon-1}\,,\quad B=\pm\frac{i}{2}\sqrt{\epsilon-1}\,, \tag{1.40}$$

we get the equation of hypergeometric type

$$x(1-x)\frac{d^2F}{dx^2}+[2A+1-(2A+2B+2)x]\,\frac{dF}{dx}-[(A+B)(A+B+1)+\Lambda]\,F=0$$

with parameters (for brevity, let $\lambda=\sqrt{4\Lambda-1}$)

$$a=\frac{1}{2}+A+B+\frac{i\lambda}{2}\,,\quad b=\frac{1}{2}+A+B-\frac{i\lambda}{2}\,,\quad c=2A+1\,,$$
$$f(x)=x^A(1-x)^B F(a,b,c;x)\,,\quad x=\frac{1-\tanh z}{2}\,. \tag{1.41}$$

We choose solutions with given behaviour in the $z\to+\infty$ $(x\to 0)$ as a plane wave moving to the right

$$A=-i\sqrt{\epsilon-1}/2,\quad B=+i\sqrt{\epsilon-1}/2,$$
$$f(x)=x^{-i\sqrt{\epsilon-1}/2}(1-x)^{+i\sqrt{\epsilon-1}/2}F(a,b,c;x),$$
$$a=\frac{1}{2}+\frac{i\lambda}{2},\quad b=\frac{1}{2}-\frac{i\lambda}{2},\quad c=-i\sqrt{\epsilon-1}+1\,,$$
$$f(x\to 0)=x^{-i\sqrt{\epsilon-1}/2}=e^{+iz\sqrt{\epsilon-1}/2}. \tag{1.42}$$

In order to find behaviour for this solution at $z\to-\infty$ $(x\to 1)$, we use the Kummer formula

$$u_1=\frac{\Gamma(c)\Gamma(c-a-b)}{\Gamma(c-a)\Gamma(c-b)}u_2+\frac{\Gamma(c)\Gamma(a+b-c)}{\Gamma(a)\Gamma(b)}u_6\,, \tag{1.43}$$

where

$$u_1=F(a,b,c;x),\quad u_2=F(a,b,a+b+1-c;1-x),$$
$$u_6=(1-x)^{c-a-b}F(c-a,c-b,c+1-a-b;1-x)\,.$$

In the region $z\to-\infty$ $(x\to 1)$, we have

$$u_1(z\to-\infty)=\frac{\Gamma(c)\Gamma(c-a-b)}{\Gamma(c-a)\Gamma(c-b)}+\frac{\Gamma(c)\Gamma(a+b-c)}{\Gamma(a)\Gamma(b)}(1-x)^{c-a-b},$$

note that $c-a-b=-i\sqrt{\epsilon-1}$. After multiplying it by

$$x^{-i\sqrt{\epsilon-1}/2}\,(1-x)^{i\sqrt{\epsilon-1}/2}$$

we obtain

$$f(z\to-\infty)=\frac{\Gamma(c)\Gamma(c-a-b)}{\Gamma(c-a)\Gamma(c-b)}(1-x)^{-i\sqrt{\epsilon-1}/2}$$
$$+\frac{\Gamma(c)\Gamma(a+b-c)}{\Gamma(a)\Gamma(b)}(1-x)^{+i\sqrt{\epsilon-1}/2}\,. \tag{1.44}$$

In other form it reads

$$f(z \to -\infty) = \frac{\Gamma(c)\Gamma(c-a-b)}{\Gamma(c-a)\Gamma(c-b)} e^{+iz\sqrt{\epsilon-1}/2} + \frac{\Gamma(c)\Gamma(a+b-c)}{\Gamma(a)\Gamma(b)} e^{-iz\sqrt{\epsilon-1}/2} . \tag{1.45}$$

This formula describes the effect of particle tunnelling which moves from the left on barrier:

$$(z \to -\infty) \;\; M e^{+iz\sqrt{\epsilon-1}/2} + N e^{-iz\sqrt{\epsilon-1}/2} \;\longrightarrow\; e^{+iz\sqrt{\epsilon-1}/2} \;\; (z \to +\infty) . \tag{1.46}$$

The corresponding transmission coefficient D is defined by the formula

$$D = \frac{|M|^2}{|N|^2} = \left| \frac{\Gamma(a)\Gamma(b)}{\Gamma(c-a)\Gamma(c-b)} \right|^2 , \tag{1.47}$$

where

$$\Gamma(a) = \Gamma(\frac{1}{2} + \frac{i\lambda}{2}), \quad \Gamma(b) = \Gamma(\frac{1}{2} - \frac{i\lambda}{2}),$$

$$\Gamma(c-a) = \Gamma(\frac{1}{2} - i(\sqrt{\epsilon-1} + \lambda/2)), \quad \Gamma(c-b) = \Gamma(\frac{1}{2} - i(\sqrt{\epsilon-1} - \lambda/2)) .$$

We could proceed with calculation, using the well-known formula for Γ-functions:

$$\Gamma(\frac{1}{2} + iZ)\Gamma(\frac{1}{2} - iZ) = \frac{\pi}{\cos i\pi Z} = \frac{\pi}{\cosh \pi Z} .$$

The situation, when a particle falls from the right on the barrier can be investigated similarly, using solutions with the following asymptotic at $z \to -\infty$ $(x \to 1)$:

$$A = -i\sqrt{\epsilon-1}/2, \quad B = +i\sqrt{\epsilon-1}/2,$$
$$g(x) = x^A(1-x)^B u_6(x) = x^{-i\sqrt{\epsilon-1}/2}(1-x)^{+i\sqrt{\epsilon-1}/2}$$
$$\times (1-x)^{c-a-b} F(c-a, c-b, c+1-a-b; 1-x) ,$$
$$a = \frac{1}{2} + \frac{i\lambda}{2}, \quad b = \frac{1}{2} - \frac{i\lambda}{2}, \quad c = -i\sqrt{\epsilon-1} + 1 ,$$
$$g(z \to -\infty(x \to 1)) = (1-x)^{-i\sqrt{\epsilon-1}/2} = e^{-iz\sqrt{\epsilon-1}/2} . \tag{1.48}$$

To find the asymptotic behaviour of solutions when $z \to +\infty$ $(x = 0)$, one needs to use the Kummer formula

$$u_6(x) = \frac{\Gamma(c+1-a-b)\Gamma(1-c)}{\Gamma(1-a)\Gamma(1-b)} u_1(x) + \frac{\Gamma(c+1-a-b)\Gamma(c-1)}{\Gamma(c-a)\Gamma(c-b)} u_5(x), \tag{1.49}$$

where

$$u_6(x) = F(a, b, a+b+1-c; 1-x), \quad u_1(x) = F(a, b, c; x),$$
$$u_5(x) = x^{1-c} F(a+1-c, b+1-c, 2-c; x)$$

in the limit $x \to 0$ this relationship gives

$$u_6(x) = \frac{\Gamma(c+1-a-b)\Gamma(1-c)}{\Gamma(1-a)\Gamma(1-b)} + \frac{\Gamma(c+1-a-b)\Gamma(c-1)}{\Gamma(c-a)\Gamma(c-b)} x^{i\sqrt{\epsilon-1}}, 1-c = +i\sqrt{\epsilon-1} .$$

Multiplying it by $x^{-i\sqrt{\epsilon-1}/2}$ we obtain

$$g(z \to +\infty \; (x \to 0))$$

$$= \frac{\Gamma(c+1-a-b)\Gamma(1-c)}{\Gamma(1-a)\Gamma(1-b)} x^{-i\sqrt{\epsilon-1}/2} + \frac{\Gamma(c+1-a-b)\Gamma(c-1)}{\Gamma(c-a)\Gamma(c-b)} x^{+i\sqrt{\epsilon-1}/2}, \tag{1.50}$$

or differently

$$g(z \to +\infty\ (x \to 0))$$

$$= \frac{\Gamma(c+1-a-b)\Gamma(c-1)}{\Gamma(c-a)\Gamma(c-b)} e^{-iz\sqrt{\epsilon-1}/2} + \frac{\Gamma(c+1-a-b)\Gamma(1-c)}{\Gamma(1-a)\Gamma(1-b)} e^{+iz\sqrt{\epsilon-1}/2}. \tag{1.51}$$

Thus, the wave moving from the right partially passes through and partially reflects from the barrier:

$$M' e^{-iz\sqrt{\epsilon-1}/2} + N' e^{+iz\sqrt{\epsilon-1}/2} \quad (z \to +\infty). \tag{1.52}$$

The transmission coefficient $D' = 1 - R'$ equals

$$D' = \frac{|M'|^2}{|N'|^2} = \left| \frac{\Gamma(1-a)\Gamma(1-b)}{\Gamma(c-a)\Gamma(c-b)} \right|^2, \tag{1.53}$$

where

$$\Gamma(1-a) = \Gamma(\frac{1}{2} - \frac{i\lambda}{2}), \quad \Gamma(1-b) = \Gamma(\frac{1}{2} + \frac{i\lambda}{2}),$$

$$\Gamma(c-a) = \Gamma(\frac{1}{2} - i(\sqrt{\epsilon-1} + \lambda/2)), \quad \Gamma(c-b) = \Gamma(\frac{1}{2} - i(\sqrt{\epsilon-1} - \lambda/2))\ ;$$

calculations can be continued with the known formula for Γ-functions:

$$\Gamma(\frac{1}{2} + iZ)\Gamma(\frac{1}{2} - iZ) = \frac{\pi}{\cos i\pi Z} = \frac{\pi}{\cosh \pi Z}.$$

1.3 Particle in the Lobachevsky space

In the special system of cylindric coordinates in the Lobachevsky space, analogue of the uniform magnetic field is determined by the relations [6] (we use dimensionless coordinates):

$$\begin{gathered} dS^2 = c^2dt^2 - \text{ch}^2 z(dr^2 + \text{sh}^2 r d\phi^2) + dz^2, \\ \sqrt{-g} = \rho^3 \text{ sh } r \text{ ch}^2 z, \quad A_\phi = -B\rho^2(\text{ch } r - 1), \ F_{r\phi} = -B\rho \text{ sh } r, \\ B_3 = -B\rho \text{ sh } r, \quad B^3 = -\frac{B}{\rho \text{ sh } r \text{ ch}^4 z}, \quad B_i B^i = B^2 \text{ch}^{-4} z. \end{gathered} \tag{1.54}$$

We start with the known form of the generalised Schrödinger equation for a Cox scalar particle

$$D_t \Psi = \frac{1}{2M\rho^2} \Big[\overset{\circ}{D}_1 \overset{*}{D}_1 + \overset{\circ}{D}_2 \frac{1}{\text{sh}^2 r} \overset{*}{D}_2 + \overset{\circ}{D}_3 \overset{*}{D}_3 \Big] \Psi,$$

where

$$D_1 = i\hbar\partial_r, \quad D_2 = i\hbar\partial_\phi + \frac{e}{c}B\rho^2(\text{ch } r - 1), \quad D_3 = i\hbar\partial_z,$$

$$\overset{\circ}{D}_1 = i\hbar(\partial_r + \frac{\text{ch } r}{\text{sh } r}), \quad \overset{\circ}{D}_2 = i\hbar\partial_\phi + \frac{e}{c}B\rho^2(\text{ch } r - 1), \quad \overset{\circ}{D}_3 = i\hbar(\partial_z + 2\frac{\text{sh } z}{\text{ch } z}),$$

$$\overset{*}{D}_1 = \frac{1}{1+\Gamma^2 B^2 \text{ch}^{-4} z}\Big[i\hbar\partial_r - \frac{\Gamma B \text{ch}^{-2} z}{\text{sh} r}(i\hbar\partial_\phi + \frac{e}{c}B\rho^2(\text{ch } r - 1))\Big],$$

$$\overset{*}{D}_2 = \frac{1}{1+\Gamma^2 B^2 \text{ch}^{-4} z}\Big[(i\hbar\partial_\phi + \frac{e}{c}B\rho^2(\text{ch } r - 1)) + i\hbar\Gamma B \text{ch}^{-2} z \text{ sh } r\partial_r\Big],$$

$$\overset{*}{D}_3 = \frac{(D_3 + \Gamma^2 B^3\ B_3 D_3)}{1+\Gamma^2 B^2 \text{ch}^{-4} z} = i\hbar\partial_z;$$

below we apply notations $B\rho^2/\hbar c = b$, $\Gamma B \text{ch}^{-2} z = \gamma(z)$. With the use of the relations

$$\frac{1}{2M\rho^2}\ \overset{\circ}{D}_1\ g^{11}\ \overset{*}{D}_1 = -\frac{\hbar^2 \text{ch}^{-2} z}{2M\rho^2(1+\gamma^2(z))}$$

$$\times\Big(\partial_r^2 + (\frac{\text{ch } r}{\text{sh } r} + i\gamma(z)b\frac{\text{ch } r - 1}{\text{sh } r})\partial_r - \frac{\gamma(z)}{\text{sh } r}\partial_r\partial_\phi + i\gamma(z)b\Big),$$

$$\frac{1}{2M\rho^2}\ \overset{\circ}{D}_2\ g^{22}\ \overset{*}{D}_2 = -\frac{\hbar^2 \text{ch}^{-2} z}{2M\rho^2(1+\gamma^2(z))}$$

$$\times\Big[\frac{1}{\text{sh}^2 r}[\partial_\phi - ib(\text{ch } r - 1)]^2 + \gamma(z)[\partial_\phi - ib(\text{ch } r - 1)]\ \frac{1}{\text{sh } r}\partial_r\Big],$$

$$\frac{1}{2M\rho^2}\ \overset{\circ}{D}_3\ g^{33}\ \overset{*}{D}_3 = -\frac{\hbar^2}{2M\rho^2}(\partial_z + 2\frac{\text{sh } z}{\text{ch } z})\partial_z,$$

and of the substitution for wave function

$$\Psi = e^{-iEt/\hbar}e^{im\phi}Z(z)R(r)\ , \quad \epsilon = \frac{E}{\hbar^2/2M\rho^2}\ ; \tag{1.55}$$

we derive the following equation (by physical reason we make the change $\gamma \Longrightarrow i\gamma$)

$$\Big(\frac{\text{ch}^{-2} z}{1-\gamma^2(z)}\Big(\partial_r^2 + \frac{\text{ch } r}{\text{sh } r}\partial_r - \frac{[m - b(\text{ch } r - 1)]^2}{\text{sh}^2 r} + b\gamma(z)\Big)$$
$$+\epsilon + (\partial_z + 2\frac{\text{sh } z}{\text{ch } z})\partial_z\Big)R(r)Z(z) = 0\ . \tag{1.56}$$

After separating the variables, we arrive at two equations (note that $\gamma = B\Gamma$):

$$\Big(\frac{d^2}{dr^2} + \frac{\text{ch } r}{\text{sh } r}\frac{d}{dr} - \frac{[m - b(\text{ch } r - 1)]^2}{\text{sh}^2 r} + \Lambda\Big)R = 0, \tag{1.57}$$

$$\Big(\frac{d^2}{dz^2} + 2\frac{\text{sh } z}{\text{ch } z}\frac{d}{dz} + \epsilon + \frac{b\gamma - \Lambda\ \text{ch}^2 z}{\text{ch}^4 z - \gamma^2}\Big)Z = 0. \tag{1.58}$$

In the following, we consider only the differential equation in the variable z. Let $Z = (\cosh z)^{-1} f(z)$, then we have the equation

$$\Big(\frac{d^2}{dz^2} + \epsilon - 1 - U(z)\Big)f(z) = 0\,,$$
$$U(z) = -\frac{b\gamma - \Lambda\,\text{ch}^2 z}{\text{ch}^4 z - \gamma^2}, \quad U(z \to \pm\infty) = +0\,, \tag{1.59}$$

which can be viewed as the Schrödinger equation in the effective potential field $U(z)$. The corresponding effective force is determined by the formula

$$F_z = -\frac{dU}{dz} = 2\,\mathrm{ch}\,z\,\mathrm{sh}\,z\,\frac{\Lambda\,\mathrm{ch}^4 z - 2b\gamma\,\mathrm{ch}^2 z + \gamma^2\Lambda}{(\mathrm{ch}^4 z - \gamma^2)^2}\,. \tag{1.60}$$

We readily find the points at which the force vanishes: those are $z = 0$ and the roots of a quadratic equation

$$\Lambda\,\mathrm{ch}^4 z - 2b\gamma\,\mathrm{ch}^2 z + \gamma^2\Lambda = 0 \implies (\mathrm{ch}^2 z)\,|_{1,2} = \frac{b}{\Lambda}\gamma \pm \sqrt{(\frac{b^2}{\Lambda^2} - 1)\gamma^2}\,. \tag{1.61}$$

It is known from considering the bound states (for motion in the variable r) that the inequality $\Lambda^2 > b^2$ must hold. This means that the square roots in eq. (1.61) are imaginary. Consequently, the points of zero force except $z = 0$ cannot exist. The situation is illustrated in Fig. 1.1.

To proceed with differential equation, we introduce the new variable

$$y = \tanh z, \quad \frac{d}{dz} = (1 - y^2)\frac{d}{dy}\,, \quad \frac{d^2}{dz^2} = (1 - y^2)^2\frac{d^2}{dy} - 2y(1 - y^2)\frac{d}{dy}\,,$$

below we will apply the notation $b\gamma = \beta$. The use of this variable allows us to put physical points $z = -\infty$ and $z = +\infty$ in different singular points of the differential equation:

$$\cosh^2 z = \frac{1}{1 - y^2}, \quad \frac{\beta - \Lambda\cosh^2 z}{\cosh^4 z - \gamma^2} = \frac{(1 - y^2)\,[\beta(1 - y^2) - \Lambda]}{1 - \gamma^2(1 - y^2)^2};$$

the equation takes the form

$$(1 - y^2)\frac{d^2 f}{dy^2} - 2y\frac{df}{dy} + \Big(\frac{1}{2}\frac{\epsilon - 1}{1 - y} + \frac{1}{2}\frac{\epsilon - 1}{1 + y}$$
$$+\frac{1}{2}\frac{-\Lambda\gamma + \beta}{[1 - \gamma(1 - y^2)]\,\gamma} + \frac{1}{2}\frac{-\Lambda\gamma - \beta}{[1 + \gamma(1 - y^2)]\,\gamma}\Big) f = 0\,. \tag{1.62}$$

We introduce the new variable x:

$$x = \frac{1 - y}{2} = \frac{1 - \tanh z}{2}, \qquad y = 1 - 2x\,,$$
$$z \to -\infty \Longrightarrow x \to 1; \qquad z \to +\infty \Longrightarrow x \to 0; \tag{1.63}$$

then the above equation transforms into

$$\Big[\frac{d^2}{dx^2} + \Big(\frac{1}{x} - \frac{1}{1 - x}\Big)\frac{d}{dx} + \frac{1}{4}\frac{\epsilon - 1}{x^2} + \frac{1}{2}\frac{-2\Lambda + \epsilon - 1}{x} + \frac{1}{4}\frac{\epsilon - 1}{(1 - x)^2} + \frac{1}{2}\frac{\epsilon - 1 - 2\Lambda}{1 - x}$$

$$+\frac{2\beta + 2\Lambda\gamma}{1 + 4\gamma x(1 - x)} + \frac{2\beta - 2\Lambda\gamma}{1 - 4\gamma x(1 - x)}\Big] f = 0\,. \tag{1.64}$$

To study the point $x = \infty$, we transform equation (1.64) to the variable $X = x^{-1}$:

$$\frac{d^2 f}{dX^2} - \frac{1}{1 - X}\frac{df}{dX} + \Big[\frac{1}{2}\frac{(\Lambda\gamma + \beta)\,(X + 4\gamma + 1)}{((4X - 4)\,\gamma + X^2)\,\gamma} + \frac{\Lambda}{1 - X}$$
$$+\frac{1}{4}\frac{\epsilon - 1}{(1 - X)^2} + \frac{1}{2}\frac{(\Lambda\gamma - \beta)\,(X - 4\gamma + 1)}{((-4X + 4)\,\gamma + X^2)\,\gamma}\Big] f = 0\,.$$

Near the point $X = 0$ it reads

$$\frac{d^2 f}{dX^2} - \frac{df}{dX} + \Big[- \frac{(\Lambda\gamma + \beta)(4\gamma + 1)}{8\gamma^2} + \Lambda + \frac{1}{4}(\epsilon - 1) + \frac{(\Lambda\gamma - \beta)(-4\gamma + 1)}{8\gamma^2} \Big] f = 0;$$

so the point $x = \infty$ is an ordinary (non-singular) one.

Two quadratic expressions in the denominators of the equation (1.64) give four regular singular points:

$$\frac{1}{1 + 4\gamma x(1-x)} = -\frac{1}{4\gamma}\frac{1}{(x - x_1)(x - x_2)}, \qquad x_{1,2} = \frac{1 \pm \sqrt{1 + \gamma^{-1}}}{2},$$

$$\frac{1}{1 - 4\gamma x(1-x)} = +\frac{1}{4\gamma}\frac{1}{(x - x_3)(x - x_4)}, \qquad x_{3,4} = \frac{1 \pm \sqrt{1 - \gamma^{-1}}}{2};$$

two identities hold

$$\frac{1}{(x - x_1)(x - x_2)} = \frac{1}{x_1 - x_2}\Big(\frac{1}{x - x_1} - \frac{1}{x - x_2}\Big),$$

$$\frac{1}{(x - x_3)(x - x_4)} = \frac{1}{x_3 - x_4}\Big(\frac{1}{x - x_3} - \frac{1}{x - x_4}\Big).$$

Equation (1.64) can be represented as

$$\Big[\frac{d^2}{dx^2} + (\frac{1}{x} - \frac{1}{1-x})\frac{d}{dx} + \frac{1}{4}\frac{\epsilon - 1}{x^2}$$

$$+\frac{1}{2}\frac{-2\Lambda + \epsilon - 1}{x} + \frac{1}{4}\frac{\epsilon - 1}{(1-x)^2} + \frac{1}{2}\frac{\epsilon - 1 - 2\Lambda}{1 - x} - \frac{2\beta + 2\Lambda\gamma}{4\gamma}\frac{1}{x_1 - x_2}(\frac{1}{x - x_1} - \frac{1}{x - x_2})$$

$$+\frac{2\beta - 2\Lambda\gamma}{4\gamma}\frac{1}{x_3 - x_4}(\frac{1}{x - x_3} - \frac{1}{x - x_4})\Big] f = 0\,. \tag{1.65}$$

Recalling that $\beta = b\gamma$, eq. (1.65) can be rewritten as

$$\Big[\frac{d^2}{dx^2} + (\frac{1}{x} - \frac{1}{1-x})\frac{d}{dx} + \frac{1}{4}\frac{\epsilon - 1}{x^2} + \frac{1}{2}\frac{-2\Lambda + \epsilon - 1}{x} + \frac{1}{4}\frac{\epsilon - 1}{(1-x)^2} + \frac{1}{2}\frac{\epsilon - 1 - 2\Lambda}{1 - x}$$

$$-\frac{2b + 2\Lambda}{4}\frac{1}{x_1 - x_2}(\frac{1}{x - x_1} - \frac{1}{x - x_2})$$

$$+\frac{2b - 2\Lambda}{4\gamma}\frac{1}{x_3 - x_4}(\frac{1}{x - x_3} - \frac{1}{x - x_4})\Big] f = 0\,. \tag{1.66}$$

Thus, we have an equation with six regular singular points:

$$0, \quad 1, \quad x_{1,2} = \frac{1 \pm \sqrt{1 + \gamma^{-1}}}{2}, \quad x_{3,4} = \frac{1 \pm \sqrt{1 - \gamma^{-1}}}{2}\,. \tag{1.67}$$

Taking in mind that the physical range of the variable is

$$z \in (-\infty, +\infty) \quad \Longleftrightarrow \quad x \in (0, 1)\,, \tag{1.68}$$

and in view of the smallness of the parameter γ, we conclude that the four singular points x_1, x_2, x_3, x_4 (two real and two complex and conjugate) do not fall inside the circle of the radius 1 about the point $x = 0$ (see Fig. 1.2).

Let us find behaviour of the solutions near two points $x=0,\ x=\infty$:

$$z\to+\infty, x\to 0, \Big(\frac{d^2}{x^2}+\frac{1}{x}\frac{d}{dx}\frac{(\epsilon-1)/4}{x^2}\Big)f=0, f=x^A, A=\pm\frac{i\sqrt{\epsilon-1}}{2};$$

$$z\to-\infty, x\to 1, \Big(\frac{d^2}{x^2}+\frac{1}{x-1}\frac{d}{dx}\frac{(\epsilon-1)/4}{(x-1)^2}\Big)f=0, f=x^B, B=\pm\frac{i\sqrt{\epsilon-1}}{2}.$$

Let us find the behaviour of solutions near singular points x_1,x_2,x_3,x_4. They are of the same type, so it suffices to consider only one case. Near the point x_1, the equation has the following structure

$$f''+\alpha f'+\frac{\beta}{x-x_1}f=0, \quad f=(x-x_1)^\rho\ ;$$

for the index ρ we obtain an algebraic equation with simple solutions:

$$\rho(\rho-1)=0 \quad\Longrightarrow\quad \rho=0,1.$$

Thus, near four points x_1,x_2,x_3,x_4 solutions behave in accordance with the relations:

$$\begin{array}{lll} x\to x_1, & f=(x-x_1)^\rho, & \rho=0,1\ ;\\ x\to x_2, & f=(x-x_2)^\rho, & \rho=0,1\ ;\\ x\to x_3, & f=(x-x_3)^\rho, & \rho=0,1\ ;\\ x\to x_4, & f=(x-x_4)^\rho, & \rho=0,1\ . \end{array} \tag{1.69}$$

For eq. (1.64)

$$\Big[\frac{d^2}{dx^2}+\Big(\frac{1}{x}-\frac{1}{1-x}\Big)\frac{d}{dx}+\frac{1}{4}\frac{\epsilon-1}{x^2}+\frac{1}{2}\frac{-2\Lambda+\epsilon-1}{x}$$
$$+\frac{1}{4}\frac{\epsilon-1}{(1-x)^2}+\frac{1}{2}\frac{\epsilon-1-2\Lambda}{1-x}+\frac{2\beta+2\Lambda\gamma}{1+4\gamma x(1-x)}+\frac{2\beta-2\Lambda\gamma}{1-4\gamma x(1-x)}\Big]f=0,$$

let us build Frobenius solutions near the point $x=0$ in the form

$$f(x)=x^A(x-1)^BF(x)=\varphi(x)F(x)\ .$$

We get an equation for $F(x)$:

$$\Big[\frac{d^2}{dx^2}+\Big(\frac{2A+1}{x}+\frac{2B+1}{x-1}\Big)\frac{d}{dx}+\frac{A^2}{x^2}+\frac{2AB+A+B}{x(x-1)}+\frac{B^2}{(x-1)^2}$$
$$+\frac{1}{4}\frac{\epsilon-1}{x^2}+\frac{1}{2}\frac{\epsilon-1-2\Lambda}{x}+\frac{1}{4}\frac{\epsilon-1}{(1-x)^2}+\frac{1}{2}\frac{\epsilon-1-2\Lambda}{1-x}$$
$$+\frac{2\beta+2\Lambda\gamma}{1+4\gamma x(1-x)}+\frac{2\beta-2\Lambda\gamma}{1-4\gamma x(1-x)}\Big]f=0.$$

Already known constraints are imposed on the parameters A and B:

$$A=\pm\frac{i\sqrt{\epsilon-1}}{2},\quad B=\pm\frac{i\sqrt{\epsilon-1}}{2}, \tag{1.70}$$

then terms x^{-2} and $(x-1)^{-2}$ vanish and we get

$$\Big[\frac{d^2}{dx^2}+\Big(\frac{2A+1}{x}+\frac{2B+1}{x-1}\Big)\frac{d}{dx}+\frac{2AB+A+B}{x(x-1)}$$

$$+\frac{1}{2}\frac{\epsilon-1-2\,\Lambda}{x}+\frac{1}{2}\frac{\epsilon-1-2\,\Lambda}{1-x}+\frac{2\,\beta+2\,\Lambda\,\gamma}{1-4\,\gamma x(x-1)}+\frac{2\,\beta-2\,\Lambda\,\gamma}{1+4\,\gamma x(x-1)}\Big]f=0\,.$$

We multiply the equation by the expression

$$x(x-1)\cdot[1-4\,\gamma x(x-1)]\cdot[1+4\,\gamma x(x-1)]\,;$$

the result is

$$\begin{gathered}
x(x-1)\;[1-16\,\gamma^2x^2(x-1)^2]F''\\
+[1-16\,\gamma^2x^2(x-1)^2]\;\;[(2A+1)(x-1)+(2B+1)x])F'\\
+\{\;(2AB+A+B)\;[1-16\,\gamma^2x^2(x-1)^2]\\
+\frac{1}{2}\,(\epsilon-1-2\,\Lambda)\;(x-1)\;[1-16\,\gamma^2x^2(x-1)^2]\\
+\frac{1}{2}(\epsilon-1-2\,\Lambda)\;x\;[1-16\,\gamma^2x^2(x-1)^2]\\
+(2\,\beta-2\,\Lambda\,\gamma)\;x(x-1)\;[1-4\,\gamma x(x-1)]\\
+\;(2\,\beta+2\,\Lambda\,\gamma)\;x(x-1)\;[1+4\,\gamma x(x-1)]\;\}\,F=0\,.
\end{gathered}\tag{1.71}$$

Note that this equation is symmetric with respect to replacement

$$x\iff(x-1)\,,\qquad A\iff B\,,\tag{1.72}$$

this allows simultaneously to build the series expansion both in the variables x and $(x-1)$.
With the use of notations

$$\begin{gathered}
\frac{\epsilon-1-2\Lambda}{2}=M,\quad 2\beta-2\Lambda\gamma=K,\quad 2\beta+2\Lambda\beta=L,\\
4\gamma=\Gamma\,,\quad 2A+2B+2=\alpha,\quad 2AB+A+B=\beta\,,
\end{gathered}\tag{1.73}$$

the above equation is written as

$$\begin{gathered}
x(x-1)\;[1-\Gamma^2x^2(x-1)^2]\frac{d^2F}{dx^2}+[1-\Gamma^2x^2(x-1)^2]\;\;(\alpha x-2A-1)\;\frac{dF}{dx}\\
+\{\beta\;[1-\Gamma^2x^2(x-1)^2]+M\;(x-1)\;[1-\Gamma^2x^2(x-1)^2]+M\;x\;[1-\Gamma^2x^2(x-1)^2]\\
+K\;x(x-1)\;[1+\Gamma x(x-1)]+L\;x(x-1)\;[1-\Gamma x(x-1)]\;\}\,F=0\,.
\end{gathered}\tag{1.74}$$

The equation for $F(x)$ can be symbolically represented as

$$PF''+QF'+RF=0,$$

where

$$\begin{gathered}
P=-\Gamma^2x^6+3\Gamma^2x^5-3\Gamma^2x^4+\Gamma^2x^3+x^2-x\\
=p_6x^6+p_5x^5+p_4x^4+p_3x^3+p_2x^2+p_1x\,,\\
Q=-\Gamma^2\alpha x^5+\Gamma^2(2A+1+2\alpha)x^4-\Gamma^2(\alpha+4A+2)x^3+\Gamma^2(2A+1)x^2+\alpha x-(2A+1)\\
=q_5x^5+q_4x^4+q_3x^3+q_2x^2+q_1x+q_0\,,\\
R=-2\Gamma^2Mx^5+(-\Gamma^2\beta+5\Gamma^2M+\Gamma K-\Gamma L)x^4+(2\Gamma^2\beta-4\Gamma^2M-2\Gamma K+2\Gamma L)x^3\\
+(-\Gamma^2\beta+\Gamma^2M+K+\Gamma K+L-\Gamma L)x^2+(2M-K-L)x+(\beta-M)
\end{gathered}$$

$$= r_5 x^5 + r_4 x^4 + r_3 x^3 + r_2 x^2 + r_1 x + r_0 \, .$$

So we can work with the equation

$$(p_6 x^6 + p_5 x^5 + p_4 x^4 + p_3 x^3 + p_2 x^2 + p_1 x) \, F''$$
$$+(q_5 x^5 + q_4 x^4 + q_3 x^3 + q_2 x^2 + q_1 x + q_0) \, F'$$
$$+(r_5 x^5 + r_4 x^4 + r_3 x^3 + r_2 x^2 + r_1 x + r_0) \, F = 0 \, .$$

Solutions are built in the form of a power series

$$F = \sum_{n=0}^{\infty} c_n x^n, \quad F' = \sum_{n=1}^{\infty} n c_n x^{n-1}, \quad F'' = \sum_{n=2}^{\infty} n(n-1) c_n x^{n-2},$$

performing necessary calculations, we obtain the equation

$$p_6 \sum_{n=6}^{\infty} (n-4)(n-5) c_{n-4} x^n + p_5 \sum_{n=5}^{\infty} (n-3)(n-4) c_{n-3} x^n + p_4 \sum_{n=4}^{\infty} (n-2)(n-3) c_{n-2} x^n$$

$$+p_3 \sum_{n=3}^{\infty} (n-1)(n-5) c_{n-1} x^n + p_2 \sum_{n=2}^{\infty} n(n-1) c_n x^n + p_1 \sum_{n=1}^{\infty} (n+1) n c_{n+1} x^n$$

$$g_5 \sum_{n=5}^{\infty} (n-4) c_{n-4} x^n + g_4 \sum_{n=4}^{\infty} (n-3) c_{n-3} x^n + g_3 \sum_{n=3}^{\infty} (n-2) c_{n-2} x^n$$

$$+g_2 \sum_{n=2}^{\infty} (n-1) c_{n-1} x^n + g_1 \sum_{n=5}^{\infty} n c_n x^n + g_0 \sum_{n=0}^{\infty} (n+1) c_{n+1} x^n$$

$$r_5 \sum_{n=5}^{\infty} c_{n-5} x^n + r_4 \sum_{n=4}^{\infty} c_{n-4} x^n + r_3 \sum_{n=3}^{\infty} c_{n-3} x^n$$

$$+r_2 \sum_{n=2}^{\infty} c_{n-2} x^n + r_1 \sum_{n=1}^{\infty} c_{n-1} x^n + r_0 \sum_{n=0}^{\infty} c_n x^n = 0 \, ;$$

whence the recurrence relations for coefficients c_n follow

$$n = 0, \qquad r_0 c_0 + g_0 c_1 = 0,$$

$$n = 1, \qquad r_1 c_0 + (r_0 + g_1) c_1 + 2 g_0 c_2 = 0,$$

$$n = 2, \qquad r_2 c_0 + (g_2 + r_1) c_1 + (r_0 + 2g_1 + 2p_2) c_2 + (6p_1 + 3g_0) c_3 = 0,$$

$$n = 3, \;\; r_3 c_0 + (g_3 + r_2) c_1 + (2p_3 + 2g_2 + r_1) c_2 + (r_0 + 6p_2 + 3g_1) c_3 + (4g_0 + 12p_1) c_4 = 0,$$

$$n = 4, \;\; r_4 c_0 + (g_4 + r_3) c_1 + (r_2 + 2p_4 + 2g_3) c_2 + (r_1 + 6p_3 + 3g_2) c_3$$
$$+(r_0 + 4g_1 + 12p_2) c_4 + (5g_0 + 20p_1) c_5 = 0,$$

$$n = 5, \qquad r_5 c_0 + (g_5 + r_4) c_1 + (r_3 + 2g_4 + 2p_5) c_2 + (r_2 + 3g_3 + 6p_4) c_3$$
$$+(r_1 + 4g_2 + 12p_3) c_4 + (r_0 + 5g_1 + 20p_2) c_5 + (6g_0 + 30p_1) c_6 = 0,$$

$$n = 6, ... \qquad r_5 c_{n-5} + [p_6(n^2 - 9n + 20) + g_5(n-4) + r_4] c_{n-4}$$

$$+[p_5(n^2-7n+12)+g_4(n-3)+r_3]c_{n-3}+[p_4(n^2-5n+6)+g_3(n-2)+r_2]c_{n-2}$$
$$+[p_3(n^2-3n+2)+g_2(n-1)+r_1]c_{n-1}+[p_2(n^2-n)+g_1n+r_0]c_n$$
$$+[p_1(n^2+n)+g_0(n+1)]c_{n+1}=0.$$

Thus, the 7-term recurrence relation is found. We divide the last expression by c_{n-5}, multiplied by n^{-2}, and tend $n\to\infty$; this results in a simple algebraic equation for the quantity associated with the possible convergence radii of the power series

$$R=\lim_{n\to\infty}\frac{c_n}{c_{n+1}},\quad R_{conv}=|R|,\; p_1+p_2R+p_3R^2+p_4R^3+p_5R^4+p_6R^5=0. \tag{1.75}$$

This equation can be rewritten differently

$$P(R)=0 \qquad \text{or} \qquad (R-1)[1-4\gamma R(R-1)]\,[1+4\gamma R(R-1)]=0\,;$$

its roots are

$$R=1,\quad R_{1,2}=\frac{1\pm\sqrt{1+\gamma^{-1}}}{2},\qquad R_{3,4}=\frac{1\pm\sqrt{1+\gamma^{-1}}}{2}. \tag{1.76}$$

The minimal radius of convergence $R^{min}_{conv}=1$, it is sufficient to cover the entire physical domain for the variable $x\in(0,1)$.

We write down solutions with the power series in the variable x (of the type I), and solutions (of the type II) with the power series in the variable $(x-1)$ (they have a simple asymptotic behaviour in different singular points):

Solutions I,

$$\begin{gathered}A=+\frac{i\sqrt{\epsilon-1}}{2},\; B=+\frac{i\sqrt{\epsilon-1}}{2},\\ f=x^{+i\sqrt{\epsilon-1}/2}(x-1)^{+i\sqrt{\epsilon-1}/2}F(x),\bar{f}=x^{+i\sqrt{\epsilon-1}/2}(x-1)^{+i\sqrt{\epsilon-1}/2}F(1-x);\end{gathered} \tag{1.77}$$

$$\begin{gathered}A=-\frac{i\sqrt{\epsilon-1}}{2},\; B=-\frac{i\sqrt{\epsilon-1}}{2},\\ f^*=x^{-i\sqrt{\epsilon-1}/2}(x-1)^{-i\sqrt{\epsilon-1}/2}F^*(x);\bar{f}^*=x^{-i\sqrt{\epsilon-1}/2}(x-1)^{-i\sqrt{\epsilon-1}/2}F^*(1-x);\end{gathered} \tag{1.78}$$

Solutions II,

$$\begin{gathered}A=+\frac{i\sqrt{\epsilon-1}}{2},\; B=-\frac{i\sqrt{\epsilon-1}}{2},\\ g=x^{+i\sqrt{\epsilon-1}/2}(x-1)^{-i\sqrt{\epsilon-1}/2}G(x),\bar{g}=x^{-i\sqrt{\epsilon-1}/2}(x-1)^{+i\sqrt{\epsilon-1}/2}G(1-x),\end{gathered} \tag{1.79}$$

$$\begin{gathered}A=-\frac{i\sqrt{\epsilon-1}}{2},\; B=+\frac{i\sqrt{\epsilon-1}}{2},\\ g^*=x^{-i\sqrt{\epsilon-1}/2}(x-1)^{+i\sqrt{\epsilon-1}/2}G^*(x),g^*=x^{+i\sqrt{\epsilon-1}/2}(x-1)^{-i\sqrt{\epsilon-1}/2}G^*(1-x).\end{gathered} \tag{1.80}$$

Remember that

$$\begin{gathered}x=\frac{1-\tanh z}{2}=\frac{1}{1+e^{2z}},\quad 1-x=\frac{1+\tanh z}{2}=\frac{1}{1+e^{-2z}}\,,\\ z\to+\infty,\quad x\to e^{-2z},\quad 1-x\to 1\,,\quad z\to-\infty,\quad x\to 1,\quad 1-x\to e^{+2z}\,.\end{gathered} \tag{1.81}$$

Analysis of the main differential equation can be performed also with the use of the variable

$$\mathrm{ch}^2 z = Z\,, \quad \frac{d}{dz} = 2\mathrm{sh}\, z\, \mathrm{ch}\, z \frac{d}{dZ}\,, \quad 2\frac{\mathrm{sh}\, z}{\mathrm{ch}\, z}\frac{d}{dz} = 4(Z-1)\frac{d}{dZ}\,, \quad \frac{d^2}{dz^2} = \frac{d}{dz} 2\mathrm{sh}\, z\, \mathrm{ch}\, z \frac{d}{dZ}$$

$$= 4\mathrm{sh}^2 z\, \mathrm{ch}^2 z \frac{d^2}{dZ^2} + 2(\mathrm{ch}^2 z + \mathrm{sh}^2 z)\frac{d}{dZ} = 4Z(Z-1)\frac{d^2}{dZ^2} + 2(2Z-1)\frac{d}{dZ}\,,$$

note that this variable does not distinguish the regions $z \to -\infty$ and $z \to +\infty$. In this variable the basic equation takes the form

$$\Big[\frac{d^2}{dZ^2} + \Big(\frac{1/2}{Z} + \frac{1/2}{Z-1}\Big)\frac{d}{dZ} + \Big(\frac{b}{4\gamma} - \frac{\epsilon - 1}{4}\Big)\frac{1}{Z} + \Big(\frac{\epsilon-1}{4} + \frac{b\gamma - \Lambda}{4(1-\gamma^2)}\Big)\frac{1}{Z-1}$$
$$+ \frac{\Lambda - b}{8\gamma(1-\gamma)}\frac{1}{Z-\gamma} - \frac{\Lambda + b}{8\gamma(1+\gamma)}\frac{1}{Z+\gamma}\Big] F(Z) = 0\,. \qquad (1.82)$$

This equation has five singular points; four of them are regular $Z = -\gamma,\ +\gamma,\ 0,\ 1$. Points $Z = 0, \pm\gamma,\ \mid \gamma \mid << 1$ lie outside the physical range of the variable $Z:\ Z = \mathrm{ch}^2 z \in [1, +\infty)$. The character of the singularity in the point $Z = \infty$ will be investigated below.

Note that the summing of the simple fractions in eq. (1.82) leads to

$$\Big(\frac{b}{4\gamma} - \frac{\epsilon-1}{4}\Big)\frac{1}{Z} + \Big(\frac{\epsilon-1}{4} + \frac{b\gamma - \Lambda}{4(1-\gamma^2)}\Big)\frac{1}{Z-1}$$

$$+ \frac{\Lambda - b}{8\gamma(1-\gamma)}\cdot\frac{1}{Z-\gamma} - \frac{\Lambda+b}{8\gamma(1+\gamma)}\cdot\frac{1}{Z+\gamma} = -\frac{1}{4}\,\frac{-(Z^2-\gamma^2)(\epsilon-1) + \Lambda Z - b\gamma}{(Z-1)\,Z\,(Z^2-\gamma^2)}\,.$$

In short form, eq. (1.82) is written as follows

$$\Big[\frac{d^2}{dZ^2} + \Big(\frac{1/2}{Z} + \frac{1/2}{Z-1}\Big)\frac{d}{dZ} + \frac{A}{Z} + \frac{B}{Z-1} + \frac{C}{Z-\gamma} + \frac{D}{Z+\gamma}\Big]F(Z) = 0, \qquad (1.83)$$

where

$$A = \frac{b}{4\gamma} - \frac{\epsilon-1}{4}, \quad B = \frac{\epsilon-1}{4} + \frac{b\gamma-\Lambda}{4(1-\gamma^2)},$$

$$C = \frac{\Lambda - b}{8\gamma(1-\gamma)}, \quad D = -\frac{\Lambda+b}{8\gamma(1+\gamma)}.$$

Since the equation does not contain complex quantities, we conclude that each complex-valued solution will be accompanied by its counterpart, the complex-conjugate one.

Now let us investigate the type of the singular point ∞. To do this, we transform the above equation to the variable $y = Z^{-1}$:

$$\Big[\frac{d^2}{dy^2} + \Big(\frac{1}{y} - \frac{1/2}{1-y}\Big)\frac{d}{dy} + \frac{A}{y^3} + \frac{B}{y^3(1-y)} + \frac{C}{y^3(1-\gamma y)} + \frac{D}{y^3(1+\gamma y)}\Big]F(y) = 0.$$

After decomposing all fractions into elementary ones:

$$\frac{B}{y^3(1-y)} = \frac{B}{y^3} + \frac{B}{y^2} + \frac{B}{y} + \frac{B}{1-y},$$

$$\frac{C}{y^3(1-y\gamma)} = \frac{C}{y^3} + \frac{C\gamma}{y^2} + \frac{C\gamma^2}{y} + \frac{C\gamma^3}{1-y\gamma},$$

$$\frac{D}{y^3(1+y\gamma)} = \frac{D}{y^3} - \frac{D\gamma}{y^2} + \frac{D\gamma^2}{y} - \frac{D\gamma^3}{1+y\gamma},$$

we arrive at the most symmetrical form

$$\Big[\frac{d^2}{dy^2} + \Big(\frac{1}{y} - \frac{1/2}{1-y}\Big)\frac{d}{dy} + \frac{A+B+C+D}{y^3}$$

$$+\frac{B+C\gamma-D\gamma}{y^2} + \frac{B+C\gamma^2+D\gamma^2}{y} + \frac{B}{1-y} + \frac{C\gamma^3}{1-y\gamma} - \frac{D\gamma^3}{1+y\gamma}\Big]F(y) = 0\,.$$

An important identity is proved:

$$A+B+C+D$$
$$= \frac{b}{4\gamma} - \frac{\epsilon}{4} + \frac{\epsilon}{4} + \frac{b\gamma - \Lambda}{4(1+\gamma)(1-\gamma)} + \frac{\Lambda - b}{8\gamma(1-\gamma)} - \frac{\Lambda + b}{8\gamma(1+\gamma)}$$
$$= \frac{b - b\gamma^2 + b\gamma^2 - \lambda\gamma}{4\gamma(1-\gamma^2)} + \frac{\lambda + \lambda\gamma - b - b\gamma - \lambda + \lambda\gamma - b + b\gamma}{8\gamma(1-\gamma^2)}$$
$$= \frac{b - b\gamma^2 + b\gamma^2 - \lambda\gamma - b + \lambda\gamma}{4\gamma(1-\gamma^2)} = 0\,.$$

Therefore, the equation simplifies

$$\Big[\frac{d^2}{dy^2} + (\frac{1}{y} - \frac{1/2}{1-y})\frac{d}{dy} + \frac{B+C\gamma-D\gamma}{y^2} + \frac{B+C\gamma^2+D\gamma^2}{y}$$

$$+\frac{B}{1-y} - \frac{C\gamma^2}{y-\gamma^{-1}} - \frac{D\gamma^2}{y+\gamma^{-1}}\Big]F(y) = 0, \tag{1.84}$$

the structure of eq. (1.84) indicates that $Z = \infty$ is the regular singular point.

We turn to eq. (1.84) and find behaviour of its solutions in the neighbourhood of the point $y = 0$:

$$\Big(\frac{d^2}{dy^2} + \frac{1}{y}\frac{d}{dy} + \frac{B+C\gamma-D\gamma}{y^2}\Big)F(y) = 0\,,$$

$$f = y^a, \qquad a^2 + B + (C-D)\gamma = 0\,, \quad a = \pm\sqrt{-B-(C-D)\gamma}\,.$$

Hence, taking into account the explicit form of the coefficients, we get

$$a_1 = \frac{+i\sqrt{\epsilon-1}}{2}, \qquad a_2 = \frac{-i\sqrt{\epsilon-1}}{2}. \tag{1.85}$$

Thus, we obtain

$$y \to 0, \quad F(y) = y^a = Z^{-a} = (\cosh z)^{-2a} = \Big(\frac{e^z+e^{-z}}{2}\Big)^{\mp i\sqrt{\epsilon-1}}. \tag{1.86}$$

We have two possibilities:

$$z \to -\infty, \qquad F \sim e^{\pm i\sqrt{\epsilon-1}\,z}\,; \qquad z \to +\infty, \qquad F \sim e^{\mp i\sqrt{\epsilon-1}\,z}\,. \tag{1.87}$$

Let us find approximate solutions near the point $y = 1$:

$$\begin{gathered}\Big(\frac{d^2}{dy^2} + \frac{1/2}{y-1}\frac{d}{dy} - \frac{B}{y-1}\Big)F(y) = 0\,,\\ F = (y-1)^b, \quad b(b-1) + \frac{b}{2} = 0, \quad b_1 = 0, \quad b_2 = \frac{1}{2}\,.\end{gathered} \tag{1.88}$$

Similarly, we find the possible behaviour of solutions near singular points $+1/\gamma$ and $-1/\gamma$:

$$y \longrightarrow +1/\gamma, \quad F - (y - 1/\gamma,)^{\rho}, \quad \rho_{1,2} = 0, 1 \; ; \\ y \longrightarrow -1/\gamma, \quad F - (y + 1/\gamma,)^{\sigma}, \quad \sigma_{1,2} = 0, 1 \; . \tag{1.89}$$

The physical domain for the variable is the interval $Z \in (1, \infty)$ or $y \in (0, 1)$.

Let us construct local Frobenius solutions of eq. (1.84)

$$\Big[\frac{d^2}{dy^2} + (\frac{1}{y} - \frac{1/2}{1-y})\frac{d}{dy} + \frac{B + C\gamma - D\gamma}{y^2} \\ + \frac{B + C\gamma^2 + D\gamma^2}{y} + \frac{B}{1-y} - \frac{C\gamma}{y - 1/\gamma} - \frac{D\gamma^2}{y + 1/\gamma}\Big] F(y) = 0 \tag{1.90}$$

near the point $Z = \infty$ $(y = 0)$. Taking into account the established asymptotics these solutions are built in the form

$$F = y^a (y-1)^b f(y); \tag{1.91}$$

we obtain the equation for $f(y)$:

$$f'' + \Big(\frac{2a+1}{y} + \frac{2b+1/2}{y-1}\Big) f' + \Big[\frac{a^2 + B + (C-D)\gamma}{y^2} + \frac{b(b-1) + b/2}{(y-1)^2} \\ + \frac{-2ab - a/2 - b + B + (C+D)\gamma^2}{y} + \frac{2ab + a/2 + b - B}{y-1} + \frac{C\gamma^3}{1 - y\gamma} - \frac{D\gamma^3}{1 + y\gamma}\Big] f = 0 \, .$$

Now apply already known restrictions

$$\frac{a^2 + B + (C-D)\gamma}{y^2} = 0 \quad \Longrightarrow \quad a = a_1, a_2 \, ,$$

$$\frac{b(b-1) + b/2}{(y-1)^2} = 0 \quad \Longrightarrow \quad b = 0, \frac{1}{2} \; ;$$

so we arrive at

$$F = y^a (y-1)^b f(y), \qquad f'' + \Big(\frac{2a+1}{y} + \frac{2b+1/2}{y-1}\Big) f' \\ + \Big(\frac{-2ab - a/2 - b + B + (C+D)\gamma^2}{y} + \frac{2ab + a/2 + b - B}{y-1} \\ - \frac{C\gamma^2}{y - 1/\gamma} - \frac{D\gamma^2}{y + 1/\gamma}\Big) f = 0 \, . \tag{1.92}$$

There arises possibility to construct four solutions (remind that $Z = \cosh^2 z$)

$$1. \quad a_1 = \frac{+i\sqrt{\epsilon - 1}}{2}, \; b_1 = 0, \quad F_{(1)} = (\cosh z)^{-i\sqrt{\epsilon-1}} \, f_{(1)}(\frac{1}{Z}) \; ; \\ 1'. \quad a_2 = \frac{-i\sqrt{\epsilon - 1}}{2}, \; b_1 = 0, \quad F_{(1')} = (\cosh z)^{+i\sqrt{\epsilon-1}} f_{(1')}(\frac{1}{Z}) \; ; \tag{1.93}$$

and

$$2. \; a_1 = \frac{+i\sqrt{\epsilon - 1}}{2}, b_2 = \frac{1}{2}, \quad F_{(2)} = +i(\cosh z)^{-i\sqrt{\epsilon-1}} \frac{\sinh z}{\cosh z} f_{(2)}(\frac{1}{Z}); \\ 2'. \; a_2 = \frac{-i\sqrt{\epsilon - 1}}{2}, b_2 = \frac{1}{2}, \quad F_{(2')} = -i(\cosh z)^{+i\sqrt{\epsilon-1}}; \frac{\sinh z}{\cosh z} f_{(2')}(\frac{1}{Z}). \tag{1.94}$$

We note that these solutions are divided into two pairs of conjugate ones.

All four functions $f_i(y)$ are subject to the equation with the general structure (see eq. (1.92))

$$f'' + \Big(\frac{\alpha_1}{y} + \frac{\beta_1}{y-1}\Big) f' + \Big(\frac{\alpha}{y} + \frac{\beta}{y-1} + \frac{c}{y-1/\gamma} + \frac{d}{y+1/\gamma}\Big) f = 0 \,. \tag{1.95}$$

Multiplying eq. (1.95) by $y\,(y-1)\,(y-1/\gamma)\,(y+1/\gamma)$, we get

$$y\,(y-1)\,(y-1/\gamma)\,(y+1/\gamma)\, f''$$
$$+\Big[\frac{\alpha_1\, y\,(y-1)\,(y-1/\gamma)\,(y+1/\gamma)}{y} + \frac{\beta_1\, y\,(y-1)\,(y-1/\gamma)\,(y+1/\gamma)}{y-1}\Big] f'$$
$$+\Big[\frac{\alpha\, y\,(y-1)\,(y-1/\gamma)\,(y+1/\gamma)}{y} + \frac{\beta\, y\,(y-1)\,(y-1/\gamma)\,(y+1/\gamma)}{y-1}$$
$$+\frac{c\, y\,(y-1)\,(y-1/\gamma)\,(y+1/\gamma)}{y-1/\gamma} + \frac{d\, y\,(y-1)\,(y-1/\gamma)\,(y+1/\gamma)}{y+1/\gamma}\Big] f = 0$$

or

$$[y^4 - y^3 - y^2/\gamma^2 + y/\gamma^2]\, f'' + \Big[(\alpha_1+\beta_1)\, y^3 - \alpha_1 y^2 - \frac{\alpha_1+\beta_1}{\gamma^2}\, y + \frac{\alpha_1}{\gamma^2}\Big] f'$$

$$+\Big[\, (\alpha+\beta+c+d)\; y^3 - \Big(\alpha + c + d - \frac{c-d}{\gamma}\Big)\, y^2 - \Big(\frac{\alpha+\beta}{\gamma^2} + \frac{c-d}{\gamma}\Big)\, y + \frac{\alpha}{\gamma^2}\Big] f = 0 \,.$$

Solutions can be constructed as power series

$$f = \sum_{n=0}^{\infty} a_n y^n, \quad f' = \sum_{n=1}^{\infty} n a_n y^{n-1}, \quad f'' = \sum_{n=2}^{\infty} n(n-1) a_n y^{n-2} \,.$$

From the equation for f we get

$$\sum_{n=2}^{\infty} n(n-1) a_n y^{n+2} - \sum_{n=2}^{\infty} n(n-1) a_n y^{n+1} - \frac{1}{\gamma^2} \sum_{n=2}^{\infty} n(n-1) a_n y^{n} + \frac{1}{\gamma^2} \sum_{n=2}^{\infty} n(n-1) a_n y^{n-1}$$

$$+(\alpha_1+\beta_1) \sum_{n=1}^{\infty} n a_n y^{n+2} - \alpha_1 \sum_{n=1}^{\infty} n a_n y^{n+1} - \frac{\alpha_1+\beta_1}{\gamma^2} \sum_{n=1}^{\infty} n a_n y^{n} + \frac{\alpha_1}{\gamma^2} \sum_{n=1}^{\infty} n a_n y^{n-1}$$

$$+(\alpha+\beta+c+d) \sum_{n=0}^{\infty} a_n y^{n+3} - \Big(\alpha + c + d - \frac{c-d}{\gamma}\Big) \sum_{n=0}^{\infty} a_n y^{n+2}$$

$$-\Big(\frac{\alpha+\beta}{\gamma^2} + \frac{c-d}{\gamma}\Big) \sum_{n=0}^{\infty} a_n y^{n+1} + \frac{\alpha}{\gamma^2} \sum_{n=0}^{\infty} a_n y^{n} = 0 \,,$$

further changing the summation indices we derive

$$\sum_{k=4}^{\infty} (k-2)(k-3)\, a_{k-2}\, y^k - \sum_{k=3}^{\infty} (k-1)(k-2)\, a_{k-1}\, y^k - \frac{1}{\gamma^2} \sum_{k=2}^{\infty} k(k-1)\, a_k\, y^k + \frac{1}{\gamma^2} \sum_{k=1}^{\infty} k(k+1)\, a_{k+1}\, y^k$$

$$+(\alpha_1+\beta_1) \sum_{k=3}^{\infty} (k-2)\, a_{k-2}\, y^k - \alpha_1 \sum_{k=2}^{\infty} (k-1)\, a_{k-1}\, y^k - \frac{\alpha_1+\beta_1}{\gamma^2} \sum_{k=1}^{\infty} k\, a_k\, y^k + \frac{\alpha_1}{\gamma^2} \sum_{k=0}^{\infty} (k+1)\, a_{k+1}\, y^k$$

$$+(\alpha+\beta+c+d)\sum_{k=3}^{\infty} a_{k-3}\, y^k - \Big(\alpha+c+d-\frac{c-d}{\gamma}\Big)\sum_{k=2}^{\infty} a_{k-2}\, y^k$$

$$-\Big(\frac{\alpha+\beta}{\gamma^2}+\frac{c-d}{\gamma}\Big)\sum_{k=1}^{\infty} a_{k-1}\, y^k + \frac{\alpha}{\gamma^2}\sum_{k=0}^{\infty} a_k\, y^k = 0\,.$$

Equate to zero the coefficients of all powers of y^k:

$$k=0,\quad \frac{\alpha_1}{\gamma^2}\, a_1 + \frac{\alpha}{\gamma^2}\, a_0 = 0\,,$$

$$k=1,\qquad \frac{2}{\gamma^2}a_2 - \frac{\alpha_1+\beta_1}{\gamma^2}a_1 + 2\frac{\alpha_1}{\gamma^2}a_2 - \Big(\frac{\alpha+\beta}{\gamma^2}+\frac{c-d}{\gamma}\Big)a_0 + \frac{\alpha}{\gamma^2}a_1 = 0\,,$$

$$k=2,\quad -\frac{1}{\gamma^2}2a_2 + \frac{1}{\gamma^2}6a_3 - \alpha_1 a_1 - \frac{\alpha_1+\beta_1}{\gamma^2}2a_2 + \frac{\alpha_1}{\gamma^2}3a_3$$

$$-\Big(\alpha+c+d-\frac{c-d}{\gamma}\Big)a_0 - \Big(\frac{\alpha+\beta}{\gamma^2}+\frac{c-d}{\gamma}\Big)a_1 + \frac{\alpha}{\gamma^2}a_2 = 0\,,$$

$$k=3\,,\quad -2a_2 - \frac{1}{\gamma^2}6a_3 + \frac{1}{\gamma^2}12a_4 + (\alpha_1+\beta_1)a_1 - \alpha_1 2a_2 - \frac{\alpha_1+\beta_1}{\gamma^2}3a_3 + \frac{\alpha_1}{\gamma^2}4a_4$$

$$+(\alpha+\beta+c+d)\,a_0 - \Big(\alpha+c+d-\frac{c-d}{\gamma}\Big)a_1 - \Big(\frac{\alpha+\beta}{\gamma^2}+\frac{c-d}{\gamma}\Big)a_2 + \frac{\alpha}{\gamma^2}a_3 = 0$$

$$k=4\,,\quad 2a_2 - 6a_3 - \frac{12}{\gamma^2}a_4 + \frac{20}{\gamma^2}a_5 + (\alpha_1+\beta_1)2a_2 - \alpha_1 3a_3 - \frac{\alpha_1+\beta_1}{\gamma^2}4a_4 + \frac{\alpha_1}{\gamma^2}5a_5$$

$$+(\alpha+\beta+c+d)\,a_1 - \Big(\alpha+c+d-\frac{c-d}{\gamma}\Big)a_2 - \Big(\frac{\alpha+\beta}{\gamma^2}+\frac{c-d}{\gamma}\Big)a_3 + \frac{\alpha}{\gamma^2}a_4 = 0,$$

$$k=5,\,6,\,7,\,\ldots\quad (k-2)(k-3)a_{k-2} - (k-1)(k-2)a_{k-1} - \frac{1}{\gamma^2}k(k-1)a_k + \frac{1}{\gamma^2}k(k+1)a_{k+1}$$

$$+(\alpha_1+\beta_1)(k-2)a_{k-2} - \alpha_1(k-1)a_{k-1} - \frac{\alpha_1+\beta_1}{\gamma^2}ka_k + \frac{\alpha_1}{\gamma^2}(k+1)a_{k+1}$$

$$+(\alpha+\beta+c+d)\,a_{k-3} - \Big(\alpha+c+d-\frac{c-d}{\gamma}\Big)a_{k-2} - \Big(\frac{\alpha+\beta}{\gamma^2}+\frac{c-d}{\gamma}\Big)a_{k-1} + \frac{\alpha}{\gamma^2}a_k = 0.$$

Thus, we arrive at the 5-term recurrence relations for coefficients a_n:

$$k=5,\,6,\,7,\,\ldots\qquad (\alpha+\beta+c+d)\;a_{k-3}$$

$$+\Big[(k-2)(k-3)+\alpha_1+\beta_1)(k-2) - \Big(\alpha+c+d-\frac{c-d}{\gamma}\Big)\Big]a_{k-2}$$

$$-\Big[(k-1)(k-2)+\alpha_1(k-1)+\Big(\frac{\alpha+\beta}{\gamma^2}+\frac{c-d}{\gamma}\Big)\Big]a_{k-1}$$

$$+\Big[-\frac{1}{\gamma^2}k(k-1) - \frac{\alpha_1+\beta_1}{\gamma^2}k + \frac{\alpha}{\gamma^2}\Big]a_k$$

$$+\Big[\frac{1}{\gamma^2}k(k+1) + \frac{\alpha_1}{\gamma^2}(k+1)\Big]a_{k+1} = 0\,. \qquad (1.96)$$

The Poincaré–Perron approach is used to analyse the convergence radii of the series. To this end, the recurrence relation is divided by a_{k-3}:

$$(\alpha+\beta+c+d)+\Big[(k-2)(k-3)+(\alpha_1+\beta_1)(k-2)-\Big(\alpha+c+d-\frac{c-d}{\gamma}\Big)\Big]\frac{a_{k-2}}{a_{k-3}}$$

$$-\Big[(k-1)(k-2)+\alpha_1(k-1)+\Big(\frac{\alpha+\beta}{\gamma^2}+\frac{c-d}{\gamma}\Big)\Big]\frac{a_{k-1}}{a_{k-2}}\frac{a_{k-2}}{a_{k-3}}$$

$$+\Big[-\frac{1}{\gamma^2}k(k-1)-\frac{\alpha_1+\beta_1}{\gamma^2}k+\frac{\alpha}{\gamma^2}\Big]\frac{a_k}{a_{k-1}}\frac{a_{k-1}}{a_{k-2}}\frac{a_{k-2}}{a_{k-3}}$$

$$+\Big[\frac{1}{\gamma^2}k(k+1)+\frac{\alpha_1}{\gamma^2}(k+1)\Big]\frac{a_{k+1}}{a_k}\frac{a_k}{a_{k-1}}\frac{a_{k-1}}{a_{k-2}}\frac{a_{k-2}}{a_{k-3}}=0\,.$$

The radius of convergence is the inverse to the quantity

$$r=\lim_{k\to\infty}\frac{a_{k+1}}{a_k}\,,\qquad R_{conv}=\frac{1}{|\,r\,|}\,.$$

To find an algebraic equation for r, we multiply the above equation by k^{-2} and tend k to ∞. This result in

$$r-r^2-\frac{1}{\gamma^2}r^3+\frac{1}{\gamma^2}r^4=0\quad\Longrightarrow\quad r(r-1)(r-\gamma)(r+\gamma)=0\,;$$

the roots are

$$r=0,\ 1,\ -\gamma,\ +\gamma; \tag{1.97}$$

the possible convergence radii are

$$R_{conv}=1,\quad\frac{1}{|\gamma|},\quad\infty\,. \tag{1.98}$$

The minimal radius of convergence $R_{conv}=1$ of the series in the variable y covers the entire physical range of the variable $y\in(0,1)$ (which corresponds to $Z\in(1,\infty)$).

It is possible to explore the most general substitution

$$F=y^a(y-1)^b(y-1/\gamma)^\rho(y+1/\gamma)^\sigma f(y)\,; \tag{1.99}$$

the function $f(y)$ obeys the equation

$$\frac{d^2f}{dy^2}+\Big[\frac{2\,\sigma}{y+\gamma^{-1}}+\frac{1/2+2\,b}{y-1}+\frac{2\,\rho}{y-\gamma^{-1}}+\frac{1+2\,a}{y}\Big]\frac{df}{dy}$$

$$+\Big[\frac{-2\gamma^2\left[1/2\,D\gamma^2+((a+\rho/2+1/2)\,\sigma+D/2)\,\gamma+\sigma\,(a+b+\rho/2+3/4)\right]}{(\gamma y+1)\,(\gamma+1)}$$

$$+\frac{1}{2}\,\frac{b\,(2\,b-1)}{(y-1)^2}$$

$$+\frac{1}{2}\frac{\left[(4a+4\rho+4\sigma+2)\,b-2B+a+\rho+\sigma\right]\gamma^2+4\,(\rho-\sigma)\,(b+1/4)\,\gamma+(-4a-2)\,b+2B-a}{(y-1)\,(\gamma-1)\,(\gamma+1)}$$

$$-\frac{\gamma^2\left[C\gamma^2+((-2a-\sigma-1)\,\rho-C)\,\gamma+2\,(a+b+\sigma/2+3/4)\,\rho\right]}{(\gamma y-1)\,(\gamma-1)}$$

$$+\frac{1}{2}\frac{(2C+2D)\,\gamma^2-4\,(a+1/2)\,(\rho-\sigma)\,\gamma+(-4b-1)\,a+2B-2b}{y}$$
$$+\frac{\rho\gamma^2\,(\rho-1)}{(\gamma y-1)^2}+\frac{C\gamma+a^2-D\gamma+B}{y^2}+\frac{\sigma\gamma^2\,(\sigma-1)}{(\gamma y+1)^2}\Big]\,f=0\,. \qquad (1.100)$$

The evident restrictions should be imposed on the parameters

$$\begin{aligned}
\frac{a^2+B+(C-D)\gamma}{y^2}=0 \quad &\Longrightarrow \quad a=a_1,a_2\;,\\
\frac{b(b-1)+b/2}{(y-1)^2}=0 \quad &\Longrightarrow \quad b=0,1/2\;,\\
\frac{\rho\,\gamma^2\,(\rho-1)}{(\gamma y-1)^2}=0 \quad &\Longrightarrow \quad \rho=0,1\;,\\
\frac{\sigma\,\gamma^2\,(\sigma-1)}{(\gamma y+1)^2}=0 \quad &\Longrightarrow \quad \sigma=0,1;
\end{aligned} \qquad (1.101)$$

as a result, we have the ability to build 16 solutions. For the function $f(y)$, we get a simple equation

$$\frac{d^2 f}{dy^2}+\Big[\frac{1+2\,a}{y}+\frac{1/2+2\,b}{y-1}+\frac{2\,\rho}{y-1/\gamma}+\frac{2\,\sigma}{y+1/\gamma}\Big]\,\frac{df}{dy}$$
$$+\Big[\frac{(C+D)\gamma^2-2(a+1/2)(\rho-\sigma)\gamma+(-2b-1/2)a+B-b}{y}$$
$$+\frac{[(4a+4\rho+4\sigma+2)b-2B+a+\rho+\sigma]\gamma^2+4(\rho-\sigma)(b+1/4)\gamma+(-4a-2)b+2B-a}{2(\gamma-1)(\gamma+1)}\,\frac{1}{y-1}$$
$$+\frac{-\gamma[D\gamma^2+((2a+\rho+1)\sigma+D)\gamma+2\sigma\,(a+b+\rho/2+3/4)]}{\gamma+1}\,\frac{1}{y+1/\gamma}$$
$$-\frac{\gamma[C\gamma^2+((-2a-\sigma-1)\rho-C)\gamma+2\rho(a+b+\sigma/2+3/4)]}{\gamma-1}\,\frac{1}{y-1/\gamma}\Big]\,f=0\,.$$

The equation can be written briefly as follows

$$\frac{d^2 f}{dy^2}+\Big(\frac{A_1}{y}+\frac{B_1}{y-1}+\frac{C_1}{y-1/\gamma}+\frac{D_1}{y+1/\gamma}\Big)\,\frac{df}{dy}$$
$$+\Big(\frac{A_2}{y}+\frac{B_2}{y-1}+\frac{C_2}{y-1/\gamma}+\frac{D_2}{y+1/\gamma}\Big)\,f=0. \qquad (1.102)$$

We give the explicit form of the coefficients $A_1, B_1, C_1, D_1, A_2, B_2, C_2, D_2$ for all 16 cases. The following method can be used: we select eight options for each case depending on $a=a_1,a_2$ being complex conjugate by-cases (we give only for a_1):

$a_1=+i\,\sqrt{\epsilon-1}/2,\ b=0,\ \rho=0,\ \sigma=0\,,$

$$F_{(1)}=(\cosh z)^{-i\sqrt{\epsilon-1}}\,f_{(1)}(y)\,,$$

$$A_1=1+i\sqrt{\epsilon-1}\,,\ B_1=\frac{1}{2}\,,\ C_1=0\,,\ D_1=0\,,$$

$$A_2=\frac{1}{4}\,(-i\sqrt{\epsilon-1}+\Lambda-\epsilon-1)\,,$$

$$B_2 = \frac{1}{4}\,\frac{(\gamma^2-1)\,i\,\sqrt{\epsilon-1}+(\epsilon+1)\,\gamma^2+\beta\,\gamma+\Lambda-\epsilon-1}{(\gamma-1)\,(\gamma+1)}\,,$$

$$C_2 = -\frac{1}{8}\,\frac{\gamma\,(\Lambda+\beta)}{\gamma-1}\,, \quad D_2 = -\frac{1}{8}\,\frac{\gamma\,(\Lambda-\beta)}{\gamma+1};$$

$a_1 = +i\,\sqrt{\epsilon-1}/2,\ b=1/2,\ \rho=0,\ \sigma=0\,,$

$$F_{(2)} = +i\,(\cosh z)^{-i\sqrt{\epsilon-1}}\,\frac{\sinh z}{\cosh z}\,f_{(2)}(y),$$

$$A_1 = 1+i\sqrt{\epsilon-1}\,,\ B_1=\frac{3}{2}\,,\ C_1=0\,,\ D_1=0\,,$$

$$A_2 = \frac{1}{4}\,\left(-3i\sqrt{\epsilon-1}+\Lambda-\epsilon-3\right)\,,$$

$$B_2 = \frac{3}{4}\,\frac{(\gamma^2-1)\,i\,\sqrt{\epsilon-1}+(1/3\epsilon+1)\,\gamma^2+1/3\,\beta\,\gamma+1/3\Lambda-1/3\,\epsilon-1}{(\gamma-1)\,(\gamma+1)}\,,$$

$$C_2 = -\frac{1}{8}\,\frac{\gamma\,(\Lambda+\beta)}{\gamma-1}\,, \quad D_2 = -\frac{1}{8}\,\frac{\gamma\,(\Lambda-\beta)}{\gamma+1};$$

$a_1 = +i\,\sqrt{\epsilon-1}/2,\ b=0,\ \rho=1,\ \sigma=1\,,$

$$F_{(3)} = (\cosh z)^{-i\sqrt{\epsilon-1}}\,\frac{\gamma-\cosh^2 z}{\gamma\cosh^2 z}\,\frac{\gamma+\cosh^2 z}{\gamma\cosh^2 z}\,f_{(3)}(y)\,,$$

$$A_1 = 1+i\sqrt{\epsilon-1}\,,\ B_1=\frac{1}{2}\,,\ C_1=2\,,\ D_1=2\,,$$

$$A_2 = \frac{1}{4}\,(-i\sqrt{\epsilon-1}+\Lambda-\epsilon-1)\,,$$

$$B_2 = \frac{1}{4}\,\frac{(\gamma^2-1)\,i\,\sqrt{\epsilon-1}+(\epsilon+5)\,\gamma^2+\beta\,\gamma+\Lambda-\epsilon-1}{(\gamma-1)\,(\gamma+1)}\,,$$

$$C_2 = \frac{\left[(\gamma-1)\,i\,\sqrt{\epsilon-1}-1/8\,\Lambda-5/2+2\,\gamma-1/8\,\beta\right]\gamma}{\gamma-1}\,,$$

$$D_2 = \frac{-8\,i\,(\gamma+1)\,\gamma\sqrt{\epsilon-1}-\gamma\,(16\,\gamma+\Lambda-\beta+20)}{8\,\gamma+8};$$

$a_1 = +i\,\sqrt{\epsilon-1}/2,\ b=1/2,\ \rho=1,\ \sigma=1\,,$

$$F_{(4)} = +i\,(\cosh z)^{-i\sqrt{\epsilon-1}}\,\frac{\sinh z}{\cosh z}\,\frac{\gamma-\cosh^2 z}{\gamma\cosh^2 z}\,\frac{\gamma+\cosh^2 z}{\gamma\cosh^2 z}\,f_{(4)}(y),$$

$$A_1 = 1+i\sqrt{\epsilon-1}\,,\ B_1=\frac{3}{2}\,,\ C_1=2\,,\ D_1=2\,,$$

$$A_2 = \frac{1}{4}\,\left(-3i\sqrt{\epsilon-1}+\Lambda-\epsilon-3\right)\,,$$

$$B_2 = \frac{3}{4}\,\frac{(\gamma^2-1)\,i\,\sqrt{\epsilon-1}+(1/3\epsilon+5)\,\gamma^2+1/3\,\beta\,\gamma+1/3\Lambda-1/3\,\epsilon-1}{(\gamma-1)\,(\gamma+1)}\,,$$

$$C_2 = \frac{\left[(\gamma-1)\, i\sqrt{\epsilon-1} - 1/8\,\Lambda - 7/2 + 2\,\gamma - 1/8\,\beta\right]\gamma}{\gamma-1},$$

$$D_2 = \frac{-8\,i\,(\gamma+1)\,\gamma\sqrt{\epsilon-1} - \gamma\,(16\,\gamma - \beta + \Lambda + 28)}{8\,\gamma+8};$$

$a_1 = +i\sqrt{\epsilon-1}/,\ b = 0,\ \rho = 0,\ \sigma = 1,$

$$F_{(5)} = (\cosh z)^{-i\sqrt{\epsilon-1}}\,\frac{\gamma+\cosh^2 z}{\gamma\cosh^2 z}\,f_{(5)}(y)\,,$$

$$A_1 = 1 + i\sqrt{\epsilon-1},\ B_1 = \frac{1}{2},\ C_1 = 0,\ D_1 = 2,$$

$$A_2 = \left(\gamma - \frac{1}{4}\right) i\sqrt{\epsilon-1} - \frac{1}{4}\,\epsilon - \frac{1}{4} + \gamma + \frac{1}{4}\,\Lambda,$$

$$B_2 = \frac{1}{4}\,\frac{(\gamma^2-1)\, i\sqrt{\epsilon-1} + (\epsilon+3)\,\gamma^2 + (\beta-2)\,\gamma + \Lambda - \epsilon - 1}{(\gamma-1)\,(\gamma+1)},$$

$$C_2 = -\frac{\gamma\,(\Lambda+\beta)}{8\,\gamma-8},$$

$$D_2 = \frac{-8\,i\gamma\,(\gamma+1)\sqrt{\epsilon-1} - \gamma\,(8\,\gamma - \beta + \Lambda + 12)}{8\,\gamma+8};$$

$a_1 = +i\sqrt{\epsilon-1}/2,\ b = 1/2,\ \rho = 0,\ \sigma = 1,$

$$F_{(6)} = +i\,(\cosh z)^{-i\sqrt{\epsilon-1}}\,\frac{\sinh z}{\cosh z}\,\frac{\gamma+\cosh^2 z}{\gamma\cosh^2 z}\,f_{(6)}(y),$$

$$A_1 = 1 + i\sqrt{\epsilon-1},\ B_1 = \frac{3}{2},\ C_1 = 0,\ D_1 = 2,$$

$$A_2 = \left(\gamma - \frac{3}{4}\right) i\sqrt{\epsilon-1} - \frac{1}{4}\,\epsilon - \frac{3}{4} + \gamma + \frac{1}{4}\,\Lambda,$$

$$B_2 = \frac{3}{4}\,\frac{(\gamma^2-1)\, i\sqrt{\epsilon-1} + (1/3\,\epsilon+3)\,\gamma^2 + (-2+1/3\,\beta)\,\gamma - 1 + 1/3\,\Lambda - 1/3\,\epsilon}{(\gamma-1)\,(\gamma+1)},$$

$$C_2 = -\frac{\gamma\,(\Lambda+\beta)}{8\,\gamma-8},$$

$$D_2 = \frac{-8\,i\gamma\,(\gamma+1)\sqrt{\epsilon-1} - \gamma\,(8\,\gamma - \beta + \Lambda + 20)}{8\,\gamma+8};$$

$a_1 = +i\sqrt{\epsilon-1}/2,\ b = 0,\ \rho = 1,\ \sigma = 0,$

$$F_{(7)} = (\cosh z)^{-i\sqrt{\epsilon-1}}\,\frac{\gamma-\cosh^2 z}{\gamma\cosh^2 z}\,f_{(7)}(y)\,,$$

$$A_1 = 1 + i\sqrt{\epsilon-1},\ B_1 = \frac{1}{2},\ C_1 = 2,\ D_1 = 0,$$

$$A_2 = -\left(\gamma + \frac{1}{4}\right) i\sqrt{\epsilon-1} - \frac{1}{4}\,\epsilon - \frac{1}{4} - \gamma + \frac{1}{4}\,\Lambda,$$

$$B_2 = \frac{1}{4}\,\frac{(\gamma^2-1)\, i\sqrt{\epsilon-1} + (\epsilon+3)\,\gamma^2 + (\beta+2)\,\gamma + \Lambda - \epsilon - 1}{(\gamma-1)\,(\gamma+1)},$$

$$C_2 = \frac{\gamma\left[(\gamma-1)\, i\sqrt{\epsilon-1} - 1/8\,\Lambda - 3/2 + \gamma - 1/8\,\beta\right]}{\gamma-1},$$

$$D_2 = \frac{\gamma\,(-\Lambda+\beta)}{8\,\gamma+8};$$

$a_1 = +i\sqrt{\epsilon-1}/2,\ b = 1/2,\ \rho = 1,\ \sigma = 0,$

$$F_{(8)} = +i\,(\cosh z)^{-i\sqrt{\epsilon-1}}\,\frac{\sinh z}{\cosh z}\,\frac{\gamma-\cosh^2 z}{\gamma\cosh^2 z}\,f_{(8)}(y),$$

$$A_1 = 1 + i\sqrt{\epsilon-1},\ B_1 = \frac{3}{2},\ C_1 = 2,\ D_1 = 0,$$

$$A_2 = -\left(\gamma+\frac{3}{4}\right) i\sqrt{\epsilon-1} - \frac{1}{4}\,\epsilon - \frac{3}{4} - \gamma + \frac{1}{4}\,\Lambda,$$

$$B_2 = \frac{3}{4}\,\frac{(\gamma^2-1)\, i\sqrt{\epsilon-1} + (1/3\,\epsilon+3)\,\gamma^2 + (2+1/3\,\beta)\,\gamma - 1 + 1/3\,\Lambda - 1/3\,\epsilon}{(\gamma-1)\,(\gamma+1)},$$

$$C_2 = \frac{\gamma\left[(\gamma-1)\, i\sqrt{\epsilon-1} - 1/8\,\Lambda - 5/2 + \gamma - 1/8\,\beta\right]}{\gamma-1},$$

$$D_2 = \frac{\gamma\,(-\Lambda+\beta)}{8\,\gamma+8}.$$

We will investigate the power series for the equations of the form (1.102):

$$\frac{d^2 f}{dy^2} + \left(\frac{A_1}{y} + \frac{B_1}{y-1} + \frac{C_1}{y-1/\gamma} + \frac{D_1}{y+1/\gamma}\right)\frac{df}{dy}$$
$$+\left(\frac{A_2}{y} + \frac{B_2}{y-1} + \frac{C_2}{y-1/\gamma} + \frac{D_2}{y+1/\gamma}\right) f = 0. \qquad (1.103)$$

Equation (1.103) is multiplied by $y\,(y-1)\,(y-1/\gamma)\,(y+1/\gamma)$:

$$y\,(y-1)\,(y-1/\gamma)\,(y+1/\gamma)\,f''$$
$$+\left[\frac{A_1\, y\,(y-1)\,(y-1/\gamma)\,(y+1/\gamma)}{y} + \frac{B_1\, y\,(y-1)\,(y-1/\gamma)\,(y+1/\gamma)}{y-1}\right.$$
$$\left.+\frac{C_1\, y\,(y-1)\,(y-1/\gamma)\,(y+1/\gamma)}{y-1/\gamma} + \frac{D_1\, y\,(y-1)\,(y-1/\gamma)\,(y+1/\gamma)}{y+1/\gamma}\right] f'$$
$$+\left[\frac{A_2\, y\,(y-1)\,(y-1/\gamma)\,(y+1/\gamma)}{y} + \frac{B_2\, y\,(y-1)\,(y-1/\gamma)\,(y+1/\gamma)}{y-1}\right.$$
$$\left.+\frac{C_2\, y\,(y-1)\,(y-1/\gamma)\,(y+1/\gamma)}{y-1/\gamma} + \frac{D_2\, y\,(y-1)\,(y-1/\gamma)\,(y+1/\gamma)}{y+1/\gamma}\right] f = 0$$

or

$$
\begin{aligned}
&[y^4 - y^3 - y^2/\gamma^2 + y/\gamma^2]\, f'' \\
&+\Big[(A_1 + B_1 + C_1 + D_1)\, y^3 - \Big(A_1 + C_1 + D_1 - \frac{C_1 - D_1}{\gamma} \Big)\, y^2 \\
&\qquad - \Big(\frac{A_1 + B_1}{\gamma^2} + \frac{C_1 - D_1}{\gamma} \Big)\, y + \frac{A_1}{\gamma^2} \Big] f' \\
&+\Big[(A_2 + B_2 + C_2 + D_2)\, y^3 - \Big(A_2 + C_2 + D_2 - \frac{C_2 - D_2}{\gamma} \Big)\, y^2 \\
&\qquad - \Big(\frac{A_2 + B_2}{\gamma^2} + \frac{C_2 - D_2}{\gamma} \Big)\, y + \frac{A_2}{\gamma^2} \Big] f = 0 .
\end{aligned} \tag{1.104}
$$

Solutions are built in the form of a power series

$$
f = \sum_{n=0}^{\infty} a_n y^n, \quad f' = \sum_{n=1}^{\infty} n a_n y^{n-1}, \quad f'' = \sum_{n=2}^{\infty} n(n-1) a_n y^{n-2} .
$$

From the equation for f we get

$$
\begin{aligned}
&\sum_{n=2}^{\infty} n(n-1) a_n y^{n+2} - \sum_{n=2}^{\infty} n(n-1) a_n y^{n+1} - \frac{1}{\gamma^2} \sum_{n=2}^{\infty} n(n-1) a_n y^n + \frac{1}{\gamma^2} \sum_{n=2}^{\infty} n(n-1) a_n y^{n-1} \\
&+ (A_1 + B_1 + C_1 + D_1) \sum_{n=1}^{\infty} n a_n y^{n+2} - \Big(A_1 + C_1 + D_1 - \frac{C_1 - D_1}{\gamma} \Big) \sum_{n=1}^{\infty} n a_n y^{n+1} \\
&- \Big(\frac{A_1 + B_1}{\gamma^2} + \frac{C_1 - D_1}{\gamma} \Big) \sum_{n=1}^{\infty} n a_n y^n + \frac{A_1}{\gamma^2} \sum_{n=1}^{\infty} n a_n y^{n-1} \\
&+ (A_2 + B_2 + C_2 + D_2) \sum_{n=0}^{\infty} a_n y^{n+3} - \Big(A_2 + C_2 + D_2 - \frac{C_2 - D_2}{\gamma} \Big) \sum_{n=0}^{\infty} a_n y^{n+2} \\
&- \Big(\frac{A_2 + B_2}{\gamma^2} + \frac{C_2 - D_2}{\gamma} \Big) \sum_{n=0}^{\infty} a_n y^{n+1} + \frac{A_2}{\gamma^2} \sum_{n=0}^{\infty} a_n y^n = 0 .
\end{aligned}
$$

The summation indices are changed

$$
\begin{aligned}
&\sum_{k=4}^{\infty} (k-2)(k-3) a_{k-2} y^k - \sum_{k=3}^{\infty} (k-1)(k-2) a_{k-1} y^k - \frac{1}{\gamma^2} \sum_{k=2}^{\infty} k(k-1) a_k y^k + \frac{1}{\gamma^2} \sum_{k=1}^{\infty} k(k+1) a_{k+1} y^k \\
&+ (A_1 + B_1 + C_1 + D_1) \sum_{k=3}^{\infty} (k-2) a_{k-2} y^k - \Big(A_1 + C_1 + D_1 - \frac{C_1 - D_1}{\gamma} \Big) \sum_{k=2}^{\infty} (k-1) a_{k-1} y^k \\
&- \Big(\frac{A_1 + B_1}{\gamma^2} + \frac{C_1 - D_1}{\gamma} \Big) \sum_{k=1}^{\infty} k a_k y^k + \frac{A_1}{\gamma^2} \sum_{k=0}^{\infty} (k+1) a_{k+1} y^k \\
&+ (A_2 + B_2 + C_2 + D_2) \sum_{k=3}^{\infty} a_{k-3} y^k - \Big(A_2 + C_2 + D_2 - \frac{C_2 - D_2}{\gamma} \Big) \sum_{k=2}^{\infty} a_{k-2} y^k \\
&- \Big(\frac{A_2 + B_2}{\gamma^2} + \frac{C_2 - D_2}{\gamma} \Big) \sum_{k=1}^{\infty} a_{k-1} y^k + \frac{A_2}{\gamma^2} \sum_{k=0}^{\infty} a_k y^k = 0 .
\end{aligned}
$$

We equate to zero the coefficients of all powers of y^k:

$$k=0\,, \qquad \frac{A_1}{\gamma^2}\,a_1+\frac{A_2}{\gamma^2}\,a_0=0\,,$$

$$k=1\,, \qquad \frac{2}{\gamma^2}\,a_2-\Big(\frac{A_1+B_1}{\gamma^2}+\frac{C_1-D_1}{\gamma}\Big)\,a_1+2\,\frac{A_1}{\gamma^2}\,a_2$$
$$-\Big(\frac{A_2+B_2}{\gamma^2}+\frac{C_2-D_2}{\gamma}\Big)a_0+\frac{A_2}{\gamma^2}a_1=0\,,$$

$$k=2\,, \qquad -\frac{2}{\gamma^2}\,a_2+\frac{6}{\gamma^2}\,a_3-\Big(A_1+C_1+D_1-\frac{C_1-D_1}{\gamma}\Big)a_1-2\Big(\frac{A_1+B_1}{\gamma^2}+\frac{C_1-D_1}{\gamma}\Big)a_2$$
$$+3\,\frac{A_1}{\gamma^2}\,a_3-\Big(A_2+C_2+D_2-\frac{C_2-D_2}{\gamma}\Big)\,a_0-\Big(\frac{A_2+B_2}{\gamma^2}+\frac{C_2-D_2}{\gamma}\Big)\,a_1+\frac{A_2}{\gamma^2}\,a_2=0\,,$$

$$k=3\,, \qquad -2\,a_2-\frac{6}{\gamma^2}\,a_3+\frac{12}{\gamma^2}\,a_4+(A_1+B_1+C_1+D_1)\,a_1$$
$$-2\Big(A_1+C_1+D_1-\frac{C_1-D_1}{\gamma}\Big)a_2-3\Big(\frac{A_1+B_1}{\gamma^2}+\frac{C_1-D_1}{\gamma}\Big)a_3+4\,\frac{A_1}{\gamma^2}\,a_4$$
$$+(A_2+B_2+C_2+D_2)\,a_0-\Big(A_2+C_2+D_2-\frac{C_2-D_2}{\gamma}\Big)\,a_1$$
$$-\Big(\frac{A_2+B_2}{\gamma^2}+\frac{C_2-D_2}{\gamma}\Big)\,a_2+\frac{A_2}{\gamma^2}\,a_3=0\,,$$

$$k=4\,, \qquad 2\,a_2-6\,a_3-\frac{12}{\gamma^2}\,a_4+\frac{20}{\gamma^2}\,a_5+2\,(A_1+B_1+C_1+D_1)\,a_2$$
$$-3\Big(A_1+C_1+D_1-\frac{C_1-D_1}{\gamma}\Big)a_3-4\Big(\frac{A_1+B_1}{\gamma^2}+\frac{C_1-D_1}{\gamma}\Big)a_4+5\,\frac{A_1}{\gamma^2}\,a_5$$
$$+(A_2+B_2+C_2+D_2)\,a_1-\Big(A_2+C_2+D_2-\frac{C_2-D_2}{\gamma}\Big)\,a_2$$
$$-\Big(\frac{A_2+B_2}{\gamma^2}+\frac{C_2-D_2}{\gamma}\Big)\,a_3+\frac{A_2}{\gamma^2}\,a_4=0\,,$$

$$k=5,\,6,\,7,\,\ldots \qquad (k-2)(k-3)a_{k-2}-(k-1)(k-2)a_{k-1}-\frac{1}{\gamma^2}\,k(k-1)a_k+\frac{1}{\gamma^2}\,k(k+1)a_{k+1}$$
$$+(A_1+B_1+C_1+D_1)\,(k-2)a_{k-2}-\Big(A_1+C_1+D_1-\frac{C_1-D_1}{\gamma}\Big)(k-1)a_{k-1}$$
$$-\Big(\frac{A_1+B_1}{\gamma^2}+\frac{C_1-D_1}{\gamma}\Big)ka_k+\frac{A_1}{\gamma^2}(k+1)a_{k+1}$$
$$+(A_2+B_2+C_2+D_2)\,a_{k-3}-\Big(A_2+C_2+D_2-\frac{C_2-D_2}{\gamma}\Big)\,a_{k-2}$$
$$-\Big(\frac{A_2+B_2}{\gamma^2}+\frac{C_2-D_2}{\gamma}\Big)\,a_{k-1}+\frac{A_2}{\gamma^2}\,a_k=0\,.$$

Thus, we arrive at the 5-term recurrence relations for the coefficients:

$$k=5,\,6,\,7,\,\ldots \qquad (A_2+B_2+C_2+D_2)\,a_{k-3}$$
$$+\Big[(k-2)(k-3)+(A_1+B_1+C_1+D_1)\,(k-2)-\Big(A_2+C_2+D_2-\frac{C_2-D_2}{\gamma}\Big)\Big]a_{k-2}$$
$$-\Big[(k-1)(k-2)+\Big(A_1+C_1+D_1-\frac{C_1-D_1}{\gamma}\Big)(k-1)+\Big(\frac{A_2+B_2}{\gamma^2}+\frac{C_2-D_2}{\gamma}\Big)\Big]a_{k-1}$$
$$+\Big[-\frac{1}{\gamma^2}\,k(k-1)-\Big(\frac{A_1+B_1}{\gamma^2}+\frac{C_1-D_1}{\gamma}\Big)k+\frac{A_2}{\gamma^2}\Big]a_k$$

$$+\Big[\frac{1}{\gamma^2}\,k(k+1)+\frac{A_1}{\gamma^2}(k+1)\Big]a_{k+1}=0\,.$$

We investigate the convergence of this series by the Poincaré–Perron method. To do this, we divide the recurrence relation by a_{k-3}:

$$(A_2+B_2+C_2+D_2)$$
$$+\Big[(k-2)(k-3)+(A_1+B_1+C_1+D_1)\,(k-2)-\Big(A_2+C_2+D_2-\frac{C_2-D_2}{\gamma}\Big)\Big]\frac{a_{k-2}}{a_{k-3}}$$
$$-\Big[(k-1)(k-2)+\Big(A_1+C_1+D_1-\frac{C_1-D_1}{\gamma}\Big)(k-1)+\Big(\frac{A_2+B_2}{\gamma^2}+\frac{C_2-D_2}{\gamma}\Big)\Big]\frac{a_{k-1}}{a_{k-2}}\frac{a_{k-2}}{a_{k-3}}$$
$$+\Big[-\frac{1}{\gamma^2}\,k(k-1)-\Big(\frac{A_1+B_1}{\gamma^2}+\frac{C_1-D_1}{\gamma}\Big)k+\frac{A_2}{\gamma^2}\Big]\frac{a_k}{a_{k-1}}\frac{a_{k-1}}{a_{k-2}}\frac{a_{k-2}}{a_{k-3}}$$
$$+\Big[\frac{1}{\gamma^2}\,k(k+1)+\frac{A_1}{\gamma^2}(k+1)\Big]\frac{a_{k+1}}{a_k}\frac{a_k}{a_{k-1}}\frac{a_{k-1}}{a_{k-2}}\frac{a_{k-2}}{a_{k-3}}=0\,.$$

Further, we multiply this relation by k^{-2} and tend $k\to\infty$, the result is

$$r-r^2-\frac{1}{\gamma^2}r^3+\frac{1}{\gamma^2}r^4=0\qquad\Longrightarrow\qquad r(r-1)(r-\gamma)(r+\gamma)=0\,;$$

the roots are

$$r=0,\ 1,\ -\gamma,\ +\gamma$$

so the convergence radii are possible

$$R_{conv}=1,\quad\frac{1}{|\gamma|},\quad\infty\,.$$

The minimal R_{conv} is enough to cover all physical domain for the variable $y:\ y\in(0,1)$.

On the basis of the used substitution, two pairs of complex conjugate solutions can be built. We write down solutions as the power series in the variable (x), and at the same time, solutions as the power series in the variable $(x-1)$ (they have a simple asymptotic behaviour at different singular points):

Two conjugate solutions,

$$A=+\frac{i\sqrt{\epsilon-1}}{2},\quad B=+\frac{i\sqrt{\epsilon-1}}{2},$$
$$f=x^{+i\sqrt{\epsilon-1}/2}(x-1)^{+i\sqrt{\epsilon-1}/2}F(x),$$
$$\bar{f}=x^{+i\sqrt{\epsilon-1}/2}(x-1)^{+i\sqrt{\epsilon-1}/2}F(1-x);\tag{1.105}$$
$$A=-\frac{i\sqrt{\epsilon-1}}{2},\quad B=-\frac{i\sqrt{\epsilon-1}}{2},$$
$$f^*=x^{-i\sqrt{\epsilon-1}/2}(x-1)^{-i\sqrt{\epsilon-1}/2}F^*(x)\,,$$
$$\bar{f}^*=x^{-i\sqrt{\epsilon-1}/2}(x-1)^{-i\sqrt{\epsilon-1}/2}F^*(1-x)\,;\tag{1.106}$$

Two conjugate solutions,

$$A=+\frac{i\sqrt{\epsilon-1}}{2},\quad B=-\frac{i\sqrt{\epsilon-1}}{2},$$
$$g=x^{+i\sqrt{\epsilon-1}/2}(x-1)^{-i\sqrt{\epsilon-1}/2}G(x),$$
$$\bar{g}=x^{-i\sqrt{\epsilon-1}/2}(x-1)^{+i\sqrt{\epsilon-1}/2}G(1-x)\,;\tag{1.107}$$
$$A=-\frac{i\sqrt{\epsilon-1}}{2},\quad B=+\frac{i\sqrt{\epsilon-1}}{2},$$
$$g^*=x^{-i\sqrt{\epsilon-1}/2}(x-1)^{+i\sqrt{\epsilon-1}/2}G^*(x)\,,$$
$$\bar{g}^*=x^{+i\sqrt{\epsilon-1}/2}(x-1)^{-i\sqrt{\epsilon-1}/2}G^*(1-x)\,.\tag{1.108}$$

1.4 Numerical study of the tunnelling effect

We start with solutions with a plane wave asymptotic which propagates from $z \to +\infty$ on the left:

$z \to +\infty$,

$$f(z) = x^A = (\frac{1 - \tanh z}{2})^A = (e^{-2z})^A = e^{+i\sqrt{\epsilon-1}z}, \; A = -i\frac{\sqrt{\epsilon - 1}}{2}. \tag{1.109}$$

In region $z \to -\infty$, behaviour of these solutions may be only of the following form

$$z \to -\infty, \quad f(z) = m \cdot e^{+i\sqrt{\epsilon-1}z} + n \cdot e^{-i\sqrt{\epsilon-1}z} . \tag{1.110}$$

Far on the left (at $z \to -\infty$), we should choose any two points z_1 and z_2, so obtaining two linear equations with respect to the complex amplitudes m and n:

$$\begin{cases} f(z_1) = x_1^A F(x_1) = m \cdot e^{+i\sqrt{\epsilon-1}\, z_1} + n \cdot e^{-i\sqrt{\epsilon-1}\, z_1}, \\ f(z_2) = x_2^A F(x_2) = m \cdot e^{+i\sqrt{\epsilon-1}\, z_2} + n \cdot e^{-i\sqrt{\epsilon-1}\, z_2}, \end{cases} \tag{1.111}$$

where

$$f(z) = x^A F(x), \qquad F = 1 + d_1 x + d_2 x^2 + ..., \qquad x \in (0, 1) . \tag{1.112}$$

For definiteness we take the values

$$z_1 = -99, \quad z_2 = -100, \tag{1.113}$$

and start with the values $\epsilon = 2$, $\Lambda = 1$; then calculate the modulus of complex m and n from (1.111), then find its ratio $r = |m|/|n|$.

We take the following values for Λ: $\Lambda = 1, 5, 10, 20$, and in each case we calculate the quantity

$$\Delta(\epsilon) = 1 + (r - 1) \cdot 10^{85} , \tag{1.114}$$

the energy values ϵ are taken from 2 to 30 with the step of argument 1 (the multiplier 10^{85} is omitted), we get the Δ interval $(-10; 10)$ – see Figs. 1.3–1.7.

1.5 Conclusion

Generalised Schrödinger equation for a scalar Cox particle is studied in the presence of a magnetic field in the background of Lobachevsky space.

The form of the effective potential along z direction curve says that we have a quantum-mechanical problem of tunnelling type. The derived equation has six regular singular points. To physical domains $z = \pm\infty$, there correspond the singular points 0 and 1 of the derived equation. The solutions of the equation are constructed with the help of power series. These series are convergent in the whole physical domain $z \in (-\infty, +\infty)$. On the basis of the analysis of the constructed solutions, we examine tunnelling effect through the potential barrier numerically.

When considering the ordinary particle in Lobachevsky space, there arises a more simple and known problem of tunnelling type, which is solvable exactly in terms of hypergeometric functions.

Visualisation of functions, some analytical transformations, and numerical study were performed with the use of the graphical, analytical, and numerical facilities of *Mathematica* 10 system. Also, the results obtained in [11–14] were used.

1.6 Figures

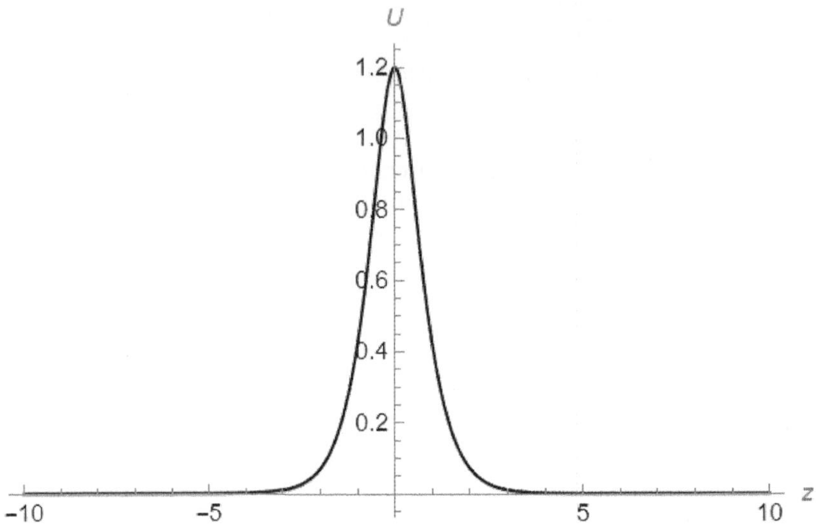

FIGURE 1.1
The graph of the potential U(z) at $\Lambda = 1, \gamma = \frac{1}{2}, b = \frac{1}{5}$.

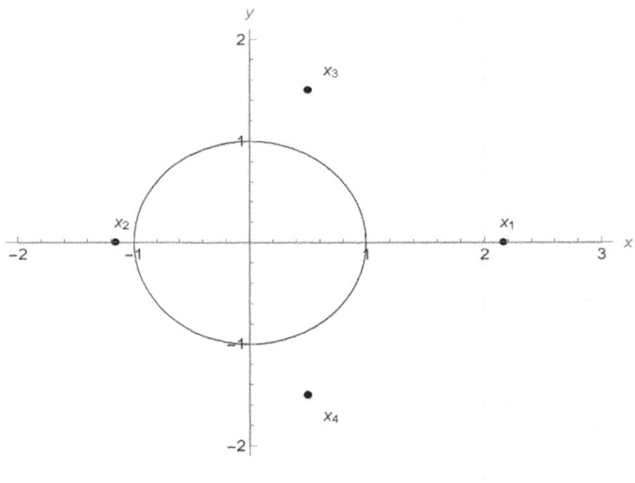

FIGURE 1.2
Location of the singular points in the complex plane.

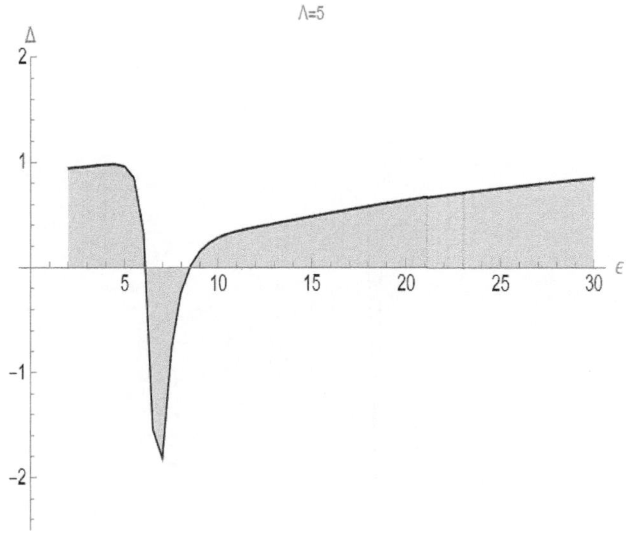

FIGURE 1.3
$\Lambda = 5, \quad \epsilon \in [2, 30]$.

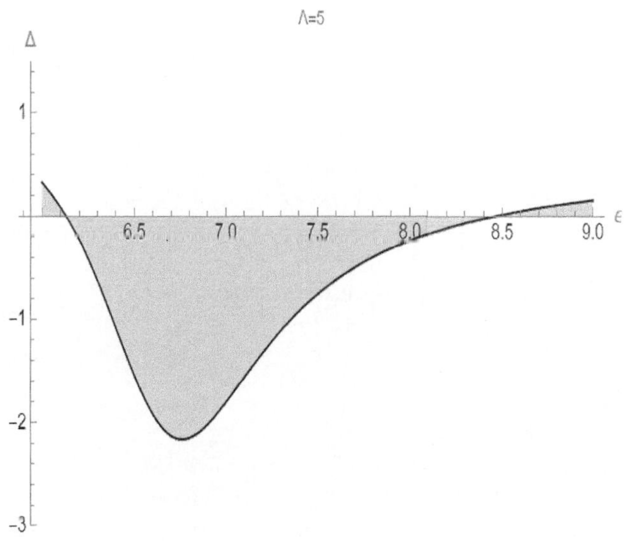

FIGURE 1.4
$\Lambda = 5, \quad \epsilon \in [6, 9]$.

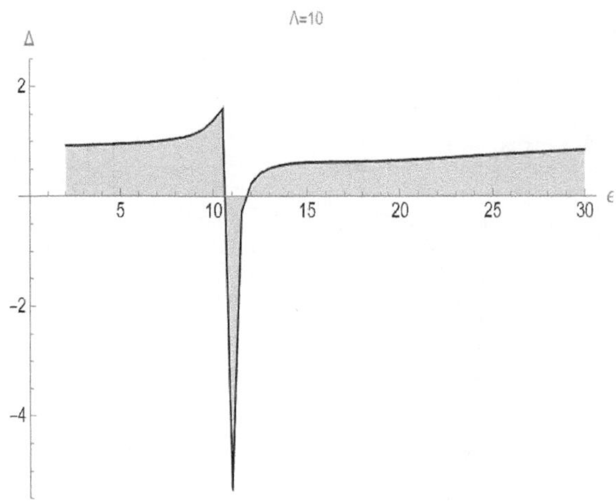

FIGURE 1.5
$\Lambda = 10, \quad \epsilon \in [2, 30]$.

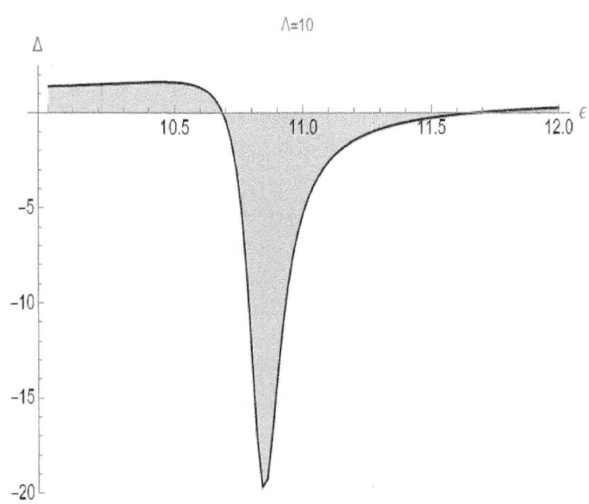

FIGURE 1.6
$\Lambda = 10, \quad \epsilon \in [10, 12]$.

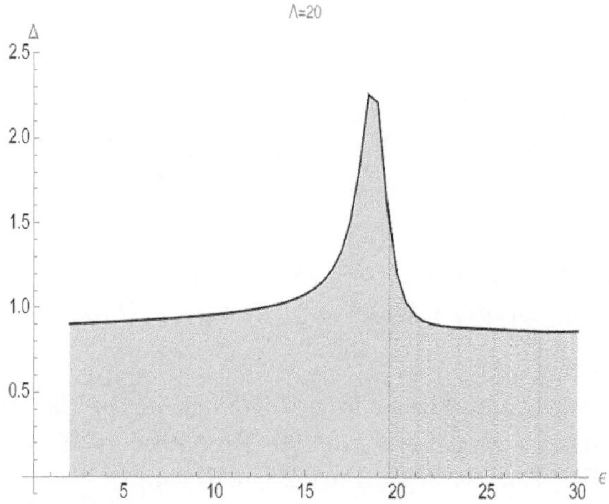

FIGURE 1.7
$\Lambda = 20, \quad \epsilon \in [2, 30]$.

Bibliography

[1] W. Cox. Higher-rank representations for zero-spin field theories. *Journal of Physics A: Mathematical and General*, 15: 627–635, 1982.

[2] V.A. Pletyukhov, V.M. Red'kov and V.A. Strazhev. *Relativistic Wave Equations and Intrinsic Degrees of Freedom.* Belorissina Science, Minsk, 2015.

[3] S.S. Schweber. *An Introduction to Relativistic Quantum Field Theory.* Harper & Row, Publishers, Inc., New York, 1961.

[4] E.M. Ovsiyuk. Spin zero Cox's particle with an intrinsic structure: general analysis in external electromagnetic and gravitational fields. *Ukrainian Journal of Physics*, 60(6): 485–496, 2015.

[5] O.V. Veko. Cox's particle in magnetic and electric fields on the background of hyperbolic Lobachevsky geometry. *Nonlinear Phenomena in Complex Systems*, 19: 50–61, 2016.

[6] V.M. Redkov and E.M. Ovsiyuk. *Quantum Mechanics in Space of Constant Curvature.* Nova Science Publishers, New York, 2012.

[7] A. Ronveaux. *Heun's Differential Equation.* Oxford University Press, Oxford, 1995.

[8] S. Yu. Slavyanov and W. Lay. *Special Functions. A Unified Theory Based on Singularities.* Oxford University Press, Oxford, 2000.

[9] L.D. Landau and E.M. Lifshitz. *Quantum Mechanics (Vol. 3 of A Course of Theoretical Physics).* Pergamon Press, New York, 1965.

[10] A.A. Bogush, V.M. Red'kov and G.G. Krylov. Schrödinger particle in magnetic and electric fields in Lobachevsky and Riemann spaces. *Nonlinear Phenomena in Complex Systems*, 11(4): 403–416, 2008.

[11] A.N. Prokopenya and A.V. Chichurin. *Application of the Mathematica System to the Solution of Ordinary Differential Equations.* BSU Publ., Minsk, 1999 (in Russian).

[12] V.M. Red'kov and A.V. Chichurin. A symbolic-numerical method for solving the differential equation describing the states of polarizable particle in coulomb potential. *Programming and Computer Software*, 40(2): 86–92, 2014.

[13] E.M. Ovsiyuk, A.V. Chichurin and V.M. Red'kov. Nonrelativistic vector particle in Coulomb field on the background of Lobachevsky geometry: analytical and numerical study, visualization. *Studia i Materialy. European University in Warsaw*, 10: 45–57, 2015.

[14] A.V. Chichurin and H.N. Shvychkina. *Application of the Mathematica System for Solving of Differential Equations and the Problems of Mathematical Modeling.* BSU Publ., Minsk, 2016, Part 1, 2017, Part 2 (in Russian).

2

Cox scalar particle in magnetic field, the spherical space

Generalised Schrödinger equation for a spin zero particle with intrinsic structure by Darwin–Cox is studied in the presence of a magnetic field on the background of 3-dimensional spherical Riemann space. The separation of the variables is done. An equation describing the motion of the particle along the axis z is studied. The form of the effective potential indicates that we have a quantum-mechanical problem with the complicated box-type potential. Frobenius solutions of the equation are constructed, and the convergence of the relevant series is proved by Poincaré-Perron method. These series are convergent in the all physical domain of the variable $z \in [-\pi/2, +\pi/2]$. Due to the compactness of the spherical space, the existence of discrete energy levels is assumed; however, any exact quantisation rule is not known. An approximate method for producing the discrete spectrum of energy is developed; it is based on the use of polynomials instead of power series involved in exact Frobenius solutions. A numerical study and visualisation of constructed solutions are performed.

2.1 The Cox equation for a scalar particle

In the frames of the theory of generalised relativistic wave equations, a special model for a spin-zero particle was proposed by Cox [1]. An updated treatment of this theory can be seen in recent books [2, 3].

Cox constructed the wave equation for a scalar particle within the model with a larger set of tensor functions than it exists in the conventional Proca's approach. Namely, he used the set of a scalar, 4-vector, antisymmetric, and symmetric tensor, thus starting with the 20-component wave function. We use Proca's type generalised system [3], obtained after elimination from the initial Cox's system of two 2nd-rank tensors:

$$\Lambda_\alpha^{\ \beta}\Phi_\beta = D_\alpha\Phi\,, \quad D^\alpha\Phi_\alpha = \mu\Phi\,, \quad \Lambda_\alpha^{\ \beta} = \mu\delta_\alpha^\beta + \lambda F_\alpha^{\ \beta}\,, \tag{2.1}$$

where $D_\alpha = i\hbar\partial_\alpha - (e/c)A_\alpha$ and $\mu = mc$; λ is a free parameter of the theory related to an additional structure of the particle, below we use the quantity $\Gamma = \lambda/\mu$. From eq. (2.1) follows a generalised Klein–Fock–Gordon equation for the scalar function Φ:

$$\left[\mu D^\rho (\Lambda^{-1})_\rho^{\ \alpha} D_\alpha - \mu^2\right]\Phi = 0\,. \tag{2.2}$$

DOI: 10.1201/9781003472377-2

Explicit expression for matrix $(\Lambda^{-1})_{\rho}^{\ \alpha}$ is known [3]:

$$(\Lambda^{-1})_{\alpha}^{\ \beta} = \frac{1}{\mu^2(\mu^2 - \frac{\lambda^2}{2} F_{\rho}^{\ \sigma} F_{\sigma}^{\ \rho}) - \lambda^4 (\frac{1}{4} F_{\alpha}^{\ \beta} F_{\beta}^{\times \rho})^2}$$
$$\times \Big\{ \mu(\mu^2 - \frac{\lambda^2}{2} F_{\rho}^{\ \sigma} F_{\sigma}^{\ \rho})\, \delta_{\alpha}^{\beta} - \lambda(\mu^2 - \frac{\lambda^2}{2} F_{\rho}^{\ \sigma} F_{\sigma}^{\ \rho}) F_{\alpha}^{\ \beta}$$
$$+ \mu\lambda^2 F_{\alpha}^{\ \sigma} F_{\sigma}^{\ \beta} - \lambda^3, F_{\alpha}^{\ \sigma} F_{\sigma}^{\ \delta} F_{\delta}^{\ \beta} \Big\}. \tag{2.3}$$

For curved space-time models, we have more complicated form of Λ^{-1}:

$$(\Lambda^{-1})_{\alpha}^{\ \beta} = \frac{1}{\mu^2(\mu^2 - \lambda^2 I) - \lambda^4 J^2}$$
$$\times \left\{ \mu\,(\,\mu^2 - \lambda^2 I\,)\, \delta_{\alpha}^{\beta} - \lambda\mu^2\, F_{\alpha}^{\ \beta} + \mu\lambda^2\, F_{\alpha}^{\ \sigma} F_{\sigma}^{\ \beta} - \lambda^3\, J(x)\, F_{\alpha}^{\times \beta} \right\}, \tag{2.4}$$

where

$$I(x) = \frac{1}{2}\,(F_{\alpha}^{\ \beta} F_{\beta}^{\ \alpha}) = -(\,g^{00} E_i E^i + B_i B^i\,)\,,$$
$$J(x) = \frac{1}{4}(F_{\alpha}^{\times \beta})(F_{\beta}^{\ \rho}) = -\frac{1}{\sqrt{-g}}\,(E_i B_i)\,,$$
$$(F_{\alpha}^{\ \beta}) = \left| \begin{array}{cccc} 0 & E^1 & E^2 & E^3 \\ -g^{00} E_1 & 0 & g^{22} B_3 & -g^{33} B_2 \\ -g^{00} E_2 & -g^{11} B_3 & 0 & g^{33} B_1 \\ -g^{00} E_3 & g^{11} B_2 & -g^{22} B_1 & 0 \end{array} \right|,$$
$$(F^{\times})^{\alpha\beta} = \frac{1}{2} \epsilon^{\alpha\beta\rho\sigma}(x) F_{\rho\sigma}, \quad \epsilon^{0123}(x) = \epsilon(x), \quad \epsilon(x) = \frac{1}{\sqrt{-g}}\,.$$

Cox's electromagnetic structure may be related to the known Darwin [4] interaction term in Schrödinger equation; this additional interaction is related to the non-point-like distribution of the electric charge in the finite volume of the sphere.

In recent papers [5–7], it was studied behaviour of such a particle in external magnetic and electric fields, in Minkowski space, and in spaces with simple non-Euclidean geometries: hyperbolic and spherical ones.

Also it was performed the non-relativistic approximation in eq. (2.2). The Schrödinger equation for the Cox particle has the form

$$D_t \Psi = \frac{1}{2m} \overset{\circ}{D}_k (-g^{kj})(K_j^{\ l} D_l + mc K_j^{\ 0}) \Psi - \frac{1}{2} [(K_0^{\ 0} - 1) mc^2 + K_0^{\ j} c D_j] \Psi\,, \tag{2.5}$$

where the notations are used:

$$dS^2 = c^2 dt^2 + g_{kl}(x)\, dx^k dx^l\,, \quad K_{\rho}^{\ \alpha} = mc(\Lambda^{-1})_{\rho}^{\ \alpha} = \mu(\Lambda^{-1})_{\rho}^{\ \alpha}\,,$$
$$i\hbar\partial_t - eA_0 = D_t\,,\ ic\hbar\partial_k - eA_k = cD_k\,,\ \frac{ic\hbar}{\sqrt{-g}} \frac{\partial}{\partial x^k} \sqrt{-g} - eA_k = c\, \overset{\circ}{D}_k\ . \tag{2.6}$$

Solutions of this non-relativistic equation in the presence of the uniform magnetic and electric fields have been found [3]. In particular, for the Cox particle in a magnetic field, we have [3] a modified energy spectrum

$$E = \frac{p^2}{2M} + \frac{\omega\hbar}{1 - (\Gamma B)^2} (n + \frac{m + \mid m \mid + 1}{2}) - \frac{\omega\hbar}{1 - (\Gamma B)^2} \frac{\Gamma B}{2}\,. \tag{2.7}$$

For Cox particle, the frequency of a quantum oscillator changes as follows:

$$\omega \quad \Longrightarrow \quad \tilde{\omega} = \frac{\omega}{1 - \Gamma^2 B^2}\,, \qquad \omega = \frac{eB}{Mc}\,. \tag{2.8}$$

It turns out that the intrinsic structure of the particle rather specifically interacts with the curved geometry of the space-time. Let us shortly discuss the problem of a particle in magnetic field on the background of the Lobachevsky space [3]. In cylindrical coordinates, the analogue of the uniform magnetic field is determined by the relations:

$$\begin{aligned} dS^2 &= c^2 dt^2 - \cosh^2 z (dr^2 + \sinh^2 r\, d\phi^2) + dz^2, \\ A_\phi &= -B\rho^2(\cosh r - 1),\ F_{r\phi} = -B\rho \sinh r\,, \\ B_3 &= -B\rho\sinh r\,,\ B^3 = -\frac{B}{\rho\sinh r\cosh^4 z}, \end{aligned} \tag{2.9}$$

the curvature radius is noted as ρ. After separating the variables in the Schrödinger equation, we get the radial equation

$$\Big(\frac{d^2}{dr^2} + \frac{\cosh r}{\sinh r}\frac{d}{dr} - \frac{[m - b(\cosh r - 1)]^2}{\sinh^2 r} + \Lambda\Big) R(r) = 0\,, \tag{2.10}$$

and the equation for $F(z)$

$$\Big(\frac{d^2}{dz^2} + 2\frac{\sinh z}{\cosh z}\frac{d}{dz} + \epsilon + \frac{b\gamma - \Lambda\cosh^2 z}{\cosh^4 z - \gamma^2}\Big) Z(z) = 0\,, \tag{2.11}$$

note the notations

$$\frac{eB\rho^2}{\hbar c} = b\,, \quad \Gamma B \cosh^{-2} z = \gamma \cosh^{-2} z\,.$$

For the radial equation, a detailed study shows [3] that here we deal with a finite series of bound states, which is described by the relations

$$\begin{aligned} m < 2B, &\quad \frac{m + \mid m \mid}{2} + n + 1/2 \leq B\,, \quad n = 0, 1, \ldots, N_B\,, \\ \Lambda - 1/4 &= 2B\Big(\frac{m + \mid m \mid}{2} + n + 1/2\Big) - \Big(\frac{m + \mid m \mid}{2} + n + \frac{1}{2}\Big)^2. \end{aligned} \tag{2.12}$$

In usual units, the last formulas read

$$\begin{aligned} \Lambda - \frac{1}{4} &= \rho^2\Lambda_0 - \frac{1}{4}, \quad \lim_{\rho\to\infty} \Lambda_0 = \frac{2M}{\hbar^2}\Big(E - \frac{P^2}{2M}\Big), \\ m < 2B,&\ m + n + 1/2 \leq \frac{eB}{\hbar c}\rho^2, \\ \rho^2\Lambda_0 - \frac{1}{4} &= 2\frac{eB}{\hbar c}\rho^2\Big(\frac{m + \mid m \mid}{2} + n + 1/2\Big) \\ &-\Big(\frac{m + \mid m \mid}{2} + n + 1/2\Big)^2, \ n = 0, 1, \ldots, N_B\,. \end{aligned} \tag{2.13}$$

At the limit of vanishing curvature, we obtain the known result for the flat space

$$E - \frac{P^2}{2M} = \frac{eB\hbar}{Mc}\Big(\frac{m + \mid m \mid}{2} + n + 1/2\Big).$$

Moreover, we have $\Lambda - 1/4 = 2BN - N^2$, $n = 0, 1, \ldots, N_B$, where

$$\frac{1}{2} \leq N = \frac{m + \mid m \mid}{2} + n + 1/2 \leq |B| = b\,,$$

and therefore Λ obeys the restriction

$$b \le \Lambda \le b^2 + \frac{1}{4} \,. \tag{2.14}$$

Eq. (2.11) may be considered as a Schrödinger equation

$$[\, \frac{d^2}{dz^2} + \epsilon - 1 - U(z)\,]\, f(z) = 0\,,$$
$$U(z) = -\frac{b\gamma - \Lambda \cosh^2 z}{\cosh^4 z - \gamma^2}\,, \qquad U(z \to \pm\infty) = +0\,, \tag{2.15}$$

with effective potential $U(z)$. We find the points of local extremum for this potential, they are $z = 0$ and the roots of a quadratic equation

$$(\cosh^2 z)\,|_{1,2} = \frac{b}{\Lambda}\gamma \pm \sqrt{(\frac{b^2}{\Lambda^2} - 1)\gamma^2}\,. \tag{2.16}$$

While considering the bound states for the radial equations, we noted inequality $\Lambda^2 > b^2$. This means that the square root in eq. (2.16) is an imaginary number. Therefore, there exist no other points of zero force except $z = 0$. The form of the effective potential $U(z)$ says that we have a quantum-mechanical problem of a tunnelling type.

Equation (2.16) reduces [3] to a differential equation with six regular singular points. To physical infinities $z = -\infty$, $+\infty$, there correspond the singular points 0 and 1. Frobenius solutions of this equation were constructed, and the convergence of the relevant series is proved by Poincaré–Perron method. These series are convergent in the whole physical domain $z \in (-\infty, +\infty)$. A numerical study of the tunnelling effect was performed in [7].

In the present chapter, we will study the problem of the Cox particle in an external magnetic field, but now on the background of the spherical Riemann space. In cylindric coordinates, we have

$$dS^2 = dt^2 - \cos^2 z\,(dr^2 + \sin^2 r\, d\phi^2) - dz^2, \qquad z \in [-\pi/2, +\pi/2]\,,$$
$$A_\phi = B\rho^2(\cos r - 1),\; F_{r\phi} = B_3 = B\rho \sin r,\; B^3 = \frac{B}{\rho \sin r \cos^4 z}\,. \tag{2.17}$$

After separating the variables in the Schrödinger equation, we derive equations in r, z-variables:

$$\Big(\frac{d^2}{dr^2} + \frac{\cos\, r}{\sin\, r}\frac{d}{dr} - \frac{[m + b(\cos\, r - 1)]^2}{\sin^2 r} + \Lambda\Big) R(r) = 0\,, \tag{2.18}$$

$$\Big(\frac{d^2}{dz^2} - 2\,\frac{\sin\, z}{\cos\, z}\frac{d}{dz} + \epsilon - \frac{b\gamma + \Lambda\, \cos^2 z}{\cos^4 z - \gamma^2}\Big) F(z) = 0\,. \tag{2.19}$$

Analysis of the radial equation gives a completely discrete energy spectrum:

$$m > 0\,, \quad \Lambda + \frac{1}{4} = (n + 1/2 + m)(n + 1/2 + m + 2b)\,;$$
$$m < -2b\,, \quad \Lambda + \frac{1}{4} = (n + 1/2 - m)(n + 1/2 - m - 2b)\,;$$
$$-2b < m \le 0\,, \quad \Lambda + \frac{1}{4} = (n + 1/2)(n + 1/2 - 2b)\,. \tag{2.20}$$

In usual measure units, these formulas read

$$\begin{aligned} m > 0\,, \quad & \rho^2 \Lambda_0 + \frac{1}{4} = +2\frac{eb}{\hbar c}\rho^2 (n+m+1/2) + (n+m+1/2)^2\,; \\ m < -2\frac{eb}{\hbar c}\rho^2\,, \quad & \rho^2 \Lambda_0 + \frac{1}{4} = -2\frac{eb}{\hbar c}\rho^2 (n-m+1/2) + (n-m+1/2)^2\,; \\ -2\frac{eb}{\hbar c}\rho^2 < m \le 0\,, \quad & \rho^2 \Lambda_0 + \frac{1}{4} = 2\frac{eb}{\hbar c}\rho^2 (n+1/2) + (n+1/2)^2\,. \end{aligned} \tag{2.21}$$

Transition to the case of the flat Minkowski space is achieved by $\rho \to \infty$; in this way we obtain the known result:

$$m < 0\,,\ \Lambda_0 = 2\frac{eb}{\hbar c}(n+1/2)\,; \quad m \ge 0\,,\ \frac{2M}{\hbar^2}\,(E - \frac{P^2}{2M}) = +2\frac{eb}{\hbar c}(n+m+1/2)\,.$$

The goal of the present chapter is to study (analytically and numerically) solutions of the equation in z-variable (2.19). In fact, here we have a quantum mechanical problem for a particle in a box with a complicated potential, a mathematical task is reduced to a differential equation with six regular singularities.

2.2 Separation of the variables

In cylindric coordinates of the Riemann spherical model of spherical Riemann space we have

$$\begin{aligned} & dS^2 = dt^2 - \cos^2 z\,(dr^2 + \sin^2 r\, d\phi^2) - dz^2, \quad \sqrt{-g} = \sin r \cos^2 z\,; \\ & g_{\alpha\beta} = (1, -\cos^2 z, -\sin^2 r \cos^2 z, -1)\,, \quad r \in [0,\pi]\,, \quad z \in [-\pi/2, +\pi/2]\,. \end{aligned} \tag{2.22}$$

An analogue of the uniform magnetic field is determined by the relations [7]:

$$\begin{aligned} & A_\phi = B\rho^2\,(\cos r - 1)\,, \quad F_{r\phi} = B\rho \sin r\,, \\ & B_3 = B\rho \sin r\,, \quad B^3 = \frac{B}{\rho \sin r \cos^4 z}\,, \quad B_i B^i = \frac{B^2}{\cos^4 z}\,. \end{aligned} \tag{2.23}$$

We start with the generalised Schrödinger equation in the form

$$D_t\,\Psi = -\frac{1}{2M\rho^2}\,\overset{\circ}{D}_k\, g^{kj}(x)\,\overset{*}{D}_j\,\Psi\,,$$

or,

$$D_t\,\Psi = \frac{1}{2M\rho^2}[\,\overset{\circ}{D}_1\,\frac{1}{\cos^2 z}\,\overset{*}{D}_1 + \overset{\circ}{D}_2\,\frac{1}{\sin^2 r\,\cos^2 z}\,\overset{*}{D}_2 + \overset{\circ}{D}_3\overset{*}{D}_3\,]\,\Psi\,,$$

where

$$D_1 = i\hbar\partial_r\,, \quad D_2 = i\hbar\partial_\phi - \frac{e}{c}\,B\rho^2(\cos r - 1)\,, \quad D_3 = i\hbar\partial_z\,,$$

$$\overset{\circ}{D}_1 = i\hbar(\partial_r + \frac{\cos r}{\sin r})\,, \quad \overset{\circ}{D}_2 = i\hbar\partial_\phi - \frac{e}{c}B\rho^2(\cos r - 1)\,, \quad \overset{\circ}{D}_3 = i\hbar(\partial_z - 2\frac{\sin z}{\cos z})\,,$$

$$\begin{aligned} \overset{*}{D}_1 &= \frac{1}{1 + \Gamma^2 B^2 \cos^{-4} z}(D_1 - \Gamma B_3 D^2) \\ &= \frac{1}{1 + \Gamma^2 B^2 \cos^{-4} z}[\,i\hbar\partial_r + \Gamma B \sin r \cos^2 z\left(i\hbar\partial_\phi - \frac{e}{c}B\rho^2(\cos r - 1)\right)\,]\,, \end{aligned}$$

$$\overset{*}{D}_2 = \frac{1}{1+\Gamma^2 B^2 \cos^{-4} z}(\,D_2 + \Gamma B_3 D^1)$$

$$= \frac{1}{1+\Gamma^2 B^2 \cos^{-4} z}[\,i\hbar\partial_\phi - \frac{e}{c}B\rho^2(\cos\, r - 1) - i\hbar\,\frac{\Gamma B \sin r}{\cos^2 z}\,\partial_r\,]\,,$$

$$\overset{*}{D}_3 = \frac{1}{1+\Gamma^2 B^2 \cos^{-4} z}(D_3 + \Gamma^2 B^3\, B_3 D_3) = i\hbar\partial_z\,.$$

We further get

$$\frac{1}{2M\rho^2}\,\overset{\circ}{D}_1\, g^{11}\,\overset{*}{D}_1 = \frac{\hbar^2}{2M\rho^2 \cos^2 z\,(1+\Gamma^2 B^2 \cos^{-4} z)}\,(\partial_r + \frac{\cos\, r}{\sin\, r})$$

$$\times[\,\partial_r + \frac{\Gamma B}{\sin r\,\cos^2 z}(\partial_\phi + i\frac{eB\rho^2}{\hbar c}(\cos\, r - 1))\,]\,,$$

or with the notations

$$\frac{eB\rho^2}{\hbar c} = b\,, \qquad \frac{\Gamma B}{\cos^2 z} = \gamma(z)\,,$$

we obtain

$$\frac{1}{2M\rho^2}\,\overset{\circ}{D}_1\, g^{11}\,\overset{*}{D}_1 = \frac{\hbar^2}{2M\rho^2 \cos^2 z\,(1+\gamma^2(z))}\,(\partial_r + \frac{\cos\, r}{\sin\, r})$$

$$\times[\,\partial_r + \frac{\gamma(z)}{\sin r}\,(\partial_\phi + i\,b\,(\cos\, r - 1))\,]$$

$$= \frac{\hbar^2}{2M\rho^2 \cos^2 z\,(1+\gamma^2(z))}[\,\partial_r^2 + \frac{\cos r}{\sin r}\,\partial_r + i\gamma(z)b\frac{\cos r - 1}{\sin r}\,\partial_r + \frac{\gamma(z)}{\sin r}\partial_r\partial_\phi - i\gamma(z)b\,]\,.$$

Similarly, we derive

$$\frac{1}{2M\rho^2}\,\overset{\circ}{D}_2\, g^{22}\,\overset{*}{D}_2 = \frac{\hbar^2}{2M\rho^2 \sin^2 r\,\cos^2 z\,(1+\gamma^2(z))}$$

$$\times\,[\partial_\phi + ib(\cos\, r - 1)]\;[\partial_\phi + ib(\cos\, r - 1) - \gamma(z)\sin r\,\partial_r]\,;$$

$$\frac{1}{2M\rho^2}\,\overset{\circ}{D}_3\, g^{33}\,\overset{*}{D}_3 = \frac{\hbar^2}{2M\rho^2}\,(\partial_z - 2\frac{\sin\, z}{\cos\, z})\,\partial_z\,.$$

By using the substitution for the wave function

$$\Psi = e^{-iEt/\hbar} e^{im\phi} F(z) R(r)\,, \qquad \epsilon = \frac{E}{\hbar^2/2M\rho^2}\,,$$

we reduce the Schrödinger equation to the form

$$\Big\{\,\frac{1}{\cos^2 z\,(1+\gamma^2(z))}\Big[\,\partial_r^2 + \frac{\cos\, r}{\sin\, r}\partial_r - \frac{[m + b(\cos\, r - 1)]^2}{\sin^2 r} - i\gamma(z)b\,\Big]$$

$$+\epsilon + (\partial_z - 2\frac{\sin\, z}{\cos\, z})\partial_z\,\Big\}\;R(r)F(z) = 0\,.$$

For physical reasons, the function $\gamma(z)$ must be imaginary, so the change $i\gamma(z) \rightsquigarrow \gamma(z)$ should be done; we obtain

$$\Big[\frac{1}{\cos^2 z\,(1-\gamma^2(z))}\Big(\partial_r^2 + \frac{\cos\, r}{\sin\, r}\,\partial_r - \frac{[m + b(\cos\, r - 1)]^2}{\sin^2 r} - b\gamma(z)\Big)$$

$$+\epsilon + (\partial_z - 2\frac{\sin\, z}{\cos\, z})\partial_z\Big] R(r)F(z) = 0\,.$$

Then by separating the variables, we get

$$\frac{1}{R(r)}\Big(\frac{d^2}{dr^2}+\frac{\cos r}{\sin r}\frac{d}{dr}-\frac{[m+b(\cos r-1)]^2}{\sin^2 r}\Big)R(r)$$

$$+\frac{1}{F(z)}\cos^2 z\,(1-\gamma^2(z))\Big(-\frac{b\,\gamma(z)}{\cos^2 z\,(1-\gamma^2(z))}+\epsilon+(\partial_z-2\frac{\sin z}{\cos z})\partial_z\Big)F(z)=0\,.$$

Hence we infer

$$\Big(\frac{d^2}{dr^2}+\frac{\cos r}{\sin r}\frac{d}{dr}-\frac{[m+b(\cos r-1)]^2}{\sin^2 r}+\Lambda\Big)R(r)=0\,, \tag{2.24}$$

and

$$\Big(\frac{d^2}{dz^2}-2\frac{\sin z}{\cos z}\frac{d}{dz}+\epsilon-\frac{b\,\gamma(z)}{\cos^2 z\,(1-\gamma^2(z))}-\frac{\Lambda}{\cos^2 z\,(1-\gamma^2(z))}\Big)F(z)=0\,.$$

The last relation explicitly looks like[1]

$$\Big(\frac{d^2}{dz^2}-2\frac{\sin z}{\cos z}\frac{d}{dz}+\epsilon-\frac{b\gamma+\Lambda\cos^2 z}{\cos^4 z-\gamma^2}\Big)F(z)=0\,. \tag{2.25}$$

From eq. (2.25), by excluding the term with the first derivative (let $Z(z)=\cos^{-1} z f(z)$), we obtain the quantum-mechanical equation with the effective potential

$$\begin{gathered}\Big[\,\frac{d^2}{dz^2}+\epsilon+1-U(z)\,\Big]\,f(z)=0\,,\quad U(z)=\frac{b\gamma+\Lambda\cos^2 z}{\cos^4 z-\gamma^2}\,,\\ U(z=0)=\frac{b\gamma+\Lambda}{1-\gamma^2}\,,\qquad U(z=\pm\frac{\pi}{2})=-\frac{b}{\gamma}\,.\end{gathered} \tag{2.26}$$

The magnetic field is directed along the axis z; quantisation of the parameter $\Lambda>0$ is known from the analysis of the equation in the transversal coordinate r. The parameter γ is associated with an additional intrinsic structure of the Cox particle; it is assumed to be sufficiently small.

The local extremum may be attended at the points

$$F_z=-\frac{dU}{dz}=-2\cos z\sin z\,\frac{\Lambda\cos^4 z+2\,b\,\gamma\cos^2 z+\gamma^2\Lambda}{(\cos^4 z-\gamma^2)^2}\,; \tag{2.27}$$

i.e., at the point $z=0$ and at the roots of the quadratic equation

$$\Lambda\cos^4 z+2,b\gamma\cos^2 z+\gamma^2\Lambda=0\Longrightarrow(\cos^2 z)\,|_{1,2}=-\frac{b}{\Lambda}\gamma\pm\sqrt{(\frac{b^2}{\Lambda^2}-1)\gamma^2}. \tag{2.28}$$

The value under the square root is negative because of the known inequality $\Lambda^2>b^2$, so there exist only one extremum point $z=0$ in the real domain of the variable.

Here we face a much different problem than in the case of Lobachevsky space. Indeed, the potential $U(z)$ becomes infinite at two physical points

$$\cos^4 z-\gamma^2=0\quad\cos^2 z=+\gamma,\quad\cos^2 z=-\gamma\,;$$

it is assumed that the parameter γ can be arbitrarily small.

Accordingly, the graph of the potential function $U(z), z\in(-[i/2,+\pi/2]$ has two symmetric vertical asymptotes defined by the equation $\cos^4 z_0=\gamma^2$; with the decrease of parameter γ^2, these asymptotes tend to the endpoints of the interval $(-\pi/2,+\pi/2)$; see Figs. 2.1, 2.2, and 2.3. The potential that enters eq. (2.26) is singular only at two points, namely $-z_0$ and $+z_0$.

[1]Let it be $\gamma=B\Gamma$.

2.3 Ordinary particle in Riemann space

Before examining the rather complicated eq. (2.26), let us consider a simpler problem that occurs for the usual scalar particle in the external magnetic field

$$\Big[\frac{d^2}{dz^2} + \epsilon + 1 - U(z) \Big] f(z) = 0\,, \quad U(z) = \frac{\Lambda}{\cos^2 z}\,, \qquad (2.29)$$
$$U(z=0) = \Lambda\,, \quad U(z = \pm\frac{\pi}{2}) = +\infty\,.$$

We introduce the variable

$$y = \tan z\,, \quad \frac{d}{dz} = (1+y^2)\frac{d}{dy}\,, \quad \frac{d^2}{dz^2} = (1+y^2)^2 \frac{d^2}{dy^2} + 2y(1+y^2)\frac{d}{dy}\,;$$

then (2.29) takes the form

$$\Big[(1+y^2)\frac{d^2}{dy^2} + 2y\frac{d}{dy} + \frac{\epsilon - 1}{(1+y^2)} - \Lambda \Big] f = 0\,.$$

To reduce the equation to the hypergeometric type, we use the variable x:

$$x = \frac{1 - iy}{2}\,, \qquad (1+y^2) = 1 - (1-2x)^2 = 4x(1-x)\,,$$

so that

$$\Big[x(1-x)\frac{d^2}{dx^2} + (1-2x)\frac{d}{dx} - \frac{\epsilon - 1}{4x(1-x)} + \Lambda \Big] f = 0\,. \qquad (2.30)$$

The singular points of the last equation are $0, 1, \infty$. The physical features are located at the imaginary infinities:

$$x = \frac{1 - i\tan z}{2}\,, \quad z \to \mp\frac{\pi}{2}\,, \quad x \to \pm i\infty\,.$$

We find behaviour of the solutions near the (non-physical) singular points:

$$x \to 0\,, \quad xf'' + f' - \frac{\epsilon - 1}{4x} f = 0 \quad f = x^A\,, \quad A = \pm\frac{\sqrt{\epsilon - 1}}{2}\,;$$

$$x \to 1\,, \quad (1-x)f'' - f' - \frac{\epsilon - 1}{4(1-x)} f = 0\,, \quad f = (1-x)^B\,, \quad B = \pm\frac{\sqrt{\epsilon - 1}}{2}\,.$$

The general solution is constructed in the form $f(x) = x^A(1-x)^B F(x)$:

$$x(1-x)F'' + [(2A+1) - (2A+2B+2)x]\; F'$$

$$-[\;(A+B)(A+B+1) - \Lambda + \frac{1}{4}\frac{-4A^2 + \epsilon - 1}{x} + \frac{1}{4}\frac{4B^2 - \epsilon + 1}{x-1}\;]\; F = 0\,.$$

The known restrictions are imposed on the parameters A and B, and the equation simplifies to hypergeometric type

$$x(1-x)F'' + [(2A+1) - (2A+2B+2)x]F' - [(A+B)(A+B+1) - \Lambda]F = 0$$

with the parameters

$$c = 2A + 1\,, \; a = \frac{1}{2} + A + B + \frac{\sqrt{1+4\Lambda}}{2}\,, \; b = \frac{1}{2} + A + B - \frac{\sqrt{1+4\Lambda}}{2}\,. \tag{2.31}$$

To choose the necessary solutions – they need to be finite and continuous – we recall the meaning of the variable x:

$$z \to +\frac{\pi}{2}\,, \quad x \to -i\infty\,, \quad z \to -\frac{\pi}{2}\,, \quad x \to +i\infty\,;$$

therefore, the most interesting are Kummer solutions of the following types:

$$u_3(x) = (-x)^{-a} F(a, a+1-c, a+1-b; \frac{1}{x})\,, \quad u_4(x) = (-x)^{-b} F(b, b+1-c, b+1-a; \frac{1}{x})\,.$$

To investigate the possibility of obtaining solutions in the form of polynomials, we consider the explicit form of the parameters a and b for four different variants depending on A and B:

$$A = -\frac{\sqrt{\epsilon-1}}{2}\,, \; B = -\frac{\sqrt{\epsilon-1}}{2}\,, \; a = \frac{1}{2} - \sqrt{\epsilon-1} + \frac{\sqrt{4\Lambda+1}}{2}\,, \; b = \frac{1}{2} - \sqrt{\epsilon-1} - \frac{\sqrt{4\Lambda+1}}{2}\,,$$

$$A = +\frac{\sqrt{\epsilon-1}}{2}\,, \; B = +\frac{\sqrt{\epsilon-1}}{2}\,, \; a = \frac{1}{2} + \sqrt{\epsilon-1} + \frac{\sqrt{4\Lambda+1}}{2}\,, \; b = \frac{1}{2} + \sqrt{\epsilon-1} - \frac{\sqrt{4\Lambda+1}}{2}\,,$$

$$A = +\frac{\sqrt{\epsilon-1}}{2}\,, \; B = -\frac{\sqrt{\epsilon-1}}{2}\,, \quad a = \frac{1}{2} + \frac{\sqrt{4\Lambda+1}}{2}\,, \; b = \frac{1}{2} - \frac{\sqrt{4\Lambda+1}}{2}\,,$$

$$A = -\frac{\sqrt{\epsilon-1}}{2}\,, \; B = +\frac{\sqrt{\epsilon-1}}{2}\,, \; a = \frac{1}{2} + \frac{\sqrt{4\Lambda+1}}{2}\,, \; b = \frac{1}{2} - \frac{\sqrt{4\Lambda+1}}{2}\,.$$

Obviously, only the first choice allows us to impose the quantisation rules $a = -n$, leading to reasonable (from the physical point of view) energy spectrum:

$$\frac{1}{2} - \sqrt{\epsilon-1} + \frac{\sqrt{4\Lambda+1}}{2} = -n \Longrightarrow \sqrt{\epsilon-1} = \frac{1}{2} + \frac{\sqrt{4\Lambda+1}}{2} + n\,. \tag{2.32}$$

We are to follow the behaviour of solutions related to u_3 and u_4 at the points $z\pm\pi/2$, $x \to \pm i\infty$. First, we examine the solution on the basis of u_3:

$$\begin{aligned} f(x \to \infty) &= x^A (1-x)^B (-x)^{-a} F(a, a+1-c, a+1-b; \frac{1}{x}) \\ &\sim x^{A+B-a} = x^{-\sqrt{\epsilon-1}+n} = x^{-1/2-\sqrt{4\Lambda+1}/2} \to 0\,. \end{aligned} \tag{2.33}$$

Here we have finite wave functions in the whole physical region.

Similarly, we examine the functions related to u_4:

$$\begin{aligned} f(x \to \infty) &= x^A (1-x)^B (-x)^{-b} F(b, b+1-c, b+1-a; \frac{1}{x}) \\ &\sim x^{A+B-b} = x^{-\sqrt{\epsilon-1}-1/2+\sqrt{\epsilon-1}+\sqrt{4\Lambda+1}/2} = x^{-1/2+\sqrt{4\Lambda+1}/2} \to \infty\,; \end{aligned} \tag{2.34}$$

such solutions are divergent at the points $z \to \pm\pi/2$, and they are not of interest for physical reasons.

Another variable $\cos^2 z = Z$ may be used to study eq. (2.29):

$$\Big[(Z(1-Z)\frac{d^2}{dZ^2} + (\frac{1}{2} - Z)\frac{d}{dZ} + \frac{\epsilon+1}{4} - \frac{\Lambda/4}{Z^2} \Big] f(Z) = 0\,; \tag{2.35}$$

the singular points $Z = 0, 1$ are physical. Let $f = Z^a (1-Z)^b F(Z)$; for $F(Z)$ we get the equation

$$Z(1-Z)\frac{d^2F}{dZ^2} + [2a + \frac{1}{2} - (2a+2b+1)Z]\frac{dF}{dZ}$$
$$+\Big[\frac{1}{4}(\epsilon+1) - (a+b)^2 + \frac{1}{2}\frac{b(2b-1)}{1-Z} - \frac{1}{4}\frac{-4a^2+2a+\Lambda}{Z}\Big]F = 0\,.$$

By imposing the following restrictions on the parameters a and b:

$$a = \frac{1}{4} \pm \frac{1}{4}\sqrt{1+4\Lambda}\,, \qquad b = 0, \ \frac{1}{2},$$

the above equation simplifies

$$Z(1-Z)\frac{d^2F}{dZ^2} + \Big[2a + \frac{1}{2} - (2a+2b+1)Z\Big]\frac{dF}{dZ} + \Big[\frac{1}{4}(\epsilon+1) - (a+b)^2\Big]F = 0\,.$$

This is an equation of hypergeometric type, with parameters

$$\alpha = a + b - \frac{1}{2}\sqrt{\epsilon+1}\,, \quad \beta = a + b + \frac{1}{2}\sqrt{\epsilon+1}\,, \quad \gamma = 2a + \frac{1}{2}\,.$$

Without loss of generality, the parameters a, b are fixed as follows

$$a = \frac{1}{4} + \frac{1}{4}\sqrt{1+4\Lambda}\,, \quad b = 0\,, \quad \gamma = 1 + \frac{1}{2}\sqrt{1+4\Lambda}\,,$$
$$\alpha = \frac{1}{4} + \frac{1}{4}\sqrt{1+4\Lambda} - \frac{1}{2}\sqrt{\epsilon+1}\,, \quad \beta = \frac{1}{4} + \frac{1}{4}\sqrt{1+4\Lambda} + \frac{1}{2}\sqrt{\epsilon+1}\,. \qquad (2.36)$$

The following Kummer functions can be selected as the two linearly independent solutions:

$$u_1(Z) = F(\alpha, \beta, \gamma, Z)\,, \quad u_5(Z) = Z^{1-\gamma}F(\alpha+1-\gamma, \beta+1-\gamma, 2-\gamma, Z)\,.$$

These hypergeometric functions are transformed into polynomials if one of the first two parameters is either zero or negative integer. From the physical point of view, the quantisation condition on the admissible energy levels is possible only for the function $u_1(Z)$:

$$\alpha = -n, \qquad \frac{1}{4} + \frac{1}{4}\sqrt{1+4\Lambda} - \frac{1}{2}\sqrt{\epsilon+1} = -n \qquad \Longrightarrow$$

$$\sqrt{\epsilon+1} = \frac{1+\sqrt{1+4\Lambda}}{2} + 2n\,; \qquad n = 0, 1, 2, \ldots \qquad (2.37)$$

the first term describes the contribution to total energy due to the motion along the transverse coordinate[2]. The term determined by the quantum number n is due to the motion along the z-axis.

The expressions for the other three combinations of parameters are[3]:

$$\beta = \frac{1}{4} + \frac{1}{4}\sqrt{1+4\Lambda} + \frac{1}{2}\sqrt{\epsilon+1} = -n\,,$$

$$\alpha + 1 - \gamma = \frac{1}{4} - \frac{\sqrt{1+4\Lambda}}{4} - \frac{\sqrt{\epsilon+1}}{2} = -n\,,$$

[2]The quantisation of this term was derived from the analysis of the radial equation.

[3]We note that these do not lead to physical spectra.

$$\beta + 1 - \gamma = \frac{1}{4} - \frac{\sqrt{1+4\Lambda}}{4} + \frac{\sqrt{\epsilon+1}}{2} = -n \, .$$

To the bound states must correspond everywhere finite solutions. At the point $Z = 0$ ($z = \pm\pi/2$), the total wave function vanishes:

$$f_1(Z) = Z^a u_1(Z) = Z^{(1+\sqrt{1+4\,\Lambda})/4} \to 0 \, . \tag{2.38}$$

To find the behaviour of the solutions at the point $Z = 1$ ($z = 0$), we use the Kummer formulas. They make it possible to decompose the above solutions in terms of other solutions depending on the argument $(1 - Z)$. There exist two such solutions:

$$u_2(Z) = F(\alpha, \beta, \alpha + \beta + 1 - c, 1 - Z) \, ,$$

$$u_6(Z) = (1 - Z)^{\gamma - \alpha - \beta} F(\gamma - \alpha, \gamma - \beta, \gamma + 1 - \alpha - \beta, 1 - Z) \, ,$$

and the needed Kummer formula is

$$u_1(Z) = \frac{\Gamma(\gamma)\Gamma(\gamma - \alpha - \beta)}{\Gamma(\gamma - \alpha)\Gamma(\beta - \beta)} u_2(Z) + \frac{\Gamma(\gamma)\Gamma(\alpha + \beta - \gamma)}{\Gamma(\alpha)\Gamma(\beta)} u_6(Z) \, .$$

At the point $(Z \to 1)$, this decomposition takes the form

$$U_1(Z \to 1) = \frac{\Gamma(\gamma)\Gamma(\gamma - \alpha - \beta)}{\Gamma(\gamma - \alpha)\Gamma(\beta - \beta)} + \frac{\Gamma(\gamma)\Gamma(\alpha + \beta - \gamma)}{\Gamma(\alpha)\Gamma(\beta)} (1 - Z)^{\gamma - \alpha - \beta} \, .$$

In view of $\gamma - \alpha - \beta = 1/2$, the previous formula is simplified to

$$f_1(Z \to 1) = u_1(Z \to 1) = \frac{\Gamma(\gamma)\Gamma(\gamma - \alpha - \beta)}{\Gamma(\gamma - \alpha)\Gamma(\beta - \beta)} \, . \tag{2.39}$$

Thus, the correct complete wave function referring to the bound states with the energy levels

$$\sqrt{\epsilon + 1} = \frac{1 + \sqrt{1 + 4\,\Lambda}}{2} + 2n \tag{2.40}$$

is given by the expression

$$f_1(Z) = Z^{(1+\sqrt{1+4\,\Lambda})/4} = F(-n, \frac{\sqrt{1+4\Lambda}}{2} + \frac{1}{2} + n, 1 + \frac{\sqrt{1+4\Lambda}}{2}, Z) \, . \tag{2.41}$$

The energy levels related to formula (2.40) are illustrated by Fig. 2.4.

2.4 Cox particle, analysis in the variable $x = \tan z$

We start with the equation

$$\Big[\frac{d^2}{dz^2} + \epsilon - 1 - \frac{\beta + \Lambda \cos^2 z}{\cos^4 z - \gamma^2} \Big] f = 0 \, . \tag{2.42}$$

With the use of the new variable

$$x = \tan z \, , \quad \frac{d}{dz} = \frac{dx}{dz}\frac{d}{dx} = \frac{1}{\cos^2 z}\frac{d}{dx} = (1 + x^2)\frac{d}{dx} \, , \tag{2.43}$$

we transform the above equation to the form

$$\Big[(1+x^2)\frac{d^2}{dx^2} + 2x\frac{d}{dx} + \frac{\epsilon - 1}{1+x^2} - \frac{\beta(1+x^2)+\Lambda}{1-\gamma^2(1+x^2)^2} \Big] f = 0\,. \tag{2.44}$$

Further we introduce the variable y:

$$y = \frac{1-ix}{2}, \qquad x^2 = -(1-2y)^2, \qquad 1+x^2 = 4y(1-y)\,; \tag{2.45}$$

the equation takes the form

$$\frac{d^2}{dy^2} + (\frac{1}{y} - \frac{1}{1-y})\frac{d}{dy} + \Big[-\frac{\epsilon-1}{4}\frac{1}{y^2} - \frac{\epsilon-1}{2}\frac{1}{y} - \frac{\epsilon-1}{2}\frac{1}{1-y} - \frac{\epsilon-1}{4}\frac{1}{(1-y)^2}$$
$$+\frac{\gamma\Lambda+\beta}{2\gamma}\frac{1}{1-4y\gamma(1-y)}\frac{1}{y(1-y)} + \frac{\gamma\Lambda-\beta}{2\gamma}\frac{1}{1+4y\gamma(1-y)}\frac{1}{y(1-y)} \Big] f = 0\,. \tag{2.46}$$

With the use of the notation, $y(1-y) = f$, we obtain the identity

$$\frac{1}{1-4\gamma y(1-y)}\frac{1}{y(1-y)} \equiv \frac{1}{4\gamma f}\frac{1}{f} = \frac{1-4\gamma f+4\gamma f}{1-4\gamma f}\frac{1}{f} = \frac{1}{f} + \frac{4\gamma}{1-4\gamma f},$$

that is

$$\frac{1}{1-4\gamma y(1-y)}\frac{1}{y(1-y)} = \frac{1}{y} + \frac{1}{1-y} + \frac{4\gamma}{1-4\gamma y(1-y)};$$

similarly derive

$$\frac{1}{1+4\gamma y(1-y)}\frac{1}{y(1-y)} = \frac{1}{y} + \frac{1}{1-y} - \frac{4\gamma}{1+4\gamma y(1-y)}.$$

Let us write down the expressions for the singular points:

$$1-4\gamma y(1-y) = 0\,, \quad y^2 - y + \frac{1}{4\gamma} = 0 \implies y_{1,2} = \frac{1 \pm \sqrt{1-\gamma^{-1}}}{2},$$

$$\frac{1}{1-4\gamma y(1-y)} = \frac{1}{4\gamma}\frac{1}{(y-y_1)(y-y_2)} = \frac{1}{4\gamma}\frac{1}{y_1-y_2}(\frac{1}{y-y_1} - \frac{1}{y-y_2}),$$

$$1+4\gamma y(1-y) = 0\,, \quad y^2 - y - \frac{1}{4\gamma} = 0 \implies y_{3,4} = \frac{1 \pm \sqrt{1+\gamma^{-1}}}{2},$$

$$\frac{1}{1+4\gamma y(1-y)} = -\frac{1}{4\gamma}\frac{1}{(y-y_3)(y-y_4)} = -\frac{1}{4\gamma}\frac{1}{y_3-y_4}(\frac{1}{y-y_3} - \frac{1}{y-y_4}).$$

Thus, the above equation may be presented as follow

$$\frac{d^2}{dy^2} + (\frac{1}{y} - \frac{1}{1-y})\frac{d}{dy} + \Big\{ -\frac{\epsilon-1}{4}\frac{1}{y^2} - \frac{\epsilon-1}{2}\frac{1}{y} - \frac{\epsilon-1}{2}\frac{1}{1-y} - \frac{\epsilon-1}{4}\frac{1}{(1-y)^2}$$
$$+\frac{\gamma\Lambda+\beta}{2\gamma}[\frac{1}{y} + \frac{1}{1-y} + \frac{1}{y_1-y_2}(\frac{1}{y-y_1} - \frac{1}{y-y_2})]$$
$$+\frac{\gamma\Lambda-\beta}{2\gamma}\Big[\frac{1}{y} + \frac{1}{1-y} + \frac{1}{y_3-y_4}(\frac{1}{y-y_3} - \frac{1}{y-y_4})\Big]\Big\} f = 0\,; \tag{2.47}$$

the singular points

$$0, \quad 1, \quad y_1, \quad y_2, \quad y_3, \quad y_4 \tag{2.48}$$

are regular.

Let us study the character of the point $y = \infty$. To do this, we recalculate the equation to the variable $Y = y^{-1}$:

$$Y = y^{-1}\,, \quad y = \frac{1}{Y}\,, \quad 1 - y = \frac{Y-1}{Y}\,, \quad y \to \infty\,, \ Y \to 0\,,$$

$$\frac{d}{dy} = -\frac{1}{y^2}\frac{d}{dY} = -Y^2\frac{d}{dY}\,, \quad \frac{d^2}{dy^2} = Y^4\frac{d}{dY} + 2Y^3\frac{d}{dY}\,,$$

so eq. (2.47) takes the form

$$\frac{d^2}{dY^2} + \frac{1}{Y-1}\frac{d}{dY}$$
$$+\Big\{ -\frac{\epsilon-1}{4}\frac{1}{Y^2} - \frac{\epsilon-1}{2}\frac{1}{Y^3} - \frac{\epsilon-1}{2}\frac{1}{Y^3}\frac{1}{Y-1} - \frac{\epsilon-1}{4}\frac{1}{Y^2}\frac{1}{(Y-1)^2}$$
$$+\frac{\gamma\Lambda+\beta}{2\gamma}\Big[\frac{1}{Y^3} + \frac{1}{Y^3}\frac{1}{Y-1} + \frac{1}{y_1-y_2}\frac{1}{Y^3}(\frac{1}{1-y_1Y} - \frac{1}{1-y_2Y})\Big]$$
$$+\frac{\gamma\Lambda-\beta}{2\gamma}\Big[\frac{1}{Y^3} + \frac{1}{Y^3}\frac{1}{Y-1} + \frac{1}{y_3-y_4}\frac{1}{Y^3}(\frac{1}{1-y_3Y} - \frac{1}{1-y_4Y})\Big]\Big\}f = 0\,. \tag{2.49}$$

In the neighbourhood of the point $y \to \infty$ $(Y \to 0)$, eq. (2.49) becomes simpler

$$\frac{d^2f}{dY^2} - \frac{df}{dY} + \Big(-\frac{\epsilon-1}{4}\frac{1}{Y^2} - \frac{\epsilon-1}{2}\frac{1}{Y^3} + \frac{\epsilon-1}{2}\frac{1}{Y^3} - \frac{\epsilon-1}{4}\frac{1}{Y^2}\Big)\, f = 0\,,$$

or differently,

$$\frac{d^2f}{dY^2} - \frac{df}{dY} = 0\,,$$

which means that the point $y = \infty$ $(Y = 0)$ is not singular.

2.5 Analysis in the variable $\cos^2 z = Z$

Let us apply the more convenient variable $\cos^2 z = Z$; then eq. (2.42) takes the form

$$\Big[\frac{d^2}{dZ^2} + (\frac{1}{2Z} + \frac{1}{2}\frac{1}{Z-1})\frac{d}{dZ} - \frac{\epsilon+1}{4Z(Z-1)}$$
$$+\frac{b\gamma + \Lambda Z}{(Z-\gamma)(Z+\gamma)4Z(Z-1)}\Big]f(Z) = 0\,. \tag{2.50}$$

The equation (2.50) can be written more symmetrically as

$$\Big[\frac{d^2}{dZ^2} + \Big(\frac{1}{2Z} + \frac{1}{2}\frac{1}{Z-1}\Big)\frac{d}{dZ} + \Big(\frac{b}{4\gamma} + \frac{\epsilon+1}{4}\Big)\frac{1}{Z}$$
$$+\Big(-\frac{\epsilon+1}{4} + \frac{b\gamma+\Lambda}{4(1-\gamma^2)}\Big)\frac{1}{Z-1}$$
$$-\frac{\Lambda+b}{8\gamma(1-\gamma)}\frac{1}{Z-\gamma} + \frac{\Lambda-b}{8\gamma(1+\gamma)}\frac{1}{Z+\gamma}\Big]f = 0\,, \tag{2.51}$$

We have four regular singularities: $Z = -\gamma,\ +\gamma,\ 0,\ 1$; the physical region of the variable is the interval $Z \in [0, +1]$. Equation (2.51) can be written in short form

$$\Big[\frac{d^2}{dZ^2} + (\frac{1/2}{Z} + \frac{1/2}{Z-1})\frac{d}{dZ} + \frac{A}{Z} + \frac{B}{Z-1} + \frac{C}{Z-\gamma} + \frac{D}{Z+\gamma}\Big] f(Z) = 0, \tag{2.52}$$

where

$$A = \frac{b}{4\gamma} + \frac{\epsilon+1}{4}, \quad B = -\frac{\epsilon+1}{4} + \frac{b\gamma+\Lambda}{4(1-\gamma^2)}, \quad C = -\frac{\Lambda+b}{8\gamma(1-\gamma)}, \quad D = \frac{\Lambda-b}{8\gamma(1+\gamma)} .$$

Since the equation does not contain complex values, we conclude that each complex-valued solution will be accompanied by a conjugate one.

Let us investigate the point $z = \infty$. To this end, we transform the equation to the variable $y = Z^{-1}$:

$$\Big[\frac{d^2}{dy^2} + (\frac{1}{y} - \frac{1/2}{1-y})\frac{d}{dy} + \frac{A}{y^3} + \frac{B}{y^3(1-y)} + \frac{C}{y^3(1-\gamma y)} + \frac{D}{y^3(1+\gamma y)}\Big] f(y) = 0 \, ,$$

or

$$\Big[\frac{d^2}{dy^2} + (\frac{1}{y} - \frac{1/2}{1-y})\frac{d}{dy} + \frac{A+B+C+D}{y^3}$$
$$+\frac{B + C\gamma - D\gamma}{y^2} + \frac{B + C\gamma^2 + D\gamma^2}{y} + \frac{B}{1-y} + \frac{C\gamma^3}{1-y\gamma} - \frac{D\gamma^3}{1+y\gamma}\Big] f = 0 \, .$$

Since we have the identity $A+B+C+D = 0$, the equation in the variable y gets simplified

$$\Big[\frac{d^2}{dy^2} + (\frac{1}{y} - \frac{1/2}{1-y})\frac{d}{dy} + \frac{B + C\gamma - D\gamma}{y^2}$$
$$+\frac{B + C\gamma^2 + D\gamma^2}{y} + \frac{B}{1-y} - \frac{C\gamma^2}{y-\gamma^{-1}} - \frac{D\gamma^2}{y+\gamma^{-1}}\Big] f(y) = 0 \, , \tag{2.53}$$

whence we conclude that the singularity at $Z = \infty$ is also regular.

Among the five singular points

$$Z = 0,\ 1,\ -\gamma,\ +\gamma,\ \infty \quad \text{or} \quad y = \infty,\ 1,\ -\frac{1}{\gamma},\ +\frac{1}{\gamma},\ 0 \, , \tag{2.54}$$

only two points are physical:

$$z \to 0,\ Z \to +1,\ y \to +1; \qquad z \to \pm\frac{\pi}{2},\ Z \to 0 \, , y \to \infty \, . \tag{2.55}$$

Now we turn to studying solutions of eq. (2.53). There is no difference which variable byTis used: $Z = \cos^2 z$ or $y = Z^{-1} = \cos^{-2} z$. For definiteness we consider the variable y.

Near the physical singular point $y = \infty$, $Z = 0$ $(z = \pm\pi/2)$, we have

$$(\frac{d^2}{dZ^2}F + \frac{1/2}{Z}\frac{d}{dZ} + \frac{A}{Z})f = 0 \, , \quad f = Z^{\nu} = y^{-\nu}, \quad \nu = 0, \quad \nu = \frac{1}{2} \, . \tag{2.56}$$

Near the physical singular point $y = 1$, $Z = 1$ $(z = 0)$ we get

$$(\frac{d^2}{dy^2} + \frac{1/2}{y-1}\frac{d}{dy} - \frac{B}{y-1}) f(y) = 0 \, ,$$
$$f = (y-1)^b = (\frac{1-Z}{Z})^b = (1-Z)^b \, , \quad b_1 = 0, \quad b_2 = \frac{1}{2} \, , \tag{2.57}$$

the solutions tend to zero $f = \sqrt{y-1}$, or tend to a constant value $f = (y-1)^0 = 1$.

Near the (non-physical) singular point $y = 0, Z = \infty$ we have the equation

$$\Big(\frac{d^2}{dy^2} + \frac{1}{y}\frac{d}{dy} + \frac{B + C\gamma - D\gamma}{y^2}\Big) f(y) = 0 \,;$$

its solutions are constructed in the form

$$f = y^a, \quad a^2 + B + (C - D)\gamma = 0 \,, \quad a_{1,2} = \pm\sqrt{-B - (C - D)\gamma} \,;$$

taking into account the expressions for B, C, D, we obtain

$$a_1 = \frac{+\sqrt{\epsilon + 1}}{2} \,, \quad a_2 = \frac{-\sqrt{\epsilon + 1}}{2} \,;$$
$$f(y) = y^{\alpha} = Z^{-\alpha} = (\cos z)^{-2a} = \Big(\frac{e^{iz} + e^{-iz}}{2}\Big)^{\mp\sqrt{\epsilon+1}} \,. \tag{2.58}$$

The possible behaviour of the solutions near the points $+1/\gamma, -1/\gamma$ is given by relations:

$$y \longrightarrow +1/\gamma \,, \quad F - (y - 1/\gamma)^{\rho} \,, \quad \rho_{1,2} = 0\,, 1 \,;$$
$$y \longrightarrow -1/\gamma \,, \quad F - (y + 1/\gamma,)^{\sigma} \,, \quad \sigma_{1,2} = 0, 1 \,. \tag{2.59}$$

At these points the solutions vanish or tend to a constant value, and it does not matter from which side to approach the point z_0.

When constructing solutions corresponding to bound states, we are primarily interested in finite functions. In view of the behaviour of solutions near the points $\pm z_0$, the potential discontinuity at the points $\pm z_0$ does not lead to any serious consequences.

Taking into account the established asymptotics, we will construct the solutions of eq. (2.53) of the form $f = y^a (y - 1)^b F(y)$; we obtain the equation

$$F'' + \Big(\frac{2a + 1}{y} + \frac{2b + 1/2}{y - 1}\Big) F' + \Big[\frac{a^2 + B + (C - D)\gamma}{y^2} + \frac{b(b - 1) + b/2}{(y - 1)^2}$$
$$+ \frac{-2ab - a/2 - b + B + (C + D)\gamma^2}{y} + \frac{2ab + a/2 + b - B}{y - 1} + \frac{C\gamma^3}{1 - y\gamma} - \frac{D\gamma^3}{1 + y\gamma}\Big] F = 0 \,.$$

We apply now the already known restrictions

$$a^2 + B + (C - D)\gamma = 0 \implies a = a_1, a_2 \,, \quad b(b - 1) + b/2 = 0 \implies b = 0, \frac{1}{2} \,,$$

further we obtain

$$F'' + \Big(\frac{2a + 1}{y} + \frac{2b + 1/2}{y - 1}\Big) F'$$
$$+ \Big[\frac{-2ab - a/2 - b + B + (C + D)\gamma^2}{y} + \frac{2ab + a/2 + b - B}{y - 1} - \frac{C\gamma^2}{y - 1/\gamma} - \frac{D\gamma^2}{y + 1/\gamma}\Big] F = 0.$$

The functions $F_i(y)$ are subjected to the equation which is the same

$$F'' + \Big(\frac{\alpha_1}{y} + \frac{\beta_1}{y - 1}\Big) F' + \Big(\frac{\alpha}{y} + \frac{\beta}{y - 1} + \frac{c}{y - 1/\gamma} + \frac{d}{y + 1/\gamma}\Big) F = 0 \,. \tag{2.60}$$

We multiply eq. (2.60) by

$$y\,(y - 1)\,(y - 1/\gamma)\,(y + 1/\gamma),$$

so we get

$$y\,(y - 1)\,(y - 1/\gamma)\,(y + 1/\gamma)\,F''$$

$$+\Big[\frac{\alpha_1\, y\,(y-1)\,(y-1/\gamma)\,(y+1/\gamma)}{y}+\frac{\beta_1\, y\,(y-1)\,(y-1/\gamma)\,(y+1/\gamma)}{y-1}\Big]F'$$

$$+\Big[\frac{\alpha\, y\,(y-1)\,(y-1/\gamma)\,(y+1/\gamma)}{y}+\frac{\beta\, y\,(y-1)\,(y-1/\gamma)\,(y+1/\gamma)}{y-1}$$

$$+\frac{c\, y\,(y-1)\,(y-1/\gamma)\,(y+1/\gamma)}{y-1/\gamma}+\frac{d\, y\,(y-1)\,(y-1/\gamma)\,(y+1/\gamma)}{y+1/\gamma}\Big]F=0\,,$$

or

$$[y^4-y^3-y^2/\gamma^2+y/\gamma^2]F''+\Big[(\alpha_1+\beta_1)y^3-\alpha_1 y^2-\frac{\alpha_1+\beta_1}{\gamma^2}y+\frac{\alpha_1}{\gamma^2}\Big]F'$$

$$+\Big[(\alpha+\beta+c+d)y^3-(\alpha+c+d-\frac{c-d}{\gamma})\,y^2-(\frac{\alpha+\beta}{\gamma^2}+\frac{c-d}{\gamma})y+\frac{\alpha}{\gamma^2}\Big]F=0\,.$$

Solutions can be built as power series

$$F=\sum_{n=0}^{\infty}a_n y^n,\quad F'=\sum_{n=1}^{\infty}n a_n y^{n-1},\quad F''=\sum_{n=2}^{\infty}n(n-1)a_n y^{n-2}\,.$$

Further we obtain

$$\sum_{n=2}^{\infty}n(n-1)a_n y^{n+2}-\sum_{n=2}^{\infty}n(n-1)a_n y^{n+1}-\frac{1}{\gamma^2}\sum_{n=2}^{\infty}n(n-1)a_n y^{n}+\frac{1}{\gamma^2}\sum_{n=2}^{\infty}n(n-1)a_n y^{n-1}$$

$$+(\alpha_1+\beta_1)\sum_{n=1}^{\infty}n a_n y^{n+2}-\alpha_1\sum_{n=1}^{\infty}n a_n y^{n+1}-\frac{\alpha_1+\beta_1}{\gamma^2}\sum_{n=1}^{\infty}n a_n y^{n}+\frac{\alpha_1}{\gamma^2}\sum_{n=1}^{\infty}n a_n y^{n-1}$$

$$+(\alpha+\beta+c+d)\sum_{n=0}^{\infty}a_n y^{n+3}-\left(\alpha+c+d-\frac{c-d}{\gamma}\right)\sum_{n=0}^{\infty}a_n y^{n+2}$$

$$-\left(\frac{\alpha+\beta}{\gamma^2}+\frac{c-d}{\gamma}\right)\sum_{n=0}^{\infty}a_n y^{n+1}+\frac{\alpha}{\gamma^2}\sum_{n=0}^{\infty}a_n y^{n}=0\,.$$

After changing the summation indices, we get

$$\sum_{k=4}^{\infty}(k-2)(k-3)\,a_{k-2}\,y^k-\sum_{k=3}^{\infty}(k-1)(k-2)\,a_{k-1}\,y^k$$

$$-\frac{1}{\gamma^2}\sum_{k=2}^{\infty}k(k-1)\,a_k\,y^k+\frac{1}{\gamma^2}\sum_{k=1}^{\infty}k(k+1)\,a_{k+1}\,y^k$$

$$+(\alpha_1+\beta_1)\sum_{k=3}^{\infty}(k-2)\,a_{k-2}\,y^k-\alpha_1\sum_{k=2}^{\infty}(k-1)\,a_{k-1}\,y^k$$

$$-\frac{\alpha_1+\beta_1}{\gamma^2}\sum_{k=1}^{\infty}k\,a_k\,y^k+\frac{\alpha_1}{\gamma^2}\sum_{k=0}^{\infty}(k+1)\,a_{k+1}\,y^k$$

$$+(\alpha+\beta+c+d)\sum_{k=3}^{\infty}a_{k-3}\,y^k-\left(\alpha+c+d-\frac{c-d}{\gamma}\right)\sum_{k=2}^{\infty}a_{k-2}\,y^k$$

$$-\Big(\frac{\alpha+\beta}{\gamma^2}+\frac{c-d}{\gamma}\Big)\sum_{k=1}^{\infty}a_{k-1}\,y^k+\frac{\alpha}{\gamma^2}\sum_{k=0}^{\infty}a_k\,y^k=0\,.$$

Equating to zero the coefficients at all powers y^k:

$$k=0\,,\qquad \frac{\alpha_1}{\gamma^2}\,a_1+\frac{\alpha}{\gamma^2}\,a_0=0\,,$$

$$k = 1\,, \qquad \frac{2}{\gamma^2}\,a_2 - \frac{\alpha_1+\beta_1}{\gamma^2}\,a_1 + 2\,\frac{\alpha_1}{\gamma^2}\,a_2 - (\frac{\alpha+\beta}{\gamma^2} + \frac{c-d}{\gamma})a_0 + \frac{\alpha}{\gamma^2}\,a_1 = 0\,,$$

$$k = 2\,, \qquad -\frac{1}{\gamma^2}\,2\,a_2 + \frac{1}{\gamma^2}\,6\,a_3 - \alpha_1\,a_1 - \frac{\alpha_1+\beta_1}{\gamma^2}\,2\,a_2 + \frac{\alpha_1}{\gamma^2}\,3\,a_3$$

$$-(\alpha + c + d - \frac{c-d}{\gamma})\;a_0 - (\frac{\alpha+\beta}{\gamma^2} + \frac{c-d}{\gamma})\,a_1 + \frac{\alpha}{\gamma^2}\,a_2 = 0\,,$$

$$k = 3\,, \qquad -2\,a_2 - \frac{1}{\gamma^2}\,6\,a_3 + \frac{1}{\gamma^2}\,12\,a_4 + (\alpha_1+\beta_1)\,a_1 - \alpha_1 2\,a_2 - \frac{\alpha_1+\beta_1}{\gamma^2}\,3\,a_3 + \frac{\alpha_1}{\gamma^2}\,4\,a_4$$

$$+(\alpha+\beta+c+d)\,a_0 - (\alpha + c + d - \frac{c-d}{\gamma})\,a_1 - (\frac{\alpha+\beta}{\gamma^2} + \frac{c-d}{\gamma})\,a_2 + \frac{\alpha}{\gamma^2}\,a_3 = 0\,,$$

$$k = 4\,, \qquad 2\,a_2 - 6\,a_3 - \frac{12}{\gamma^2}\,a_4 + \frac{20}{\gamma^2}\,a_5 + (\alpha_1+\beta_1)\,2\,a_2 - \alpha_1 3\,a_3 - \frac{\alpha_1+\beta_1}{\gamma^2}\,4\,a_4 + \frac{\alpha_1}{\gamma^2}\,5\,a_5$$

$$+(\alpha+\beta+c+d)\,a_1 - (\alpha + c + d - \frac{c-d}{\gamma})\,a_2 - (\frac{\alpha+\beta}{\gamma^2} + \frac{c-d}{\gamma})\,a_3 + \frac{\alpha}{\gamma^2}\,a_4 = 0\,,$$

$$k = 5,\ 6,\ 7,\ \ldots \qquad (k-2)(k-3)\,a_{k-2} - (k-1)(k-2)\,a_{k-1} - \frac{1}{\gamma^2}k(k-1)\,a_k + \frac{1}{\gamma^2}k(k+1)\,a_{k+1}$$

$$+(\alpha_1+\beta_1)(k-2)\,a_{k-2} - \alpha_1(k-1)\,a_{k-1} - \frac{\alpha_1+\beta_1}{\gamma^2}k\,a_k + \frac{\alpha_1}{\gamma^2}(k+1)\,a_{k+1}$$

$$+(\alpha+\beta+c+d)\,a_{k-3} - (\alpha + c + d - \frac{c-d}{\gamma})\,a_{k-2} - (\frac{\alpha+\beta}{\gamma^2} + \frac{c-d}{\gamma})\,a_{k-1} + \frac{\alpha}{\gamma^2}\,a_k = 0\,,$$

we derive the 5-term recurrence relation for the coefficients

$$k = 5,\ 6,\ 7,\ \ldots \qquad (\alpha+\beta+c+d)\,a_{k-3}$$

$$+[(k-2)(k-3) + \alpha_1 + \beta_1)(k-2) - (\alpha + c + d - \frac{c-d}{\gamma})]a_{k-2}$$

$$-[(k-1)(k-2) + \alpha_1(k-1) + (\frac{\alpha+\beta}{\gamma^2} + \frac{c-d}{\gamma})]a_{k-1}$$

$$+[-\frac{1}{\gamma^2}k(k-1) - \frac{\alpha_1+\beta_1}{\gamma^2}k + \frac{\alpha}{\gamma^2}]a_k + [\frac{1}{\gamma^2}k(k+1) + \frac{\alpha_1}{\gamma^2}(k+1)]a_{k+1} = 0\,.$$

The Poincaré–Perron method is used to analyse the convergence radii of the series. We divide the recurrence relation by a_{k-3}:

$$(\alpha+\beta+c+d)$$

$$+[(k-2)(k-3) + (\alpha_1+\beta_1)(k-2) - (\alpha + c + d - \frac{c-d}{\gamma})]\frac{a_{k-2}}{a_{k-3}}$$

$$-[(k-1)(k-2) + \alpha_1(k-1) + (\frac{\alpha+\beta}{\gamma^2} + \frac{c-d}{\gamma})]\frac{a_{k-1}}{a_{k-2}}\frac{a_{k-2}}{a_{k-3}}$$

$$+[-\frac{1}{\gamma^2}k(k-1) - \frac{\alpha_1+\beta_1}{\gamma^2}k + \frac{\alpha}{\gamma^2}]\frac{a_k}{a_{k-1}}\frac{a_{k-1}}{a_{k-2}}\frac{a_{k-2}}{a_{k-3}}$$

$$+[\frac{1}{\gamma^2}k(k+1) + \frac{\alpha_1}{\gamma^2}(k+1)]\frac{a_{k+1}}{a_k}\frac{a_k}{a_{k-1}}\frac{a_{k-1}}{a_{k-2}}\frac{a_{k-2}}{a_{k-3}} = 0\,.$$

The convergence radius of the power series is the inverse of $|r|$:

$$r = \lim_{k\to\infty}\frac{a_{k+1}}{a_k}\,, \qquad R_{conv} = \frac{1}{|\,r\,|}\,.$$

In order to find an algebraic equation for r, the resulting equation is multiplied by k^{-2}, and we let k tend to $k \to \infty$. In this way we obtain

$$r - r^2 - \frac{1}{\gamma^2} r^3 + \frac{1}{\gamma^2} r^4 = 0 \qquad \Longrightarrow \qquad r(r-1)(r-\gamma)(r+\gamma) = 0 \, .$$

The following roots are $r = 0,\ 1,\ -\gamma,\ +\gamma$, so for convergence radii we get

$$R_{conv} = 1, \ \frac{1}{|\gamma|}, \ \infty \, . \tag{2.61}$$

We recall that the physical domain for the variable y is $y = \cos^{-2} z, y \in (1, \infty)$.

2.6 Analysis in the variable $\sin^2 z = x$

In the equation

$$\Big[\frac{d^2}{dZ^2} + (\frac{1/2}{Z} + \frac{1/2}{Z-1})\frac{d}{dZ} + \frac{A}{Z} + \frac{B}{Z-1} + \frac{C}{Z-\gamma} + \frac{D}{Z+\gamma}\Big] f(Z) = 0,$$

we make the change of the variable to[4]

$$Z = 1 - x, \quad x = 1 - Z = \sin^2 z \, , \quad \frac{d}{dZ} = -\frac{d}{dx} \, . \tag{2.62}$$

The equation takes the form

$$\Big[\frac{d^2}{dx^2} + (\frac{1/2}{x-1} + \frac{1/2}{x})\frac{d}{dx} - \frac{A}{x-1} - \frac{B}{x} - \frac{C}{x-(1-\gamma)} - \frac{D}{x-(1+\gamma)}\Big] f(Z) = 0 \, .$$

Let $1 - \gamma = s, 1 + \gamma = t$; then the above equation can be written as

$$\Big(\frac{d^2}{dx^2} + (\frac{1/2}{x-1} + \frac{1/2}{x})\frac{d}{dx} - \frac{A}{x-1} - \frac{B}{x} - \frac{C}{x-s} - \frac{D}{x-t}\Big) f(Z) = 0 \, . \tag{2.63}$$

Its solutions are built in the form

$$f(x) = x^a (x-1)^b G(x), \qquad a = 0, \frac{1}{2}, \quad b = 0, \frac{1}{2} \, .$$

For the function G, we have the equation

$$\frac{d^2 G}{dx^2} + (\frac{1}{2}\frac{4b+1}{x-1} + \frac{1}{2}\frac{1+4a}{x})\frac{dG}{dx} + \Big(\frac{1}{2}\frac{4ab - 2A + a + b}{x-1}$$

$$+ \frac{1}{2}\frac{b(2b-1)}{(x-1)^2} + \frac{1}{2}\frac{a(2a-1)}{x^2} + \frac{1}{2}\frac{-4ab - 2B - a - b}{x} - \frac{C}{x-s} - \frac{D}{x-t}\Big) G = 0 \, .$$

We apply now the already known restrictions $a = 0, 1/2, \ \ b = 0, 1/2$. The choice $a = 0$ gives symmetrical under the replacement $z \rightsquigarrow -z$ solutions, with the property $f(z \to 0) = \text{const}$;

[4]This new variable is more convenient, because the physical domain is $Z \in (0, 1)$.

the choice $a = 1/2$ gives anti-symmetric under the replacement $z \rightsquigarrow -z$ solutions, with the property $f(z \to 0) = \sin z \to 0$. In this way, we obtain the equation

$$\frac{d^2G}{dx^2} + \Big(\frac{1}{2}\frac{1+4a}{x} + \frac{1}{2}\frac{4b+1}{x-1}\Big)\frac{dG}{dx}$$
$$+\Big(\frac{1}{2}\frac{-4ab-2B-a-b}{x} + \frac{1}{2}\frac{4ab-2A+a+b}{x-1} - \frac{C}{x-s} - \frac{D}{x-t}\Big)G = 0\,.$$

We shall follow all four possibilities of the last equation. The functions $G_i(x)$ are subject to the equation with the general structure

$$G'' + \Big(\frac{\alpha_1}{x} + \frac{\beta_1}{x-1}\Big)G' + \Big(\frac{\alpha}{x} + \frac{\beta}{x-1} + \frac{c}{x-s} + \frac{d}{x-t}\Big)G = 0\,, \tag{2.64}$$

where

$$\alpha_1 = \frac{1}{2}\,(1+4\,a)\,, \quad \beta_1 = \frac{1}{2}\,(1+4\,b)\,,$$

$$\alpha = \frac{1}{2}\,(-4\,ab - 2\,B - a - b)\,, \quad \beta = \frac{1}{2}\,(4\,ab - 2\,A + a + b)\,, \quad c = -C\,, \quad d = -D\,.$$

We multiply the equation by $x\,(x-1)\,(x-s)\,(x-t)$, so we get

$$\big[x^4 - (1+s+t)\,x^3 + (st+s+t)\,x^2 - stx\big]\;G''$$
$$+\big\{(\alpha_1+\beta_1)\,x^3 - [(1+s+t)\,\alpha_1 + (s+t)\,\beta_1]\,x^2 + [(st+s+t)\,\alpha_1 + st\beta_1]\,x - \alpha_1 st\big\}\,G'$$
$$+\big\{(\alpha+\beta+c+d)\,x^3 - [(1+s+t)\,\alpha + (s+t)\,\beta + (1+t)\,c + (1+s)\,d\,]\,x^2$$
$$+\;[(st+s+t)\,\alpha + st\beta + ct + ds]\,x - \alpha\,st\big\}\,G = 0\,.$$

The solutions can be built as power series

$$G = \sum_{n=0}^{\infty} a_n x^n, \quad G' = \sum_{n=1}^{\infty} n a_n x^{n-1}, \quad G'' = \sum_{n=2}^{\infty} n(n-1)a_n x^{n-2}\,.$$

Further we get

$$\sum_{n=2}^{\infty} n(n-1)a_n x^{n+2} - (1+s+t)\sum_{n=2}^{\infty} n(n-1)a_n x^{n+1} + (st+s+t)\sum_{n=2}^{\infty} n(n-1)a_n x^{n}$$
$$-st\sum_{n=2}^{\infty} n(n-1)a_n x^{n-1} + (\alpha_1+\beta_1)\sum_{n=1}^{\infty} n a_n x^{n+2} - [(1+s+t)\,\alpha_1 + (s+t)\;\beta_1]\sum_{n=1}^{\infty} n a_n x^{n+1}$$
$$+[(st+s+t)\,\alpha_1 + st\beta_1]\sum_{n=1}^{\infty} n a_n x^{n} - \alpha_1\,st\sum_{n=1}^{\infty} n a_n x^{n-1} + (\alpha+\beta+c+d)\sum_{n=0}^{\infty} a_n x^{n+3}$$
$$-[(1+s+t)\,\alpha + (s+t)\,\beta + (1+t)\,c + (1+s)\,d\,]\sum_{n=0}^{\infty} a_n x^{n+2}$$
$$+[(st+s+t)\,\alpha + st\beta + ct + ds]\sum_{n=0}^{\infty} a_n x^{n+1} - \alpha\,st\sum_{n=0}^{\infty} a_n x^{n} = 0\,.$$

By changing the summation indices, we get

$$\sum_{k=4}^{\infty}(k-2)(k-3)a_{k-2}x^k - (1+s+t)\sum_{k=3}^{\infty}(k-1)(k-2)a_{k-1}x^k + (st+s+t)\sum_{k=2}^{\infty} k(k-1)a_k x^k$$

$$-st\sum_{k=1}^{\infty} k(k+1)a_{k+1}x^k + (\alpha_1+\beta_1)\sum_{k=3}^{\infty}(k-2)a_{k-2}x^k$$

$$-\left[(1+s+t)\,\alpha_1 + (s+t)\;\beta_1\right]\sum_{k=2}^{\infty}(k-1)a_{k-1}x^k$$

$$+\left[(st+s+t)\,\alpha_1 + st\beta_1\right]\sum_{k=1}^{\infty} ka_kx^k - \alpha_1\,st\sum_{k=0}^{\infty}(k+1)a_{k+1}x^k + (\alpha+\beta+c+d)\sum_{k=3}^{\infty} a_{k-3}x^k$$

$$-\left[(1+s+t)\,\alpha + (s+t)\,\beta + (1+t)\,c + (1+s)\,d\right]\sum_{k=2}^{\infty} a_{k-2}x^k$$

$$+\left[(st+s+t)\,\alpha + st\beta + ct + ds\right]\sum_{k=1}^{\infty} a_{k-1}x^k - \alpha\,st\sum_{k=0}^{\infty} a_kx^k = 0\,.$$

We equate to zero the coefficients at the powers of y^k, and we get a system of recurrence relations:

$$k=0\,,\qquad -\alpha_1\,st\,a_1 - \alpha\,st\,a_0 = 0\,,$$

$$k=1\,,\qquad -2st\,a_2 + \left[(st+s+t)\,\alpha_1 + st\beta_1\right]a_1 - 2\alpha_1\,st\,a_2$$
$$+\left[(st+s+t)\,\alpha + st\beta + ct + ds\right]a_0 - \alpha\,st\,a_1 = 0\,,$$

$$k=2\,,\qquad 2\;(st+s+t)\;a_2 - 6\,st\,a_3 - \left[(1+s+t)\,\alpha_1 + (s+t)\;\beta_1\right]\;a_1$$
$$+2\;\left[(st+s+t)\,\alpha_1 + st\beta_1\right]\;a_2 - 3\,\alpha_1\,st\,a_3$$
$$-\left[(1+s+t)\,\alpha + (s+t)\,\beta + (1+t)\,c + (1+s)\,d\right]a_0$$
$$+\left[(st+s+t)\,\alpha + st\beta + ct + ds\right]a_1 - \alpha\,st\,a_2 = 0\,,$$

$$k=3\,,\qquad -2\;(1+s+t)\;a_2 + 6\;(st+s+t)\;a_3 - 12\,st\,a_4 + (\alpha_1+\beta_1)\;a_1$$
$$-2\;\left[(1+s+t)\,\alpha_1 + (s+t)\;\beta_1\right]\;a_2 + 3\;\left[(st+s+t)\,\alpha_1 + st\beta_1\right]\;a_3$$
$$-4\,\alpha_1\,st\,a_4 + (\alpha+\beta+c+d)\;a_0 -$$
$$-\left[(1+s+t)\,\alpha + (s+t)\,\beta + (1+t)\,c + (1+s)\,d\right]\;a_1$$
$$+\left[(st+s+t)\,\alpha + st\beta + ct + ds\right]a_2 - \alpha\,st\,a_3 = 0\,,$$

$$k=4\,,\qquad 2\,a_2 - 6\;(1+s+t)\;a_3 + 12\;(st+s+t)\;a_4 - 20\,st\,a_5 + 2\;(\alpha_1+\beta_1)\;a_2$$
$$-3\;\left[(1+s+t)\,\alpha_1 + (s+t)\;\beta_1\right]\;a_3 + 4\;\left[(st+s+t)\,\alpha_1 + st\beta_1\right]\;a_4$$
$$-5\,\alpha_1\,st\,a_5 + (\alpha+\beta+c+d)\;a_1$$
$$-\left[(1+s+t)\,\alpha + (s+t)\,\beta + (1+t)\,c + (1+s)\,d\right]a_2$$
$$+\left[(st+s+t)\,\alpha + st\beta + ct + ds\right]a_3 - \alpha\,st\,a_4 = 0\,,$$

$k = 5,\ 6,\ 7,\ \ldots$

$$(k-2)(k-3)\,a_{k-2} - (1+s+t)\;(k-1)(k-2)\,a_{k-1} + (st+s+t)\;k(k-1)\,a_k$$
$$-stk(k+1)a_{k+1} + (\alpha_1+\beta_1)\,(k-2)a_{k-2} - \left[(1+s+t)\,\alpha_1 + (s+t)\,\beta_1\right]\,(k-1)a_{k-1}$$
$$+\left[(st+s+t)\,\alpha_1 + st\beta_1\right]\;k\,a_k - \alpha_1\,st\,(k+1)\,a_{k+1} + (\alpha+\beta+c+d)\;a_{k-3}$$
$$-\left[(1+s+t)\,\alpha + (s+t)\,\beta + (1+t)\,c + (1+s)\,d\right]a_{k-2}$$
$$+\left[(st+s+t)\,\alpha + st\beta + ct + ds\right]a_{k-1} - \alpha\,st\,a_k = 0\,.$$

Thus, we get the 5-term recurrence relations for the coefficients

$$k = 5,\, 6,\, 7,\, \ldots \quad (\alpha+\beta+c+d)\; a_{k-3} + \{(k-2)(k-3) + (\alpha_1+\beta_1)\,(k-2)$$
$$-\,[(1+s+t)\,\alpha + (s+t)\,\beta + (1+t)\,c + (1+s)\,d]\}a_{k-2}$$
$$+\{-\,(1+s+t)\;(k-1)(k-2) - [(1+s+t)\,\alpha_1 + (s+t)\;\beta_1]\;(k-1)$$
$$+\,(st+s+t)\,\alpha + st\beta + ct + ds\}\; a_{k-1} +$$
$$+\{(st+s+t)\,k(k-1) + [(st+s+t)\,\alpha_1 + st\beta_1]\;k - \alpha\,st\}\,a_k$$
$$+\{-st\,k(k+1) - \alpha_1\,st\,(k+1)\}\,a_{k+1} = 0\,.$$

The Poincaré–Perron approach is used to analyse the question of the convergence radius of the series. The recurrence relation is divided by a_{k-3}:

$$(\alpha+\beta+c+d) + \{(k-2)(k-3) + (\alpha_1+\beta_1)\,(k-2)$$
$$-\,[(1+s+t)\,\alpha + (s+t)\,\beta + (1+t)\,c + (1+s)\,d]\}\frac{a_{k-2}}{a_{k-3}}$$
$$+\,\{-\,(1+s+t)\;(k-1)(k-2) - [(1+s+t)\,\alpha_1 + (s+t)\,\beta_1]\,(k-1)$$
$$+\,(st+s+t)\,\alpha + st\beta + ct + ds\}\;\frac{a_{k-1}}{a_{k-2}}\frac{a_{k-2}}{a_{k-3}}$$
$$+\{(st+s+t)\;k(k-1) + [(st+s+t)\,\alpha_1 + st\beta_1]\;k - \alpha\,st\}\frac{a_k}{a_{k-1}}\frac{a_{k-1}}{a_{k-2}}\frac{a_{k-2}}{a_{k-3}}$$
$$+\{-st\,k(k+1) - \alpha_1\,st\,(k+1)\}\frac{a_{k+1}}{a_k}\frac{a_k}{a_{k-1}}\frac{a_{k-1}}{a_{k-2}}\frac{a_{k-2}}{a_{k-3}} = 0\,.$$

The convergence radius of power series is the inverse to $|r|$:

$$r = \lim_{k\to\infty} \frac{a_{k+1}}{a_k}\,, \qquad R_{conv} = \frac{1}{|\,r\,|}\,.$$

To find an algebraic equation for r, the resulting equation is multiplied by k^{-2}, and we let k tend to $k \to \infty$. The result is

$$r - (1+s+t)r^2 + (st+s+t)r^3 - str^4 = 0 \quad\Longrightarrow\quad (-st)r(r-\frac{1}{st})(r-\frac{1}{t})(r-\frac{1}{s}) = 0\,;$$

that is, the roots are

$$r = 0\,, \quad r = 1\,, \quad r = \frac{1}{t}\,, \quad r = \frac{1}{t}\,.$$

Accordingly, the following radii of convergence are possible:

$$R_{conv} = \infty,\; 1\,,\; |t| = |1+\gamma|\,,\; |s| = |1-\gamma|\,. \tag{2.65}$$

Thus, the power series near the point $x = 0$ is guaranteed to converge in a circle of radius $R_{conv} = 1 - |\gamma|$; this is the most interesting area from the physical standpoint. We note that the convergence may be further extended due to the non-singular behaviour of the solutions near the points $|x| = |1 \pm \gamma|$.

2.7 Figures

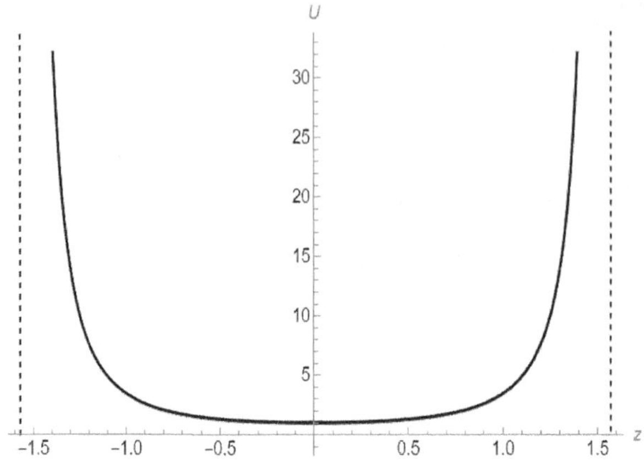

FIGURE 2.1
Potential $U(z)$ at $\gamma = 0$.

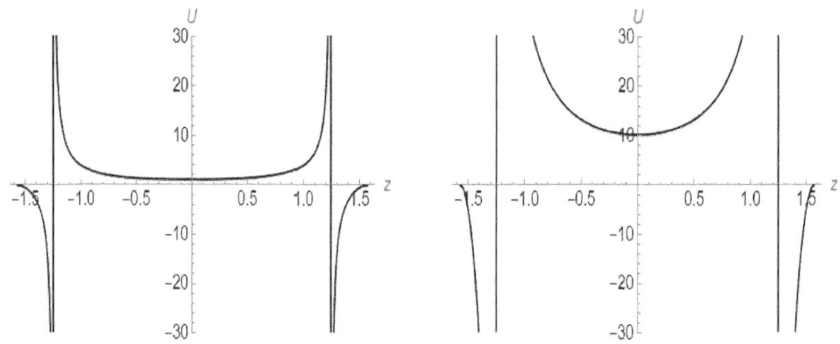

FIGURE 2.2
Potential $U(z)$ at $\gamma = \frac{1}{10}, b = \frac{1}{100}$; two cases: $\Lambda = 1$ and $\Lambda = 10$.

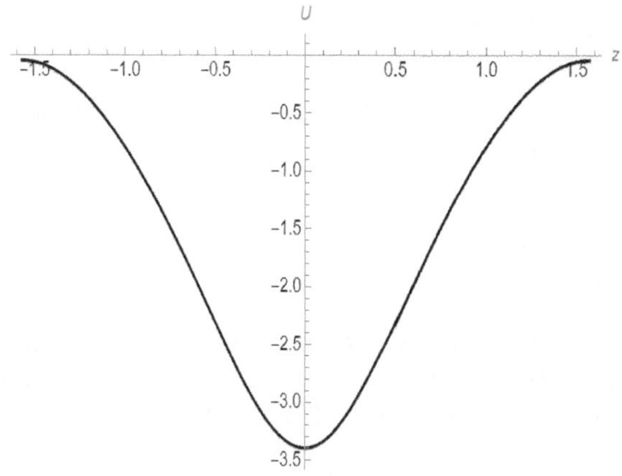

FIGURE 2.3
Potential $U(z)$ *at* $\gamma = 2, b = \frac{1}{10}, \Lambda = 10$.

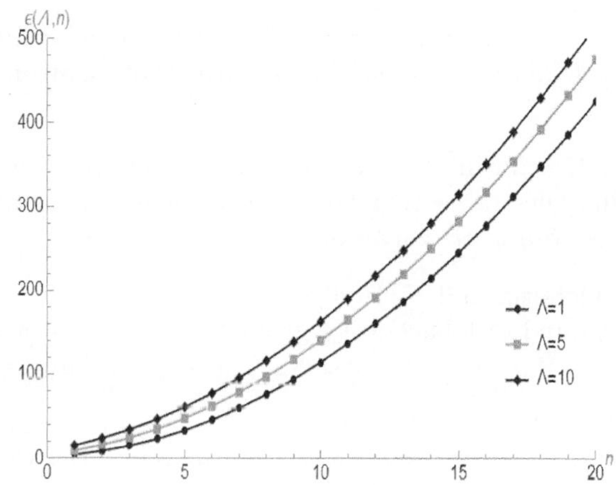

FIGURE 2.4
Three series of energy levels at $\Lambda = 1, 5, 10$.

Bibliography

[1] W. Cox. Higher-rank representations for zero-spin field theories. *Journal of Physics A: Mathematical and General*, 15: 627–635, 1982.

[2] V. Pletyukhov, V. Red'kov and V. Strazhev. *Relativistic Wave Equations and Intrinsic Degrees of Freedom.* Belorissian Science, Minsk, 2015.

[3] V. Kisel, E. Ovsiyuk, O. Veko, Y. Voynova, V. Balan and V. Red'kov. *Elementary Particles with Internal Structure in External Fields.* Nova Science Publishers Inc., New York, 2018.

[4] S. Schweber. *An Introduction to Relativistic Quantum Field Theory.* Harper & Row, Publishers, Inc., New York, 1961.

[5] E. Ovsiyuk. Spin zero cox's particle with an intrinsic structure: general analysis in external electromagnetic and gravitational fields. *Ukrainian Journal of Physics*, 60(6): 485–496, 2015.

[6] O. Veko. Cox's particle in magnetic and electric fields on the background of hyperbolic Lobachevsky geometry. *Nonlinear Phenomena in Complex Systems*, 19(1): 50–61, 2016.

[7] A. Chichurin, E. Ovsiyuk and V. Redkov. Modeling the quantum tunnelling effect for a particle with intrinsic structure in presence of external magnetic field in the Lobachevsky space. *Computers and Mathematics with Applications*, 75: 1550–1565, 2018.

[8] L. Landay and E.M. Lifshitz. *Quantum Mechanics. Volume 3. A Course of Theoretical Physics.* Pergamon Press, Oxford, 1965.

[9] G. Beytmen and A. Erdeyli. *Higher Transcendantle Functions.* Volume 1, Nauka, Moscow, 1973.

[10] A. Prokopenya and A. Chichurin. *Application of the Mathematica System to the Solution of Ordinary Differential Equations.* Belarus State University Press, Minsk, 1999 [in Russian].

[11] V. Red'kov and A. Chichurin. A symbolic-numerical method for solving the differential equation describing the states of polarizable particle in coulomb potential. *Programming and Computer Software*, 40(2): 86–92, 2014.

[12] E. Ovsiyuk, A. Chichurin and V. Red'kov. Nonrelativistic vector particle in coulomb field on the background of lobachevsky geometry: analytical and numerical study, visualisation. *Studia i Materialy. European University in Warsaw*, 10: 45–57, 2015.

[13] A. Chichurin and H. Shvychkina. *Application of the Mathematica System for Solving of Differential Equations and the Problems of Mathematical Modeling.* Part 2, Belarus State University Press, Minsk, 2017 [in Russian].

3

Cox particle in the Coulomb field

Generalised Klein-Fock-Gordon equation for a scalar particle with the Darwin-Cox structure, which takes into account the distribution of the electric charge of the particle inside a finite spherical region is studied in the presence of the external Coulomb field. There are constructed exact Frobenius type solutions of the derived equations, convergence of the relevant power series with 8-term recurrent relations is studied. As analytical quantisation rule is taken so-called transcendency conditions. It provides us with a 4th-order algebraic equation with respect to energy values, which has four sets of roots. One set of roots, $0 < E_{n,k} < 1$, depending on the angular momentum $n = 0, 1, 2, ...$ and the main quantum number $n = 0, 1, 2, ...$, may be interpreted as corresponding to some bound states of the particle in the Coulomb field. In the same manner, a generalised nonrelativistic Schrödinger equation for such a particle is studied, the final results are similar.

3.1 Setting the problem

We start with the tensor system of equations for scalar Cox particle which charge is distributed in finite volume (see [1–8])

$$\frac{mc}{\hbar}\Big(\delta_{\alpha}^{\beta} + \frac{\lambda}{mc/\hbar} F_{\alpha}^{\ \beta}\Big)\Phi_{\beta} = D_{\alpha}\Phi\,, \quad D^{\alpha}\Phi_{\alpha} = \frac{mc}{\hbar}\Phi\,, \tag{3.1}$$

where $D_{\alpha} = i\nabla_{\alpha} + (e/\hbar c)A_{\alpha}$. Nonzero parameter λ corresponds to an additional Darwin–Cox structure of the particle. In short form, eq. (3.1) reads

$$\frac{mc}{\hbar}\Lambda_{\alpha}^{\ \beta}\Phi_{\beta} = D_{\alpha}\Phi\,, \quad D^{\alpha}\Phi_{\alpha} = \frac{mc}{\hbar}\Phi\,. \tag{3.2}$$

Multiplying the first equation by the inverse matrix $(\Lambda^{-1})_{\rho}^{\ \alpha}$, we get

$$\frac{mc}{\hbar}\Phi_{\rho} = (\Lambda^{-1})_{\rho}^{\ \alpha} D_{\alpha}\Phi\,, \quad D^{\alpha}\Phi_{\alpha} = \frac{mc}{\hbar}\Phi\,, \tag{3.3}$$

whence after eliminating the vector component we obtain a generalised Klein–Fock–Gordon equation for scalar $\Phi(x)$

$$\Big(D_{\rho}(\Lambda^{-1})^{\rho\alpha} D_{\alpha}\Phi \; - \frac{m^2c^2}{\hbar^2}\Big)\Phi = 0\,. \tag{3.4}$$

In arbitrary curvilinear coordinates with metrics $g_{\alpha\beta}(x)$, the above equation is written as follows

$$\Big[\Big(\frac{i}{\sqrt{-g}}\frac{\partial}{\partial x^{\rho}}\sqrt{-g} + \frac{e}{c\hbar}A_{\rho}\Big)(\Lambda^{-1})^{\rho\alpha}\Big(i\frac{\partial}{\partial x^{\alpha}} + \frac{e}{c\hbar}A_{\alpha}\Big) \; - \frac{m^2c^2}{\hbar^2}\Big]\Phi = 0\,. \tag{3.5}$$

DOI: 10.1201/9781003472377-3

3.2 Separating the variables

Consider the Cox particle in the presence of external Coulomb field

$$A_0 = \frac{e}{r}, \; e > 0 \quad F_{r0} = -\frac{e}{r^2}, \qquad dS^2 = c^2dt^2 - dr^2 - r^2d\theta^2 - r^2\sin^2\theta d\phi^2 \,. \tag{3.6}$$

In this case, the above matrix ${\Lambda_\alpha}^\beta$ takes the form

$$\Lambda = ({\Lambda_\alpha}^\beta) = \left| \begin{array}{cccc} 1 & \frac{\mu}{r^2} & 0 & 0 \\ \frac{\mu}{r^2} & 1 & 0 & 0 \\ 0 & 0 & 1 & 0 \\ 0 & 0 & 0 & 1 \end{array} \right|, \quad \mu = -\frac{\lambda e}{mc/\hbar}\,; \tag{3.7}$$

its inverse matrix equals

$$\Lambda^{-1} = K = ({K_\beta}^\rho) = \left| \begin{array}{cccc} \frac{r^4}{r^4-\mu^2} & -\frac{\mu r^2}{r^4-\mu^2} & 0 & 0 \\ -\frac{\mu r^2}{r^4-\mu^2} & \frac{r^4}{r^4-\mu^2} & 0 & 0 \\ 0 & 0 & 1 & 0 \\ 0 & 0 & 0 & 1 \end{array} \right|, \tag{3.8}$$

or differently (note the location of indices)

$$(K^{\beta\rho}) = \left| \begin{array}{cccc} \frac{r^4}{r^4-\mu^2} & -\frac{\mu r^2}{r^4-\mu^2} & 0 & 0 \\ +\frac{\mu r^2}{r^4-\mu^2} & -\frac{r^4}{r^4-\mu^2} & 0 & 0 \\ 0 & 0 & -1/r^2 & 0 \\ 0 & 0 & 0 & -1/r^2\sin^2\theta \end{array} \right|. \tag{3.9}$$

Let us apply notations

$$\bar{D}_\alpha = \frac{i}{\sqrt{-g}}\frac{\partial}{\partial x^\rho}\sqrt{-g} + \frac{e}{c\hbar}A_\rho\,, \quad D_\beta = i\frac{\partial}{\partial x^\alpha} + \frac{e}{c\hbar}A_\alpha\,,$$

then eq. (3.5) takes the form

$$\Big[\bar{D}_0 K^{00} D_0 + \bar{D}_0 K^{0r} D_r + \bar{D}_r K^{r0} D_0 + \bar{D}_r K^{rr} D_r$$
$$+\bar{D}_\theta K^{\theta\theta} D_\theta + \bar{D}_\phi K^{\phi\phi} D_\phi - \frac{m^2c^2}{\hbar^2}\Big]\Phi = 0\,. \tag{3.10}$$

Further, we obtain (let $\alpha = e^2/\hbar c$)

$$\Big\{(i\partial_0 + \frac{\alpha}{r})K^{00}(i\partial_0 + \frac{\alpha}{r}) + (i\partial_0 + \frac{\alpha}{r})K^{0r}i\partial_r$$
$$+\frac{i}{r^2}\partial_r r^2 K^{r0}(i\partial_0 + \frac{\alpha}{r}) + \frac{i}{r^2}\partial_r r^2 K^{rr} i\partial_r$$
$$-\frac{1}{r^2\sin\theta}\partial_\theta\sin\theta\partial_\theta - \frac{1}{r^2\sin^2\theta}\partial_\phi\partial_\phi - \frac{m^2c^2}{\hbar^2}\Big\}\Phi = 0\,,$$

or

$$\Big\{\left(i\partial_0 + \frac{\alpha}{r}\right)^2\frac{r^4}{r^4-\mu^2} - \left(i\partial_0 + \frac{\alpha}{r}\right)\frac{\mu r^2}{r^4-\mu^2}i\partial_r$$
$$+\frac{i}{r^2}\partial_r r^2\frac{\mu r^2}{r^4-\mu^2}\left(i\partial_0 + \frac{\alpha}{r}\right) - \frac{i}{r^2}\partial_r r^2\frac{r^4}{r^4-\mu^2}i\partial_r$$
$$+\frac{1}{r^2}\left(\frac{1}{\sin\theta}\partial_\theta\sin\theta\partial_\theta + \frac{1}{\sin^2\theta}\partial_\phi\partial_\phi\right) - \frac{m^2c^2}{\hbar^2}\Big\}\Phi = 0\,. \tag{3.11}$$

If $\mu = 0$, then from eq. (3.11) follows an expected equation

$$\Big\{\Big(i\partial_0 + \frac{\alpha}{r}\Big)^2 + \frac{1}{r^2}\partial_r r^2 \partial_r + \frac{1}{r^2}\Big(\frac{1}{\sin\theta}\partial_\theta \sin\theta \partial_\theta + \frac{1}{\sin^2\theta}\partial_\phi\partial_\phi\Big) - \frac{m^2c^2}{\hbar^2}\Big\}\Phi = 0 .$$

After separating the variables with the use of the substitution

$$\Phi = e^{-iE'x^0/c\hbar}\, Y_{lm}(\theta\phi)\, R(r)\,, \qquad \epsilon = E'/c\hbar, \quad [\epsilon] = \text{meter}^{-1}$$

from eq. (3.10) we get the radial equation

$$\Big\{(\epsilon + \frac{\alpha}{r})^2 \frac{r^4}{r^4+\Gamma^2} - \Gamma(\epsilon + \frac{\alpha}{r})\frac{r^2}{r^4+\Gamma^2}\frac{d}{dr}$$

$$+\Gamma(\frac{d}{dr} + \frac{2}{r})\frac{r^2}{r^4+\Gamma^2} + \frac{\alpha}{r}) + (\frac{d}{dr} + \frac{2}{r})\frac{r^4}{r^4+\Gamma^2}\frac{d}{dr} - \frac{l(l+1)}{r^2} - M^2\Big\}R = 0\,, \tag{3.12}$$

in the following we will apply the notations

$$M = \frac{mc}{\hbar}, \quad e^2 = \alpha = \frac{1}{137}, \quad L = l(l+1)\,, \quad i\mu = \Gamma,\ \Gamma^* = \Gamma\,. \tag{3.13}$$

Equation (3.12) can be transformed to the form

$$\Big\{(\epsilon + \frac{\alpha}{r})^2 \frac{r^4}{r^4+\Gamma^2} - \Gamma(\epsilon + \frac{\alpha}{r})\frac{r^2}{r^4+\Gamma^2}\frac{d}{dr}$$

$$+\Gamma(\epsilon + \frac{\alpha}{r})\frac{r^2}{r^4+\Gamma^2}\frac{d}{dr} + \Big(\Gamma\frac{r^2}{r^4+\Gamma^2}(\epsilon + \frac{\alpha}{r})\Big)' + \Gamma\frac{2}{r}\frac{r^2}{r^4+\Gamma^2}(\epsilon + \frac{\alpha}{r})$$

$$+\frac{r^4}{r^4+\Gamma^2}\frac{d^2}{dr^2} + \Big(\frac{r^4}{r^4+\Gamma^2}\Big)'\frac{d}{dr} + \frac{2}{r}\frac{r^4}{r^4+\Gamma^2}\frac{d}{dr} - \frac{l(l+1)}{r^2} - M^2\Big\}R = 0\,,$$

that is

$$\Big\{\frac{r^4}{r^4+\Gamma^2}\frac{d^2}{dr^2} + \Big[\Big(\frac{r^4}{r^4+\Gamma^2}\Big)' + \frac{2}{r}\frac{r^4}{r^4+\Gamma^2}\Big]\frac{d}{dr} + (\epsilon + \frac{\alpha}{r})^2\frac{r^4}{r^4+\Gamma^2}$$

$$+\Big[\Gamma\frac{r^2}{r^4+\Gamma^2}\Big(\epsilon + \frac{\alpha}{r}\Big)\Big]' + \Gamma\frac{2}{r}\frac{r^2}{r^4+\Gamma^2}(\epsilon + \frac{\alpha}{r}) - \frac{l(l+1)}{r^2} - M^2\Big\}R = 0\,. \tag{3.14}$$

Allowing for identities

$$\frac{2}{r}\frac{r^4}{r^4+\Gamma^2} + \Big[\frac{r^4}{r^4+\Gamma^2}\Big]' = \frac{r^4}{r^4+\Gamma^2}\Big(\frac{2}{r} + \frac{4\Gamma^2}{r(r^4+\Gamma^2)}\Big),$$

$$\Big(\Gamma\frac{r^2}{r^4+\Gamma^2}(\epsilon + \frac{\alpha}{r})\Big)' = \frac{r^4}{r^4+\Gamma^2}\Big(\frac{2\epsilon\Gamma}{r^3} + \frac{\alpha\Gamma}{r^4} - \frac{4(\epsilon\alpha r + \Gamma\alpha)}{r^4+\Gamma^2}\Big),$$

we reduce eq. (3.14) to the form

$$\Big\{\frac{r^4}{r^4+\Gamma^2}\frac{d^2}{dr^2} + \frac{r^4}{r^4+\Gamma^2}\Big(\frac{2}{r} + \frac{4\Gamma^2}{r(r^4+\Gamma^2)}\Big)\frac{d}{dr}$$

$$+(\epsilon + \frac{\alpha}{r})^2\frac{r^4}{r^4+\Gamma^2} + \frac{r^4}{r^4+\Gamma^2}\Big(\frac{2\epsilon\Gamma}{r^3} + \frac{\alpha\Gamma}{r^4} - \frac{4(\epsilon\Gamma r + \Gamma\alpha)}{r^4+\Gamma^2}\Big)$$

$$+\frac{2\Gamma}{r^3}\frac{r^4}{r^4+\Gamma^2}(\epsilon + \frac{\alpha}{r}) - \frac{l(l+1)}{r^2} - M^2\Big\}R = 0\,,$$

or

$$\Big\{\frac{d^2}{dr^2}+\Big(\frac{2}{r}+\frac{4\Gamma^2}{r(r^4+\Gamma^2)}\Big)\frac{d}{dr}$$
$$+(\epsilon+\frac{\alpha}{r})^2+\frac{2\epsilon\Gamma}{r^3}+\frac{\alpha\Gamma}{r^4}-\frac{4}{r^4+\Gamma^2}(\epsilon\Gamma r+\Gamma\alpha)$$
$$+\frac{2\Gamma}{r^3}(\epsilon+\frac{\alpha}{r})-\frac{l(l+1)}{r^2}(1+\frac{\Gamma^2}{r^4})-M^2(1+\frac{\Gamma^2}{r^4})\Big\}R=0\,.$$

Finally, we arrive at the equation

$$\frac{d^2R}{dr^2}+\Big[\frac{2}{r}+\frac{4\Gamma^2}{r(r^4+\Gamma^2)}\Big]\frac{dR}{dr}+\Big[\epsilon^2-M^2+\frac{2\alpha\epsilon}{r}+\frac{\alpha^2-l(l+1)}{r^2}$$
$$+\frac{4\Gamma\epsilon}{r^3}+\frac{3\alpha\Gamma-\Gamma^2M^2}{r^4}-\frac{l(l+1)\Gamma^2}{r^6}-\frac{4(\Gamma\epsilon r+\Gamma\alpha)}{r^4+\Gamma^2}\Big]R=0\,. \tag{3.15}$$

Equation (3.15) has four regular and two irregular points

$$r^4+\Gamma^2=\Big(r-e^{+i\pi/4}\sqrt{\Gamma}\Big)\Big(r+e^{+i\pi/4}\sqrt{\Gamma}\Big)\Big(r-e^{-i\pi/4}\sqrt{\Gamma}\Big)\Big(r+e^{-i\pi/4}\sqrt{\Gamma}\Big),$$
$$r=0,\ \text{Rank}=3\,,\qquad r=\infty,\ \text{Rank}=2\,; \tag{3.16}$$

note identities

$$\sigma\equiv(-\Gamma^2)^{1/4},\quad \sigma^2=\sqrt{-\Gamma^2}=i\Gamma,\quad \sigma^4=-\Gamma^2,$$
$$\frac{1}{r^4+\Gamma^2}=\frac{1}{4\sigma^3}\Big(\frac{1}{r-\sigma}-\frac{1}{r+\sigma}+\frac{i}{r-i\sigma}-\frac{i}{r+i\sigma}\Big). \tag{3.17}$$

Solutions near regular points behave as follows

$$r\to+\sigma, R\sim(r-\sigma)^\rho,\rho=0,2;\quad r\to-\sigma,\ R\sim(r+\sigma)^\rho,\rho=0,2;$$
$$r\to+i\sigma, R\sim(r-i\sigma)^\rho,\rho=0,2;\quad r\to-i\sigma, R\sim(r+i\sigma)^\rho,\rho=0,2. \tag{3.18}$$

Because the point $r=0$ has the rank 3, Frobenius type solutions (in vicinity of the point $r=0$) are searched in the form

$$R(r)=r^Ce^{Ar}e^{B/r}e^{D/r^2}f(r)\,. \tag{3.19}$$

3.3 States with zero angular momentum, $l=0$

If $l=0$, eq. (3.15) simplifies

$$\frac{d^2R}{dr^2}+\Big(\frac{6}{r}-\frac{4r^3}{r^4+\Gamma^2}\Big)\frac{dR}{dr}$$
$$+\Big[\epsilon^2-M^2+\frac{2\alpha\epsilon}{r}+\frac{\alpha^2}{r^2}+\frac{4\Gamma\epsilon}{r^3}+\frac{3\alpha\Gamma-\Gamma^2M^2}{r^4}-\frac{4\Gamma(\epsilon r+\alpha)}{r^4+\Gamma^2}\Big]R=0\,, \tag{3.20}$$

it has the same four regular points, and two irregular, $r=0,r=\infty$, both of the rank 2. Allowing for identities (recall that $\sqrt[4]{-\Gamma^2}=\sigma$):

$$-\frac{4r^3}{(r^4+\Gamma^2)}=-\frac{1}{r-\sigma}-\frac{1}{r-i\sigma}-\frac{1}{r+\sigma}-\frac{1}{r+i\sigma}\,, \tag{3.21}$$

$$\frac{-4\Gamma\epsilon r - 4\alpha\Gamma}{r^4+\Gamma^2}$$

$$= \frac{1}{\sigma^3}\Big\{\frac{-\Gamma\epsilon\sigma - \alpha\Gamma}{r-\sigma} + \frac{i(-i\Gamma\epsilon\sigma - \alpha\Gamma)}{r - i\sigma} - \frac{\Gamma\epsilon\sigma - \alpha\Gamma}{r+\sigma} - \frac{i(i\Gamma\epsilon\sigma - \alpha\Gamma)}{r+i\sigma}\Big\}, \tag{3.22}$$

we transform eq. (3.20) to the form

$$\frac{d^2R}{dr^2} + \Big[\frac{6}{r} - \frac{1}{r-\sigma} - \frac{1}{r-i\sigma} - \frac{1}{r+\sigma} - \frac{1}{r+i\sigma}\Big]\frac{dR}{dr}$$

$$+\Big[\epsilon^2 - M^2 + \frac{2\alpha\epsilon}{r} + \frac{\alpha^2}{r^2} + \frac{4\Gamma\epsilon}{r^3}$$

$$+\frac{3\alpha\Gamma - \Gamma^2M^2}{r^4} - \frac{\Gamma\epsilon}{\sigma^2}\Big(\frac{1}{r-\sigma} - \frac{1}{r-i\sigma} + \frac{1}{r+\sigma} - \frac{1}{r+i\sigma}\Big)$$

$$-\frac{\alpha\Gamma}{\sigma^3}\Big(\frac{1}{r-\sigma} - \frac{1}{r+\sigma} + \frac{i}{r-i\sigma} - \frac{i}{r+i\sigma}\Big)\Big]R = 0$$

or

$$\frac{d^2R}{dr^2} + \Big(\frac{6}{r} - \frac{1}{r-\sigma} - \frac{1}{r-i\sigma} - \frac{1}{r+\sigma} - \frac{1}{r+i\sigma}\Big)\frac{dR}{dr}$$

$$+\Big[\epsilon^2 - M^2 + \frac{2\alpha\epsilon}{r} + \frac{\alpha^2}{r^2} + \frac{4\Gamma\epsilon}{r^3} + \frac{3\alpha\Gamma - \Gamma^2M^2}{r^4}$$

$$-\frac{\Gamma(\epsilon\sigma+\alpha)}{(r-\sigma)\sigma^3} + \frac{\Gamma(\epsilon\sigma+i\alpha)}{(r+i\sigma)\sigma^3} + \frac{\Gamma(-\epsilon\sigma+\alpha)}{(r+\sigma)\sigma^3} - \frac{\Gamma(-\epsilon\sigma+i\alpha)}{(r-i\sigma)\sigma^3}\Big]R = 0\,. \tag{3.23}$$

Let us specify eq. (3.23) near the point $r = 0$:

$$R'' + \frac{6}{r}R' + \frac{(3\Gamma\alpha - \Gamma^2M^2)}{r^4}R = 0\,. \tag{3.24}$$

We can see that the region in the vicinity of $r = 0$ is forbidden for the classical motion of the particle only if parameter Γ is negative. Indeed, inequality $(3\Gamma\alpha - \Gamma^2M^2) < 0$ yields

$$\Gamma(\Gamma - \frac{3\alpha}{M^2}) > 0 \quad \Longrightarrow \quad a)\;\; \Gamma < 0; \qquad b)\;\; \Gamma > \frac{3\alpha}{M^2}\,. \tag{3.25}$$

The variant b) seems to be nonphysical because we assume that parameter Γ may be as small as we wish, including the value $\Gamma = 0$. The inverse of eq. (3.25) restriction $0 < \Gamma < \frac{3\alpha}{M^2}$ implies the existence of the above boundary.

In accordance with the structure of singularity in $r = 0$, the Frobenius type solutions of eq. (3.23) are searched in the form

$$R(r) = r^C e^{Ar} e^{B/r}\, f(r), \tag{3.26}$$

this results in the following equation for $f(r)$:

$$\frac{d^2f}{dr^2} + \Big(\frac{2C+6}{r} - \frac{2B}{r^2} + 2A - \frac{1}{r+i\sigma} - \frac{1}{r+\sigma} - \frac{1}{r-\sigma} - \frac{1}{r-i\sigma}\Big)\frac{df}{dr}$$

$$+\Big[\frac{2\alpha\epsilon + 6A + 2AC}{r} + \frac{\alpha^2 - 2AB + 5C + C^2}{r^2}$$

$$+\frac{4\Gamma\epsilon - 4B - 2BC}{r^3} + \frac{3\alpha\Gamma - \Gamma^2 M^2 + B^2}{r^4}$$

$$+A^2 - M^2 + \epsilon^2 + \frac{-\Gamma\epsilon\sigma + \alpha\Gamma - A\sigma^3 + B\sigma + C\sigma^2}{\sigma^3}\frac{1}{(r+\sigma)}$$

$$+\frac{-\Gamma\epsilon\sigma - \alpha\Gamma - A\sigma^3 + B\sigma - C\sigma^2}{\sigma^3}\frac{1}{(r-\sigma)}$$

$$+\frac{\Gamma\epsilon\sigma + i\Gamma\alpha - A\sigma^3 - B\sigma - iC\sigma^2}{\sigma^3}\frac{1}{(r+i\sigma)}$$

$$+\frac{\Gamma\epsilon\sigma - i\Gamma\alpha - A\sigma^3 - B\sigma + iC\sigma^2}{\sigma^3}\frac{1}{(r-i\sigma)}\Big] f = 0. \tag{3.27}$$

Impose restrictions on A, B, C:

$$A^2 - M^2 + \epsilon^2 = 0 \quad \Longrightarrow \quad A = \pm\sqrt{M^2 - \epsilon^2}, \tag{3.28}$$

$$3\alpha\Gamma - \Gamma^2 M^2 + B^2 = 0 \quad \Longrightarrow \quad B = \delta\sqrt{\Gamma(\Gamma M^2 - 3\alpha)},$$

$$C = 2(\frac{\Gamma\epsilon}{B} - 1) = -2 + \delta\frac{2\Gamma\epsilon}{\sqrt{\Gamma(\Gamma M^2 - 3\alpha)}}, \tag{3.29}$$

here arise four possibilities: $\pm 1, \delta = \pm 1$.

For describing the bound states, of interest are solutions tending to zero in infinity, so parameter A should be negative

$$A = -\sqrt{M^2 - \epsilon^2}. \tag{3.30}$$

Solutions tend to zero at the point $r = 0$, only at negative B,

$$B = -\sqrt{\Gamma(\Gamma M^2 - 3\alpha)}, \quad \Gamma < 0 \quad (R > 0). \tag{3.31}$$

Let us study the possibility of imaginary B:

$$\Gamma(\Gamma M^2 - 3\alpha) < 0, \quad e^{B/r} = e^{\pm iR/r}, \quad R > 0, \quad \Gamma \in (0, \frac{3\alpha}{M^2}), \ \Gamma > 0. \tag{3.32}$$

This implies the following behaviour in vicinity of $r = 0$:

$$r \to 0, \qquad R \sim \frac{1}{r^2} e^{\mp i\frac{2\Gamma\epsilon}{R}\ln r} e^{\pm iR/r}; \tag{3.33}$$

which is hardly consistent with any bound states. For this reason, in the following, we assume that $\Gamma < 0$.

Taking in mind restrictions (3.29), we simplify eq. (3.27) to the form

$$\frac{d^2 f}{dr^2} + \Big(\frac{2C+6}{r} - \frac{2B}{r^2} + 2A - \frac{1}{r+\sigma} - \frac{1}{r-\sigma} - \frac{1}{r+i\sigma} - \frac{1}{r-i\sigma}\Big)\frac{df}{dr}$$

$$+\Big[\frac{2\alpha\epsilon + 6A + 2AC}{r} + \frac{\alpha^2 - 2AB + 5C + C^2}{r^2}$$

$$+\frac{-\Gamma\epsilon\sigma + \alpha\Gamma - A\sigma^3 + B\sigma + C\sigma^2}{\sigma^3(r+\sigma)} + \frac{-\Gamma\epsilon\sigma - \alpha\Gamma - A\sigma^3 + B\sigma - C\sigma^2}{\sigma^3(r-\sigma)}$$

$$+\frac{\Gamma\epsilon\sigma + i\Gamma\alpha - A\sigma^3 - B\sigma - iC\sigma^2}{\sigma^3(r+i\sigma)} + \frac{\Gamma\epsilon\sigma - i\Gamma\alpha - A\sigma^3 - B\sigma + iC\sigma^2}{\sigma^3(r-i\sigma)}\Big] f = 0. \tag{3.34}$$

Further, we will apply its short form

$$f'' + \Big(a + \frac{a_1}{r} + \frac{a_2}{r^2} - \frac{1}{r+\sigma} - \frac{1}{r-\sigma} - \frac{1}{r+i\sigma} - \frac{1}{r-i\sigma}\Big) f'$$
$$+ \Big(\frac{b_1}{r} + \frac{b_2}{r^2} + \frac{\beta_1}{r+\sigma} + \frac{\beta_2}{r-\sigma} + \frac{\beta_3}{r+i\sigma} + \frac{\beta_4}{r-i\sigma}\Big) f = 0 . \tag{3.35}$$

Multiplying it by $r^2(r+\sigma)(r-\sigma)(r+i\sigma)(r-i\sigma) = r^2(r^4 - \sigma^4)$, we obtain

$$\Big(r^6 - \sigma^4 r^2\Big)\frac{d^2 f}{dr^2} + \Big[a r^6 + (a_1 - 4)\, r^5 + a_2\, r^4 - \sigma^4 a r^2 - \sigma^4 a_1\, r - \sigma^4 a_2\Big]\frac{df}{dr}$$
$$+\Big[\,(b_1 + \beta_1 + \beta_2 + \beta_3 + \beta_4)\, r^5 + ((-\beta_1 + \beta_2 - i\beta_3 + i\beta_4)\,\sigma + b_2)\, r^4$$
$$+\sigma^2\,(\beta_1 + \beta_2 - \beta_3 - \beta_4)\, r^3$$
$$+ (-\beta_1 + \beta_2 + i\beta_3 - i\beta_4)\,\sigma^3 r^2 - b_1\,\sigma^4 r - b_2\,\sigma^4\Big] f = 0 . \tag{3.36}$$

Solutions can be searched as a power series,

$$f = \sum_{n=0}^{\infty} c_n r^n, \qquad \frac{df}{dr} = \sum_{n=1}^{\infty} n c_n r^{n-1}, \qquad \frac{d^2 f}{dr^2} = \sum_{n=2}^{\infty} n(n-1) c_n r^{n-2},$$

$$\sum_{n=2}^{\infty} n(n-1) c_n r^{n+4} - \sigma^4 \sum_{n=2}^{\infty} n(n-1) c_n r^n + a \sum_{n=1}^{\infty} n c_n r^{n+5} + (a_1 - 4) \sum_{n=1}^{\infty} n c_n r^{n+4}$$
$$+ a_2 \sum_{n=1}^{\infty} n c_n r^{n+3} - \sigma^4 a \sum_{n=1}^{\infty} n c_n r^{n+1} - \sigma^4 a_1 \sum_{n=1}^{\infty} n c_n r^n - \sigma^4 a_2 \sum_{n=1}^{\infty} n c_n r^{n-1}$$
$$+ (b_1 + \beta_1 + \beta_2 + \beta_3 + \beta_4) \sum_{n=0}^{\infty} c_n r^{n+5} + ((-\beta_1 + \beta_2 - i\beta_3 + i\beta_4)\,\sigma + b_2) \sum_{n=0}^{\infty} c_n r^{n+4}$$
$$+ \sigma^2\,(\beta_1 + \beta_2 - \beta_3 - \beta_4) \sum_{n=0}^{\infty} c_n r^{n+3} + (-\beta_1 + \beta_2 + i\beta_3 - i\beta_4)\,\sigma^3 \sum_{n=0}^{\infty} c_n r^{n+2}$$
$$- b_1\,\sigma^4 \sum_{n=0}^{\infty} c_n r^{n+1} - b_2\,\sigma^4 \sum_{n=0}^{\infty} c_n r^n = 0 .$$

After changing the summation indices, we obtain

$$\sum_{k=6}^{\infty} (k-4)(k-5) c_{k-4} r^k - \sigma^4 \sum_{k=2}^{\infty} k(k-1) c_k r^k + a \sum_{k=6}^{\infty} (k-5) c_{k-5} r^k + (a_1 - 4) \sum_{k=5}^{\infty} (k-4) c_{k-4} r^k$$
$$+ a_2 \sum_{k=4}^{\infty} (k-3) c_{k-3} r^k - \sigma^4 a \sum_{k=2}^{\infty} (k-1) c_{k-1} r^k - \sigma^4 a_1 \sum_{k=1}^{\infty} k c_k r^k - \sigma^4 a_2 \sum_{k=0}^{\infty} (k+1) c_{k+1} r^k$$
$$+ (b_1 + \beta_1 + \beta_2 + \beta_3 + \beta_4) \sum_{k=5}^{\infty} c_{k-5} r^k + [(-\beta_1 + \beta_2 - i\beta_3 + i\beta_4)\,\sigma + b_2] \sum_{k=4}^{\infty} c_{k-4} r^k$$
$$+ \sigma^2\,(\beta_1 + \beta_2 - \beta_3 - \beta_4) \sum_{k=3}^{\infty} c_{k-3} r^k + (-\beta_1 + \beta_2 + i\beta_3 - i\beta_4)\,\sigma^3 \sum_{k=2}^{\infty} c_{k-2} r^k$$
$$- b_1\,\sigma^4 \sum_{k=1}^{\infty} c_{k-1} r^k - b_2\,\sigma^4 \sum_{k=0}^{\infty} c_k r^k = 0 .$$

Thus, we get the recurrent formulas

$$k = 0, \qquad a_2 c_1 + b_2 c_0 = 0,$$

$$k = 1, \quad 2 a_2 c_2 + a_1 c_1 + b_1 c_0 + b_2 c_1 = 0,$$

$$k = 2, \quad -2\sigma c_2 - \sigma a c_1 - 2\sigma a_1 c_2 - 3\sigma a_2 c_3 + (-\beta_1 + \beta_2 + i\beta_3 - i\beta_4) c_0 - b_1 \sigma c_1 - b_2 \sigma c_2 = 0,$$

$$k = 3, \quad -6\sigma^2 c_3 - 2\sigma^2 a c_2 - 3\sigma^2 a_1 c_3 - 4\sigma^2 a_2 c_4 + (\beta_1 + \beta_2 - \beta_3 - \beta_4) c_0$$
$$+ (-\beta_1 + \beta_2 + i\beta_3 - i\beta_4) \sigma c_1 - b_1 \sigma^2 c_2 - b_2 \sigma^2 c_3 = 0,$$

$$k = 4, \quad -12\sigma^4 c_4 + a_2 c_1 - 3\sigma^4 a c_3 - \sigma^4 a_1 4 c_4 - 5\sigma^4 a_2 c_5 + [(-\beta_1 + \beta_2 - i\beta_3 + i\beta_4)\sigma + b_2] c_0$$
$$+\sigma^2 (\beta_1 + \beta_2 - \beta_3 - \beta_4) c_1 + (-\beta_1 + \beta_2 + i\beta_3 - i\beta_4) \sigma^3 c_2 - b_1 \sigma^4 c_3 - b_2 \sigma^4 c_4 = 0,$$

$$k = 5, \qquad -20\sigma^4 c_5 + (a_1 - 4) c_1 + 2a_2 c_2 - 4\sigma^4 a c_4 - \sigma^4 a_1 5 c_5 - 6\sigma^4 a_2 c_6$$
$$+ (b_1 + \beta_1 + \beta_2 + \beta_3 + \beta_4) c_0 + [(-\beta_1 + \beta_2 - i\beta_3 + i\beta_4)\sigma + b_2] c_1$$
$$+\sigma^2 (\beta_1 + \beta_2 - \beta_3 - \beta_4) c_2 + (-\beta_1 + \beta_2 + i\beta_3 - i\beta_4) \sigma^3 c_3 - b_1 \sigma^4 c_4 - b_2 \sigma^4 c_5 = 0,$$

$$k = 6, \quad 2c_2 - 30\sigma^4 c_6 + a c_1 + 2(a_1 - 4) c_2 + 3a_2 c_3 - 5\sigma^4 a c_5 - \sigma^4 a_1 6 c_6$$
$$-7\sigma^4 a_2 c_7 + (b_1 + \beta_1 + \beta_2 + \beta_3 + \beta_4) c_1 + [(-\beta_1 + \beta_2 - i\beta_3 + i\beta_4)\sigma + b_2] c_2$$
$$+\sigma^2 (\beta_1 + \beta_2 - \beta_3 - \beta_4) c_3 + (-\beta_1 + \beta_2 + i\beta_3 - i\beta_4) \sigma^3 c_4 - b_1 \sigma^4 c_5 - b_2 \sigma^4 c_6 = 0,$$

the basic 7-term recurrent formula reads

$$k = 5,\ 6,\ 7, \ldots, \qquad [a(k-5) + (b_1 + \beta_1 + \beta_2 + \beta_3 + \beta_4)]\, c_{k-5}$$
$$+ [(k-4)(k-5) + (a_1 - 4)(k-4) + \{(-\beta_1 + \beta_2 - i\beta_3 + i\beta_4)\sigma + b_2\}]\, c_{k-4}$$
$$+ \left[a_2(k-3) + \sigma^2 (\beta_1 + \beta_2 - \beta_3 - \beta_4)\right] c_{k-3} + (-\beta_1 + \beta_2 + i\beta_3 - i\beta_4)\sigma^3 c_{k-2}$$
$$+ \left[-\sigma^4 a(k-1) - b_1 \sigma^4\right] c_{k-1} + \left[-\sigma^4 k(k-1) - \sigma^4 a_1 k - b_2 \sigma^4\right] c_k - \sigma^4 a_2 (k+1) c_{k+1} = 0.$$

Let us divide the last relation by $k^2 c_{k-5}$:

$$\frac{1}{k^2}\left[a(k-5) + (b_1 + \beta_1 + \beta_2 + \beta_3 + \beta_4)\right]$$
$$+\frac{1}{k^2}\left[(k-4)(k-5) + (a_1 - 4)(k-4) + \{(-\beta_1 + \beta_2 - i\beta_3 + i\beta_4)\sigma + b_2\}\right] \frac{c_{k-4}}{c_{k-5}}$$
$$+\frac{1}{k^2}\left[a_2(k-3) + \sigma^2(\beta_1 + \beta_2 - \beta_3 - \beta_4)\right] \frac{c_{k-3}}{c_{k-4}} \frac{c_{k-4}}{c_{k-5}}$$
$$+\frac{1}{k^2}(-\beta_1 + \beta_2 + i\beta_3 - i\beta_4)\sigma^3 \frac{c_{k-2}}{c_{k-3}} \frac{c_{k-3}}{c_{k-4}} \frac{c_{k-4}}{c_{k-5}}$$
$$+\frac{1}{k^2}\left[-\sigma^4 a(k-1) - b_1 \sigma^4\right] \frac{c_{k-1}}{c_{k-2}} \frac{c_{k-2}}{c_{k-3}} \frac{c_{k-3}}{c_{k-4}} \frac{c_{k-4}}{c_{k-5}}$$
$$+\frac{1}{k^2}\left[-\sigma^4 k(k-1) - \sigma^4 a_1 k - b_2 \sigma^4\right] \frac{c_k}{c_{k-1}} \frac{c_{k-1}}{c_{k-2}} \frac{c_{k-2}}{c_{k-3}} \frac{c_{k-3}}{c_{k-4}} \frac{c_{k-4}}{c_{k-5}}$$
$$-\frac{1}{k^2}\sigma^4 a_2 (k+1) \frac{c_{k+1}}{c_k} \frac{c_k}{c_{k-1}} \frac{c_{k-1}}{c_{k-2}} \frac{c_{k-2}}{c_{k-3}} \frac{c_{k-3}}{c_{k-4}} \frac{c_{k-4}}{c_{k-5}} = 0$$

and tend $k \longrightarrow \infty$. This results in an algebraic equation which determines possible convergence radii

$$R_{conv} = \frac{1}{|r|}, \qquad \lim_{k \to \infty} \frac{c_{k-4}}{c_{k-5}} = r, \qquad r - \sigma^4 r^5 = 0.$$

Whence it follows

$$R_{conv} = \frac{1}{|\sigma|} = |\sqrt{\Gamma}|\,,\ \infty\,. \tag{3.37}$$

Because on the boundary of the circle of radius $|\Gamma|$, behaviour of solutions is quite regular, we can assume that the series converges for any finite r.

To get some quantisation rules, we will apply restriction which determines so called transcendental Frobenius solutions [8,9]. To this end, let us turn to recurrent formula

$$P_{k-5}c_{k-5} + P_{k-4}c_{k-4} + \ldots + P_k c_k + P_{k+1}c_{k+1} = 0, \quad k = 6, 7, 8, \ldots \tag{3.38}$$

and impose restriction of vanishing the coefficient P_{k-5} at c_{k-5}. It should be noted that if in addition to constraint $P_{k-5} = 0$, we require six coefficients be vanished

$$c_{k-4} = 0, \quad c_{k-3} = 0, \quad c_{k-2} = 0, \quad c_{k-1} = 0, \quad c_k = 0, \quad c_{k+1} = 0, \tag{3.39}$$

then due to the recurrent formula, the series terminates as a polynomial. Of course, it is assumed that equations (3.38) and (3.39) permit consistent solutions. The study below will show that for the problem under consideration such consistent solutions do not exist.

3.4 Studying the states with $l = 1, 2, 3, \ldots$

Now we turn to a more complex case when $l = 1, 2, 3, \ldots$. Equation (3.15) has the form (two additional terms arise in comparison with the case $l = 0$)

$$\frac{d^2R}{dr^2} + \Big(\frac{6}{r} - \frac{1}{r-\sigma} - \frac{1}{r-i\sigma} - \frac{1}{r+\sigma} - \frac{1}{r+i\sigma}\Big)\frac{dR}{dr}$$
$$+\Big[\epsilon^2 - M^2 + \frac{2\alpha\epsilon}{r} + \frac{\alpha^2}{r^2} - \frac{L}{r^2} + \frac{4\Gamma\epsilon}{r^3} + \frac{3\alpha\Gamma - \Gamma^2 M^2}{r^4} - \frac{\Gamma^2 L}{r^6}$$
$$-\frac{\Gamma(\epsilon\sigma+\alpha)}{(r-\sigma)\sigma^3} + \frac{\Gamma(\epsilon\sigma+i\alpha)}{(r+i\sigma)\sigma^3} + \frac{\Gamma(-\epsilon\sigma+\alpha)}{(r+\sigma)\sigma^3} - \frac{\Gamma(-\epsilon\sigma+i\alpha)}{(r-i\sigma)\sigma^3}\Big]R = 0\,. \tag{3.40}$$

In the vicinity of the point $r = 0$, the form of the equation differs substantially from that when $l = 0$,

$$R'' + \frac{2}{r}R' - \frac{L\Gamma^2}{r^6}R = 0,$$

whence we conclude that the region near the point $r = 0$ is forbidden for classical motion of the particle, and no constraint for parameter Γ arises. Now, Frobenius-type solutions are searched in the form

$$R(r) = r^C e^{Ar} e^{B/r} e^{D/r^2} f(r)\,, \tag{3.41}$$

for function $f(r)$ we get equation

$$\frac{d^2f}{dr^2} + \Big[2\,A + \frac{2\,C+6}{r} - \frac{2\,B}{r^2} - \frac{4\,D}{r^3} - \frac{1}{r-\sigma} - \frac{1}{r-i\sigma} - \frac{1}{r+\sigma} - \frac{1}{r+i\sigma}\Big]\frac{df}{dr}$$
$$+\Big[\frac{2\,\alpha\,\epsilon + 6\,A + 2\,AC}{r} + \frac{\alpha^2 + 5\,C + C^2 - 2\,AB - L}{r^2} + \frac{4\,\Gamma\epsilon - 4\,B - 2\,CB - 4\,AD}{r^3}$$
$$+\frac{B^2 - 6\,D - 4\,CD + 3\,\alpha\,\Gamma - \Gamma^2 M^2}{r^4} + \frac{4\,BD}{r^5} + \frac{-\Gamma^2 L + 4\,D^2}{r^6} + A^2 + \epsilon^2 - M^2$$

$$+\frac{-\Gamma\epsilon\sigma+\alpha\Gamma-A\sigma^3+B\sigma+C\sigma^2-2D}{(r+\sigma)\sigma^3}+\frac{\Gamma\epsilon\sigma-i\Gamma\alpha-B\sigma+iC\sigma^2+2iD-A\sigma^3}{(r-i\sigma)\sigma^3}$$
$$+\frac{-\Gamma\epsilon\sigma-\alpha\Gamma-A\sigma^3+B\sigma-C\sigma^2+2D}{(r-\sigma)\sigma^3}+\frac{\Gamma\epsilon\sigma+i\Gamma\alpha-2iD-iC\sigma^2-B\sigma-A\sigma^3}{(r+i\sigma)\sigma^3}\Big]f=0\,.$$

Imposing four restrictions, we find possible values for parameters A, B, C, D:

$$A^2-M^2+\epsilon^2=0 \implies A=\pm\sqrt{M^2-\epsilon^2}\,,$$
$$\frac{1}{r^6},\quad -\Gamma^2 L+4D^2=0 \implies D=\pm\frac{1}{2}\sqrt{L}|\Gamma|, \tag{3.42}$$
$$\frac{1}{r^5},\quad 4BD=0 \implies B=0\,, \qquad \frac{1}{r^4},\quad C=-\frac{1}{4D}\left(6D-3\alpha\Gamma+\Gamma^2M^2\right).$$

To bound states there correspond the values

$$A=-\sqrt{M^2-\epsilon^2}\,,\quad B=0\,,\quad D=-\frac{1}{2}\sqrt{L}|\Gamma|\,,$$
$$C=\frac{1}{2\sqrt{L}|\Gamma|}\Big(-3\sqrt{L}|\Gamma|-3\alpha\Gamma+\Gamma^2M^2\Big)\,. \tag{3.43}$$

According to this, the previous equation simplifies

$$\frac{d^2f}{dr^2}+\Big[2A+\frac{2C+6}{r}-\frac{2B}{r^2}-\frac{4D}{r^3}-\frac{1}{r-\sigma}-\frac{1}{r-i\sigma}-\frac{1}{r+\sigma}-\frac{1}{r+i\sigma}\Big]\frac{df}{dr}$$
$$+\Big[\frac{2\alpha\epsilon+6A+2AC}{r}+\frac{\alpha^2+5C+C^2-2AB-L}{r^2}+\frac{4\Gamma\epsilon-4B-2CB-4AD}{r^3}$$
$$+\frac{-\Gamma\epsilon\sigma+\alpha\Gamma-A\sigma^3+B\sigma+C\sigma^2-2D}{(r+\sigma)\sigma^3}+\frac{\Gamma\epsilon\sigma-i\Gamma\alpha-B\sigma+iC\sigma^2+2iD-A\sigma^3}{(r-i\sigma)\sigma^3}$$
$$+\frac{-\Gamma\epsilon\sigma-\alpha\Gamma-A\sigma^3+B\sigma-C\sigma^2+2D}{(r-\sigma)\sigma^3}+\frac{\Gamma\epsilon\sigma+i\Gamma\alpha-2iD-iC\sigma^2-B\sigma-A\sigma^3}{(r+i\sigma)\sigma^3}\Big]f=0;$$

below we apply its short form

$$f''+\Big(a+\frac{a_1}{r}+\frac{a_2}{r^2}+\frac{a_3}{r^3}-\frac{1}{r+\sigma}-\frac{1}{r-\sigma}-\frac{1}{r+i\sigma}-\frac{1}{r-i\sigma}\Big)f'$$
$$+\Big(\frac{b_1}{r}+\frac{b_2}{r^2}+\frac{b_3}{r^3}+\frac{\beta_1}{r+\sigma}+\frac{\beta_2}{r-\sigma}+\frac{\beta_3}{r+i\sigma}+\frac{\beta_4}{r-i\sigma}\Big)f=0\,. \tag{3.44}$$

Multiplying this equation by $r^3(r+\sigma)(r-\sigma)(r+i\sigma)(r-i\sigma)=r^3(r^4-\sigma^4)$, we get

$$\left(r^7-\sigma^4r^3\right)f''$$
$$+\Big[ar^7+(a_1-4)r^6+a_2r^5+a_3r^4-\sigma^4ar^3-\sigma^4a_1r^2-\sigma^4a_2r-a_3\sigma^4\Big]f'$$
$$+\Big\{(b_1+\beta_1+\beta_2+\beta_3+\beta_4)r^6+[b_2+(-\beta_1+\beta_2-i\beta_3+i\beta_4)\sigma]r^5$$
$$+\left[b_3+(\beta_1+\beta_2-\beta_3-\beta_4)\sigma^2\right]r^4$$
$$+\sigma^3(-\beta_1+\beta_2+i\beta_3-i\beta_4)r^3-b_1\sigma^4r^2-b_2\sigma^4r-b_3\sigma^4\Big\}f=0\,. \tag{3.45}$$

Solutions are constructed as power series,

$$f=\sum_{n=0}^{\infty}c_nr^n,\qquad \frac{df}{dr}=\sum_{n=1}^{\infty}nc_nr^{n-1},\qquad \frac{d^2f}{dr^2}=\sum_{n=2}^{\infty}n(n-1)c_nr^{n-2}\,;$$

$$\sum_{n=2}^{\infty} n(n-1)c_n r^{n+5} - \sigma^4 \sum_{n=2}^{\infty} n(n-1)c_n r^{n+1}$$

$$+a \sum_{n=1}^{\infty} nc_n r^{n+6} + (a_1 - 4) \sum_{n=1}^{\infty} nc_n r^{n+5} + a_2 \sum_{n=1}^{\infty} nc_n r^{n+4} + a_3 \sum_{n=1}^{\infty} nc_n r^{n+3}$$

$$-\sigma^4 a \sum_{n=1}^{\infty} nc_n r^{n+2} - \sigma^4 a_1 \sum_{n=1}^{\infty} nc_n r^{n+1} - \sigma^4 a_2 \sum_{n=1}^{\infty} nc_n r^{n} - a_3\, \sigma^4 \sum_{n=1}^{\infty} nc_n r^{n-1}$$

$$+ (b_1 + \beta_1 + \beta_2 + \beta_3 + \beta_4) \sum_{n=0}^{\infty} c_n r^{n+6} + \{b_2 + (-\beta_1 + \beta_2 - i\beta_3 + i\beta_4)\, \sigma\} \sum_{n=0}^{\infty} c_n r^{n+5}$$

$$+ \{b_3 + (\beta_1 + \beta_2 - \beta_3 - \beta_4)\, \sigma^2\} \sum_{n=0}^{\infty} c_n r^{n+4} + \sigma^3 (-\beta_1 + \beta_2 + i\beta_3 - i\beta_4) \sum_{n=0}^{\infty} c_n r^{n+3}$$

$$-b_1\, \sigma^4 \sum_{n=0}^{\infty} c_n r^{n+2} - b_2\, \sigma^4 \sum_{n=0}^{\infty} c_n r^{n+1} - b_3\, \sigma^4 \sum_{n=0}^{\infty} c_n r^{n} = 0\,.$$

Changing the summation indices, we get

$$\sum_{k=7}^{\infty} (k-5)(k-6)c_{k-5} r^k - \sigma^4 \sum_{k=3}^{\infty} (k-1)(k-2)c_{k-1} r^k$$

$$+a \sum_{k=7}^{\infty} (k-6)c_{k-6} r^k + (a_1 - 4) \sum_{k=6}^{\infty} (k-5)c_{k-5} r^k + a_2 \sum_{k=5}^{\infty} (k-4)c_{k-4} r^k + a_3 \sum_{k=4}^{\infty} (k-3)c_{k-3} r^k$$

$$-\sigma^4 a \sum_{k=3}^{\infty} (k-2)c_{k-2} r^k - \sigma^4 a_1 \sum_{k=2}^{\infty} (k-1)c_{k-1} r^k - \sigma^4 a_2 \sum_{k=1}^{\infty} kc_k r^k - a_3\, \sigma^4 \sum_{k=0}^{\infty} (k+1)c_{k+1} r^k$$

$$+ (b_1 + \beta_1 + \beta_2 + \beta_3 + \beta_4) \sum_{k=6}^{\infty} c_{k-6} r^k + [b_2 + (-\beta_1 + \beta_2 - i\beta_3 + i\beta_4)\, \sigma] \sum_{k=5}^{\infty} c_{k-5} r^k$$

$$+ \left[b_3 + (\beta_1 + \beta_2 - \beta_3 - \beta_4)\, \sigma^2\right] \sum_{k=4}^{\infty} c_{k-4} r^k + \sigma^3 (-\beta_1 + \beta_2 + i\beta_3 - i\beta_4) \sum_{k=3}^{\infty} c_{k-3} r^k$$

$$-b_1\, \sigma^4 \sum_{k=2}^{\infty} c_{k-2} r^k - b_2\, \sigma^4 \sum_{k=1}^{\infty} c_{k-1} r^k - b_3\, \sigma^4 \sum_{k=0}^{\infty} c_k r^k = 0\,.$$

Thus, we obtain the recurrent relations

$$k = 0, \qquad a_3 c_1 + b_3 c_0 = 0,$$

$$k = 1, \quad a_2 c_1 + 2a_3 c_2 + b_2 c_0 + b_3 c_1 = 0,$$

$$k = 2, \quad a_1 c_1 + 2a_2 c_2 + 3a_3 c_3 + b_1 c_0 + b_2 c_1 + b_3 c_2 = 0,$$

$$k = 3, \quad 2\sigma c_2 + \sigma a c_1 + 2\sigma a_1 c_2 + 3\sigma a_2 c_3 + 4a_3 \sigma c_4$$
$$+ (\beta_1 - \beta_2 - i\beta_3 + i\beta_4)\, c_0 + b_1 \sigma c_1 + b_2 \sigma c_2 + b_3 \sigma c_3 = 0,$$

$$k = 4, \qquad 6\sigma^4 c_3 - a_3 c_1 + 2\sigma^4 a c_2 + 3\sigma^4 a_1 c_3 + \sigma^4 a_2 4 c_4 + 5a_3 \sigma^4 c_5$$
$$- \left[b_3 + (\beta_1 + \beta_2 - \beta_3 - \beta_4)\, \sigma^2\right] c_0$$
$$-\sigma^3 (-\beta_1 + \beta_2 + i\beta_3 - i\beta_4)\, c_1 + b_1 \sigma^4 c_2 + b_2 \sigma^4 c_3 + b_3 \sigma^4 c_4 = 0,$$

$$k = 5, \quad -12\sigma^4 c_4 + a_2 c_1 + 2a_3 c_2 - 3\sigma^4 a c_3 - 4\sigma^4 a_1 c_4 - 5\sigma^4 a_2 c_5$$
$$-6a_3 \sigma^4 c_6 + [b_2 + (-\beta_1 + \beta_2 - i\beta_3 + i\beta_4)\, \sigma]\, c_0 + \left[b_3 + (\beta_1 + \beta_2 - \beta_3 - \beta_4)\, \sigma^2\right] c_1$$
$$+\sigma^3 (-\beta_1 + \beta_2 + i\beta_3 - i\beta_4)\, c_2 - b_1 \sigma^4 c_3 - b_2 \sigma^4 c_4 - b_3 \sigma^4 c_5 = 0,$$

$$k=6,\quad -20\sigma^4 c_5+(a_1-4)\,c_1+2a_2c_2+3a_3c_3-4\sigma^4 ac_4-5\sigma^4 a_1c_5-6\sigma^4 a_2c_6-7a_3\sigma^4 c_7$$
$$+(b_1+\beta_1+\beta_2+\beta_3+\beta_4)\,c_0+[b_2+(-\beta_1+\beta_2-i\beta_3+i\beta_4)\,\sigma]\,c_1$$
$$+\left[b_3+(\beta_1+\beta_2-\beta_3-\beta_4)\,\sigma^2\right]c_2+\sigma^3\,(-\beta_1+\beta_2+i\beta_3-i\beta_4)\,c_3$$
$$-b_1\sigma_4^4-b_2\sigma^4 c_5-b_3\sigma^4 c_6=0,$$
$$k=7,\quad 2c_2-30\sigma^4 c_6+ac_1+2\,(a_1-4)\,c_2+3a_2c_3+4a_3c_4-5\sigma^4 ac_5$$
$$-6\sigma^4 a_1c_6-7\sigma^4 a_2c_7-8a_3\sigma^4 c_8+(b_1+\beta_1+\beta_2+\beta_3+\beta_4)\,c_1$$
$$+[b_2+(-\beta_1+\beta_2-i\beta_3+i\beta_4)\,\sigma]\,c_2+\left[b_3+(\beta_1+\beta_2-\beta_3-\beta_4)\,\sigma^2\right]c_3$$
$$+\sigma^3\,(-\beta_1+\beta_2+i\beta_3-i\beta_4)\,c_4-b_1\sigma^4 c_5-b_2\sigma^4 c_6-b_3\sigma^4 c_7=0\,.$$

So the basic 8-term recurrent formula reads

$$k=6,\,7,\,8,\ldots,\qquad [a\,(k-6)+(b_1+\beta_1+\beta_2+\beta_3+\beta_4)]\;c_{k-6}$$
$$+\{(k-5)(k-6)+(a_1-4)\;(k-5)+[b_2+(-\beta_1+\beta_2-i\beta_3+i\beta_4)\,\sigma]\}\;c_{k-5}$$
$$+\left\{a_2\,(k-4)+\left[b_3+(\beta_1+\beta_2-\beta_3-\beta_4)\,\sigma^2\right]\right\}\;c_{k-4}$$
$$+\left[a_3\,(k-3)+\sigma^3\,(-\beta_1+\beta_2+i\beta_3-i\beta_4)\right]\;c_{k-3}$$
$$-\sigma^4\,[a\,(k-2)+b_1\,]\;c_{k-2}-\sigma^4\,[(k-1)(k-2)+a_1\,(k-1)+b_2]\;c_{k-1}$$
$$-\sigma^4\,[a_2\,k+b_3]\;c_k-a_3\,\sigma^4\,(k+1)c_{k+1}=0\,, \tag{3.46}$$

or shortly

$$k=6,7,8,\ldots,\qquad P_{k-6}c_{k-6}+P_{k-5}c_{k-5}+P_{k-4}c_{k-4}+P_{k-3}c_{k-3}$$
$$+P_{k-2}c_{k-2}+P_{k-1}c_{k-1}+P_kc_k+P_{k+1}c_{k+1}=0\,. \tag{3.47}$$

In accordance with the Poincaré–Perron method, we divide the last relation by k^2c_{k-6}:

$$\frac{1}{k^2}\,[a\,(k-6)+(b_1+\beta_1+\beta_2+\beta_3+\beta_4)]$$
$$+\frac{1}{k^2}\,\{(k-5)(k-6)+(a_1-4)\;(k-5)+[b_2+(-\beta_1+\beta_2-i\beta_3+i\beta_4)\,\sigma]\}\;\frac{c_{k-5}}{c_{k-6}}$$
$$+\frac{1}{k^2}\left\{a_2\,(k-4)+\left[b_3+(\beta_1+\beta_2-\beta_3-\beta_4)\,\sigma^2\right]\right\}\;\frac{c_{k-4}}{c_{k-5}}\frac{c_{k-5}}{c_{k-6}}$$
$$+\frac{1}{k^2}\left[a_3\,(k-3)+\sigma^3\,(-\beta_1+\beta_2+i\beta_3-i\beta_4)\right]\;\frac{c_{k-3}}{c_{k-4}}\frac{c_{k-4}}{c_{k-5}}\frac{c_{k-5}}{c_{k-6}}$$
$$-\frac{\sigma^4}{k^2}\,[a\,(k-2)+b_1\,]\;\frac{c_{k-2}}{c_{k-3}}\frac{c_{k-3}}{c_{k-4}}\frac{c_{k-4}}{c_{k-5}}\frac{c_{k-5}}{c_{k-6}}$$
$$-\frac{\sigma^4}{k^2}\,[(k-1)(k-2)+a_1\,(k-1)+b_2]\;\frac{c_{k-1}}{c_{k-2}}\frac{c_{k-2}}{c_{k-3}}\frac{c_{k-3}}{c_{k-4}}\frac{c_{k-4}}{c_{k-5}}\frac{c_{k-5}}{c_{k-6}}$$
$$-\frac{\sigma^4}{k^2}\,[a_2\,k+b_3]\;\frac{c_k}{c_{k-1}}\frac{c_{k-1}}{c_{k-2}}\frac{c_{k-2}}{c_{k-3}}\frac{c_{k-3}}{c_{k-4}}\frac{c_{k-4}}{c_{k-5}}\frac{c_{k-5}}{c_{k-6}}-$$
$$-\frac{\sigma^4}{k^2}a_3\,(k+1)\frac{c_{k+1}}{c_k}\frac{c_k}{c_{k-1}}\frac{c_{k-1}}{c_{k-2}}\frac{c_{k-2}}{c_{k-3}}\frac{c_{k-3}}{c_{k-4}}\frac{c_{k-4}}{c_{k-5}}\frac{c_{k-5}}{c_{k-6}}=0\,,$$

and tend $k \longrightarrow \infty$. This results in an algebraic equation

$$\lim_{k\to\infty} \frac{c_{k-5}}{c_{k-6}} = r, \quad r - \sigma^4 r^5 = 0\,.$$

So two possible convergence radii are

$$R_{conv} = \frac{1}{|\sigma|} = |\sqrt{\Gamma}|\,,\ \infty\,. \tag{3.48}$$

Because on the boundary of the circle of radius $|\Gamma|$, behaviour of solutions is quite regular, we can assume that the series converge for any finite r. In fact, this result coincides with that found for the case of $l = 0$. The deference consists in more complicated recurrent formula, 8-term instead of 6-term. For states with of $l = 1, 2, \ldots$, the transcendency condition has the form

$$P_{k-6} = a\,(k-6) + (b_1 + \beta_1 + \beta_2 + \beta_3 + \beta_4) = 0, \quad k = 6, 7, 8, \ldots\,; \tag{3.49}$$

we will need this relationship in the following.

3.5 Qualitative study

Let us examine behaviour of the effective squared momentum for the problem. First consider the case of $l = 0$:

$$\frac{d^2R}{dr^2} + \Big(\frac{6}{r} - \frac{4r^3}{r^4+\Gamma^2}\Big)\frac{dR}{dr}$$
$$+\Big[\epsilon^2 - M^2 + \frac{2\alpha\epsilon}{r} + \frac{\alpha^2}{r^2} + \frac{4\Gamma\epsilon}{r^3} + \frac{3\alpha\Gamma - \Gamma^2 M^2}{r^4} - \frac{4\Gamma(\epsilon r + \alpha)}{r^4+\Gamma^2}\Big] R = 0\,. \tag{3.50}$$

Eliminating the first derivative term by simple substitution

$$R(r) = \varphi(r)\, f(r), \quad R' = \varphi' f + \varphi f', \quad R'' = \varphi'' f + 2\varphi' f' + \varphi f''\,,$$

we get

$$f'' + \Big(\frac{6}{r} - \frac{4r^3}{r^4+\Gamma^2} + \frac{\varphi'}{\varphi}\Big) f' + \Big\{\frac{\varphi''}{\varphi} + \frac{\varphi'}{\varphi}\Big(\frac{6}{r} - \frac{4r^3}{r^4+\Gamma^2}\Big)$$
$$+\epsilon^2 - M^2 + \frac{2\alpha\epsilon}{r} + \frac{\alpha^2}{r^2} + \frac{4\Gamma\epsilon}{r^3} + \frac{3\alpha\Gamma - \Gamma^2 M^2}{r^4} - \frac{4\Gamma(\epsilon r + \alpha)}{r^4+\Gamma^2}\Big\} f = 0\,.$$

The needed function $\varphi(r)$ is

$$\Big(\frac{6}{r} - \frac{4r^3}{r^4+\Gamma^2} + \frac{\varphi'}{\varphi}\Big) = 0 \quad \Longrightarrow \quad \varphi = \sqrt{\frac{r^4+\Gamma^2}{r^6}}\,. \tag{3.51}$$

Thus, we reduce the above equation to the form

$$\frac{d^2 f}{dr^2} + P^2(r) f = 0\,,$$

where the squared momentum is determined by the formula

$$P^2(r) = \epsilon^2 - M^2 + \frac{2\alpha\epsilon}{r} + \frac{\alpha^2 - 6}{r^2}$$
$$+\frac{4\Gamma\epsilon}{r^3} + \frac{\Gamma\,(3\alpha - \Gamma M^2)}{r^4} + \frac{6r^2 - 4\Gamma\epsilon r - 4\alpha\Gamma}{r^4+\Gamma^2} + \frac{12\Gamma^2 r^2}{(r^4+\Gamma^2)^2}\,. \tag{3.52}$$

Near the point $r = 0$, this quantity behaves

$$\Gamma < 0, \quad P^2(r)_{r\to 0} \sim +\frac{\Gamma\,(3\alpha - \Gamma M^2)}{r^4} \to -\infty\,, \tag{3.53}$$

$$0 < \Gamma < \frac{3\alpha}{M^2}, \quad P^2(r)_{r\to 0} \sim +\frac{\Gamma\,(3\alpha - \Gamma M^2)}{r^4} \to +\infty\,. \tag{3.54}$$

In infinity, it behaves as follows:

$$r \to \infty, \qquad P^2(r) \sim \epsilon^2 - M^2\,. \tag{3.55}$$

For states with $l = 1, 2, 3, \ldots$, the main equation is

$$\frac{d^2R}{dr^2} + \Big(\frac{6}{r} - \frac{4r^3}{r^4+\Gamma^2}\Big)\frac{dR}{dr}$$
$$+\Big[\epsilon^2 - M^2 + \frac{2\alpha\epsilon}{r} + \frac{\alpha^2}{r^2} - \frac{L}{r^2} + \frac{4\Gamma\epsilon}{r^3} + \frac{3\alpha\Gamma - \Gamma^2 M^2}{r^4} - \frac{\Gamma^2 L}{r^6} - \frac{4\Gamma(\epsilon r + \alpha)}{r^4+\Gamma^2}\Big]R = 0$$

after eliminating the first derivative term, it takes the form

$$R(r) = \varphi(r)\, f(r), \qquad \varphi = \sqrt{\frac{r^4+\Gamma^2}{r^6}}, \qquad \frac{d^2 f}{dr^2} + P^2(r) f = 0\,,$$

where

$$P^2(r) = \epsilon^2 - M^2 + \frac{2\,\alpha\,\epsilon}{r} + \frac{\alpha^2 - 6 - L}{r^2} + \frac{4\,\Gamma\epsilon}{r^3} - \frac{\Gamma\,(-3\,\alpha + \Gamma M^2)}{r^4}$$
$$-\frac{\Gamma^2 L}{r^6} + \frac{6\,r^2 - 4\,\Gamma\epsilon\, r - 4\,\alpha\,\Gamma}{r^4+\Gamma^2} + \frac{12\,\Gamma^2 r^2}{(r^4+\Gamma^2)^2}\,. \tag{3.56}$$

It behaves near the singular points as follows

$$P^2(r)_{r\to 0} \sim -\Gamma^2 \frac{l(l+1)}{r^6} \to -\infty\,, \quad P^2(r)_{r\to\infty} \sim \epsilon^2 - M^2 < 0\,. \tag{3.57}$$

3.6 Dimensionless quantities, qualitative study

Let us take the Compton wave length λ of the particle as a unit length

$$M = \frac{mc}{\hbar} = \frac{1}{\lambda}, \tag{3.58}$$

then there arise a number of dimensionless quantities

$$Mr = x\,, \quad \frac{\epsilon}{M} = E, \quad \Gamma M^2 \Longrightarrow \Gamma\,. \tag{3.59}$$

Then equation at $l = 0$ (see eq. (3.20)) is re-written as follows

$$\frac{d^2R}{dx^2} + \Big(\frac{6}{x} - \frac{4x^3}{x^4+\Gamma^2}\Big)\frac{dR}{dx}$$
$$+\Big[E^2 - 1 + \frac{2\alpha E}{x} + \frac{\alpha^2}{x^2} + \frac{4\Gamma E}{x^3} + \frac{3\alpha\Gamma - \Gamma^2}{x^4} - \frac{4\Gamma(Ex+\alpha)}{x^4+\Gamma^2}\Big]R = 0\,. \tag{3.60}$$

In fact, such a transition is reached through formal changes

$$r \Longrightarrow x, \quad M \Longrightarrow 1, \quad \epsilon \Longrightarrow E, \quad \Gamma \Longrightarrow \Gamma \,. \tag{3.61}$$

After eliminating the first derivative term in eq. (3.60) we obtain the expression for squared momentum (see eq. (3.52))

$$\begin{gathered} P^2(x) = E^2 - 1 + \frac{2\alpha E}{x} + \frac{\alpha^2 - 6}{x^2} \\ + \frac{4\Gamma E}{x^3} + \frac{\Gamma(3\alpha - \Gamma)}{x^4} + \frac{6x^2 - 4\Gamma E x - 4\alpha\Gamma}{x^4 + \Gamma^2} + \frac{12\Gamma^2 x^2}{(x^4 + \Gamma^2)^2} \,. \end{gathered} \tag{3.62}$$

The above noticed possibilities for the Cox parameter are now read

$$I, \quad \Gamma < 0 \,, \qquad II, \quad 0 < \Gamma < 3\alpha \,. \tag{3.63}$$

The substitution for Frobenius solutions is

$$\begin{gathered} l = 0, \quad R(x) = x^C e^{Ax} e^{B/x} \, f(x), \quad A = \pm\sqrt{1 - E^2} \,, \\ B = \pm\sqrt{-\Gamma(3\,\alpha - \Gamma)} \,, \quad C = 2(\frac{\Gamma E}{B} - 1) \,. \end{gathered} \tag{3.64}$$

For two possibilities in (3.63), we have corresponding expressions for parameters (assuming the description of the bound states)

$I, \quad \Gamma < 0,$

$$A = -\sqrt{1 - E^2}, \qquad B = -\sqrt{-\Gamma(3\alpha - \Gamma)}, \quad C = \frac{-2\Gamma E}{\sqrt{-\Gamma(3\alpha - \Gamma)}} - 2; \tag{3.65}$$

$II, \quad 0 < \Gamma < 3\alpha \,,$

$$A = -\sqrt{1 - E^2}, \qquad B = \pm i\sqrt{\Gamma(3\alpha - \Gamma)}, \quad C = \frac{2\Gamma E}{\pm i\sqrt{\Gamma(3\alpha - \Gamma)}} - 2. \tag{3.66}$$

Let us study the function $P^2(x)$ numerically. We fix the basic parameters as

$$I, \qquad E = 0.999 \,, \quad \alpha = \frac{1}{137} \,, \quad \Gamma = -0.001 \,.$$

The roots of equation $P^2(x) = 0$ are

$$\begin{gathered} x_1 = +0.0319933, \quad x_2 = +7.299278, \\ -0.0313352, \quad -0.2118184, \\ 0.1690023 \pm 0.1267129 \cdot i, \quad 0.0003370 \pm 0.0315803 \cdot i, \\ -0.0003330 \pm 0.0019250 \cdot i, \quad -0.0652559 \pm 0.1981235 \cdot i, \\ -0.0652559 \pm 0.1981235 \cdot i, \quad -0.0003330 \pm 0.0019250 \cdot i, \\ 0.0003370 \pm 0.0315803 \cdot i, \quad 0.1690023 \pm 0.1267129 \cdot i \,. \end{gathered}$$

The existence of the positive roots x_1, x_2 means that the classical motion of the particle is possible in the region $x \in (x_1, x_2)$.

Let us find the roots of equation $P^2(x) = 0$ at positive Γ:

$$II, \qquad E = 0.999 \,, \quad \alpha = \frac{1}{137} \,, \quad \Gamma = +0.001 \,.$$

$$x_1 = 0.00222882, \quad x_2 = 0.03125030, \quad x_3 = 7.2992596,$$
$$x_4 = -0.00156276, \quad x_5 = -0.03192458, \quad x_6 = -0.20709844,$$
$$x_7 = 0.17076664 + 0.12237871\, i, \quad x_8 = 0.170766640 - 0.12237871\, i,$$
$$x_9 = -0.06870490 + 0.20066378\, i, \quad x_{10} = -0.06870490 - 0.20066378\, i,$$
$$x_{11} = -0.00032889 - 0.03165733\, i, \quad x_{12} = -0.00032889 + 0.03165733\, i\,.$$

We may note the existence of three positive roots, and the region near the point $r = 0$ is permitted for classical motion. This means that the whole region for variable x is divided into parts, permitted (+) and forbidden (−) for classical motion:

$$(+)\ x \in (0, x_1), \quad (-)\ x \in (x_1, x_2), \quad (+)\ x \in (x_2, x_3), \quad (-)\ x \in (x_3, +\infty).$$

This implies a completely different physical system than in the case when $\Gamma < 0$.

For the remaining values of the orbital momentum $l = 1, 2, 3, ...$, the main equation after performing the formal changes

$$r \Longrightarrow x, \quad M \Longrightarrow 1, \quad \epsilon \Longrightarrow E, \quad \Gamma \Longrightarrow \Gamma\,, \quad \sigma \Longrightarrow \Sigma\ \ (\Gamma^2 = -\Sigma^4) \tag{3.67}$$

takes on the form

$$\frac{d^2R}{dx^2} + \left(\frac{6}{x} - \frac{1}{x-\Sigma} - \frac{1}{x-i\Sigma} - \frac{1}{x+\Sigma} - \frac{1}{x+i\Sigma}\right)\frac{dR}{dx}$$
$$+ \left[E^2 - 1 + \frac{2\alpha E}{x} + \frac{\alpha^2}{x^2} - \frac{L}{x^2} + \frac{4\Gamma E}{x^3} + \frac{3\alpha\Gamma - \Gamma^2}{x^4} - \frac{\Gamma^2 L}{x^6}\right.$$
$$\left. - \frac{\Gamma\,(E\Sigma + \alpha)}{(x-\Sigma)\,\Sigma^3} + \frac{\Gamma\,(E\Sigma + i\alpha)}{(x+i\Sigma)\,\Sigma^3} + \frac{\Gamma\,(-E\Sigma + \alpha)}{(x+\Sigma)\,\Sigma^3} - \frac{\Gamma\,(-E\Sigma + i\alpha)}{(x-i\Sigma)\,\Sigma^3}\right] R = 0\,. \tag{3.68}$$

After eliminating the first derivative term, we get an equation with the effective squared momentum

$$P^2(x) = E^2 - 1 + \frac{2\,\alpha\,E}{x} + \frac{\alpha^2 - 6 - L}{x^2}$$
$$+ \frac{4\,\Gamma E}{x^3} + \frac{\Gamma\,(3\,\alpha - \Gamma)}{x^4} - \frac{\Gamma^2 L}{x^6} + \frac{6\,x^2 - 4\,\Gamma E x - 4\,\alpha\,\Gamma}{x^4 + \Gamma^2} + \frac{12\,\Gamma^2 x^2}{(x^4+\Gamma^2)^2}\,. \tag{3.69}$$

Let us study behaviour of the function $P^2(x)$ numerically. Let it be

$$I, \qquad E = 0.99999\,, \quad \alpha = \frac{1}{137}\,, \quad \Gamma = -0.001\,, \quad L = 2\ \ (l = 1)\,.$$

The roots of the equation $P^2(x) = 0$ are

$$x_1 = +182.75523, \quad x_2 = +547.16813,$$
$$0.03675161 \pm 0.01494842 \cdot i, \quad 0.01418804 \pm 0.01428520 \cdot i,$$
$$0.01486661 \pm 0.03657386 \cdot i, \quad -0.01480131 \pm 0.03678462 \cdot i,$$
$$-0.01432517 \pm 0.01428510 \cdot i, \quad -0.03667977 \pm 0.01473772 \cdot i,$$
$$-0.03667977 \pm 0.01473772 \cdot i, \quad -0.01432517 \pm 0.01428510 \cdot i,$$
$$-0.01480131 \pm 0.03678462 \cdot i, \quad 0.01486661 \pm 0.0365739 \cdot i,$$
$$0.01418804 \pm 0.01428520 \cdot i, \quad 0.03675161 \pm 0.01494842 \cdot i\,.$$

There exists the region $x \in (x_1, x_2)$ permissible for classical motion.

3.7 Quantisation of energy, the case of minimal $l = 0$

Frobenius solutions are searched in the form

$$R(r) = x^C e^{Ax} e^{B/x} e^{D/x^2} f(x)\,, \tag{3.70}$$

we will follow both possibilities (assuming description of bound states)
I, $\Gamma < 0$,

$$A = -\sqrt{1-E^2}\,,\; B = 0\,,\; D = +\frac{\sqrt{L}}{2}\Gamma\,,\; C = -\frac{3}{2} - \frac{3\alpha - \Gamma}{2\sqrt{L}}\,; \tag{3.71}$$

II, $0 < \Gamma < 3\alpha$,

$$A = -\sqrt{1-E^2}\,, \quad B = 0\,,\; D = -\frac{\sqrt{L}}{2}\Gamma\,,\; C = -\frac{3}{2} + \frac{3\alpha - \Gamma}{2\sqrt{L}}\,. \tag{3.72}$$

The bound states are possible only at negative Γ. We have the following expressions for parameters

$$A = -\sqrt{1-E^2}\,, \quad B = -\sqrt{-\Gamma(3\alpha - \Gamma)}\,, \quad C = \frac{-2\Gamma E}{\sqrt{-\Gamma(3\alpha-\Gamma)}} - 2 \tag{3.73}$$

and the equation for $f(x)$,

$$\frac{d^2 f}{dx^2} + \Big(\frac{2C+6}{x} - \frac{2B}{x^2} + 2A - \frac{1}{x+\Sigma} - \frac{1}{x-\Sigma} - \frac{1}{x+i\Sigma} - \frac{1}{x-i\Sigma}\Big)\frac{df}{dx}$$

$$+\Big[\frac{2\alpha E + 6A + 2AC}{x} + \frac{\alpha^2 - 2AB + 5C + C^2}{x^2}$$

$$+\frac{-\Gamma E\Sigma + \alpha\Gamma - A\Sigma^3 + B\Sigma + C\Sigma^2}{\Sigma^3}\frac{1}{(x+\Sigma)}$$

$$+\frac{-\Gamma E\Sigma - \alpha\Gamma - A\Sigma^3 + B\Sigma - C\Sigma^2}{\Sigma^3}\frac{1}{(x-\Sigma)}$$

$$+\frac{\Gamma E\Sigma + i\Gamma\alpha - A\Sigma^3 - B\Sigma - iC\Sigma^2}{\Sigma^3}\frac{1}{(x+i\Sigma)}$$

$$+\frac{\Gamma E\Sigma - i\Gamma\alpha - A\Sigma^3 - B\Sigma + iC\Sigma^2}{\Sigma^3}\frac{1}{(x-i\Sigma)}\Big]f = 0,$$

or shortly (recall that $\Gamma^2 = -\Sigma^4$)

$$f'' + \Big(a + \frac{a_1}{x} + \frac{a_2}{x^2} - \frac{1}{x+\Sigma} - \frac{1}{x-\Sigma} - \frac{1}{x+i\Sigma} - \frac{1}{x-i\Sigma}\Big)f'$$
$$+\Big(\frac{b_1}{x} + \frac{b_2}{x^2} + \frac{\beta_1}{x+\Sigma} + \frac{\beta_2}{x-\Sigma} + \frac{\beta_3}{x+i\Sigma} + \frac{\beta_4}{x-i\Sigma}\Big)f = 0\,. \tag{3.74}$$

Let us consider the transcendency condition (3.38):

$$k = 5, 6, 7, ..., \qquad P_{k-5} = a\,(k-5) + (b_1 + \beta_1 + \beta_2 + \beta_3 + \beta_4) = 0\,. \tag{3.75}$$

Taking into account identities

$$a = 2\,A = -2\,\sqrt{1-E^2}\,, \quad b_1 = 2\,\alpha\,E + 6\,A + 2\,AC\,,$$

$$\beta_1+\beta_2+\beta_3+\beta_4$$
$$= \frac{-\Gamma E\Sigma + \alpha\Gamma - A\Sigma^3 + B\Sigma + C\Sigma^2}{\Sigma^3} + \frac{-\Gamma E\Sigma - \alpha\Gamma - A\Sigma^3 + B\Sigma - C\Sigma^2}{\Sigma^3}$$
$$+\frac{\Gamma E\Sigma + i\Gamma\alpha - A\Sigma^3 - B\Sigma - iC\Sigma^2}{\Sigma^3} + \frac{\Gamma E\Sigma - i\Gamma\alpha - A\Sigma^3 - B\Sigma + iC\Sigma^2}{\Sigma^3} = -4A\,,$$

we find analytical form of the above condition

$$a\,(k-5) + (b_1+\beta_1+\beta_2+\beta_3+\beta_4)$$
$$= 2\,A\,k - 10A + 2\,\alpha\,E + 6\,A + 2\,AC - 4A = 2\,A\,k - 8A + 2\,\alpha\,E + 2\,AC = 0\,,$$

whence it follows equation for energy

$$k = 6,7,8,9,10,..., \qquad 2\,\alpha\,E - 2\,\sqrt{1-E^2}\Big(k - 6 - \frac{2\,\Gamma E}{\sqrt{-\Gamma\,(3\,\alpha-\Gamma)}}\Big) = 0 \qquad (3.76)$$

or differently

$$-\frac{4\Gamma E^4}{3\alpha-\Gamma} + E^2\Big(\alpha^2 - \frac{12\alpha}{\Gamma-3\alpha} + (k-12)k + 32\Big) - \frac{4\Gamma(k-6)E^3}{\sqrt{\Gamma(\Gamma-3\alpha)}} + \frac{4\Gamma(k-6)E}{\sqrt{\Gamma(\Gamma-3\alpha)}} - (k-6)^2 = 0$$

The roots at $k = 5,\ldots,28$ are: one positive, one negative, and two complex.

TABLE 3.1
The energy roots, positive, negative, and complex

k	E
5	-0.99998675, 0.99992138, 2.3926155 ± 0.0192232 i
6	0, 0, -0.99984749, 0.99984749
7	-0.99992138, 0.99998675, -2.3926155 ± 0.0192232 i
8	-0.99998936, 0.99999544, -4.7851687 ± 0.0178584 i
9	-0.99999600, 0.99999772, -7.1777493 ± 0.0176361 i
10	-0.99999792, 0.99999864, -9.5703316 ± 0.0175602 i
11	-0.99999873, 0.99999909, -11.9629143 ± 0.0175254 i
12	-0.99999915, 0.99999935, -14.355497 ± 0.017507 i
13	-0.99999939, 0.99999952, -16.748080 ± 0.017495 i
14	-0.99999954, 0.99999962, -19.140663 ± 0.017488 i
15	-0.99999964, 0.99999970, -21.533245 ± 0.017483 i
16	-0.99999971, 0.99999975, -23.925828 ± 0.017479 i
17	-0.99999976, 0.99999980, -26.318411 ± 0.017477 i
18	-0.99999980, 0.99999983, -28.710994 ± 0.017475 i
19	-0.99999983, 0.99999985, -31.103577 ± 0.017473 i
20	-0.99999986, 0.99999987, -33.496160 ± 0.017472 i
21	-0.99999987, 0.99999989, -35.888742 ± 0.017471 i
22	-0.99999989, 0.99999990, -38.281325 ± 0.017470 i
23	-0.99999990, 0.99999991, -40.673908 ± 0.017469 i
24	-0.99999991, 0.99999992, -43.066491 ± 0.017469 i
25	-0.99999992, 0.99999993, -45.459074 ± 0.017468 i
26	-0.99999993, 0.99999994, -47.851656 ± 0.017468 i
27	-0.99999994, 0.99999994, -50.244239 ± 0.017468 i
28	-0.99999994, 0.99999995, -52.636822 ± 0.017467 i

Retain only positive roots

TABLE 3.2
The positive energy roots

k	E	k	E
5	0.99992138	6	0.99984749
7	0.99998675	8	0.99999544
9	0.99999772	10	0.99999864
11	0.99999909	12	0.99999935
13	0.99999952	14	0.99999962
15	0.99999970	16	0.99999975
17	0.99999980	18	0.99999983
19	0.99999985	20	0.99999987
21	0.99999989	22	0.99999990
23	0.99999991	24	0.99999992
25	0.99999993	26	0.99999994
27	0.99999994	28	0.99999995

All these roots belong to interval $0 < E_1 < 1$.

3.8 Quantisation of energy at $l = 1, 2, 3, ...$

Now let $l = 1, 2, 3, ...$. The transcendency condition $P_{k-6} = 0$ takes the form

$$(2k - 13)\sqrt{1 - E^2} = 2\alpha E + \frac{\sqrt{1 - E^2}(3\alpha + \Gamma)}{\sqrt{L}}.$$

Taking the values parameters

$$\alpha = \frac{1}{137}, \ \Gamma = -10^{-3}, l = 1 \ (L = 2), \ k = 6, ..., 29, \tag{3.77}$$

we get the following the energy levels

TABLE 3.3
The positive energy roots, parameters (3.77)

k	E	k	E
6	0.99989654	7	0.99989024
8	0.99998804	9	0.99999571
10	0.99999782	11	0.99999868
12	0.99999912	13	0.99999937
14	0.99999953	15	0.99999963
16	0.99999970	17	0.99999976
18	0.99999980	19	0.99999983
20	0.99999985	21	0.99999987
22	0.99999989	23	0.99999990
24	0.99999991	25	0.99999992
26	0.99999993	27	0.99999994
28	0.99999994	29	0.99999995

At other set of parameters

$$\alpha = \frac{1}{137}, \ \Gamma = -10^{-3}, \ l = 2 \ (L = 6), \ k = 6, ..., 29, \tag{3.78}$$

we get the energy levels.

TABLE 3.4
The positive energy roots, parameters (3.78)

k	E	k	E
6	0.99989525	7	0.99989162
8	0.99998809	9	0.99999572,
10	0.99999782	11	0.99999868
12	0.99999912	13	0.99999937
14	0.99999953	15	0.99999963
16	0.99999970	17	0.99999976
18	0.99999980	19	0.99999983
20	0.99999985	21	0.99999987
22	0.99999989	23	0.99999990
24	0.99999991	25	0.99999992
26	0.99999993	27	0.99999994
28	0.99999994	29	0.99999995

At other set of parameters

$$\alpha = \frac{1}{137}, \ \Gamma = -10^{-3}, \ l = 5 \ (L = 30), \ k = 6, ..., 29, \tag{3.79}$$

we get the energy levels.

TABLE 3.5
The positive energy roots, parameters (3.79)

k	E	k	E
6	0.99989427	7	0.99989264
8	0.99998813	9	0.99999573
10	0.99999782	11	0.99999868
12	0.99999912	13	0.99999937
14	0.99999953	15	0.99999963
16	0.99999970	17	0.99999976
18	0.99999980	19	0.99999983
20	0.99999985	21	0.99999987
22	0.99999989	23	0.99999990
24	0.99999991	25	0.99999992
26	0.99999993	27	0.99999994
28	0.99999994	29	0.99999995

For comparison of the different series of energy levels we use the quantity $k - 5 = n = 0, 1, 2, ...,$

$$\Delta = E_n(l = 2) - E_n(l = 5).$$

TABLE 3.6
Differences between the energy values

n	Δ	n	Δ
0	$9.864 \cdot 10^{-7}$	1	$-1.024 \cdot 10^{-6}$
2	$-3.745 \cdot 10^{-8}$	3	$-8.070 \cdot 10^{-9}$
4	$-2.938 \cdot 10^{-9}$	5	$-1.382 \cdot 10^{-9}$
6	$-7.564 \cdot 10^{-10}$	7	$-4.5813 \cdot 10^{-10}$
8	$-2.982 \cdot 10^{-10}$	9	$-2.048 \cdot 10^{-10}$
10	$-1.467 \cdot 10^{-10}$	11	$-1.086 \cdot 10^{-10}$
12	$-8.267 \cdot 10^{-11}$	13	$-6.437 \cdot 10^{-11}$
14	$-5.110 \cdot 10^{-11}$	15	$-4.124 \cdot 10^{-11}$
16	$-3.376 \cdot 10^{-11}$	17	$-2.798 \cdot 10^{-11}$
18	$-2.345 \cdot 10^{-11}$	19	$-1.985 \cdot 10^{-11}$
20	$-1.695 \cdot 10^{-11}$	21	$-1.459 \cdot 10^{-11}$
23	$-1.265 \cdot 10^{-11}$	24	$-1.103 \cdot 10^{-11}$

3.9 Nonrelativistic approximation

Nonrelativistic equation for Darwin–Cox particle has the form [4]

$$\Big[D_t + \frac{1}{2}[(K_0{}^0 - 1)mc^2 + K_0{}^j cD_j]\Big]\Psi$$
$$= -\Big[\frac{1}{2m}\bar{D}_k g^{kj}(x)\,(K_j{}^l D_l + mcK_j{}^0)\Big]\Psi, \tag{3.80}$$

where

$$(K_\beta{}^\rho) = \begin{vmatrix} \frac{r^4}{r^4-\mu^2} & -\frac{\mu r^2}{r^4-\mu^2} & 0 & 0 \\ -\frac{\mu r^2}{r^4-\mu^2} & \frac{r^4}{r^4-\mu^2} & 0 & 0 \\ 0 & 0 & 1 & 0 \\ 0 & 0 & 0 & 1 \end{vmatrix}, \quad [\mu] = \text{meter}^2 \tag{3.81}$$

and

$$D_t = i\hbar\partial_t + \frac{e}{c}A_0, \quad D_k = i\hbar\partial_k + \frac{e}{c}A_k, \quad \bar{D}_k = i\hbar\frac{1}{\sqrt{-g}}\partial_k\sqrt{-g} + \frac{e}{c}A_k, \tag{3.82}$$

$$g^{\alpha\beta}(x) = \begin{vmatrix} 1 & 0 & 0 & 0 \\ 0 & -1 & 0 & 0 \\ 0 & 0 & -1/r^2 & 0 \\ 0 & 0 & 0 & -1/r^2\sin^2\theta \end{vmatrix}, \quad \sqrt{-g} = r^2\sin\theta\,.$$

Taking into account the presence of the Coulomb field $A_0 = \frac{e}{r}$, we can obtain

$$(K_0{}^0 - 1)mc^2 + K_0{}^j cD_j = \frac{\mu^2}{r^4-\mu^2}\,mc^2 - i\hbar c\frac{\mu r^2}{r^4-\mu^2}\partial_r\,,$$

whence allowing for $i\mu = \Gamma,\ \Gamma^* = \Gamma$ we get

$$(K_0{}^0 - 1)mc^2 + K_0{}^j cD_j = -\frac{\Gamma^2}{r^4+\Gamma^2}\,mc^2 - \hbar c\frac{\Gamma r^2}{r^4+\Gamma^2}\partial_r\,. \tag{3.83}$$

Then, we find expression for operator from the right

$$-\Big\{\frac{1}{2m}\bar{D}_k g^{kj}(x)\,(K_j{}^l D_l + mcK_j{}^0)\Big\}$$

$$= -\frac{\hbar^2}{2m}\Big\{\frac{1}{r^2}\frac{\partial}{\partial r}r^2\Big(K_r{}^r\frac{\partial}{\partial r} - i\frac{mc}{\hbar}K_r{}^0\Big) + \frac{1}{\sin\theta}\frac{\partial}{\partial\theta}\sin\theta\frac{1}{r^2}\frac{\partial}{\partial\theta} + \frac{\partial}{\partial\phi}\frac{1}{r^2\sin^2\theta}\frac{\partial}{\partial\phi}\Big\}$$

$$= -\frac{\hbar^2}{2m}\Big\{\frac{1}{r^2}\frac{\partial}{\partial r}r^2\Big(\frac{r^4}{r^4-\mu^2}\frac{\partial}{\partial r} + i\frac{mc}{\hbar}\frac{\mu r^2}{r^4-\mu^2}\Big) + \frac{1}{r^2}\Big(\frac{1}{\sin\theta}\frac{\partial}{\partial\theta}\sin\theta\frac{\partial}{\partial\theta} + \frac{1}{\sin^2\theta}\frac{\partial^2}{\partial\phi^2}\Big)\Big\}.$$

Thus, the Schrödinger-like equation takes the form

$$\begin{gathered}\Big\{\Big(i\hbar\frac{\partial}{\partial t} + \frac{e^2}{r}\Big) + \frac{1}{2}\Big[-\frac{\Gamma^2}{r^4+\Gamma^2}\,mc^2 - \hbar c\frac{\Gamma r^2}{r^4+\Gamma^2}\partial_r\Big]\Big\}\Psi \\ = -\frac{\hbar^2}{2m}\Big\{\frac{1}{r^2}\frac{\partial}{\partial r}r^2\Big(\frac{r^4}{r^4+\Gamma^2}\frac{\partial}{\partial r} + \frac{mc}{\hbar}\frac{\Gamma r^2}{r^4+\Gamma^2}\Big) - \frac{1}{r^2}\hat{l}^2\Big\}\Psi\,.\end{gathered} \tag{3.84}$$

Using the usual substitution for the wave function

$$\Psi = e^{-i\epsilon t/\hbar}\,Y_{lm}(\theta\phi)\,R(r)\,,$$

we get the radial equation

$$\begin{gathered}\Big[\Big(\epsilon + \frac{e^2}{r}\Big) + \frac{1}{2}\Big(-\frac{\Gamma^2}{r^4+\Gamma^2}\,mc^2 - \hbar c\frac{\Gamma r^2}{r^4+\Gamma^2}\frac{d}{dr}\Big)\Big]R(r) \\ = -\frac{\hbar^2}{2m}\Big\{\frac{1}{r^2}\frac{d}{dr}r^2\Big(\frac{r^4}{r^4+\Gamma^2}\frac{d}{dr} + \frac{mc}{\hbar}\frac{\Gamma r^2}{r^4+\Gamma^2}\Big) - \frac{l(l+1)}{r^2}\Big\}R(r)\,.\end{gathered} \tag{3.85}$$

Divide the equation by mc^2:

$$\Big[\frac{1}{mc^2}\Big(\epsilon + \frac{e^2}{r}\Big) + \frac{1}{2}\Big(-\frac{\Gamma^2}{r^4+\Gamma^2} - \frac{\hbar}{mc}\frac{\Gamma r^2}{r^4+\Gamma^2}\frac{d}{dr}\Big)\Big]R(r)$$

$$= -\frac{1}{2}\frac{\hbar^2}{m^2c^2}\Big[\frac{1}{r^2}\frac{d}{dr}r^2\Big(\frac{r^4}{r^4+\Gamma^2}\frac{d}{dr} + \frac{mc}{\hbar}\frac{\Gamma r^2}{r^4+\Gamma^2}\Big) - \frac{l(l+1)}{r^2}\Big]R(r)\,,$$

and transform it to dimensionless variables

$$E = \frac{\epsilon}{mc^2},\; r\frac{mc}{\hbar} = x,\; \frac{1}{mc^2}\frac{e^2}{r} = \frac{\alpha}{x},\; d\alpha = \frac{e^2}{\hbar c} = \frac{1}{137},\; \Gamma\frac{m^2c^2}{\hbar^2} \Longrightarrow \Gamma. \tag{3.86}$$

Then, we obtain the following equation

$$\begin{gathered}\Big[2\Big(E + \frac{\alpha}{x}\Big) + \Big(-\frac{\Gamma^2}{x^4+\Gamma^2} - \frac{\Gamma x^2}{x^4+\Gamma^2}\frac{d}{dx}\Big) \\ + \frac{1}{x^2}\frac{d}{dx}x^2\Big(\frac{x^4}{x^4+\Gamma^2}\frac{d}{dx} + \frac{\Gamma x^2}{x^4+\Gamma^2}\Big) - \frac{l(l+1)}{x^2}\Big]R(r) = 0\,,\end{gathered} \tag{3.87}$$

or

$$\begin{gathered}\frac{d^2R}{dx^2} + \Big[-\frac{4\,x^3}{x^4+\Gamma^2} + \frac{6}{x}\Big]\frac{dR}{dx} \\ + \Big[2\,E + \frac{2\,\alpha}{x} - \frac{l(l+1)}{x^2} + \frac{4\Gamma}{x^3} + \frac{\Gamma^2(2\,E-1)}{x^4} + \frac{2\,\alpha\,\Gamma^2}{x^5} - \frac{\Gamma^2 l(l+1)}{x^6} - \frac{4\,x\Gamma}{x^4+\Gamma^2}\Big]R = 0.\end{gathered} \tag{3.88}$$

Here, we have an equation with regular four singularities

$$r^4 + \Gamma^2 = (r - e^{+i\pi/4}\sqrt{\Gamma})\,(r + e^{+i\pi/4}\sqrt{\Gamma})\,(r - e^{-i\pi/4}\sqrt{\Gamma})\,(r + e^{-i\pi/4}\sqrt{\Gamma})\,,$$
$$\{\, r_1,\, r_2,\, r_3,\, r_4 \,\} = \{\, e^{+i\pi/4}\sqrt{\Gamma}\,, \quad -e^{+i\pi/4}\sqrt{\Gamma}\,, \quad e^{-i\pi/4}\sqrt{\Gamma}\,, \quad -e^{-i\pi/4}\sqrt{\Gamma}\,\}\,, \tag{3.89}$$

and two irregular

$$x = 0, \ \ \text{Rank} = 3, \qquad x = \infty, \ \ \text{Rank} = 2\,.$$

Behaviour of solutions near the regular points is

$$r \to +\sigma, \quad R \sim (r-\sigma)^{\rho}, \quad \rho = 0, 2\,; \quad r \to -\sigma, \quad R \sim (r+\sigma)^{\rho}, \quad \rho = 0, 2\,;$$
$$r \to +i\sigma, \quad R \sim (r-i\sigma)^{\rho}, \quad \rho = 0, 2\,; \quad r \to -i\sigma, \quad R \sim (r+i\sigma)^{\rho}, \quad \rho = 0, 2\,. \tag{3.90}$$

In the case $l = 0$, the above equation simplifies

$$\frac{d^2 R}{dx^2} + \left[-\frac{4\,x^3}{x^4 + \Gamma^2} + \frac{6}{x} \right] \frac{dR}{dx}$$
$$+ \left[2\,E + \frac{2\,\alpha}{x} + \frac{4\,\Gamma}{x^3} + \frac{\Gamma^2\,(2\,E - 1)}{x^4} + \frac{2\,\alpha\,\Gamma^2}{x^5} - \frac{4\,x\Gamma}{x^4 + \Gamma^2} \right] R = 0\,; \tag{3.91}$$

here we have an equation with four regular points and two irregular,

$$x = 0, \ \ \text{Rank} = \frac{5}{2}, \qquad x = \infty, \ \ \text{Rank} = 2\,.$$

3.10 Frobenius solutions at $l = 1, 2, ...$

Let us construct the Frobenius solutions of eq. (3.88) in the form

$$R(r) = e^{Ax} x^C e^{B/x} e^{D/x^2} f(r)\,. \tag{3.92}$$

For $f(x)$ we get equation

$$\frac{d^2 f}{dx^2} + \Big[-\frac{4x^3}{(x^4 + \Gamma^2)} + \frac{6 + 2C}{x} - \frac{2B}{x^2} - \frac{4D}{x^3} + 2A \Big] \frac{df}{dx}$$
$$+ \Big[\frac{2CA + 6A + 2\alpha}{x} + \frac{-2AB + C^2 + 5C - l(l+1)}{x^2} + \frac{-2CB - 4AD - 4B + 4\Gamma}{x^3}$$
$$+ \frac{-4x^2 C - 4x^3 A + 4xB + 8D - 4x\Gamma}{(x^4 + \Gamma^2)} + (2E + A^2)$$
$$+ \frac{-4CD + B^2 - 6D - \Gamma^2 + 2\Gamma^2 E}{x^4}$$
$$+ \frac{4BD + 2\alpha\Gamma^2}{x^5} + \frac{-\Gamma^2 l(l+1) + 4D^2}{x^6} \Big] f = 0\,. \tag{3.93}$$

Let us fix parameters C, A, B, D:

$$2E + A^2 = 0 \quad \Longrightarrow \quad A = \pm\sqrt{-2E}, \ E < 0\,; \tag{3.94}$$

$$-\Gamma^2 l(l+1)+4D^2=0 \quad \Longrightarrow \quad D=\pm\frac{|\Gamma|}{2}\sqrt{l(l+1)}=\pm\frac{|\Gamma|\,L}{2}\,; \tag{3.95}$$

$$4BD+2\alpha\Gamma^2=0 \quad \Longrightarrow \quad B=-\frac{\alpha\Gamma^2}{2D}=\mp\frac{\alpha|\Gamma|}{L}\,; \tag{3.96}$$

$$-4CD-6D+B^2-\Gamma^2(1-2E)=0 \quad \Longrightarrow$$

$$C=-\frac{3}{2}+\frac{B^2-\Gamma^2(1-2E)}{4D}=-\frac{3}{2}\pm|\Gamma|\,\frac{\alpha^2/L^2-1+2E}{2L}\,. \tag{3.97}$$

To describe bound states, we are to use the following parameters

$$\begin{gathered} A=-\sqrt{-2E}, \quad e^{-\sqrt{-2E}\,r}\to 0, \quad r\to+\infty\,; \\ D=-\frac{|\Gamma|L}{2}, \quad e^{D/x^2}\to 0, \quad r\to 0\,; \\ B=+\frac{\alpha|\Gamma|}{L}, \quad C=-\frac{3}{2}-|\Gamma|\,\frac{\alpha^2/L^2-1+2E}{2L}\,. \end{gathered} \tag{3.98}$$

The index C might be positive or negative, but near the point $x=0$, the main factor $e^{D/x^2}\to 0$ if $D<0$. Further, we follow both positive and negative values for Γ:

$\Gamma>0$,

$$A=-\sqrt{-2E}, D=-\frac{\Gamma L}{2}, B=\frac{\alpha\Gamma}{L}, C=-\frac{3}{2}-\Gamma\frac{\alpha^2/L^2-1+2E}{2L}; \tag{3.99}$$

$\Gamma<0$,

$$A=-\sqrt{-2E}, D=+\frac{\Gamma L}{2}, B=-\frac{\alpha\Gamma}{L}, C=-\frac{3}{2}+\Gamma\frac{\alpha^2/L^2-1+2E}{2L}. \tag{3.100}$$

Equation (3.93) simplifies

$$\begin{gathered} \frac{d^2 f}{dx^2}+\Big[2A+\frac{6+2C}{x}-\frac{2B}{x^2}-\frac{4D}{x^3}-\frac{4x^3}{(x^4+\Gamma^2)}\Big]\frac{df}{dx} \\ +\Big[\frac{2CA+6A+2\alpha}{x}+\frac{-2AB+C^2+5C-L^2}{x^2}+\frac{-2CB-4AD-4B+4\Gamma}{x^3} \\ +\frac{-4Ax^3-4x^2C+4x(B-\Gamma)+8D}{(x^4+\Gamma^2)}\Big]f=0\,, \end{gathered} \tag{3.101}$$

or in short form

$$\begin{gathered} f''+\Big(a_0+\frac{a_1}{x}+\frac{a_2}{x^2}+\frac{a_3}{x^3}-\frac{4x^3}{x^4+\Gamma^2}\Big)f' \\ +\Big(\frac{b_1}{x}+\frac{b_2}{x^2}+\frac{b_3}{x^3}+\frac{c_3x^3+c_2x^2+c_1x+c_0}{x^4+\Gamma^2}\Big)f=0. \end{gathered} \tag{3.102}$$

Let us write down the coefficients for both cases:

$\Gamma>0\,$,

$$a_0=-\sqrt{-8E}, \quad a_1=3-\Gamma\frac{\alpha^2/L^2-1+2E}{L}, \quad a_2=-\frac{2\alpha\Gamma}{L}, \quad a_3=2\Gamma L,$$

$$b_1 = -3\sqrt{-2E} + 2\alpha + \Gamma\sqrt{-2E}\left(\frac{-1+2E}{L} + \frac{\alpha^2}{L^3}\right),$$

$$b_2 = -\frac{21}{4} - L^2 + \frac{\Gamma(2\sqrt{-2E}\alpha - 2E + 1)}{L} + \frac{\Gamma^2(1/4 - E + E^2)}{L^2}$$
$$-\frac{\Gamma\alpha^2}{L^3} + \frac{\Gamma^2\alpha^2(-1/2 + E)}{L^4} + \frac{1}{4}\frac{\Gamma^2\alpha^4}{L^6},$$

$$b_3 = -2\sqrt{-2E}\Gamma L + 4\Gamma - \frac{\alpha\Gamma}{L} + \frac{\Gamma^2\alpha(2E-1)}{L^2} + \frac{\Gamma^2\alpha^3}{L^4},$$

$$c_0 = -4\Gamma L, c_1 = -4\Gamma + \frac{4\alpha\Gamma}{L},$$
$$c_2 = 6 + \frac{2\Gamma(\alpha^2/L^2 - 1 + 2E)}{L}, c_3 = 4\sqrt{-2E}; \quad (3.103)$$

$\Gamma < 0$,

$$a_0 = -\sqrt{-8E}, \quad a_1 = 3 + \Gamma\frac{\alpha^2/L^2 - 1 + 2E}{L}, \quad a_2 = \frac{2\alpha\Gamma}{L}, \quad a_3 = -2\Gamma L,$$

$$b_1 = -3\sqrt{-2E} + 2\alpha + \Gamma\sqrt{-2E}\left(\frac{1-2E}{L} - \frac{\alpha^2}{L^3}\right),$$

$$b_2 = -\frac{21}{4} - L^2 + \frac{\Gamma(-2\sqrt{-2E}\alpha + 2E - 1)}{L} + \frac{\Gamma^2(1/4 - E + E^2)}{L^2}$$
$$+\frac{\Gamma\alpha^2}{L^3} + \frac{\Gamma^2\alpha^2(-1/2 + E)}{L^4} + \frac{1}{4}\frac{\Gamma^2\alpha^4}{L^6},$$

$$b_3 = 2\sqrt{-2E}\Gamma L + 4\Gamma + \frac{\alpha\Gamma}{L} + \frac{\Gamma^2\alpha(2E-1)}{L^2} + \frac{\Gamma^2\alpha^3}{L^4},$$

$$c_0 = 4\Gamma L, c_1 = -4\Gamma - \frac{4\alpha\Gamma}{L},$$
$$c_2 = 6 - \frac{2\Gamma(\alpha^2/L^2 - 1 + 2E)}{L}, c_3 = 4\sqrt{-2E}. \quad (3.104)$$

Multiply eq. (3.102) by $x^3(x^4 + \Gamma^2)$, we obtain

$$x^7 f'' + x^3\Gamma^2 f'' + (x^7 a_0 + x^6 a_1 - 4x^6 + x^5 a_2 + x^4 a_3 + x^3 a_0\Gamma^2 + x^2 a_1\Gamma^2 + x a_1\Gamma^2 + a_3\Gamma^2)f'$$
$$+(x^6 b_1 + x^6 c_3 + x^5 b_2 + x^5 c_2 + x^4 b_3 + x^4 c_1 + x^2\Gamma^2 b_1 + x^3 c_0 + x\Gamma^2 b_2 + \Gamma^2 b_3)f = 0\,.$$

Solutions are searched as power series:

$$f(x) = \sum_{n=0}^{\infty} d_n x^n\,, \quad f'(x) = n\sum_{n=1}^{\infty} d_n x^{n-1}\,, \quad f''(x) = n(n-1)\sum_{n=2}^{\infty} d_n x^{n-2}\,,$$

in this way we get

$$n(n-1)\sum_{n=2}^{\infty} d_n x^{n+5} + \Gamma^2 n(n-1)\sum_{n=2}^{\infty} d_n x^{n+1} + a_0 n\sum_{n=1}^{\infty} d_n x^{n+6} + a_1 n\sum_{n=1}^{\infty} d_n x^{n+5} - 4n\sum_{n=1}^{\infty} d_n x^{n+5}$$

$$+a_2 n\sum_{n=1}^{\infty} d_n x^{n+4} + a_3 n\sum_{n=1}^{\infty} d_n x^{n+3} + a_0\Gamma^2 n\sum_{n=1}^{\infty} d_n x^{n+2} + a_1\Gamma^2 n\sum_{n=1}^{\infty} d_n x^{n+1} + a_1\Gamma^2 n\sum_{n=1}^{\infty} d_n x^{n}$$

$$+a_3\Gamma^2 n\sum_{n=1}^{\infty} d_n x^{n-1} + b_1\sum_{n=0}^{\infty} d_n x^{n+6} + c_3\sum_{n=0}^{\infty} d_n x^{n+6} + b_2\sum_{n=0}^{\infty} d_n x^{n+5} + c_2\sum_{n=0}^{\infty} d_n x^{n+5} + b_3\sum_{n=0}^{\infty} d_n x^{n+4}$$

$$+c_1\sum_{n=0}^{\infty} d_n x^{n+4} + b_1\Gamma^2\sum_{n=0}^{\infty} d_n x^{n+2} + c_0\sum_{n=0}^{\infty} d_n x^{n+3} + b_2\Gamma^2\sum_{n=0}^{\infty} d_n x^{n+1} + b_3\Gamma^2\sum_{n=0}^{\infty} d_n x^{n} = 0,$$

which after changing summation indices leads to

$$(n-5)(n-6)\sum_{n=7}^{\infty} d_{n-5}x^n + \Gamma^2(n-1)(n-2)\sum_{n=3}^{\infty} d_{n-1}x^n + a_0(n-6)\sum_{n=7}^{\infty} d_{n-6}x^n$$

$$+a_1(n-5)\sum_{n=6}^{\infty} d_{n-5}x^n - 4(n-5)\sum_{n=6}^{\infty} d_{n-5}x^n + +a_2(n-4)\sum_{n=5}^{\infty} d_{n-4}x^n$$

$$+a_3(n-3)\sum_{n=4}^{\infty} d_{n-3}x^n + a_0\Gamma^2(n-2)\sum_{n=3}^{\infty} d_{n-2}x^n + a_1\Gamma^2(n-1)\sum_{n=2}^{\infty} d_{n-1}x^n + a_1\Gamma^2 n\sum_{n=1}^{\infty} d_n x^n$$

$$+a_3\Gamma^2(n+1)\sum_{n=0}^{\infty} d_{n+1}x^n + b_1\sum_{n=6}^{\infty} d_{n-6}x^n + c_3\sum_{n=6}^{\infty} d_{n-6}x^n + b_2\sum_{n=5}^{\infty} d_{n-5}x^n$$

$$+c_2\sum_{n=5}^{\infty} d_{n-5}x^n + b_3\sum_{n=4}^{\infty} d_{n-4}x^n + c_1\sum_{n=4}^{\infty} d_{n-4}x^n$$

$$+b_1\Gamma^2\sum_{n=2}^{\infty} d_{n-2}x^n + c_0\sum_{n=3}^{\infty} d_{n-3}x^n + b_2\Gamma^2\sum_{n=1}^{\infty} d_{n-1}x^n + b_3\Gamma^2\sum_{n=0}^{\infty} d_n x^n = 0\,.$$

So, we get the recurrent relations

$$n=0, \qquad a_3\Gamma^2 d_1 + b_3\Gamma^2 d_0 = 0,$$

$$n=1, \qquad 2a_3\Gamma^2 d_2 + b_3\Gamma^2 d_1 + a_1\Gamma^2 d_1 + b_2\Gamma^2 d_0 = 0,$$

$$n=2, \qquad a_1\Gamma^2 d_1 + 2a_1\Gamma^2 d_2 + 3a_3\Gamma^2 d_3 + b_1\Gamma^2 d_0 + b_2\Gamma^2 d_1 + b_3\Gamma^2 d_2 = 0,$$

$$n=3, \qquad 2\Gamma^2 d_2 + a_0\Gamma^2 d_1 + 2a_1\Gamma^2 d_2 + 3a_1\Gamma^2 d_3 + 4a_3\Gamma^2 d_4$$
$$+b_1\Gamma^2 d_1 + c_0 d_0 + b_2\Gamma^2 d_2 + b_3\Gamma^2 d_3 = 0,$$

$$n=4, \ (6\Gamma^2 + 3a_1\Gamma^2 + b_2\Gamma^2)d_3 + (a_3 + c_0)d_1 + (2a_0\Gamma^2 + b_1\Gamma^2)d_2$$
$$+4a_1\Gamma^2 + b_3\Gamma^2)d_4 + 5a_3\Gamma^2 d_5 + (b_3 + c_1)d_0 = 0,$$

$$n=5, \qquad 6a_3\Gamma^2 d_6 + (5a_1\Gamma^2 + b_3\Gamma^2)d_5 + (12\Gamma^2 + 4a_1\Gamma^2 + b_2\Gamma^2)d_4$$
$$+(3a_0\Gamma^2 + b_1\Gamma^2)d_3 + (2a_3 + c_0)d_2 + (a_2 + b_3 + c_1)d_1 + +(b_2 + c_2)d_0 = 0,$$

$$n=6, \qquad 7a_3\Gamma^2 d_7 + (6a_1\Gamma^2 + b_3\Gamma^2)d_6 + (20\Gamma^2 + 5a_1\Gamma^2 + b_2\Gamma^2)d_5$$
$$+(4a_0\Gamma^2 + b_1\Gamma^2)d_4 + (3a_3 + c_0)d_3 + (2a_2 + b_3 + c_1)d_2 + (a_1 - 4 + b_2 + c_2)d_1 + (b_1 + c_3)d_0 = 0,$$

$$n=7, \qquad 8a_3\Gamma^2 d_8 + (7a_1\Gamma^2 + b_3\Gamma^2)d_7 + (30\Gamma^2 + 6a_1\Gamma^2 + b_2\Gamma^2)d_6 + (5a_0\Gamma^2 + b_1\Gamma^2)d_5$$
$$+(4a_3 + c_0)d_4 + (3a_2 + b_3 + c_1)d_3 + (2 + 2a_1 - 8 + b_2 + c_2)d_2 + (a_0 + b_1 + c_3)d_1 = 0,$$

..................................;

the main 8-term recurrent formula has the form

$$n = 6, 7, 8, 9, 10...$$

$$[a_0(n-6) + b_1 + c_3]d_{n-6} + [(n-5)(n-6) + a_1(n-5) - 4(n-5) + b_2 + c_2]d_{n-5}$$
$$+[a_2(n-4) + b_3 + c_1]d_{n-4} + [a_3(n-3) + c_0]d_{n-3}$$

$$+[a_0\Gamma^2(n-2)+b_1\Gamma^2]d_{n-2}+[\Gamma^2(n-1)(n-2)+a_1\Gamma^2(n-1)+b_2\Gamma^2]d_{n-1}$$
$$+[a_1\Gamma^2 n+b_3\Gamma^2]d_n+a_3\Gamma^2(n+1)d_{n+1}=0\,,$$

in symbolical form it reads

$$P_{n-6}d_{n-6}+P_{n-5}d_{n-5}...+P_nd_n+P_{n+1}d_{n+1}=0. \tag{3.105}$$

In accordance with the Poincaré–Perron method, we transform the recurrent relation to the form

$$\begin{aligned}
&[a_0(n-6)+b_1+c_3]\\
&+[(n-5)(n-6)+a_1(n-5)-4(n-5)+b_2+c_2]\frac{d_{n-5}}{d_{n-6}}\\
&+[a_2(n-4)+b_3+c_1]\frac{d_{n-4}}{d_{n-5}}\frac{d_{n-5}}{d_{n-6}}\\
&+[a_3(n-3)+c_0]\frac{d_{n-3}}{d_{n-4}}\frac{d_{n-4}}{d_{n-5}}\frac{d_{n-5}}{d_{n-6}}\\
&+[a_0\Gamma^2(n-2)+b_1\Gamma^2]\frac{d_{n-2}}{d_{n-3}}\frac{d_{n-3}}{d_{n-4}}\frac{d_{n-4}}{d_{n-5}}\frac{d_{n-5}}{d_{n-6}}\\
&+[\Gamma^2(n-1)(n-2)+a_1\Gamma^2(n-1)+b_2\Gamma^2]\frac{d_{n-1}}{d_{n-2}}\frac{d_{n-2}}{d_{n-3}}\frac{d_{n-3}}{d_{n-4}}\frac{d_{n-4}}{d_{n-5}}\frac{d_{n-5}}{d_{n-6}}\\
&+[a_1\Gamma^2n+b_3\Gamma^2]\frac{d_n}{d_{n-1}}\frac{d_{n-1}}{d_{n-2}}\frac{d_{n-2}}{d_{n-3}}\frac{d_{n-3}}{d_{n-4}}\frac{d_{n-4}}{d_{n-5}}\frac{d_{n-5}}{d_{n-6}}\\
&a_3\Gamma^2(n+1)\frac{d_{n+1}}{d_n}\frac{d_n}{d_{n-1}}\frac{d_{n-1}}{d_{n-2}}\frac{d_{n-2}}{d_{n-3}}\frac{d_{n-3}}{d_{n-4}}\frac{d_{n-4}}{d_{n-5}}\frac{d_{n-5}}{d_{n-6}}=0\,.
\end{aligned} \tag{3.106}$$

Let it be

$$\frac{d_{n-1}}{d_{n-2}}=R_{n-1},\qquad \lim_{n\to\infty}R_{n-1}=R,\qquad R_{conv}=\mid\frac{1}{R}\mid\,;$$

for R it follows a simple algebraic equation

$$R+\Gamma^2R^5=0\quad\Longrightarrow\quad R=0,\frac{i}{\sqrt{i\Gamma}},-\frac{i}{\sqrt{i\Gamma}},-\frac{i}{\sqrt{-i\Gamma}},\frac{i}{\sqrt{-i\Gamma}}\,,$$

so the possible convergence radii are

$$R_{conv}=+\infty,|\Gamma|. \tag{3.107}$$

We may assume convergence of the series for all values of the variable x.

3.11 Qualitative study of the differential equation

Let us turn again to eq. (3.88)

$$\begin{aligned}
&\frac{d^2R}{dx^2}+\Big(-\frac{4\,x^3}{x^4+\Gamma^2}+\frac{6}{x}\Big)\frac{dR}{dx}+\Big[2\,E+\frac{2\,\alpha}{x}-\frac{l\,(l+1)}{x^2}\\
&+\frac{4\,\Gamma}{x^3}+\frac{\Gamma^2\,(2\,E-1)}{x^4}+\frac{2\,\alpha\,\Gamma^2}{x^5}-\frac{\Gamma^2 l\,(l+1)}{x^6}-\frac{4\,x\Gamma}{x^4+\Gamma^2}\Big]R=0\,,
\end{aligned} \tag{3.108}$$

and eliminate the first derivative term

$$R = \varphi \bar{R}, \quad R' = \varphi' \bar{R} + \varphi \bar{R}', \quad R'' = \varphi'' \bar{R} + 2\varphi' \bar{R}' + \varphi \bar{R}''.$$

The above equation takes the form

$$\frac{\varphi''}{\varphi} \bar{R} + 2\frac{\varphi'}{\varphi} \bar{R}' + \bar{R}'' + (-\frac{4x^3}{x^4+\Gamma^2} + \frac{6}{x})(\frac{\varphi'}{\varphi} \bar{R} + \bar{R}')$$

$$+ \Big[2E + \frac{2\alpha}{x} - \frac{l(l+1)}{x^2} + \frac{4\Gamma}{x^3} + \frac{\Gamma^2(2E-1)}{x^4} + \frac{2\alpha\Gamma^2}{x^5} - \frac{\Gamma^2 l(l+1)}{x^6} - \frac{4x\Gamma}{x^4+\Gamma^2}\Big] \bar{R} = 0,$$

or differently

$$\bar{R}'' + \Big(2\frac{\varphi'}{\varphi} - \frac{4x^3}{x^4+\Gamma^2} + \frac{6}{x}\Big) \bar{R}' + \Big[\frac{\varphi''}{\varphi} + (-\frac{4x^3}{x^4+\Gamma^2} + \frac{6}{x})\frac{\varphi'}{\varphi}$$

$$+ 2E + \frac{2\alpha}{x} - \frac{l(l+1)}{x^2} + \frac{4\Gamma}{x^3} + \frac{\Gamma^2(2E-1)}{x^4} + \frac{2\alpha\Gamma^2}{x^5} - \frac{\Gamma^2 l(l+1)}{x^6} - \frac{4x\Gamma}{x^4+\Gamma^2}\Big] \bar{R} = 0.$$

We readily find the needed function φ:

$$2\frac{\varphi'}{\varphi} - \frac{4x^3}{x^4+\Gamma^2} + \frac{6}{x} = 0, \quad \varphi(x) = \frac{\sqrt{x^4+\Gamma^2}}{x^3}, \tag{3.109}$$

so the last equation takes the form

$$\bar{R}'' + \Big[\frac{\varphi''}{\varphi} + (-\frac{4x^3}{x^4+\Gamma^2} + \frac{6}{x})\frac{\varphi'}{\varphi} + 2E + \frac{2\alpha}{x}$$

$$- \frac{l(l+1)}{x^2} + \frac{4\Gamma}{x^3} + \frac{\Gamma^2(2E-1)}{x^4} + \frac{2\alpha\Gamma^2}{x^5} - \frac{\Gamma^2 l(l+1)}{x^6} - \frac{4x\Gamma}{x^4+\Gamma^2}\Big] \bar{R} = 0,$$

explicitly it reads

$$\frac{d^2}{dx^2}\bar{R} + \Big[2E + \frac{2\alpha}{x} - \frac{l(l+1)+6}{x^2} + \frac{4\Gamma}{x^3} + \frac{\Gamma^2(2E-1)}{x^4} + \frac{2\alpha\Gamma^2}{x^5} - \frac{\Gamma^2 l(l+1)}{x^6}$$

$$- \frac{4x\Gamma - 18x^2}{x^4+\Gamma^2} - \frac{12x^6}{(x^4+\Gamma^2)^2}\Big] \bar{R} = 0, \tag{3.110}$$

or shortly

$$\Big[\frac{d^2}{dx^2} + P^2(x)\Big] \bar{R} = 0.$$

Near two physical singularities, the function $P^2(x)$ behaves as follows:

$$\begin{aligned} x \to 0, \quad & P^2(x) \approx -\frac{\Gamma^2 l(l+1)}{x^6} \to -\infty; \\ x \to \infty, \quad & P^2(x) \approx 2E + \frac{2\alpha}{x} + \ldots \to 2E. \end{aligned} \tag{3.111}$$

3.12 Energy quantisation, non-relativistic case

Let $l = 1, 2, 3, ...$ The transcendency condition takes the form

$$2\alpha + \frac{\sqrt{2}\sqrt{-E}(-2\Gamma E + \Gamma - (2k-13)l(l+1))}{l(l+1)} = \frac{\sqrt{2}\alpha^2\Gamma\sqrt{-E}}{l^3(l+1)^3}. \tag{3.112}$$

At $l = 1,\ \Gamma = -10^{-3}$, and $k = 6, ..., 29$ we have

TABLE 3.7
The energy values at $l = 1,\ \Gamma = -10^{-3}$, non-relativistic case

k	E	k	E
6	-0.000106665347	7	-0.00010645219
8	-0.000011835908	9	$-4.2614952 \cdot 10^{-6}$
10	$-2.1743565 \cdot 10^{-6}$	11	$-1.3153932 \cdot 10^{-6}$
12	$-8.8057025 \cdot 10^{-7}$	13	$-6.3047628 \cdot 10^{-7}$
14	$-4.7356259 \cdot 10^{-7}$	15	$-3.6869349 \cdot 10^{-7}$
16	$-2.9516088 \cdot 10^{-7}$	17	$-2.4161817 \cdot 10^{-7}$
18	$-2.0142543 \cdot 10^{-7}$	19	$-1.7048708 \cdot 10^{-7}$
20	$-1.4616563 \cdot 10^{-7}$	21	$-1.2670037 \cdot 10^{-7}$
22	$-1.1087955 \cdot 10^{-7}$	23	$-9.7847069 \cdot 10^{-8}$
24	$-8.6984198 \cdot 10^{-8}$	25	$-7.7834775 \cdot 10^{-8}$
26	$-7.0056512 \cdot 10^{-8}$	27	$-6.3388512 \cdot 10^{-8}$
28	$-5.7629102 \cdot 10^{-8}$	29	$-5.2620405 \cdot 10^{-8}$

At $l = 2$ and $k = 6, ..., 29$ we have

TABLE 3.8
The energy values at $l = 2,\ \Gamma = -10^{-3}$, non-relativistic case

k	E	k	E
6	-0.000106594223	7	-0.00010652317
8	-0.000011838539	9	$-4.2620633 \cdot 10^{-6}$
10	$-2.1745635 \cdot 10^{-6}$	11	$-1.3154906 \cdot 10^{-6}$
12	$-8.8062362 \cdot 10^{-7}$	13	$-6.3050861 \cdot 10^{-7}$
14	$-4.7358364 \cdot 10^{-7}$	15	$-3.6870795 \cdot 10^{-7}$
16	$-2.9517124 \cdot 10^{-7}$	17	$-2.4162584 \cdot 10^{-7}$
18	$-2.0143127 \cdot 10^{-7}$	19	$-1.7049163 \cdot 10^{-7}$
20	$-1.4616923 \cdot 10^{-7}$	21	$-1.2670328 \cdot 10^{-7}$
22	$-1.1088194 \cdot 10^{-7}$	23	$-9.7849046 \cdot 10^{-8}$
24	$-8.6985855 \cdot 10^{-8}$	25	$-7.7836178 \cdot 10^{-8}$
26	$-7.0057710 \cdot 10^{-8}$	27	$-6.3389542 \cdot 10^{-8}$
28	$-5.7629995 \cdot 10^{-8}$	29	$-5.2621184 \cdot 10^{-8}$

For $l = 5$ and $k = 6, ..., 29$ we have (Fig. 3.1)

TABLE 3.9
The energy values at $l = 5$, $\Gamma = -10^{-3}$, non-relativistic case

k	E	k	E
6	-0.000106565793	7	-0.00010655158
8	-0.000011839591	9	$-4.2622907 \cdot 10^{-6}$
10	$-2.1746464 \cdot 10^{-6}$	11	$-1.3155296 \cdot 10^{-6}$
12	$-8.8064497 \cdot 10^{-7}$	13	$-6.3052154 \cdot 10^{-7}$
14	$-4.7359206 \cdot 10^{-7}$	15	$-3.6871373 \cdot 10^{-7}$
16	$-2.9517538 \cdot 10^{-7}$	17	$-2.4162891 \cdot 10^{-7}$
18	$-2.0143361 \cdot 10^{-7}$	19	$-1.7049344 \cdot 10^{-7}$
20	$-1.4617068 \cdot 10^{-7}$	21	$-1.2670445 \cdot 10^{-7}$
22	$-1.1088289 \cdot 10^{-7}$	23	$-9.7849836 \cdot 10^{-8}$
24	$-8.6986518 \cdot 10^{-8}$	25	$-7.7836739 \cdot 10^{-8}$
26	$-7.0058189 \cdot 10^{-8}$	27	$-6.3389955 \cdot 10^{-8}$
28	$-5.7630353 \cdot 10^{-8}$	29	$-5.2621496 \cdot 10^{-8}$

Comparison of the energy series

$$\Delta = E_n\ (l = 2) - E_n\ (l = 5), \qquad n = 0, 1, 2, \ldots . \tag{3.113}$$

TABLE 3.10
The differences between energy values, non-relativistic case

n	Δ	n	Δ
0	$2.8430186 \cdot 10^{-8}$	1	$-2.8413125 \cdot 10^{-8}$
2	$-1.0523492 \cdot 10^{-9}$	3	$-2.2731308 \cdot 10^{-10}$
4	$-8.2841118 \cdot 10^{-11}$	5	$-3.8977675 \cdot 10^{-11}$
6	$-2.1348515 \cdot 10^{-11}$	7	$-1.2933536 \cdot 10^{-11}$
8	$-8.4192760 \cdot 10^{-12}$	9	$-5.7836591 \cdot 10^{-12}$
10	$-4.1427564 \cdot 10^{-12}$	11	$-3.0682656 \cdot 10^{-12}$
12	$-2.3354351 \cdot 10^{-12}$	13	$-1.8185771 \cdot 10^{-12}$
14	$-1.4436463 \cdot 10^{-12}$	15	$-1.1650872 \cdot 10^{-12}$
16	$-9.5382260 \cdot 10^{-13}$	17	$-7.9069886 \cdot 10^{-13}$
18	$-6.6274890 \cdot 10^{-13}$	19	$-5.6098102 \cdot 10^{-13}$
20	$-4.7902667 \cdot 10^{-13}$	21	$-4.1228934 \cdot 10^{-13}$
22	$-3.5739498 \cdot 10^{-13}$	29	$-3.1182893 \cdot 10^{-13}$

At $l = 0$, the first energy level is (let $\Gamma = -10^{-3}$) $E = 0.9999213823$. We can construct any finite number of terms in the series; for definiteness, let us restrict ourselves by polynomials of 10th-order

$$f(x) \approx g(x)P_{10}(x), \quad g(x) = x^C e^{Ax} e^{B/x}.$$

where the coefficient of the series equal

$$\begin{gathered} d_0 = 1,\ d_1 = 565.0253196,\ d_2 = 75914.6222,\ d_3 = -5.9869325 \cdot 10^6, \\ d_4 = 1.423294 \cdot 10^9, d_5 = -5.33496 \cdot 10^{11},\ d_6 = 2.67316 \cdot 10^{14}, \\ d_7 = -1.66026 \cdot 10^{17}, d_8 = 1.22398 \cdot 10^{20}, d_9 = -1.0413 \cdot 10^{23},\ d_{10} = 1.0023 \cdot 10^{26}. \end{gathered} \tag{3.114}$$

The plot of g is given in Fig. 3.2. The plot of the complete function $f(x)$ is given in Fig. 3.3. Similarly, examine behaviour of the functions $g(x)$ and $f(x)$ with $\Gamma = -10^{-3}$ for the second energy level $E = 0.9998474908$. See Figs. 3.4 and 3.5.

3.13 Relativistic problem: polynomial solutions?

Let us examine the possibility to get exact solutions in polynomials. Recall the structure of the recurrent formula

$$
\begin{aligned}
&P_{k-6}c_{k-6} + P_{k-5}c_{k-5} + P_{k-4}c_{k-4} + P_{k-3}c_{k-3} \\
&+P_{k-2}c_{k-2} + P_{k-1}c_{k-1} + P_k c_k + P_{k+1}c_{k+1} = 0\,.
\end{aligned}
\tag{3.115}
$$

To this end, we equate the coefficients of the power series $P_{10}(x) = 1 + c_1 x + c_2 x^2 + \ldots$ ($k = \overline{1,10}$), obtaining equations with respect to energies E. In this way, we consequently obtain the following sets of energies[1]:

$c_1 = 0$,

$$E = -7.17766 \pm 0.0325504i, 4.78508 \pm 0.02143i;$$

$c_2 = 0$,

$$
\begin{aligned}
E = {} & -9.57069 \pm 0.0871618i, -7.17729 \pm 0.0000147199i, \\
& 2.3923 \pm 0.0199044i, 4.78535 \pm 3.18263 \cdot 10^{-6}i;
\end{aligned}
$$

$c_3 = 0$,

$$E = -0.0138482, 0.0136254, -11.9655 \pm 0.16361i, -9.56799 \pm 0.000165449i,$$

$$-7.17755 \pm 3.62678 \cdot 10^{-6}i, 2.3928 \pm 3.0992 \cdot 10^{-6}i, 4.78514 \pm 1.02337 \cdot 10^{-6}i;$$

$c_4 = 0$,

$$E = -0.000198405, -0.000197379, -14.364 \pm -0.261447i, -11.9562 \pm 0.000825818i,$$

$$-9.56864 \pm 0.0000803654i, -7.17839 \pm 0.0000379228i, -2.39173 \pm 0.0398022i,$$

$$2.39266 \pm 4.10168 \cdot 10^{-7}i, 4.78514;$$

$c_5 = 0$,

$$
\begin{aligned}
E = {} & -0.000139544, -0.000139344, -16.7682 \pm 0.379763i, \\
& -14.3411 \pm 0.00272001i, -11.9557 \pm 0.000633883i, \\
& -9.57531 \pm 0.000507849i, -7.17506 \pm 0.000282867i, -4.78491 \pm 0.107225i, \\
& -2.39205 \pm 7.19976 \cdot 10^{-6}i, 2.39266, 4.78514;
\end{aligned}
$$

$c_6 = 0$,

$$E = -0.000139444, -19.1804 \pm 0.517021i, -16.7221 \pm 0.00690396i,$$

$$-14.3345 \pm 0.00291035i, -11.9805 \pm 0.00286058i, -9.5589 \pm 0.00225326i,$$

$$-7.18462 \pm 0.196333i, -7.17507 \pm 0.000230603i, -4.78201 \pm 0.000212772i,$$

$$-2.39251 \pm 2.05079 \cdot 10^{-6}i, 2.39266, 4.78514;$$

$c_7 = 0$,

$$E = -0.000139444, -21.6027 \pm 0.671048i, -19.0996 \pm 0.0145641i,$$

[1] In calculations we use the command Chops with default tolerance of 10^{-10}, so we ignore imaginary additions which are smaller than 10^{-10} in modules.

$$-16.6995 \pm 0.00979856i, -14.4002 \pm 0.0106758i, -11.9325 \pm 0.00972476i,$$
$$-9.59771 - 0.310631i, -9.55899 - 0.0019359i, -7.1766 - 0.0000928371i,$$
$$-7.16875 \pm 0.00146393i, -4.78519 \pm 0.0000358196i, -2.39251,$$
$$2.39266, 4.78514;$$

$c_8 = 0,$

$$E = -0.000139444, -24.0367 \pm 0.839191i, -21.4744 \pm 0.0266521i, -19.0442 \pm 0.0271154i,$$
$$-16.8411 \pm 0.0311392i, -14.2926 \pm 0.0303364i, -12.0306 \pm 0.456284i,$$
$$-11.9332 \pm 0.00820475i, -9.56708 \pm 0.000531903i, -9.54983 \pm 0.00576288i,$$
$$-7.18192 \pm 0.000166772i, -7.17358 \pm 0.0000347737i, -4.78519,$$
$$-2.39251, 2.39266, 4.78514;$$

$c_9 = 0,$

$$E = -0.000139444, -0.000139444, -26.4832 \pm 1.01858i,$$
$$-23.8478 \pm 0.0435067i, -21.3617 \pm 0.0659823i, -19.3099 \pm 0.0773246i, -16.6408 \pm 0.0761181i,$$
$$-14.4848 \pm 0.642902i, -14.2954 \pm 0.0238801i, -11.9585 \pm 0.000974102i, -11.921 \pm 0.015833i,$$
$$-9.58514 \pm 0.000664145i, -9.5556 \pm 0.0000700813i, -7.18193 \pm 5.31255 \cdot 10^{-8}i,$$
$$-7.17359 \pm 9.49047 \cdot 10^{-8}i, -4.78519 \pm 7.98381 \cdot 10^{-10}i, -2.39251, 2.39266, 4.78514.$$

Evidently, we can see that among the roots know physically interpretable energy levels arise, so we conclude that the polynomial solutions do not exist in the problem under consideration.

3.14 Figures

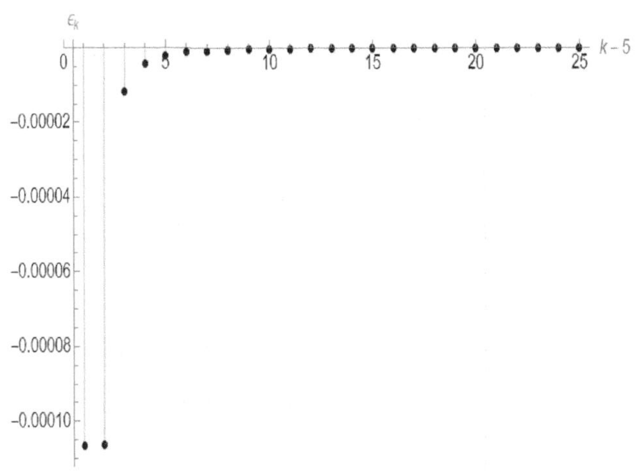

FIGURE 3.1
Energy levels at $l = 5,\ \Gamma = -10^{-3}$*, and* $k = \overline{6, 29}$.

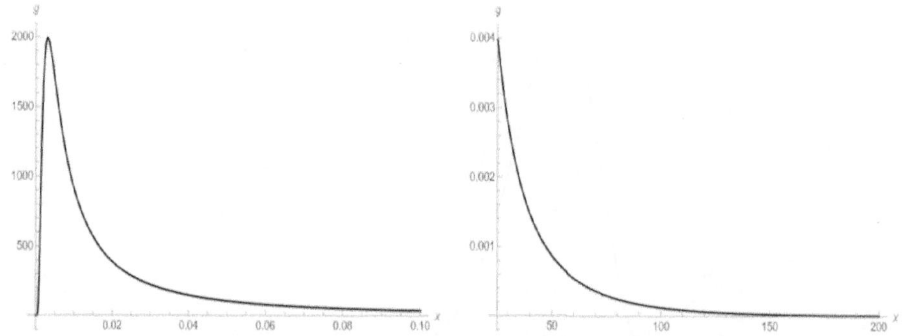

FIGURE 3.2
Plot of $g(x)$ for $l = 0,\ \Gamma = -10^{-3}$, the first energy level.

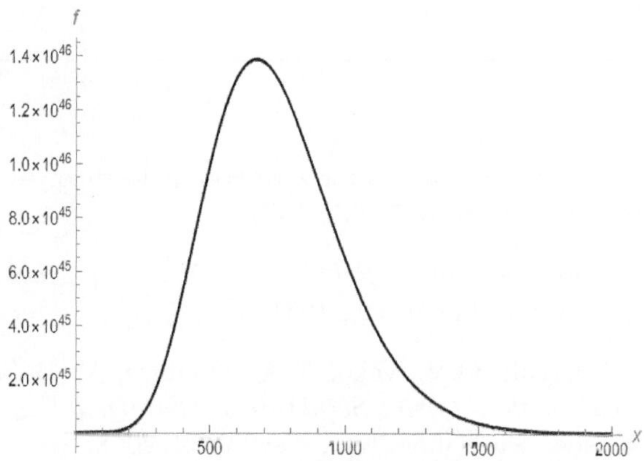

FIGURE 3.3
Plot of $f(x)$ for $l = 0,\ \Gamma = -10^{-3}$, the first energy level.

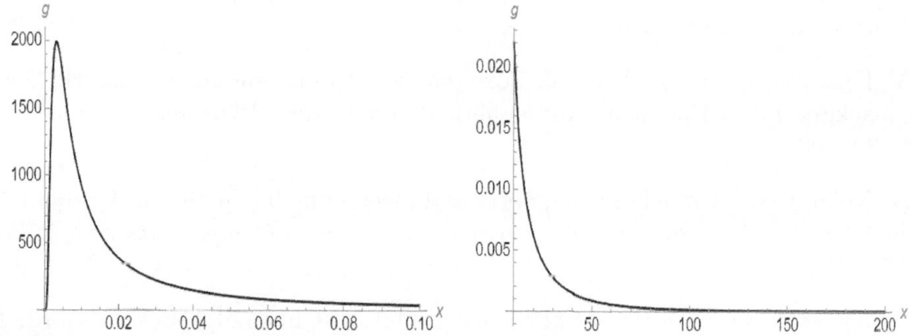

FIGURE 3.4
Plot of $g(x)$ for $l = 0,\ \Gamma = -10^{-3}$, the second energy level.

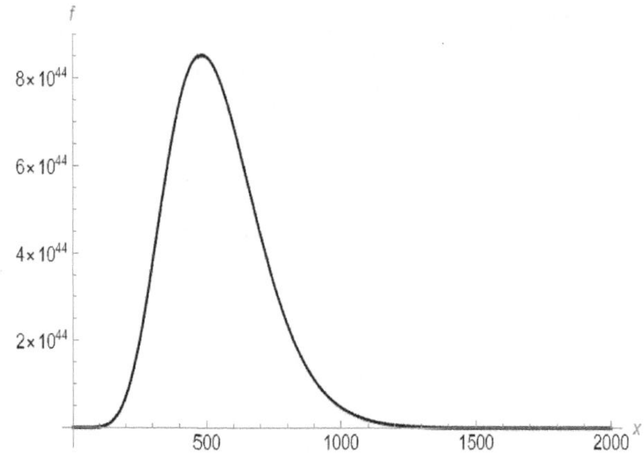

FIGURE 3.5
Plot of $f(x)$ for $l = 0$, $\Gamma = -10^{-3}$, the second energy level.

Bibliography

[1] W. Cox. Higher-rank representations for zero-spin field theories. *Journal of Physics A: Mathematical and General*, 15: 627–635, 1982.

[2] S.S. Schweber. *An Introduction to Relativistic Quantum Field Theory.* Series: Dover Books on Physics, Dover Publications, 1961.

[3] V.V. Kisel, E.M. Ovsiyuk, O.V. Veko, Y.A. Voynova, V. Balan and V.M. Red'kov. *Elementary Particles with Internal Structure in External Fields.* Vols I, II. Physical Problems. Nova Science Publishers Inc., New York, 2018.

[4] E.M. Ovsiyuk. Spin zero Cox's particle with an intrinsic structure: general analysis in external electromagnetic and gravitational fields. *Ukrainian Journal of Physics*, 60: 485–496, 2015.

[5] E.M. Ovsiyuk, O.V. Veko and K.V. Kazmerchuk. Scalar particle with intrinsic structure in electromagnetic field in curved space-time. *Problems of Physics, Mathematics and Technics*, 20(3): 32–36, 2014.

[6] K.V. Kazmerchuk, E.M. Ovsiyuk. Cox's particlele in magnetic and electric field against the background of Euclidean and spherical geometries. *Ukrainian Journal of Physics*, 60: 389–400, 2015.

[7] O.V. Veko. Cox's particle in magnetic and electric fields on the background of hyperbolic Lobachevsky geometry. *Nonlinear Phenomena in Complex Systems*, 19(1): 50–61, 2016.

[8] A. Ronveaux. *Heun's differential equation.* Oxford University Press, Oxford, 1995.

[9] S.Yu. Slavyanov and W. Lay. *Special Functions. A Unified Theory Based on Singularities.* Oxford University Press, Oxford, 2000.

4

Tunnelling Dirac particles through Schwarzschild barrier

For massless Dirac particles, the general mathematical and numerical study of the tunnelling process through the effective potential barrier generated by Schwarzschild black hole geometry is done. The study will be based on the use of eight Frobenius solutions of related 2nd-order differential equations with nonregular singularities of the rank 2. We construct these solutions in explicit form and prove that power series involved in them are converged in all physical regions of the physical region of the variable $r \in (1, +\infty)$. Results for tunnelling effect significantly differ for two situations: one when the particle falls on the barrier from within and another when the particle falls from outside. The mathematical structure of the derived asymptotic relations is exact; however, analytical expressions for involved convergent power series are not known, and further study is based on numerical summing the series. The calculations are implemented using *Mathematica* system.

4.1 Basic facts

Starting idea on which the present chapter is based appeared many years ago in Regge and Wheeler's work [1], where the stability of the Schwarzschild metric [2] was studied. In [1], a linearised wave equation for the spin 2 field on the background of the Schwarzschild metric was derived. It was established that a corresponding radial equation is of Schrödinger form with potential of barrier type. Similar results have been obtained for particles with different spins [3–12].

For Schwarzschild metric in the static coordinates $x^\alpha = (t, \theta, \phi, r)$

$$dS^2 = \Phi dt^2 - r^2 d\theta^2 - r^2 \sin^2\theta d\phi^2 - \frac{1}{\Phi} dr^2, \quad \Phi = 1 - \frac{1}{r}, \; r \in (1, +\infty), \tag{4.1}$$

the generally covariant Dirac equation [13] takes the form (we separate a special factor in the wave function, $\Psi = r^{-1}\Phi^{-1/4}(r)\psi$)

$$\left[\frac{\gamma^0}{\sqrt{\Phi}}\partial_t + i\sqrt{\Phi}\gamma^3\partial_r + \frac{1}{r}(i\gamma^1\partial_\theta + \gamma^2\frac{i\partial_\phi + i\sigma^{12}\cos\theta}{\sin\theta}) - M\right]\psi(x) = 0\,. \tag{4.2}$$

Solutions with spherical symmetry are constructed within the following substitution (instead of commonly used formalisms of spinor spherical harmonics [14] and spin-weigh harmonics [15], we apply a more simple formalism based on the use of the Wigner

DOI: 10.1201/9781003472377-4

D-functions [16, 17])

$$\psi(x)_{\epsilon jm\delta} = e^{-i\epsilon t} \left| \begin{array}{c} f_1(r)\, D_{-1/2} \\ f_2(r)\, D_{+1/2} \\ \delta\, f_2(r)\, D_{-1/2} \\ \delta\, f_1(r)\, D_{+1/2} \end{array} \right| . \tag{4.3}$$

This substitution corresponds to diagonalisation of the total angular momentum $\vec{J}^2$, J_3, and spatial reflection operator. The parity eigenvalues are $\Pi = \delta(-1)^{j+1}$, $\delta = \pm 1$. Wigner D-functions are designated as follows: $D_\sigma = D^j_{-m,\sigma}(\phi, \theta, 0)$; in this notation, the fixed parameters $j = 1/2, 3/2, ...$ and $m \in \{-j, ..., +j\}$ are omitted for brevity.

After needed calculation we produce the system of two equations

$$\Big(\Phi \frac{d}{dr} + \frac{\nu\sqrt{\Phi}}{r}\Big) f = -\,(\epsilon + M\sqrt{\Phi})\, g\,, \qquad \Big(\Phi \frac{d}{dr} - \frac{\nu\sqrt{\Phi}}{r}\Big) g = +(\epsilon - M\sqrt{\Phi})\, f\,, \tag{4.4}$$

where

$$f = (f_1 + f_2)\,, \quad g = -i(f_1 - f_2), \quad \nu = j + 1/2, \quad \nu = 1, 2, 3, \ldots .$$

Let us transform eq. (4.4) to the new variable

$$\sqrt{\Phi} = +\sqrt{1 - 1/r} = x\,, \quad r \to 1\,,\ x \to 0\,, \quad r \to +\infty\,,\ x \to +1\,; \tag{4.5}$$

the physical region for the variable is the interval $x \in (0, 1)$. Equation (4.4) take the form

$$\begin{gathered} \Big(\frac{x(1-x^2)^2}{2}\frac{d}{dx} + \nu x(1-x^2)\Big) f = -(\epsilon - Mx) g\,, \\ \Big(\frac{x(1-x^2)^2}{2}\frac{d}{dx} - \nu x(1-x^2)\Big) g = +(\epsilon + Mx) g\,. \end{gathered} \tag{4.6}$$

Whence it follows a 2nd-order equation for $f(x)$:

$$\begin{gathered} \Big\{ \frac{d^2}{dx^2} + \Big(\frac{1}{x} + \frac{2}{x+1} + \frac{2}{x-1} - \frac{1}{x+c}\Big)\frac{d}{dx} \\ -\nu \frac{2(1-3x^2)}{x(1-x^2)^2} - \frac{4\nu^2}{(1-x^2)^2} + \frac{1}{x+c}\frac{2\nu}{1-x^2} + \frac{4(\epsilon^2 - M^2x^2)}{x^2(1-x^2)^4} \Big\} f = 0\,, \end{gathered} \tag{4.7}$$

an equation for the function $g(x)$ follows from eq. (4.7) by the formal changes $f \Longrightarrow g$, $\nu \Longrightarrow -\nu$, $c \Longrightarrow -c$.

Note that $c = \epsilon/M > 1$; to the massless case, there corresponds the limit $c \to \infty$, so such a singular point vanishes and we have a simpler equation

$$\begin{gathered} \Big\{ \frac{d^2}{dx^2} + \Big(\frac{1}{x} + \frac{2}{x+1} + \frac{2}{x-1}\Big)\frac{d}{dx} \\ -\nu \frac{2(1-3x^2)}{x(1-x^2)^2} - \frac{4\nu^2}{(1-x^2)^2} + \frac{4\epsilon^2}{x^2(1-x^2)^4} \Big\} f = 0\,. \end{gathered} \tag{4.8}$$

The problem under consideration becomes clearer from a physical standpoint if one transforms equations to other radial variables r_*:

$$\Phi \frac{d}{dr} = \frac{d}{dr_*}, \quad r_* = r + \ln(r - 1), \quad r_* \in (-\infty, +\infty)\,; \tag{4.9}$$

to points $r = +1, +\infty$ there correspond the following values of r_*:

$$r \to +1,\ r_* \to -\infty; \qquad r \to +\infty,\ r_* \to +\infty\,. \tag{4.10}$$

In this variable, the system (4.4) reads

$$\Big[\frac{d}{dr_*} + \nu\varphi(r_*)\Big] f = -(\epsilon + M\sqrt{\Phi})g, \; \Big[\frac{d}{dr_*} - \nu\varphi(r_*)\Big] g = +(\epsilon - M\sqrt{\Phi})f, \tag{4.11}$$

where the function $\varphi(r_*)$ is determined as follows

$$\varphi(r_*) = \frac{\sqrt{\Phi}}{r}, \quad r_* \to \pm\infty, \quad \varphi(r_*) \to 0\,. \tag{4.12}$$

Several typical graphs for potential $\varphi(r_*)$ are given in Fig. 4.1. Below we show the location of maximum values of the potentials $\varphi(r)$ and $\varphi(r_*)$:

$$\begin{array}{lll}
\nu = 1, & r = 0.186736, & r_* = -0.349256;\\
\nu = 2, & r = 0.633808, & r_* = 0.134404;\\
\nu = 3, & r = 1.37398, & r_* = 0.351799;\\
\nu = 4, & r = 2.41039, & r_* = 0.466459;\\
\nu = 5, & r = 3.74324, & r_* = 0.535869;\\
\nu = 6, & r = 5.37252, & r_* = 0.582072;\\
\nu = 7, & r = 7.29817, & r_* = 0.614938;\\
\nu = 8, & r = 9.52018, & r_* = 0.639478;\\
\nu = 9, & r = 12.0385, & r_* = 0.658483;\\
\nu = 10, & r = 14.8532, & r_* = 0.673629;\\
\nu = 11, & r = 17.9642, & r_* = 0.685979;\\
\nu = 12, & r = 21.3715, & r_* = 0.696239;\\
\nu = 13, & r = 25.0751, & r_* = 0.704898;\\
\nu = 14, & r = 29.0751, & r_* = 0.712303;\\
\nu = 15, & r = 33.3713, & r_* = 0.718706;\\
\nu = 16, & r = 37.9638, & r_* = 0.724298;\\
\nu = 17, & r = 42.8527, & r_* = 0.729225;\\
\nu = 18, & r = 48.0378, & r_* = 0.733597;\\
\nu = 19, & r = 53.5192, & r_* = 0.737503;\\
\nu = 20, & r = 59.297, & r_* = 0.741014.
\end{array}$$

The corresponding 2nd-order equations may be presented as follows

$$\Big\{(\epsilon - M\sqrt{\Phi})(\frac{d}{dr_*} + \nu\varphi)\frac{1}{(\epsilon - M\sqrt{\Phi})}\Big\}(\frac{d}{dr_*} - \nu\varphi)g + (\epsilon^2 - M^2\Phi)g = 0\,, \tag{4.13}$$

$$\Big\{(\epsilon + M\sqrt{\Phi})\;(\frac{d}{dr_*} - \nu\varphi)\frac{1}{(\epsilon + M\sqrt{\Phi})}\Big\}(\frac{d}{dr_*} + \nu\varphi)f + (\epsilon^2 - M^2\Phi)f = 0\,. \tag{4.14}$$

Equations (4.13)–(4.14) can be reduced to the form

$$\Big(\frac{d^2}{dr_*^2} + P^2(r_*)\Big)f = 0\,, \quad \Big(\frac{d^2}{dr_*^2} + Q^2(r_*)\Big)g = 0\,; \tag{4.15}$$

near the points $r \to 1, +\infty$ they become simpler

$r \to +1$,

$$(\frac{d^2}{dr_*^2} + \epsilon^2)\;f = 0, \quad (\frac{d^2}{dr_*^2} + \epsilon^2)g = 0\,, \quad f, g \sim e^{\pm i\epsilon r_*}\,; \tag{4.16}$$

$r \to \infty$,

$$(\frac{d^2}{dr_*^2} + \epsilon^2 - M^2)f = 0, \quad (\frac{d^2}{dr_*^2} + \epsilon^2 - M^2)g = 0\,, \quad f, g \sim e^{\pm i\sqrt{\epsilon^2 - M^2} r_*}\,. \tag{4.17}$$

The quantities $P^2(r_*)$ and $Q^2(r_*)$ represent squared effective linear momentums. Here we have a problem where the quantum-mechanical tunnelling effect is possible.

4.2 Analytical study of Frobenius solutions

Let us turn to consideration of possible solutions for eq. (4.7). It is convenient to apply the following form of this equation

$$\frac{d^2}{dx^2}f + \Big(\frac{1}{x} + \frac{2}{x+1} + \frac{2}{x-1} - \frac{1}{x+c}\Big)\frac{df}{dx}f$$

$$+\Big\{ -\frac{2\nu}{x} + \frac{4\epsilon^2}{x^2} + \frac{D}{x+c} + \frac{A}{(x+1)} + \frac{A'}{(x-1)}$$

$$+\frac{B}{(x+1)^2} + \frac{B'}{(x-1)^2} + \frac{\epsilon^2 - M^2/2}{(x+1)^3} - \frac{\epsilon^2 - M^2/2}{(x-1)^3}$$

$$+ + \frac{\epsilon^2 - M^2}{4(x+1)^4} + \frac{\epsilon^2 - M^2}{4(x-1)^4}\Big\}f = 0\,, \tag{4.18}$$

where the notations are used

$$A = \frac{-8\nu^2 + 35\epsilon^2 + 8\nu - 5M^2 + 8\nu/(c-1)}{8}\,,$$

$$A' = \frac{+8\nu^2 - 35\epsilon^2 + 8\nu + 5M^2 - 8\nu/(c+1)}{8}\,,$$

$$B = \frac{-8\nu^2 + 19\epsilon^2 - 8\nu - 5M^2}{8}\,,$$

$$B' = \frac{-8\nu^2 + 19\epsilon^2 + 8\nu - 5M^2}{8}\,, \qquad D = -\frac{2\nu}{c^2 - 1}\,.$$

Recall that restriction to massless case is straightforward: $M \to 0, c \to \infty$.

In (4.18) we have an equation with three regular singular points $x = 0, -c, \infty$ and two irregular singular points $-1, +1$ of rank 2. Let us detail the asymptotics near singular points. Most interesting are the physical points $x = 0, +1$:

$$x \to 0, \quad f \sim x^\gamma\,, \quad \gamma = \pm 2i\epsilon\,; \tag{4.19}$$

$$x \to +1\,, \quad f = (x-1)^\alpha \exp\,(\frac{\beta}{x-1})\,,$$

$$\beta = \pm i \cdot \frac{\sqrt{\epsilon^2 - M^2}}{2}\,, \qquad \alpha = \pm i \cdot \frac{(\epsilon^2 - M^2) + M^2/2}{\sqrt{\epsilon^2 - M^2}}\,; \tag{4.20}$$

$$x \to -1\,, \quad f = (x+1)^{\alpha'} \exp\,(\frac{\beta'}{x+1})\,,$$

$$\beta' = \pm i \cdot \frac{\sqrt{\epsilon^2 - M^2}}{2}\,, \qquad \alpha' = \mp i \cdot \frac{(\epsilon^2 - M^2) + M^2/2}{\sqrt{\epsilon^2 - M^2}}\,; \tag{4.21}$$

$$x \to -c, \quad f = (x+c)^{\rho}, \quad \rho = 0, 2\,; \tag{4.22}$$

$$x \to \infty\ (y = x^{-1})\,, \quad (\frac{d^2 f}{dy^2} - \frac{2}{y}\frac{df}{dy}) f = 0\,, \quad f(y) \sim \frac{1}{x^{\sigma}}\,, \sigma = 0, 3\,. \tag{4.23}$$

Let us construct Frobenius solutions for eq. (4.18). It is convenient to introduce shortening notations

$$(\epsilon^2 - M^2)/4 = E, \quad \epsilon^2 - M^2/2 = E'\,,$$

then eq. (4.18) is written as

$$\frac{d^2 f}{dx^2} + (\frac{1}{x} + \frac{2}{x+1} + \frac{2}{x-1} - \frac{1}{x+c})\frac{df}{dx}$$

$$+\Big\{ -\frac{2\nu}{x} + \frac{4\epsilon^2}{x^2} + \frac{D}{x+c} + \frac{A}{(x+1)} + \frac{A'}{(x-1)} + \frac{B}{(x+1)^2} + \frac{B'}{(x-1)^2}$$

$$+ \frac{E'}{(x+1)^3} - \frac{E'}{(x-1)^3} + \frac{E}{(x+1)^4} + \frac{E}{(x-1)^4} \Big\} f = 0\,. \tag{4.24}$$

We construct solutions using the substitution

$$f(x) = x^{\gamma}(x-1)^{\alpha} \exp\,(\frac{\beta}{x-1})(x+1)^{\alpha'} \exp\,(\frac{\beta'}{x+1}) F(x)\,. \tag{4.25}$$

We get an equation for $F(x)$:

$$\frac{d^2 F}{dx^2} + \left[\frac{1+2\gamma}{x} + \frac{2+2\alpha'}{x+1} + \frac{2+2\alpha}{x-1} - \frac{1}{x+c} - \frac{2\beta'}{(x+1)^2} - \frac{2\beta}{(x-1)^2} \right] \frac{dF}{dx}$$

$$+ \Big[\frac{1}{2}\, \frac{3\beta - \beta' + 4\alpha + 2\alpha' - \beta\beta' + 2\alpha\alpha' + 4\gamma + 4\gamma\alpha + 4\gamma\beta - \alpha\beta' + \beta\alpha' + 2A'}{x-1}$$

$$- \frac{\beta}{(1+c)^2\,(x-1)} \quad \frac{\alpha}{(1+c)\,(x-1)} + \frac{E + \beta^2}{(x-1)^4} + \frac{-E' - 2\alpha\beta}{(x-1)^3}$$

$$+ \frac{E' - 2\alpha'\beta'}{(x+1)^3} + \frac{1}{2}\, \frac{2B' + 2\alpha^2 - 4\beta + 2\alpha - 4\gamma\beta - 2\beta\alpha' + \beta\beta'}{(x-1)^2} + \frac{\beta}{(x-1)^2\,(1+c)} + \frac{E + \beta'^2}{(x+1)^4}$$

$$+ \frac{-2B - 2\alpha - 2\beta - 2\alpha^2 + 4c\beta - 2\alpha\beta' - 4\gamma\beta + 2Bc + 2\alpha'^2 c - \beta\beta + 2c\alpha + 4\gamma\beta' c + 2\alpha\beta' c + \beta\beta' c}{(x+1)^2 2(c-1)}$$

$$+ \frac{-c\alpha + 2\gamma\alpha' c - 2\gamma\beta' c - c\beta - 2\nu c - \gamma - 2\gamma\beta c - 2\gamma\alpha c + c\alpha' - c\beta'}{cx} + \frac{4\epsilon^2 + \gamma^2}{x^2}$$

$$+ \frac{1}{2}\, \frac{-4\alpha' - 2\alpha\alpha' - 4\gamma\alpha' - \beta\alpha' + \beta\beta' + 3\beta' + \alpha\beta' + 4\gamma\beta' - \beta - 4\gamma - 2\alpha + 2A}{x+1}$$

$$- \frac{\beta'}{(-1+c)^2\,(x+1)} - \frac{\alpha'}{(-1+c)\,(x+1)}$$

$$+ \frac{1}{c(c-1)^2(c+1)^2}\, \frac{1}{x+c}\, [Dc^5 + (\gamma + \alpha + \alpha')\, c^4 + (-2D + \alpha' + \beta + \beta' - \alpha)\, c^3$$

$$+ (-\alpha - 2\gamma + 2\beta' - 2\beta - \alpha')c^2 + (\alpha + \beta' + D - \alpha' + \beta)c + \gamma] \Big] F = 0\,.$$

Imposing needed constraints

$$\begin{gathered}\gamma^2 + 4\epsilon^2 = 0 \Longrightarrow \gamma = \pm 2i\epsilon\,;\\ -2\alpha\beta - E' = 0,\ \beta^2 + E = 0 \Longrightarrow \beta = \pm i\sqrt{E}\,,\ \alpha = \pm i\frac{E'}{2\sqrt{E}}\,;\\ -2\alpha'\beta' + E' = 0\,,\ \beta'^2 + E = 0 \Longrightarrow \beta' = \pm i\sqrt{E}\,,\ \alpha' = \mp i\frac{E'}{2\sqrt{E}}\end{gathered} \tag{4.26}$$

(here we have eight variants), we arrive at the equation

$$\begin{gathered}\frac{d^2F}{dx^2} + \Big(\frac{2\gamma+1}{x} + \frac{2\alpha'+2}{x+1} + \frac{2\alpha+2}{x-1} - \frac{2\beta'}{(x+1)^2} - \frac{2\beta}{(x-1)^2} - \frac{1}{x+c}\Big)\frac{dF}{dx}\\ +\Big\{\frac{2\gamma\alpha' - 2\gamma\alpha - 2\gamma\beta' - 2\gamma\beta + \alpha' - \alpha - \beta' - \beta - 2\nu - \frac{\gamma}{c}}{x}\\ +\frac{4\alpha\gamma + 4\beta\gamma + 2\alpha\alpha' + \alpha'\beta - \alpha\beta' - \beta\beta' + 4\alpha + 3\beta + 4\gamma + 2\alpha' - \beta' + 2A' - \frac{2\alpha}{c+1} - \frac{2\beta}{(c+1)^2}}{2(x-1)}\\ +\frac{-4\gamma\beta + 2\alpha^2 - 2\alpha'\beta + \beta\beta' - 4\beta + 2\alpha + 2B' + \frac{2\beta}{c+1}}{2(x-1)^2}\\ +\frac{-4\alpha'\gamma + 4\beta'\gamma - 2\alpha\alpha' - \alpha'\beta + \alpha\beta' + \beta\beta' - 4\alpha' + 3\beta' - 4\gamma - 2\alpha - \beta + 2A - \frac{2\alpha'}{c-1} - \frac{2\beta'}{(c-1)^2}}{2(x+1)}\\ +\frac{4\gamma\beta' + 2\alpha'^2 + 2\alpha\beta' + \beta\beta' + 4\beta' + 2\alpha' + 2B + \frac{2\beta'}{c-1}}{2(x+1)^2}\\ +\Big(\frac{\gamma}{c} + \frac{\alpha'}{c-1} + \frac{\alpha}{c+1} + \frac{\beta'}{(c-1)^2} + \frac{\beta}{(c+1)^2} + D\Big)\frac{1}{x+c}\Big\}F = 0.\end{gathered} \tag{4.27}$$

For theoretical consideration, it is enough to use its shortened form

$$\begin{gathered}F'' + \Big(\frac{n}{x} + \frac{n_1}{x-1} + \frac{n_2}{(x-1)^2} + \frac{n_3}{x+1} + \frac{n_4}{(x+1)^2} + \frac{n_5}{x+c}\Big)F'\\ +\Big(\frac{m}{x} + \frac{m_1}{x-1} + \frac{m_2}{(x-1)^2} + \frac{m_3}{x+1} + \frac{m_4}{(x+1)^2} + \frac{m_5}{x+c}\Big)F = 0.\end{gathered} \tag{4.28}$$

Multiplying this by $x(x+c)(x-1)^2(x+1)^2$, we get

$$\begin{gathered}\big[x^6 + cx^5 - 2\,x^4 - 2\,cx^3 + x^2 + cx\big]\,F''\\ +\big[(n + n_1 + n_3 + n_5)\,x^5 + ((n + n_1 + n_3)\,c + n_1 + n_2 - n_3 + n_4)\,x^4\\ +((n_1 + n_2 - n_3 + n_4)\,c - 2\,n - n_1 + 2\,n_2 - n_3 - 2\,n_4 - 2\,n_5)\,x^3\\ +((-2\,n - n_1 + 2\,n_2 - n_3 - 2\,n_4)\,c - n_1 + n_2 + n_3 + n_4)\,x^2\\ +((-n_1 + n_2 + n_3 + n_4)\,c + n + n_5)\,x + nc\big]\,F'\\ +\big[(m + m_1 + m_3 + m_5)\,x^5 + ((m + m_1 + m_3)\,c + m_1 + m_2 - m_3 + m_4)\,x^4\\ +((m_1 + m_2 - m_3 + m_4)\,c - 2\,m - m_1 + 2\,m_2 - m_3 - 2\,m_4 - 2\,m_5)\,x^3\\ +((-2\,m - m_1 + 2\,m_2 - m_3 - 2\,m_4)\,c - m_1 + m_2 + m_3 + m_4)\,x^2\\ +((-m_1 + m_2 + m_3 + m_4)\,c + m + m_5)\,x + mc\big]\,F = 0\,.\end{gathered}$$

Solutions of the last equation may be searched in the form of power series

$$F = \sum_{k=0}^{\infty} b_k x^k, \qquad F' = \sum_{k=1}^{\infty} k b_k x^{k-1}, \qquad F'' = \sum_{k=2}^{\infty} k(k-1) b_k x^{k-2}\,,$$

further, we produce 7-term recurrent relations:

$$i = 0, \qquad n\, b_1 + m\, b_0 = 0\,,$$

$$\begin{aligned}
i = 1, \qquad & 2\, cb_2 + 2\, ncb_2 + [(-n_1 + n_2 + n_3 + n_4)\, c + n + n_5]\, b_1 + mcb_1 \\
& + [(-m_1 + m_2 + m_3 + m_4)\, c + m + m_5]\, b_0 = 0\,,
\end{aligned}$$

$$\begin{aligned}
i = 2, \qquad & +3\, ncb_3 + 6\, cb_3 + 2b_2 + 2\, [(-n_1 + n_2 + n_3 + n_4)\, c + n + n_5]\, b_2 + mcb_2 \\
& + [(-2\, n - n_1 + 2\, n_2 - n_3 - 2\, n_4)\, c - n_1 + n_2 + n_3 + n_4]\, b_1 \\
& + [(-m_1 + m_2 + m_3 + m_4)\, c + m + m_5]\, b_1 \\
& + [(-2\, m - m_1 + 2\, m_2 - m_3 - 2\, m_4)\, c - m_1 + m_2 + m_3 + m_4]\, b_0 = 0\,,
\end{aligned}$$

$$\begin{aligned}
i = 3, \qquad & 12\, cb_4 + 4\, ncb_4 + 6b_3 + 3\, [(-n_1 + n_2 + n_3 + n_4)\, c + n + n_5]\, b_3 + mcb_3 \\
& -4\, cb_2 + 2\, [(-2\, n - n_1 + 2\, n_2 - n_3 - 2\, n_4)\, c - n_1 + n_2 + n_3 + n_4]\, b_2 \\
& + [(-m_1 + m_2 + m_3 + m_4)\, c + m + m_5]\, b_2 \\
& + [(n_1 + n_2 - n_3 + n_4)\, c - 2\, n - n_1 + 2\, n_2 - n_3 - 2\, n_4 - 2\, n_5]\, b_1 \\
& + [(-2\, m - m_1 + 2\, m_2 - m_3 - 2\, m_4)\, c - m_1 + m_2 + m_3 + m_4]\, b_1 \\
& + [(m_1 + m_2 - m_3 + m_4)\, c - 2\, m - m_1 + 2\, m_2 - m_3 - 2\, m_4 - 2\, m_5]\, b_0 = 0\,,
\end{aligned}$$

$$\ldots$$

$$\begin{aligned}
i = 6, 7, 8... \qquad & (m + m_1 + m_3 + m_5)\, b_{i-5} \\
& + [(i-4)(i-5) + (n + n_1 + n_3 + n_5)\, (i-4) \\
& + (m + m_1 + m_3)\, c + m_1 + m_2 - m_3 + m_4] b_{i-4} \\
& + \{c(i-3)(i-4) + [(n + n_1 + n_3)\, c + n_1 + n_2 - n_3 + n_4]\, (i-3) \\
& + [(m_1 + m_2 - m_3 + m_4)\, c - 2\, m - m_1 + 2\, m_2 - m_3 - 2\, m_4 - 2\, m_5]\}\, b_{i-3} \\
& + \{-2(i-2)(i-3) + [(n_1 + n_2 - n_3 + n_4)\, c - 2\, n - n_1 + 2\, n_2 - n_3 - 2\, n_4 - 2\, n_5]\, (i-2) \\
& + [(-2\, m - m_1 + 2\, m_2 - m_3 - 2\, m_4)\, c - m_1 + m_2 + m_3 + m_4]\}\, b_{i-2} \\
& + \{-2\, c(i-1)(i-2) + [(-2\, n - n_1 + 2\, n_2 - n_3 - 2\, n_4)\, c - n_1 + n_2 + n_3 + n_4]\, (i-1) \\
& + [(-m_1 + m_2 + m_3 + m_4)\, c + m + m_5]\}\, b_{i-1} \\
& + \{i(i-1) + [(-n_1 + n_2 + n_3 + n_4)\, c + n + n_5]\, i + mc\}\; b_i \\
& + [ci(i+1) + nc\, (i+1)]\, b_{i+1} = 0\,.
\end{aligned}$$

In accordance with Poincaré–Perron method, we divide the main recurrent relation by b_{i-5}:

$$\begin{aligned}
& (m + m_1 + m_3 + m_5) \\
& + [(i-4)(i-5) + (n + n_1 + n_3 + n_5)\, (i-4) \\
& + (m + m_1 + m_3)\, c + m_1 + m_2 - m_3 + m_4]\, \frac{b_{i-4}}{b_{i-5}} \\
& + \{c(i-3)(i-4) + [(n + n_1 + n_3)\, c + n_1 + n_2 - n_3 + n_4]\, (i-3) \\
& + [(m_1 + m_2 - m_3 + m_4)\, c - 2\, m - m_1 + 2\, m_2 - m_3 - 2\, m_4 - 2\, m_5]\}\, \frac{b_{i-3}}{b_{i-4}} \frac{b_{i-4}}{b_{i-5}} \\
& + \{-2(i-2)(i-3) + [(n_1 + n_2 - n_3 + n_4)\, c - 2\, n - n_1 + 2\, n_2 - n_3 - 2\, n_4 - 2\, n_5]\, (i-2)
\end{aligned}$$

$$+[(-2m - m_1 + 2m_2 - m_3 - 2m_4)c - m_1 + m_2 + m_3 + m_4]\} \frac{b_{i-2}}{b_{i-3}} \frac{b_{i-3}}{b_{i-4}} \frac{b_{i-4}}{b_{i-5}}$$

$$+\{-2c(i-1)(i-2) + [(-2n - n_1 + 2n_2 - n_3 - 2n_4)c - n_1 + n_2 + n_3 + n_4](i-1)$$

$$+[(-m_1 + m_2 + m_3 + m_4)c + m + m_5]\} \frac{b_{i-1}}{b_{i-2}} \frac{b_{i-2}}{b_{i-3}} \frac{b_{i-3}}{b_{i-4}} \frac{b_{i-4}}{b_{i-5}}$$

$$+\{i(i-1) + [(-n_1 + n_2 + n_3 + n_4)c + n + n_5]i + mc\} \frac{b_i}{b_{i-1}} \frac{b_{i-1}}{b_{i-2}} \frac{b_{i-2}}{b_{i-3}} \frac{b_{i-3}}{b_{i-4}} \frac{b_{i-4}}{b_{i-5}}$$

$$+[ci(i+1) + nc(i+1)] \frac{b_{i+1}}{b_i} \frac{b_i}{b_{i-1}} \frac{b_{i-1}}{b_{i-2}} \frac{b_{i-2}}{b_{i-3}} \frac{b_{i-3}}{b_{i-4}} \frac{b_{i-4}}{b_{i-5}} = 0 .$$

After that we multiply the last relation by i^{-2} and tend $i \to \infty$, so deriving an algebraic equation for R, which determines possible convergence radii

$$R = \lim_{i \to \infty} \frac{b_i}{b_{i-1}}, \qquad R_{conv} = \frac{1}{|R|},$$
$$R + cR^2 - Rr^3 - 2cR^4 + R^5 + cR^6 = 0 \implies (1 + cR)\, R\, (R^2 - 1)^2 = 0 ; \tag{4.29}$$

the roots are $R = 0$, $R = \pm 1$, and $R = -1/c$. Therefore, possible convergence radii are $1, c > 1, \infty$. Note that the convergence radius $R_{conv} = 1$ covers all physical region for the variable $x \in (0, 1)$. In accordance with the above-mentioned symmetry, solutions for function $g(x)$ are constructed in a similar way, it is enough to make formal changes $\nu \implies -\nu$, $c \implies -c$.

4.3 Numerical study

Let us list the above eight solutions (note the change $(x-1) < 0$ to $(1-x) > 0$, it makes the involved functions single-valued in physical region $x \in (0, 1)$)

$$f(x) = x^{\gamma}(1-x)^{\alpha} \exp\left(\frac{\beta}{1-x}\right)(x+1)^{\alpha'} \exp\left(\frac{\beta'}{x+1}\right) F(x) , \tag{4.30}$$

where

$$\beta = \pm i\Gamma, \quad \alpha = \pm i\Sigma, \quad \beta' = \pm i\Gamma, \quad \alpha' = \pm i\Sigma,$$
$$\Gamma = \frac{\sqrt{\epsilon^2 - M^2}}{2}, \quad \Sigma = \frac{\epsilon^2 - M^2/2}{\sqrt{\epsilon^2 - M^2}}, \quad \gamma = \pm 2i\epsilon . \tag{4.31}$$

In order to get the massless particle, it suffices to make the formal changes in parameters:

$$c \to \infty, \quad \Gamma = \frac{1}{2}\epsilon, \quad \Sigma = \epsilon . \tag{4.32}$$

In the following we study the most simple massless case. Correspondingly, we use the following substitutions

$$g_1(x) \to h(x) e^{\frac{i\epsilon}{2(x-1)} + \frac{i\epsilon}{2(x+1)}} (1-x)^{i\epsilon} x^{2i\epsilon} (x+1)^{-i\epsilon},$$
$$g_2(x) \to h(x) e^{-\frac{i\epsilon}{2(x-1)} - \frac{i\epsilon}{2(x+1)}} (1-x)^{-i\epsilon} x^{-2i\epsilon} (x+1)^{i\epsilon}; \tag{4.33}$$

$$\begin{aligned} g_3(x) &\to h(x)e^{-\frac{i\epsilon}{2(x-1)}-\frac{i\epsilon}{2(x+1)}}(1-x)^{-i\epsilon}x^{2i\epsilon}(x+1)^{i\epsilon},\\ g_4(x) &\to h(x)e^{\frac{i\epsilon}{2(x-1)}+\frac{i\epsilon}{2(x+1)}}(1-x)^{i\epsilon}x^{-2i\epsilon}(x+1)^{-i\epsilon}; \end{aligned} \tag{4.34}$$

$$\begin{aligned} g_5(x) &\to h(x)e^{\frac{i\epsilon}{2(x-1)}-\frac{i\epsilon}{2(x+1)}}(1-x)^{i\epsilon}x^{2i\epsilon}(x+1)^{i\epsilon},\\ g_6(x) &\to h(x)e^{\frac{i\epsilon}{2(x+1)}-\frac{i\epsilon}{2(x-1)}}(1-x)^{-i\epsilon}x^{-2i\epsilon}(x+1)^{-i\epsilon}; \end{aligned} \tag{4.35}$$

$$\begin{aligned} g_7(x) &\to h(x)e^{\frac{i\epsilon}{2(x+1)}-\frac{i\epsilon}{2(x-1)}}(1-x)^{-i\epsilon}x^{2i\epsilon}(x+1)^{-i\epsilon},\\ g_8(x) &\to h(x)e^{\frac{i\epsilon}{2(x-1)}-\frac{i\epsilon}{2(x+1)}}(1-x)^{i\epsilon}x^{-2i\epsilon}(x+1)^{i\epsilon}. \end{aligned} \tag{4.36}$$

We need an explicit form of 2nd-order equations for all $h_i(x)$; it suffices to write down only four cases, h_1, h_3, h_5, and h_7, because all equations and solutions are divided into pairs of conjugated ones:

$$\begin{aligned} &\Big[-\frac{35\epsilon^2}{4(x-1)}+\frac{12\epsilon^2}{x}-\frac{13\epsilon^2}{4(x+1)}+\frac{11\epsilon^2}{4(x-1)^2}-\frac{5\epsilon^2}{4(x+1)^2}+\frac{11i\epsilon}{2(x-1)}\\ &\quad-\frac{3i\epsilon}{x}-\frac{5i\epsilon}{2(x+1)}+\frac{\nu^2}{x-1}-\frac{\nu^2}{x+1}-\frac{\nu^2}{(x-1)^2}-\frac{\nu^2}{(x+1)^2}\\ &\quad+\frac{\nu}{x-1}-\frac{2\nu}{x}+\frac{\nu}{x+1}+\frac{\nu}{(x-1)^2}-\frac{\nu}{(x+1)^2}\Big]h_1(x)\\ &+\Big[\frac{2i\epsilon}{x-1}+\frac{4i\epsilon}{x}-\frac{2i\epsilon}{x+1}-\frac{i\epsilon}{(x-1)^2}-\frac{i\epsilon}{(x+1)^2}+\frac{2}{x-1}+\frac{1}{x}+\frac{2}{x+1}\Big]h_1'(x)+h_1''(x)=0, \end{aligned}$$

$$\begin{aligned} &\Big[\frac{13\epsilon^2}{4(x-1)}-\frac{12\epsilon^2}{x}+\frac{35\epsilon^2}{4(x+1)}-\frac{5\epsilon^2}{4(x-1)^2}+\frac{11\epsilon^2}{4(x+1)^2}+\frac{5i\epsilon}{2(x-1)}+\frac{3i\epsilon}{x}-\frac{11i\epsilon}{2(x+1)}\\ &+\frac{\nu^2}{x-1}-\frac{\nu^2}{x+1}-\frac{\nu^2}{(x-1)^2}-\frac{\nu^2}{(x+1)^2}+\frac{\nu}{x-1}-\frac{2\nu}{x}+\frac{\nu}{x+1}+\frac{\nu}{(x-1)^2}-\frac{\nu}{(x+1)^2}\Big]h_3(x)\\ &+\Big[-\frac{2i\epsilon}{x-1}+\frac{4i\epsilon}{x}+\frac{2i\epsilon}{x+1}+\frac{i\epsilon}{(x-1)^2}+\frac{i\epsilon}{(x+1)^2}+\frac{2}{x-1}+\frac{1}{x}+\frac{2}{x+1}\Big]h_3'(x)+h_3''(x)=0, \end{aligned}$$

$$\begin{aligned} &\Big[-\frac{12\epsilon^2}{x-1}+\frac{12\epsilon^2}{x+1}+\frac{4\epsilon^2}{(x-1)^2}+\frac{4\epsilon^2}{(x+1)^2}+\frac{8i\epsilon}{x-1}-\frac{8i\epsilon}{x+1}\\ &+\frac{\nu^2}{x-1}-\frac{\nu^2}{x+1}-\frac{\nu^2}{(x-1)^2}-\frac{\nu^2}{(x+1)^2}+\frac{\nu}{x-1}-\frac{2\nu}{x}+\frac{\nu}{x+1}+\frac{\nu}{(x-1)^2}-\frac{\nu}{(x+1)^2}\Big]h_5(x)\\ &+\Big[\frac{2i\epsilon}{x-1}+\frac{4i\epsilon}{x}+\frac{2i\epsilon}{x+1}-\frac{i\epsilon}{(x-1)^2}+\frac{i\epsilon}{(x+1)^2}+\frac{2}{x-1}+\frac{1}{x}+\frac{2}{x+1}\Big]h_5'(x)+h_5''(x)=0, \end{aligned}$$

$$\begin{aligned} &\Big[\frac{\nu^2}{x-1}-\frac{\nu^2}{x+1}-\frac{\nu^2}{(x-1)^2}-\frac{\nu^2}{(x+1)^2}+\frac{\nu}{x-1}-\frac{2\nu}{x}+\frac{\nu}{x+1}+\frac{\nu}{(x-1)^2}-\frac{\nu}{(x+1)^2}\Big]h_7(x)\\ &+\Big[-\frac{2i\epsilon}{x-1}+\frac{4i\epsilon}{x}-\frac{2i\epsilon}{x+1}+\frac{i\epsilon}{(x-1)^2}-\frac{i\epsilon}{(x+1)^2}+\frac{2}{x-1}+\frac{1}{x}+\frac{2}{x+1}\Big]h_7'(x)+h_7''(x)=0. \end{aligned}$$

Below symbols $R_i(x)$ and $I_i(x)$ $(i = 1, 3, 5, 7)$ will designate real and imaginary parts of four converging series (they depend on quantum numbers $\epsilon, \nu = j + 1/2$). Several typical graphs of series are given in Figs. 4.2–4.5.

Numerical study shows that real and imaginary parts of the series sums $R_i(r_*) + I_i(r_*)$ demonstrate evident asymptotical behaviour at $r_* \to +\infty$:

$$\begin{aligned}
&\text{Re}\, h_1: \frac{h_1(500)}{h_1(250)} = 1.0133, \quad \frac{h_1(750)}{h_1(500)} = 1.0043, \quad \frac{h_1(1000)}{h_1(750)} = 1.0021, \\
&\frac{h_1(1250)}{h_1(1000)} = 1.0012, \quad \frac{h_1(1500)}{h_1(1250)} = 1.0008, \quad h_1(1500) = 12092.5427, \\
&\text{Im}\, h_1: \frac{h_1(500)}{h_1(250)} = 1.0349, \quad \frac{h_1(750)}{h_1(500)} = 1.0112, \quad \frac{h_1(1000)}{h_1(750)} = 1.0055, \\
&\frac{h_1(1250)}{h_1(1000)} = 1.0033, \quad \frac{h_1(1500)}{h_1(1250)} = 1.0021, \quad h_1(1500) = 279.9890;
\end{aligned} \tag{4.37}$$

$$\begin{aligned}
&\text{Re}\, h_3: \frac{h_3(500)}{h_3(250)} = 1.0138, \quad \frac{h_3(750)}{h_3(500)} = 1.0045, \quad \frac{h_3(1000)}{h_3(750)} = 1.00224, \\
&\frac{h_3(1250)}{h_3(1000)} = 1.0013, \quad \frac{h_3(1500)}{h_3(1250)} = 1.0008, \quad h_3(1500) = 22583.1144, \\
&\text{Im}\, h_3: \frac{h_3(500)}{h_3(250)} = 0.9290, \quad \frac{h_3(750)}{h_3(500)} = 0.9744, \quad \frac{h_3(1000)}{h_3(750)} = 0.9868, \\
&\frac{h_3(1250)}{h_3(1000)} = 0.9920, \quad \frac{h_3(1500)}{h_3(1250)} = 0.9946, \quad h_3(1500) = 284.8543;
\end{aligned} \tag{4.38}$$

$$\begin{aligned}
&\text{Re}\, h_5: \frac{h_5(500)}{h_5(250)} = 1.0170, \frac{h_5(750)}{h_5(500)} = 1.0055, \frac{h_5(1000)}{h_5(750)} = 1.0027, \\
&\frac{h_5(1250)}{h_5(1000)} = 1.0016, \frac{h_5(1500)}{h_5(1250)} = 1.0011, \quad h_5(1500) = 1206.6537, \\
&\text{Im}\, h_5: \frac{h_5(500)}{h_5(250)} = 1.0132, \frac{h_5(750)}{h_5(500)} = 1.0043, \frac{h_5(1000)}{h_5(750)} = 1.0021, \\
&\frac{h_5(1250)}{h_5(1000)} = 1.0012, \frac{h_5(1500)}{h_5(1250)} = 1.0000, \quad h_5(1500) = -11203.6223;
\end{aligned} \tag{4.39}$$

$$\begin{aligned}
&\text{Re}\, h_7: \frac{h_7(500)}{h_7(250)} = 1.0170, \quad \frac{h_7(750)}{h_7(500)} = 1.0055, \quad \frac{h_7(1000)}{h_7(750)} = 1.0027, \\
&\frac{h_7(1250)}{h_7(1000)} = 1.0016, \quad \frac{h_7(1500)}{h_7(1250)} = 1.0011, \quad h_7(1500) = 1206.6537, \\
&\text{Im}\, h_7: \frac{h_7(500)}{h_7(250)} = 1.0132, \quad \frac{h_7(750)}{h_7(500)} = 1.0043, \quad \frac{h_7(1000)}{h_7(750)} = 1.0021, \\
&\frac{h_7(1250)}{h_7(1000)} = 1.0012, \quad \frac{h_7(1500)}{h_7(1250)} = 1.0009, \quad h_7(1500) = -11203.62231.
\end{aligned} \tag{4.40}$$

4.4 Tunnelling process

Let us examine tunnelling effect for the particle, which moves from the right of the barrier. To this end, we start with the solution $g_2(x)$, its asymptotic behaviour at $x \to 0$ ($r_* \to -\infty$)

is given by the formula

$$g_2(x) = e^{-2i\epsilon \ln x} e^{-i\epsilon} = e^{-i\epsilon} e^{-i\epsilon r_*} . \tag{4.41}$$

Let us use notation

$$G_2(x) = e^{+i\epsilon} g_2(x), \quad G_2(r \to -\infty) = e^{-i\epsilon r_*} . \tag{4.42}$$

We may formulate the following Cauchy problem:

$$G_2(x) , \quad \frac{d}{dx} G_2(x) , \quad x_0 = 10^{-5} , \quad (r_*)_0 = -22 . \tag{4.43}$$

The tunnelling process is described by the general formula

$$e^{-i\epsilon r_*} \Longleftrightarrow A e^{-i\epsilon r_*} + B e^{+i\epsilon r_*}, \text{ or } \frac{1}{A} e^{-i\epsilon r_*} \Longleftrightarrow A e^{-i\epsilon r_*} + \frac{B}{A} e^{+i\epsilon r_*} ; \tag{4.44}$$

reflection and transmission coefficients are defined as

$$R = |\frac{B}{A}|^2, \quad D = |\frac{1}{A}|^2 . \tag{4.45}$$

Let us take in the region $r_* \to +\infty$ two close points, $s_1 = \epsilon r_{1*}, s_2 = \epsilon r_{2*}$. They give two linear equations

$$e^{-is_1} A + e^{+is_1} B = N_1, \quad e^{-is_2} A + e^{+is_2} B = N_2. \tag{4.46}$$

The values N_1 and N_2 are known from results of solving the Cauchy problem, so this linear system may be resolved with respect to variables A and B. Numerical study yields

$$\begin{aligned}
& s_1 = 100, \ s_2 = 101, \\
& |D| = 2.5871172808253399483182956090916 \times 10^{-9} \\
& |R| = 0.999999999999330 \\
& s_1 = 200, \ s_2 = 201, \\
& |D| = 2.5877017935952845053539711845416 \times 10^{-9} \\
& |R| = 0.999999999999330 \\
& s_1 = 300, \ s_2 = 301, \\
& |D| = 2.5880854478636695997420096062116 \times 10^{-9} \\
& |R| = 0.999999999999330 \\
& s_1 = 400, \ s_2 = 401, \\
& |D| = 2.5882480599386513075141952768516 \times 10^{-9} \\
& |R| = 0.999999999999330 \\
& s_1 = 500, \ s_2 = 501, \\
& |D| = 2.5882683602272274835321197825816 \times 10^{-9} \\
& 0.999999999999330 \\
& s_1 = 600, \ s_2 = 601, \\
& |D| = 2.5882297117443874845102673598216 \times 10^{-9} \\
& |R| = 0.999999999999330
\end{aligned} \tag{4.47}$$

We have calculated coefficients D and R at different values of energy:

$$
\begin{array}{lll}
\epsilon = & D = & R = \\
1 & 2.5879781045664890751285537294 \times 10^{-9} & 0.999999999999330 \\
3/2 & 1.758238744829617894686948354 2314 \times 10^{-7} & 0.999999996908314 \\
2 & 6.5645375757383672067916623723 5681 \times 10^{-6} & 0.999995690201453 \\
5/2 & 9.8942877050231482632828007936 7695 \times 10^{-6} & 0.999990208695562 \\
3 & 9.9975388921676661320561153011 2792 \times 10^{-6} & 0.999989999254247 \\
7/2 & 0.000010000098762059780 & 0.999989999540272 \\
4 & 9.9998963457682276217788142383 8663 \times 10^{-6} & 0.999990000509748
\end{array}
\tag{4.48}
$$

In a similar manner, we could examine tunnelling effect for the particle, which moves from the left on the barrier.

4.5 Conclusions

For the Dirac particle, the general mathematical and numerical study of the tunnelling process through the potential barrier generated by the Schwarzschild black hole metric has been done. We construct solutions in explicit form and prove that the power series involved in them are converged in all physical regions of the physical region of the variable $r \in (1, +\infty)$. Results for tunnelling effect significantly differ for two situations: one when the particle falls on the barrier from within and another when the particle falls from outside. The mathematical structure of the derived asymptotic relations is exact; however, analytical expressions for involved convergent power series are not known, and further study is based on numerical summing of the series. The calculations are implemented using *Mathematica* system.

4.6 Figures

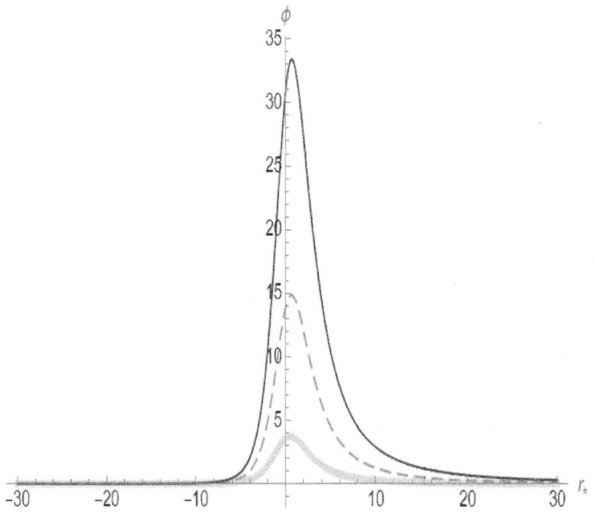

FIGURE 4.1
Potential function $\phi(r_*)$.

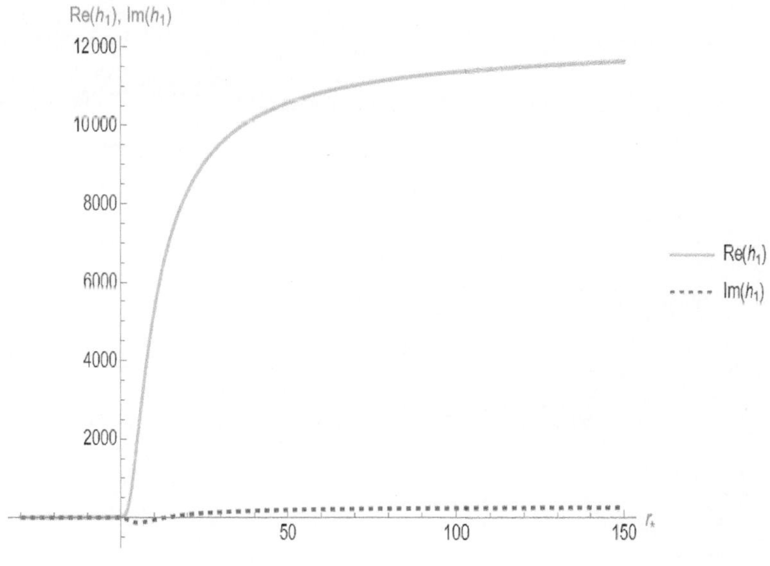

FIGURE 4.2
The graph of the series $h_1(r_*)$, $\epsilon = 1$, $\nu = 5$.

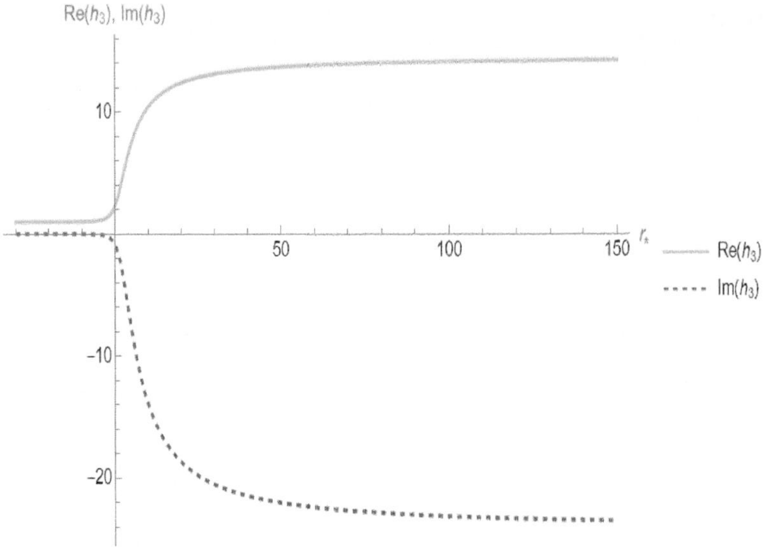

FIGURE 4.3
The graph of the series $h_3(r_*)$, $\epsilon = 1$, $\nu = 5$.

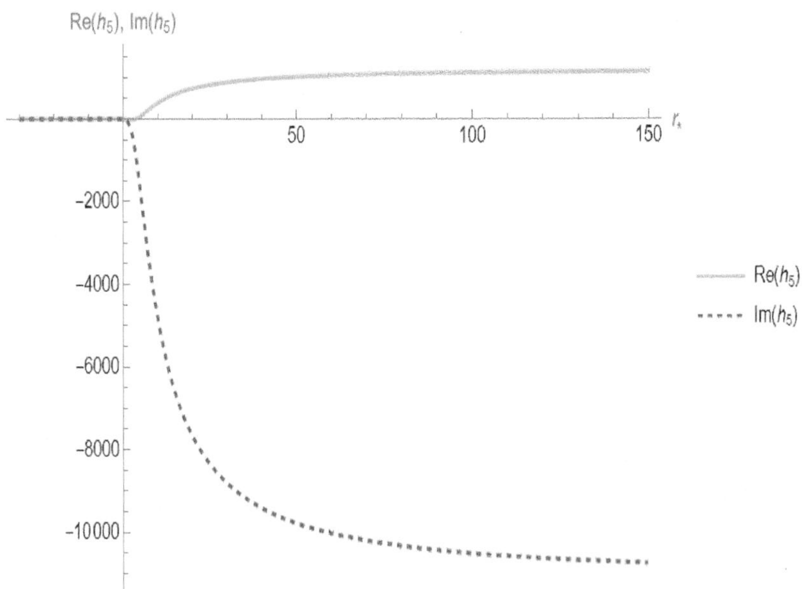

FIGURE 4.4
The graph of the series $h_5(r_*)$, $\epsilon = 1$, $\nu = 5$.

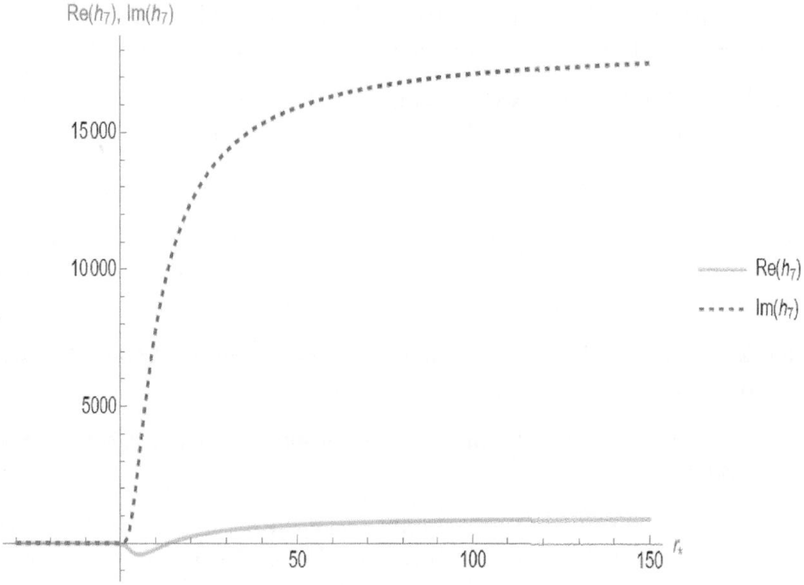

FIGURE 4.5
The graph of the series $h_7(r_*)$, $\epsilon = 1$, $\nu = 5$.

Bibliography

[1] T. Regge and John A. Wheeler. Stability of a Schwarzschild Singularity. *Physical Review*, 108(4): 1063–1069, 1957.

[2] K. Schwarzschild. Über das Gravitationsfeld eines Massenpunktes nach der Einsteinschen Theorie. *Sitzungsberichte der Königlich Preußischen Akademie der Wissenschaften zu Berlin*, K1: 189–196, 1916.

[3] D.R. Brill and John A. Wheeler. Interaction of Neutrinos and Gravitational Fields. *Reviews of Modern Physics*, 29: 465–479, 1957.

[4] J.M. Bardeen and W.H. Press. Radiation fields in the Schwarzschild background. *Journal of Mathematical Physics*, 14: 7–19, 1973.

[5] S.W. Hawking. Particle creation by black holes. *Communications in Mathematical Physics*, 43: 199–220, 1975.

[6] S. Chandrasekhar. *The Mathematical Theory of Black Holes.* Oxford University Press, Oxford, 1983.

[7] V.P. Frolov. Physical effects in graviational field of blacj holes. *FIAN Proceedings*, 169: 3–131, 1986.

[8] D.V. Gal'tsov. *Particles and Fields in Vivinity of Black Holes.* Publ. Moscow State University, Moscow, 1986 (in Russian).

[9] D.N. Page. Particle emission rates from a black hole: Massless particles from an uncharged, non-rotating hole. *Journals of Physical Review D*, 13: 198–206, 1976.

[10] Joel Smoller and Chunjing Xie. Asymptotic behaviour of massless dirac waves in schwarzschild geometry. *Annales Henri Poincare*, 13: 943–989, 2012.

[11] P. Fiziev in: https://www.researchgate.net/profile/Plamen-Fiziev/publications

[12] E.M. Ovsiyuk, O.V. Veko, Yu. A. Rusak, A.V. Chichurin and V.M. Red'kov. To analysis of the Dirac and Majorana particle solutions in Schwarzschild field. *Nonlinear Phenomena in Complex System*, 20(1): 56–72, 2017.

[13] V.M. Red'kov. *Particle Fields in Riemannian Space and the Lorent Group.* Belarussian Science, Minsk, 2009.

[14] V.V. Berestetskiy, E.M. Lifshitz and L.P. Pitaewskiy. *Quntum Electrodynamics.* Nauka, Moskow, 1980.

[15] R. Penrose and W. Rindler. *Spinors and Space-time. Volume I: Two-spinor Calculus and Relativistic Fields.* Cambridge University Press, 1984.

[16] D.A. Varshalovich, A.N. Moskalev and V.K. Hersonskiy. *Quantum Theory of Angular Moment.* Nauka, Leningrad, 1975 (in Russian).

[17] V.M. Red'kov. *Tetrad Formalism, Spherical Symmetry and Schrödinger Basis.* Publishing House "Belarusian Science", Minsk, 2011 (in Russian).

[18] V.M. Redkov and E.M. Ovsiyuk. *Quantum Mechanics in Space of Constant Curvature.* Nova Science Publishers, New York, 2012.

[19] E.M. Ovsiyuk, A.V. Chichurin and V.M. Red'kov. Nonrelativistic vector particle in Coulomb field on the background of Lobachevsky geometry: analytical and numerical study, visualization. *Studia i Materialy EWSIE*, 10: 45–57, 2015.

[20] A.V. Chichurin, E.M. Ovsiyuk and V.M. Red'kov. The quantum tunnelling effect for a particle with intrinsic structure in presence of external magnetic field in the Lobachevsky space. *Computers and Mathematics with Applications*, 75: 1550–1565, 2018.

5

On Maxwell equations in Schwarzschild space-time

It is shown that the generally covariant extended method of Riemann–Silberstein–Majorana–Oppenheimer in electrodynamics, specified in Schwarzschild metrics, after separating the variables, provides us with the possibility of reducing the problem to a differential equation similar to that arising in the case of a scalar field in the Schwarzschild space-time. This differential equation is recognised as a confluent Hein equation.

We have considered the electromagnetic field on the basis of the 10-dimensional Duffin–Kemmer approach, when in addition to six components of the strength tensor, one uses four components of an electromagnetic potential. After separation of the variable, we have arrived at a system of ten radial equations, which were simplified by the use of additional constraints followed by an eigenvalue equation for the spatial parity operator $\hat{\Pi}\Psi = P\Psi$; the radial system has been divided into two subsystems of four and six equations, respectively. In this second approach, the problem of electromagnetic field has been reduced to the confluent Hein differential equation as well.

In particular, we have shown explicitly how solutions found in complex form are embedded in matrix 10-dimensional formalism; besides, we determine radial functions that are responsible for gauge degrees of freedom.

The chapter is based on [1]–[24].

5.1 Introduction

Usually, when treating electromagnetic field in a curved space-time background [1, 2], for instance, in Schwarzschild space-time geometry [3], they use a real vector description of an electromagnetic tensor [4]. In the frames of Newman–Penrose formalism [5] for description of components of the electromagnetic tensor a spinor technique is used.

In [6], a general covariant approach to the Maxwell theory based on the use of the Riemann–Silberstein–Majorana–Oppenheimer complex representation was elaborated. Here it is used to treat Maxwell field in the background of Schwarzschild black hole. It is shown that this technique provides us with possibilities after separating the variables to reduce the problem of the Maxwell field to a differential equation similar to that arising in the case of a scalar field in the Schwarzschild space-time. This differential equation is recognised as a confluent Hein equation.

After that we turn to consideration of the electromagnetic field on the basis of a 10-dimensional description, when in addition to six components of the strength tensor, one uses four components of electromagnetic potential. Such a description of the electromagnetic field

DOI: 10.1201/9781003472377-5

is more informative because it includes gauge degrees of freedom. However, this method to describe electromagnetic field is more complicated. We use it in the matrix form of Duffin–Kemmer–Peteau (recent consideration and big list of references see in [7]). After separation of the variable, we arrive at a system of ten radial equations, which can be simplified by the use of additional constraints followed by an eigenvalue equation for the spatial parity operator $\hat{\Pi}\Psi = P\Psi$; the radial system is divided into two subsystems of four and six equations, respectively. In this second approach, the problem of electromagnetic field reduces to the confluent Hein differential equation as well. In particular, we show explicitly how previously found solutions in complex form are embedded into matrix Duffin–Kemmer–Peteau formalism; besides, we determine radial functions that are responsible for gauge degrees of freedom.

5.2 Separating the variables, Wigner functions

Matrix Maxwell equation in Schwarzschild space-time

$$
\begin{gathered}
dS^2 = \Phi\, dt^2 - \frac{dr^2}{\Phi} - r^2(d\theta^2 + \sin^2\theta d\phi^2), \quad \Phi = 1 - M/r, \\
e^{\alpha}_{(0)} = (\frac{1}{\sqrt{\Phi}}, 0, 0, 0), \qquad e^{\alpha}_{(3)} = (0, \sqrt{\Phi}, 0, 0), \\
e^{\alpha}_{(1)} = (0, 0, \frac{1}{r}, 0), \qquad e^{\alpha}_{(2)} = (1, 0, 0, \frac{1}{r\sin\theta}), \\
\gamma_{030} = \frac{\Phi'}{2\sqrt{\Phi}}, \ \gamma_{311} = \frac{\sqrt{\Phi}}{r}, \ \gamma_{322} = \frac{\sqrt{\Phi}}{r}, \ \gamma_{122} = \frac{\cos\theta}{r\sin\theta}
\end{gathered}
\tag{5.1}
$$

has the following form (assume the use of cyclic basis when the matrix S_3 is diagonal)

$$
\begin{gathered}
[-\frac{i\partial_t}{\sqrt{\Phi}} + \sqrt{\Phi}(\alpha^3\partial_r + \frac{\alpha^1 s_2 - \alpha^2 s_1}{r} + \frac{\Phi'}{2\Phi}s_3) + \frac{1}{r}\Sigma_{\theta,\phi}] \left| \begin{array}{c} 0 \\ \psi \end{array} \right| = 0, \\
\Sigma_{\theta,\phi} = \frac{\alpha^1}{r}\partial_\theta + \alpha^2\, \frac{\partial_\phi + s_3\cos\theta}{\sin\theta}, \qquad \psi = \mathbf{E} + i\mathbf{B}.
\end{gathered}
\tag{5.2}
$$

Let us diaginalise the square and the third projection of the total angular momentum of the electromagnetic field, $\mathbf{J}^2$, J^3; correspondingly, we apply the substitution

$$
\psi = e^{-i\omega t} \left| \begin{array}{c} 0 \\ \varphi_1(r) D_{-1} \\ \varphi_2(r) D_0 \\ \varphi_3(r) D_{+1} \end{array} \right| , \tag{5.3}
$$

where Wigner function are used, $D_\sigma = D^j_{-m,\sigma}(\phi, \theta, 0), \sigma = -1, 0, +1$. With the use of the known recurrent formulas

$$
\begin{gathered}
\partial_\theta D_{-1} = \frac{1}{2}(a D_{-2} - \nu D_0), \ \frac{m - \cos\theta}{\sin\theta} D_{-1} = \frac{1}{2}(a D_{-2} + \nu D_0), \\
\partial_\theta D_0 = \frac{1}{2}(\nu D_{-1} - \nu D_{+1}), \ \frac{m}{\sin\theta} D_0 = \frac{1}{2}(\nu D_{-1} + \nu D_{+1}), \\
\partial_\theta D_{+1} = \frac{1}{2}(\nu D_0 - a D_{+2}), \ \frac{m + \cos\theta}{\sin\theta} D_{+1} = \frac{1}{2}\,(\nu D_0 +\ a D_{+2})\,, \\
\nu = \sqrt{j(j+1)}, \qquad a = \sqrt{(j-1)(j+2)}
\end{gathered}
\tag{5.4}
$$

we obtain (the factor $e^{-i\omega t}$ is omitted)

$$\Sigma_{\theta\phi}\Psi' = \frac{\nu}{\sqrt{2}} \left| \begin{array}{r} (\varphi_1 + \varphi_3) D_0 \\ -i\,\varphi_2 D_{-1} \\ i\,(\varphi_1 - \varphi_3) D_0 \\ +i\,\varphi_2 D_{+1} \end{array} \right| . \tag{5.5}$$

To simplify the formula, we change the notation

$$\frac{\nu}{\sqrt{2}} = \sqrt{\frac{j(j+1)}{2}} \qquad \Longrightarrow \qquad \nu. \tag{5.6}$$

Turning to the matrix eq. (5.2), we derive the radial system of four equations

$$\begin{array}{ll} 1) & \sqrt{\Phi}\,(\frac{d}{dr} + \frac{2}{r})\,\varphi_2 + \frac{\nu}{r}\,(\varphi_1 + \varphi_3) = 0, \\ 2) & (-\frac{\omega}{\sqrt{\Phi}} - i\,\sqrt{\Phi}\,\frac{d}{dr} - i\frac{\sqrt{\Phi}}{r} - i\frac{\Phi'}{2\sqrt{\Phi}})\,\varphi_1 - \frac{i\nu}{r}\,\varphi_2 = 0, \\ 3) & -\frac{\omega}{\sqrt{\Phi}}\varphi_2 + \frac{i\nu}{r}\,(\varphi_1 - \varphi_3) = 0, \\ 4) & (-\frac{\omega}{\sqrt{\Phi}} + i\,\sqrt{\Phi}\,\frac{d}{dr} + i\frac{\sqrt{\Phi}}{r} + i\frac{\Phi'}{2\sqrt{\Phi}})\,\varphi_3 + \frac{i\nu}{r}\,\varphi_2 = 0. \end{array} \tag{5.7}$$

Combining eqs. 2) and 4), instead of eq. (5.7) we obtain

$$\begin{array}{ll} 2)+4), & -\frac{\omega}{\sqrt{\Phi}}(\varphi_1 + \varphi_3) - i(\sqrt{\Phi}\frac{d}{dr} + \frac{\sqrt{\Phi}}{r} + \frac{\Phi'}{2\sqrt{\Phi}})(\varphi_1 - \varphi_3) = 0, \\ 2)-4), & -\frac{\omega}{\sqrt{\Phi}}(\varphi_1 - \varphi_3) - i(\sqrt{\Phi}\frac{d}{dr} + \frac{\sqrt{\Phi}}{r} + \frac{\Phi'}{2\sqrt{\Phi}})(\varphi_1 + \varphi_3) - \frac{2i\nu}{r}\,\varphi_2 = 0, \\ 3) & -\frac{\omega}{\sqrt{\Phi}}\varphi_2 + \frac{i\nu}{r}\,(\varphi_1 - \varphi_3) = 0, \qquad 1)\;\; \sqrt{\Phi}(\frac{d}{dr} + \frac{2}{r})\varphi_2 + \frac{\nu}{r}\,(\varphi_1 + \varphi_3) = 0. \end{array} \tag{5.8}$$

It is readily checked that eq. 1) turns out to be identity when taking into account three remaining equations. Therefore, we have only three independent equations

$$\begin{array}{r} \frac{\omega}{\sqrt{\Phi}}\varphi_2 + \frac{i\nu}{r}\,(\varphi_1 - \varphi_3) = 0, \\ -\frac{\omega}{\sqrt{\Phi}}(\varphi_1 + \varphi_3) - i\,(\sqrt{\Phi}\,\frac{d}{dr} + \frac{\sqrt{\Phi}}{r} + \frac{\Phi'}{2\sqrt{\Phi}})(\varphi_1 - \varphi_3) = 0, \\ -\frac{\omega}{\sqrt{\Phi}}(\varphi_1 - \varphi_3) - i\,(\sqrt{\Phi}\frac{d}{dr} + \frac{\sqrt{\Phi}}{r} + \frac{\Phi'}{2\sqrt{\Phi}})(\varphi_1 + \varphi_3) - \frac{2i\nu}{r}\,\varphi_2 = 0. \end{array} \tag{5.9}$$

Introducing new variables

$$f = \varphi_1 + \varphi_3, \qquad g = \varphi_1 - \varphi_3,$$

we transform eq. (5.9) to

$$\begin{array}{l} \varphi_2 = \frac{i\nu}{\omega}\frac{\sqrt{\Phi}}{r}g, \qquad -\frac{\omega}{\Phi}f - i(\frac{d}{dr} + \frac{1}{r} + \frac{\Phi'}{2\Phi})g = 0, \\ -\frac{\omega^2}{\Phi}g - i\omega(\frac{d}{dr} + \frac{1}{r} + \frac{\Phi'}{2\Phi})\,f + \frac{2\nu^2}{r^2}\,g = 0. \end{array} \tag{5.10}$$

By using the substitutions,

$$g = \frac{1}{r\sqrt{\Phi}}\, G(r), \qquad f = \frac{1}{r\sqrt{\Phi}}\, F(r);$$

we get more simple equations

$$\varphi_2 = \frac{i\nu}{\omega}\,\frac{1}{r^2}\, G(r), \qquad i\omega\, F = \Phi \frac{d}{dr} G, \qquad +i\omega\, \frac{d}{dr}\, F + \frac{\omega^2}{\Phi}\, G - \frac{2\nu^2}{r^2}\, G = 0. \tag{5.11}$$

The 2nd-order equation for the primary variable $G(r)$ reads

$$\frac{d^2G}{dr^2} + \frac{\Phi'}{\Phi}\,\frac{dG}{dr} + \Big(\frac{\omega^2}{\Phi^2} - \frac{j(j+1)}{r^2\Phi}\Big) G = 0 \tag{5.12}$$

or

$$\frac{d^2G}{dr^2} + \frac{M}{r(r-M)}\,\frac{dG}{dr} + \Big(\frac{\omega^2 r^2}{(r-M)^2} - \frac{j(j+1)}{r(r-M)}\Big) G = 0. \tag{5.13}$$

It is convenient to apply the variable $x = r/M$, then we have

$$\frac{d^2G}{dx^2} + \Big(\frac{1}{x-1} - \frac{1}{x}\Big)\,\frac{dG}{dx}$$
$$+\Big(M^2\omega^2 + \frac{j(j+1)}{x} + \frac{2M^2\omega^2 - j(j+1)}{x-1} + \frac{M^2\omega^2}{(x-1)^2}\Big) G = 0. \tag{5.14}$$

With the use of the substitution $G = (x-1)^{\alpha}\, x^{\beta} e^{\gamma x} g(x)$, from (5.14) we get

$$\frac{d^2g}{dx^2} + \Big[\frac{1+2\alpha}{x-1} - \frac{1-2\beta}{x} + 2\gamma\Big]\,\frac{dg}{dx}$$
$$+\Big[M^2\omega^2 + \gamma^2 + \frac{M^2\omega^2+\alpha^2}{(x-1)^2} + \frac{\beta(\beta-2)}{x^2} + \frac{j(j+1)+\alpha-\beta-\gamma-2\alpha\beta+2\beta\gamma}{x}$$
$$+\frac{2M^2\omega^2 - j(j+1) - \alpha+\beta+\gamma+2\alpha\beta+2\alpha\gamma}{x-1}\Big] g = 0. \tag{5.15}$$

Imposing restrictions on parameters α, β, γ

$$\alpha = \pm iM\omega\,, \qquad \beta = 0,\ 2\,, \qquad \gamma = \pm iM\omega, \tag{5.16}$$

we simplify eq. (5.15):

$$\frac{d^2g}{dx^2} + \Big[\frac{1+2\alpha}{x-1} - \frac{1-2\beta}{x} + 2\gamma\Big]\,\frac{dg}{dx} + \Big[\frac{j(j+1)+\alpha-\beta-\gamma-2\alpha\beta+2\beta\gamma}{x}$$
$$+\frac{2M^2\omega^2 - j(j+1) - \alpha+\beta+\gamma+2\alpha\beta+2\alpha\gamma}{x-1}\Big] g = 0, \tag{5.17}$$

which is identified as the confluent Heun equation

$$\frac{d^2Z}{dz^2} + \Big[A + \frac{1+B}{z} + \frac{1+C}{z-1}\Big]\,\frac{dZ}{dz} +$$
$$+\Big[\frac{1}{2}\,\frac{A-B-C+AB-BC-2F}{z} + +\frac{1}{2}\,\frac{A+B+C+AC+BC+2D+2F}{z-1}\Big] f = 0$$

with parameters

$$A = 2\gamma\,, \quad B = 2\beta - 2\,, \quad C = 2\alpha\,, \quad D = 2M^2\omega^2\,, \quad F = 1 - j(j+1)\,. \tag{5.18}$$

5.3 Duffin–Kemmer formalism

Matrix 10-dimensional Duffin–Kemmer equation in Schwarzschild space-time takes the form (I_6 stands for the projective operator on tensor subspace)

$$\Big[i\,\beta^0 \partial_t \;+\; i\,\Phi\,(\beta^3 \partial_r \;+\; \frac{1}{r}\,(\beta^1 j^{31} \;+\; \beta^2 j^{32})$$

$$+\frac{\Phi'}{2\Phi}\,\beta^0 J^{03}) \;+\; \frac{\sqrt{\Phi}}{r}\,\Sigma^{\kappa}_{\theta,\phi} \;-I_6\,\sqrt{\Phi}\;\Big]\;\Phi(x) = 0, \tag{5.19}$$

$$\Sigma_{\theta,\phi} = \; i\,\beta^1 \partial_\theta \;+\; \beta^2\,\frac{i\partial_\phi \;+\; i\,j^{12}\cos\theta}{\sin\theta}.$$

We use the following substitution for solutions in the form of spherical waves

$$\Phi_{\omega jm}(x) = e^{-i\omega t}\,\Big[\; f_1(r) D_0, f_2(r) D_{-1}, f_3(r) D_0, f_4(r) D_{+1},$$

$$f_5(r)\; D_{-1}, f_6(r)\; D_0,\; f_7(r)\; D_{+1}, f_8(r)\; D_{-1},\; f_9(r)\; D_0,\; f_{10}(r)\; D_{+1}\Big]. \tag{5.20}$$

After separating the variables, we get ten radial equations (where $\nu = \sqrt{j(j+1)/2}$)

$$\begin{gathered}
-\Phi(\frac{d}{dr} + \frac{2}{r}) f_6 - \nu \frac{\sqrt{\Phi}}{r} (f_5 + f_7) = 0,\\
i\omega f_5 + i\Phi(\frac{d}{dr} + \frac{1}{r} + \frac{\Phi'}{2\Phi}) f_8 + i\nu \frac{\sqrt{\Phi}}{r}\; f_9 = 0,\\
i\omega f_6 + i\nu \frac{\sqrt{\Phi}}{r}\; (-f_8 + f_{10}) = 0,\\
i\omega f_7 - i\Phi(\frac{d}{dr} + \frac{1}{r} + \frac{\Phi'}{2\Phi}) f_{10} - i\nu \frac{\sqrt{\Phi}}{r} f_9 = 0,\\
-i\omega f_2 + \nu \frac{\sqrt{\Phi}}{r} f_1 - \sqrt{\Phi} f_5 = 0,\\
-i\omega f_3 - \Phi(\frac{d}{dr} + \frac{\Phi'}{2\Phi}) f_1 - \sqrt{\Phi} f_6 = 0,\\
-i\omega f_4 + \nu \frac{\sqrt{\Phi}}{r}\; f_1 \;\; \sqrt{\Phi} f_7 - 0,\\
-i\Phi(\frac{d}{dr} + \frac{1}{r})\; f_2 - i\nu \frac{\sqrt{\Phi}}{r} f_3 - \sqrt{\Phi} f_8 = 0,\\
i\nu \frac{\sqrt{\Phi}}{r} (f_2 - f_4) - \sqrt{\Phi}\; f_9 = 0,\\
i\Phi(\frac{d}{dr} + \frac{1}{r}) f_4 + i\nu \frac{\sqrt{\Phi}}{r} f_3 - \sqrt{\Phi} f_{10} = 0.
\end{gathered} \tag{5.21}$$

Let us additionally diagonalise the space reflection operator

$$\hat{P}^{cycl.}_{sph.} = \left| \begin{array}{cccccccccc} 1 & 0 & 0 & 0 & 0 & 0 & 0 & 0 & 0 & 0 \\ 0 & 0 & 0 & 1 & 0 & 0 & 0 & 0 & 0 & 0 \\ 0 & 0 & 1 & 0 & 0 & 0 & 0 & 0 & 0 & 0 \\ 0 & 1 & 0 & 0 & 0 & 0 & 0 & 0 & 0 & 0 \\ 0 & 0 & 0 & 0 & 0 & 0 & 1 & 0 & 0 & 0 \\ 0 & 0 & 0 & 0 & 0 & 1 & 0 & 0 & 0 & 0 \\ 0 & 0 & 0 & 0 & 1 & 0 & 0 & 0 & 0 & 0 \\ 0 & 0 & 0 & 0 & 0 & 0 & 0 & 0 & 0 & -1 \\ 0 & 0 & 0 & 0 & 0 & 0 & 0 & 0 & -1 & 0 \\ 0 & 0 & 0 & 0 & 0 & 0 & 0 & -1 & 0 & 0 \end{array} \right| \otimes \hat{P},$$

from the eigenvalue equation $\hat{P}^{cycl.}_{sph.} \ \Phi_{jm} = P \ \Phi_{jm}$, we get two sets of restrictions

$P = (-1)^{j+1}$,

$$f_1 = f_3 = f_6 = 0, \ f_4 = -f_2, \ f_7 = -f_5, \ f_{10} = +f_8; \tag{5.22}$$

$P = (-1)^{j}$,

$$f_9 = 0, \ f_4 = +f_2, \ f_7 = +f_5, \ f_{10} = -f_8. \tag{5.23}$$

Allowing for eqs. (5.22) and (5.23), we get two more simple subsystems. The first is

$P = (-1)^{j+1}$,

$$\begin{gathered} i\omega \ f_5 \ + \ i\Phi(\frac{d}{dr} + \frac{1}{r} \ + \ \frac{\Phi'}{2\Phi}) \ f_8 \ + \ i\nu \frac{\sqrt{\Phi}}{r} \ f_9 = 0, \qquad -i\omega \ f_2 \ - \sqrt{\Phi} \ f_5 = 0, \\ -i\Phi \ (\frac{d}{dr} + \frac{1}{r}) \ f_2 \ - \ \sqrt{\Phi} \ f_8 = 0, \qquad i2\nu \frac{\sqrt{\Phi}}{r} \ f_2 \ - \ \sqrt{\Phi} \ f_9 = 0. \end{gathered} \tag{5.24}$$

Whence it follows a 2nd-order equation for the primary variable f_2:

$$\frac{d^2 f_2}{dr^2} + \Big(\frac{\Phi'}{\Phi} + \frac{2}{r}\Big) \frac{df_2}{dr} + \Big(\frac{\omega^2}{\Phi^2} - \frac{2\nu^2}{r^2\Phi} + \frac{\Phi'}{\Phi}\frac{1}{r}\Big) f_2 = 0,$$

it yields (let $f_2 = r^{-1} F_2$)

$$\frac{d^2 F_2}{dr^2} + \frac{\Phi'}{\Phi}\frac{dF_2}{dr} + \Big(\frac{\omega^2}{\Phi^2} - \frac{j(j+1)}{r^2\Phi}\Big) F_2 = 0$$

which coincides with eq. (5.12) for G.

In the case $P = (-1)^j$, we get six equations

$$\begin{gathered} \Phi \ (\frac{d}{dr} + \frac{2}{r}) F_6 + \frac{2\nu}{r} F_5 = 0, \quad i\omega F_5 + i\Phi(\frac{d}{dr} + \frac{1}{r}) F_8 = 0, \\ i\omega F_6 - i\frac{2\nu}{r} F_8 = 0, \quad -i\omega F_2 + \frac{\nu}{r} \ F_1 - F_5 = 0, \\ i\omega F_3 + \Phi \frac{d}{dr} F_1 + \Phi F_6 = 0, \quad i\Phi(\frac{d}{dr} + \frac{1}{r}) F_2 + i\frac{\nu}{r} F_3 + F_8 = 0, \end{gathered} \tag{5.25}$$

where

$$\begin{gathered} F_1 = \sqrt{\Phi} \ f_1, \ F_2 = f_2, \ F_3 = \sqrt{\Phi} \ f_3, \\ F_5 = \sqrt{\Phi} \ f_5, \ F_6 = f_6, \ F_8 = \sqrt{\Phi} \ f_8. \end{gathered} \tag{5.26}$$

5.4 Relation between two formalisms

We start with the following identities

$$E_{(1)} = F_{(0)(1)}, \quad E_{(2)} = F_{(0)(2)}, \quad E_{(3)} = F_{(0)(3)},$$
$$B_{(1)} = -F_{(2)(3)}, \quad B_{(2)} = -F_{(3)(1)}, \quad B_{(3)} = -F_{(1)(2)} \tag{5.27}$$

and the known expressions for complex 3-vector and tensor:

$$\Psi = e^{-i\omega t} \left| \begin{array}{l} 0 \\ \varphi_1 D_{-1} \\ \varphi_2 D_0 \\ \varphi_3 D_{+1} \end{array} \right|, \qquad \Phi = e^{-i\omega t} \, [f_1 D_0, f_2 D_{-1}, f_3 D_0, f_4 D_{+1},$$

$$f_5 D_{-1}, f_6 D_0, f_7 D_{+1}, f_8 D_{-1}, f_9 D_0, f_{10} D_{+1}], \tag{5.28}$$

thus we find three relations

$$D_0 \varphi_2(r) = E_{(2)} + iB_{(2)} = F_{(0)(2)} - iF_{(3)(1)} = f_6(r)\, D_0 + i f_9(r)\, D_0,$$

$$D_{-1}\varphi_1(r) = E_{(1)} + iB_{(1)} = F_{(0)(1)} - iF_{(2)(3)} = f_5(r)\, D_{-1} - i\, f_8(r)\, D_{-1},$$

$$D_{+1}\varphi_3(r = E_{(3)} + iB_{(3)} = F_{03} - iF_{12} = f_7(r)\, D_{+1} - i f_{10}(r)\, D_{+1},$$

whence it follows

$$\varphi_2 = f_6 + i\, f_9, \qquad \varphi_1 = f_5 - i\, f_8, \qquad \varphi_3 = f_7 - i\, f_{10}. \tag{5.29}$$

Taking into account the spatial parity restrictions (5.22) and (5.23), we obtain

$$P = (-1)^{j+1}, \quad f_6 = 0, \quad f_7 = -f_5, \; f_{10} = +f_8 \quad \Longrightarrow$$
$$\varphi_2 = +i\, f_9, \quad \varphi_1 = f_5 - i\, f_8, \quad \varphi_3 = -f_5 - i\, f_8; \tag{5.30}$$

$$P = (-1)^j, \quad f_9 = 0, \quad f_7 = +f_5, \; f_{10} = -f_8 \quad \Longrightarrow$$
$$\varphi_2 = f_6, \quad \varphi_1 = f_5 - i\, f_8, \quad \varphi_3 = f_5 + i\, f_8; \tag{5.31}$$

the inverse relations are

$$P = (-1)^{j+1}, \quad f_9 = -i\varphi_2, \quad f_8 = \frac{i}{2}(\varphi_1 + \varphi_3), \quad f_5 = \frac{1}{2}(\varphi_1 - \varphi_3); \tag{5.32}$$

$$P = (-1)^j, \quad F_6 = \varphi_2, \quad F_5 = \frac{\sqrt{\Phi}}{2}(\varphi_1 + \varphi_3), \quad F_8 = i\frac{\sqrt{\Phi}}{2}(\varphi_1 - \varphi_3). \tag{5.33}$$

5.5 Studying equations for states with $P = (-1)^j$

First consider the case $P = (-1)^j$. In six equations (5.25), we are to take into account (5.33). Three first equations (containing F_5, F_6, and F_8)

$$i\omega F_6 - i\frac{2\nu}{r}F_8 = 0, \quad \Phi(\frac{d}{dr} + \frac{2}{r})F_6 + \frac{2\nu}{r}F_5 = 0, \quad i\omega F_5 + i\Phi(\frac{d}{dr} + \frac{1}{r})F_8 = 0$$

in the new variables read

$$
\begin{aligned}
i\omega\,\varphi_2 \;-\; i\frac{\nu}{r}\, i\sqrt{\Phi}\,(\varphi_1-\varphi_3)=0,\\
\Phi\,(\frac{d}{dr}+\frac{2}{r})\varphi_2+\frac{\nu}{r}\,\sqrt{\Phi}\,(\varphi_1+\varphi_3)=0,\\
i\omega\,\sqrt{\Phi}\,(\varphi_1+\varphi_3)+i\Phi\,(\frac{d}{dr}+\frac{1}{r})\,i\sqrt{\Phi}\,(\varphi_1-\varphi_3)=0.
\end{aligned}
\tag{5.34}
$$

Equations (5.34) may be compared with (5.8)

$$
\begin{aligned}
3)\quad -\frac{\omega}{\sqrt{\Phi}}\varphi_2+\frac{i\nu}{r}\,(\varphi_1-\varphi_3)=0,\\
1)\quad \sqrt{\Phi}(\frac{d}{dr}+\frac{2}{r})\varphi_2+\frac{\nu}{r}(\varphi_1+\varphi_3)=0,\\
2)+4)\quad -\frac{\omega}{\sqrt{\Phi}}(\varphi_1+\varphi_3)-i(\sqrt{\Phi}\frac{d}{dr}+\frac{\sqrt{\Phi}}{r}+\frac{\Phi'}{2\sqrt{\Phi}})(\varphi_1-\varphi_3)=0,\\
2)-4)\quad -\frac{\omega}{\sqrt{\Phi}}(\varphi_1-\varphi_3)-i(\sqrt{\Phi}\frac{d}{dr}+\frac{\sqrt{\Phi}}{r}+\frac{\Phi'}{2\sqrt{\Phi}})(\varphi_1+\varphi_3)-\frac{2i\nu}{r}\varphi_2=0.
\end{aligned}
\tag{5.35}
$$

Three first equations coincide, but the last is the consequence of these three. Recall that φ_1, φ_2, and φ_3 are determined by the primary function G:

$$
\varphi_2=-\frac{1}{i\omega}\frac{\nu}{r^2}\,G(r),\quad \varphi_1+\varphi_3=\frac{1}{i\omega}\frac{\sqrt{\Phi}}{r}\frac{d}{dr}G,\quad \varphi_1-\varphi_3=\frac{1}{r\sqrt{\Phi}}\,G(r),
\tag{5.36}
$$

where G obeys the equation

$$
\frac{d^2G}{dr^2}+\frac{M}{r(r-M)}\frac{dG}{dr}+\Big(\frac{\omega^2r^2}{(r-M)^2}-\frac{j(j+1)}{r(r-M)}\Big)G=0.
\tag{5.37}
$$

Remaining three equation in eq. (5.25) relate the radial functions F_1, F_2, and F_3 of the electromagnetic 4-vector with the radial functions F_5, F_6, and F_8 of the electromagnetic tensor by equations

$$
\begin{aligned}
i\omega F_3+\Phi\frac{d}{dr}F_1+\Phi\;F_6=0,\quad -i\omega F_2+\frac{\nu}{r}F_1-F_5=0,\\
i\Phi(\frac{d}{dr}+\frac{1}{r})F_2+i\frac{\nu}{r}F_3+F_8=0;
\end{aligned}
\tag{5.38}
$$

whence with the help of eq. (5.33) we derive

$$
\begin{aligned}
i\omega F_3+\Phi\frac{d}{dr}F_1+\Phi\varphi_2=0,\quad -i\omega F_2+\frac{\nu}{r}F_1-\frac{\sqrt{\Phi}}{2}(\varphi_1+\varphi_3)=0,\\
i\Phi(\frac{d}{dr}+\frac{1}{r})F_2+i\frac{\nu}{r}F_3+i\frac{\sqrt{\Phi}}{2}(\varphi_1-\varphi_3)=0.
\end{aligned}
\tag{5.39}
$$

In turn, from eq. (5.39), with the help of eq. (5.36), we obtain

$$
\begin{aligned}
\frac{i\omega}{\Phi}F_3+\frac{d}{dr}F_1-\frac{1}{i\omega}\frac{\nu}{r^2}G(r)=0,\\
-i\omega F_2+\frac{\nu}{r}F_1-\frac{1}{i\omega}\frac{\Phi}{\sqrt{2}r}\frac{d}{dr}G(r)=0,\\
\Phi(\frac{d}{dr}+\frac{1}{r})F_2+\frac{\nu}{r}F_3+\frac{1}{\sqrt{2}r}G(r)=0.
\end{aligned}
\tag{5.40}
$$

It should be emphasised that all terms with F_1 cancel each other. This means that the function F_1 may be arbitrary. This fact is the result of the known gauge symmetry in the electromagnetic 4-vector at the fixed electromagnetic tensor. The system (5.40) is equivalent to

$$F_3 = \frac{i\Phi}{\omega}\frac{dF_1}{dr} - \frac{\Phi\,\nu}{\omega^2 r^2}G(r)\,, \quad F_2 = -\frac{i\nu}{\omega r}F_1 + \frac{\Phi}{\sqrt{2}\,\omega^2 r}\frac{dG(r)}{dr}\,,$$
$$\frac{d^2G(r)}{dr^2} + \frac{\Phi'}{\Phi}\frac{dG(r)}{dr} - \frac{\sqrt{2}\,\nu^2}{r^2\,\Phi}G(r) + \frac{\omega^2}{\Phi^2}G(r) = 0\,. \qquad (5.41)$$

5.6 Studying the case $P = (-1)^{j+1}$

In the system (5.24), let us take into account (5.32):

$$P = (-1)^{j+1}, \quad -i\omega f_2 - \sqrt{\Phi} f_5 = 0,$$
$$i\omega f_5 + i\Phi(\frac{d}{dr} + \frac{1}{r} + \frac{\Phi'}{2\Phi})f_8 + i\nu\frac{\sqrt{\Phi}}{r}f_9 = 0,$$
$$-i\Phi(\frac{d}{dr} + \frac{1}{r})f_2 - \sqrt{\Phi}f_8 = 0, \quad i2\nu\frac{\sqrt{\Phi}}{r}f_2 - \sqrt{\Phi}f_9 = 0,$$

which yields

$$\frac{\omega}{\sqrt{\Phi}}\varphi_2 + \frac{i\nu}{r}(\varphi_1 - \varphi_3) = 0,$$
$$-\frac{\omega}{\sqrt{\Phi}}(\varphi_1 - \varphi_3) - i\sqrt{\Phi}(\frac{d}{dr} + \frac{1}{r} + \frac{\Phi'}{2\Phi})(\varphi_1 + \varphi_3) + \frac{2i\,\nu}{r}\varphi_2 = 0, \qquad (5.42)$$
$$-\sqrt{\Phi}(\frac{d}{dr} + \frac{2}{r})\varphi_2 + \frac{\nu}{r}(\varphi_1 + \varphi_3) = 0, \quad f_2 = -\frac{r}{2\,\nu}\varphi_2\,.$$

Note that the variable F_2 referring to the electromagnetic 4-vector is determined uniquely by the components of the electromagnetic tensor. This means that this class of solutions does not contain any gauge degrees of freedom.

Let us compare eq. (5.42) with the first three equations in eq. (5.8) (the last equation 2)+4) in eq. (5.8) turns out to be identity)

$$3) \quad -\frac{\omega}{\sqrt{\Phi}}\varphi_2 + \frac{i\nu}{r}(\varphi_1 - \varphi_3) = 0, \ 1) \ \sqrt{\Phi}(\frac{d}{dr} + \frac{2}{r})\varphi_2 + \frac{\nu}{r}(\varphi_1 + \varphi_3) = 0,$$
$$2) - 4), \ -\frac{\omega}{\sqrt{\Phi}}(\varphi_1 - \varphi_3) - i(\sqrt{\Phi}\frac{d}{dr} + \frac{\sqrt{\Phi}}{r} + \frac{\Phi'}{2\sqrt{\Phi}})(\varphi_1 + \varphi_3) - \frac{2i\nu}{r}\varphi_2 = 0. \qquad (5.43)$$

The systems (5.42) and (5.43) differ only in notation.

In eq. (5.42), let us introduce new variables

$$f = \varphi_1 + \varphi_3, \qquad g = \varphi_1 - \varphi_3,$$

then we get the system

$$\varphi_2 = -\frac{i\nu}{\omega}\frac{\sqrt{\Phi}}{r}g, \quad i\sqrt{\Phi}(\frac{d}{dr} + \frac{2}{r})\frac{\sqrt{\Phi}}{r}g + \frac{\omega}{r}f = 0,$$
$$-\frac{\omega^2}{\Phi}g - i\omega(\frac{d}{dr} + \frac{1}{r} + \frac{\Phi'}{2\Phi})\,f + \frac{2\nu^2}{r^2}\,g = 0. \qquad (5.44)$$

The system (5.44) is simplified by the substitution

$$g = \frac{1}{r\sqrt{\Phi}}\, G(r), \qquad f = \frac{1}{r\sqrt{\Phi}}\, F(r),$$

as a result we obtain

$$\varphi_2 = -\frac{i\nu}{\omega r^2}\, G, \quad \omega F = -i\Phi\, \frac{d}{dr} G, \quad -\frac{\omega^2}{\Phi} G - i\frac{d}{dr}\omega F + \frac{2\nu^2}{r^2} G = 0\,. \tag{5.45}$$

So, the problem reduces to the equation

$$\frac{d^2G}{dr^2} + \frac{M}{r(r-M)}\, \frac{dG}{dr} + \Big(\frac{\omega^2 r^2}{(r-M)^2} - \frac{j(j+1)}{r(r-M)} \Big) G = 0. \tag{5.46}$$

All concomitant functions are determined from eq. (5.46) as follows
$P = (-1)^{j+1}$,

$$\varphi_2 = -i\frac{\nu}{\omega r^2} G, \qquad \varphi_1 + \varphi_3 = -i\frac{\sqrt{\Phi}}{\omega r}\frac{d}{dr} G, \quad \varphi_1 - \varphi_3 = \frac{1}{r\sqrt{\Phi}} G(r). \tag{5.47}$$

In turn, for solutions with opposite parity we have
$P = (-1)^{j}$,

$$\varphi_2 = +i\frac{\nu}{\omega r^2} G, \quad \varphi_1 + \varphi_3 = -i\frac{\sqrt{\Phi}}{\omega r}\frac{d}{dr} G, \quad \varphi_1 - \varphi_3 = \frac{1}{r\sqrt{\Phi}} G(r). \tag{5.48}$$

Relations between two methods to describe the Maxwell field may be expressed as follows

$$\begin{array}{lcccc} \text{Majorana–Oppenheimer} & \Longrightarrow & \varphi_2, & +\varphi_1 & \varphi_3, \\ P = (-1)^j, \ \text{Duffin–Kemmer} & \Longrightarrow & +\varphi_2, & \varphi_1, & \varphi_3, \\ P = (-1)^{j+1}, \ \text{Duffin–Kemmer} & \Longrightarrow & -\varphi_2, & \varphi_1, & \varphi_3. \end{array}$$

5.7 The gauge degrees of freedom

Maxwell equations permit the existence of pure gauge solutions, for which the following initial substitution should be used

$$\Phi_{\omega jm}(x) = e^{-i\omega t}(\, f_1 D_0, f_2 D_{-1}, f_3 D_0, f_4 D_{+1}, 0, 0, 0, 0, 0, 0\,). \tag{5.49}$$

For states with parity $P = (-1)^{j+1}$ we have

$$P = (-1)^{j+1}, \qquad f_1 = f_3 = 0,\ f_4 = -f_2,$$

$$0 = 0, \quad -i\omega f_2 = 0, \quad -i\Phi\, (\frac{d}{dr} + \frac{1}{r})\, f_2 \ = 0, \qquad i2\nu\frac{\sqrt{\Phi}}{r}\, f_2 \ = 0,$$

whence it follows $f_2 = 0$. This means that any pure gauge states do not exist with the parity $P = (-1)^{j+1}$.

For states with the parity $P = (-1)^j$, we have

$$P = (-1)^j, \qquad f_4 = +f_2,$$

$$(F_1 = \sqrt{\Phi}\, f_1,\; F_2 = f_2,\; F_3 = \sqrt{\Phi}\, f_3\,)\,,$$

$$0 = 0, \qquad 0 = 0, \qquad 0 = 0,$$

$$-i\omega F_2 + \frac{\nu}{r} F_1 = 0, \quad i\omega F_3 + \Phi \frac{d}{dr} F_1 = 0, \quad i\Phi(\frac{d}{dr} + \frac{1}{r}) F_2 + i\frac{\nu}{r} F_3 = 0,$$

whence it follows

$$i\omega F_2 = \frac{\nu}{r} F_1, \quad i\omega F_3 = -\Phi \frac{d}{dr} F_1, \quad i\Phi(\frac{d}{dr} + \frac{1}{r}) F_2 + i\frac{\nu}{r} F_3 = 0.$$

Eliminating the variables F_2, F_3, we get an identity of $0 \equiv 0$. Therefore, the function $F_1(r)$ may be arbitrary, and the concomitant functions are defined by relations

$$i\omega F_2 = \frac{\nu}{r} F_1, \quad i\omega F_3 = -\Phi \frac{d}{dr} F_1. \tag{5.50}$$

The last relations may be verified with the use of the Lorentz gauge. In radial form, it reads

$$\frac{-i\omega}{\sqrt{\Phi}}\, f_1 - \sqrt{\Phi}\, (\frac{d}{dr} + \frac{2}{r} + \frac{\Phi'}{2\Phi})\, f_3 - \frac{\nu}{r}\, (f_2 + f_4) = 0. \tag{5.51}$$

This constraint turns to be an identity for states with the parity $P = (-1)^{j+1}$; for states with the parity $P = (-1)^j$, it takes the form

$$\frac{-i\omega}{\sqrt{\Phi}}\, f_1 - \sqrt{\Phi}\, (\frac{d}{dr} + \frac{2}{r} + \frac{\Phi'}{2\Phi})\, f_3 - \frac{2\nu}{r}\, f_2 = 0\,; \tag{5.52}$$

whence, taking into account (5.26), we obtain

$$\frac{-i\omega}{\Phi}\, F_1 - (\frac{d}{dr} + \frac{2}{r})\, F_3 - \frac{2\nu}{r}\, F_2 = 0. \tag{5.53}$$

From this, allowing for (5.50), we derive an equation

$$\Big[\frac{d^2}{dr^2} + (\frac{2}{r} + \frac{\Phi'}{\Phi})\frac{d}{dr} + \frac{\omega^2}{\Phi^2} - \frac{j(j+1)}{\Phi r^2}\Big] F_1 = 0. \tag{5.54}$$

Let us compare this result with the radial form of the wave equation for massless scalar field in Schwarzschild space-time, $\nabla^\alpha \nabla_\alpha \Psi = 0$. This wave equation reads

$$\nabla^\alpha \nabla_\alpha \Psi\Psi = \frac{1}{\sqrt{-g}} \partial_\alpha \sqrt{-g} g^{\alpha\beta} \partial_\beta \Psi$$

$$= \Big(\frac{1}{\Phi}\frac{\partial^2}{\partial t^2} - \frac{1}{r^2}\frac{\partial}{\partial r} r^2 \Phi \frac{\partial}{\partial r} - \frac{1}{\sin\theta}\frac{\partial}{\partial\theta}\sin\theta \frac{1}{r^2}\frac{\partial}{\partial\theta} - \frac{1}{r^2 \sin^2\theta}\frac{\partial^2}{\partial\phi^2}\Big)\Psi = 0;$$

so the radial equation is

$$\Big(-\frac{\omega^2}{\Phi} - \frac{1}{r^2}\frac{\partial}{\partial r} r^2 \Phi \frac{\partial}{\partial r} + \frac{j(j+1)}{r^2}\Big) f(r) = 0,$$

which coincides with eq. (5.54).

5.8 Conclusions

Let us summarise results. It is shown that the generally covariant extended method of Riemann–Silberstein–Majorana–Oppenheimer in electrodynamics, specified in Schwarzschild metrics, after separating the variables, provide us with the possibility to have reduce the problem to a differential equation similar to that arising in case of a scalar filed in the Schwarzschild space-time. This differential equation is recognised as a confluent Heun equation.

We have considered the electromagnetic field on the basis of 10-dimensional Duffin–Kemmer approach, when in addition to six components of the strength tensor, one uses four components of an electromagnetic potential. After separation of the variables, we have arrive at a system of ten radial equations, which were simplified by the use of additional constraints followed from eigenvalue equation for spatial parity operator $\hat{\Pi}\Psi = P\Psi$; the radial system has been divided into two subsystems of four and six equations, respectively. In this second approach, the problem of the electromagnetic field has been reduced to the confluent Heun differential equation as well.

In particular, we have shown explicitly how solutions found in complex form are embedded into matrix 10-dimensional formalism; besides, we determine radial functions that are responsible for gauge degrees of freedom.

Bibliography

[1] C. Misner, K.S. Thorne and J.A. Wheeler. *Gravitation.* W.H. Freeman, San Francisco, 1973.

[2] L.D. Landau and E.M. Lifshitz. *Classical Theory of Fields (Fourth Revised English Edition).* Pergamon, Oxford, 1975.

[3] K. Schwarzschild. Über das Gravitationsfeld eines Massenpunktes nach der Einsteinschen Theorie. *Sitzungsber. Preuss. Akad. Wiss. Berlin, Math. Phys.* 189–196 (1916).

[4] A. Bachelot. Gravitational scattering of electromagnetic field by Schwarzschild black-hole. *Annales de l'institut Henri Poincarée. Physique Theorique*, 54(3): 261–320, 1991.

[5] S. Chandrasekhar. *The Mathematical Theory of Black Holes.* Oxford University Press, Oxford, 1983.

[6] A.A. Bogush, G.G. Krylov, E.M. Ovsiyuk and V.M. Red'kov. Maxwell equations in complex form of Majorana-Oppenheimer, solutions with cylindric symmetry in Riemann S_3 and Lobachevsky H_3 spaces. *Ricerche di matematica*, 59(1): 59–96, 2010.

[7] A.A. Bogush, V.V. Kisel, N.G. Tokarevskaya and V.M. Red'kov. Duffin-Kemmer-Petiau formalism reexamined: non-relativistic approximation for spin 0 and spin 1 particles in a Riemannian space-time. *Annales de la Fondation Louis de Broglie*, 32(2-3): 355–381, 2007.

[8] D.A. Varshalovich, A.N. Moskalev and V.K. Hersonskiy. *Quantum Theory of Angular Moment.* Nauka, Leningrad, 1975.

[9] A.A. Bogush, V.S. Otchik and V.M. Red'kov. *General Covariant Duffin–Kemmer Formalism and Spherical Waves for a Vector Field in de Sitter Space.* Preprint 426, Institute of Physics, NANB, Minsk, 1986.

[10] V.M. Red'kov and E.A. Tolkachev. The Lorentz group, non-commutative space-time, and nonlinear electrodynamics in Majorana–Oppenheimer formalism. *Proceedings of the 14th International School & Conference "Foundation & Advances in Nonlinear Science"*, pages 1–19. Minsk, 2008.

[11] V.M. Red'kov. *Fields in Riemannian Space and the Lorentz Group.* Publishing House "Belarusian Science", Minsk, 2009.

[12] V.M. Red'kov, A.A. Bogush, N.G. Tokarevskaya and George J. Spix. Majorana-Oppenheimer approach to Maxwell electrodynamics in Riemannian space-time. *Proceedings of the 14th International School & Conference "Foundation & Advances in Nonlinear Science"*, pages 20–49. Minsk, 2008.

[13] A.A. Bogush, G.G. Krylov, E.M. Ovsiyuk and V.M. Red'kov. Maxwell equations in complex form of Majorana-Oppenheimer, solutions with cylindric symmetry in Riemann S_3 and Lobachevsky H_3 spaces. *Ricerche di matematica*, 59(1): 59–96, 2010.

[14] V. Red'kov and E. Tolkachev. The Lorentz Group, Noncommutative Space-Time, and Nonlinear Electrodynamics in Majorana-Oppenheimer Formalism. *Nonlinear Phenomena in Complex Systems*, 13(3): 249–266, 2010.

[15] V.M. Red'kov, N.G. Tokarevskaya and George. J. Spix. Majorana-Oppenheimer approach to Maxwell electrodynamics. Part I. Minkowski space. *Advances in Applied Clifford Algebras*, 22(4): 1129–1149, 2012.

[16] V.M. Red'kov, N.G. Tokarevskaya and George. J. Spix. Majorana-Oppenheimer approach to Maxwell electrodynamics. Part II. Curved Riemannian space. *Advances in Applied Clifford Algebras*, 23(1): 165–178, 2013.

[17] E.M. Ovsiyuk, V.M. Red'kov and N.G. Tokarevskaya. Majorana-Oppenheimer approach to Maxwell electrodynamics. Part III. Electromagnetic spherical waves in spaces of constant curvature. *Advances in Applied Clifford Algebras*, 23(1): 153–163, 2013.

[18] E. Ovsiyuk, O. Florea, A. Chichurin and V. Red'kov. Electromagnetic Field in Schwarzschild Black Hole Background. Analytical Treatment and Numerical Simulation. *Computer Algebra Systems in Teaching and Research*, IV(1): 204–214, 2013.

[19] E.M. Ovsiyuk, V.V. Kisel and V.M. Red'kov. *Maxwell Electrodynamics and Boson Fields in Spaces of Constant Curvature.* Nova Science Publishers Inc., New York, 2014.

[20] O.V. Veko, N.D. Vlasii, E.M. Ovsiyuk, V.M. Red'kov and Yu.A. Sitenko. Electromagnetic Field on de Sitter Expanding Universe: Majorana–Oppenheimer Formalism, Exact Solutions in non-Static Coordinates. *Nonlinear Phenomena in Complex Systems*, 17(1): 17–39, 2014.

[21] O.V. Veko and V.M. Red'kov. Peculiarities of squaring method applied to construct solutions of the Dirac, Majorana, and Weyl equations. *Nonlinear Phenomena in Complex Systems*, 18(1): 44–62, 2015.

[22] E.M. Ovsiyuk, O.V. Veko, Yu.A. Rusak, A.V. Chichurin and V.M. Red'kov. To Analysis of the Dirac and Majorana Particle Solutions in Schwarzschild Field. *Nonlinear Phenomena in Complex Systems*, 20(1): 56–72, 2016.

[23] E.M. Ovsiyuk, O.V. Veko, Yu.A. Rusak, A.V. Chichurin and V.M. Red'kov. To Analysis of the Dirac and Majorana Particle Solutions in Schwarzschild Field. *Nonlinear Phenomena in Complex Systems*, 20(1): 56–72, 2017.

[24] A.V. Chichurin, E.M. Ovsiyuk and V.M. Red'kov. The tunneling effect through Schwarzschild barrier for spin 1/2 particle, analytical and numerical study. *Romanian Reports in Physics*, 76 (3): 110, 2024.

6

Particle with polarisability in the Coulomb field

Methods for solving the differential equation describing the wave functions of a polarisable particle in the Coulomb potential are discussed. Relations between the coefficients under which the general solutions of this equation can be found in analytical form are detailed. For the case of zero polarisability, the general solution to this equation in terms of special functions is obtained; for the first values of the parameter j, plots of the corresponding solutions are presented. For nonzero polarisability and certain specially chosen values of the parameters, solutions possessing the required physical properties are constructed with the use of numerical methods and functional objects of the type DifferentialRoot. Instructions in Mathematica are presented which permit to apply elaborated methods in studying other problems in physics and mathematics.

The chapter is based on [1–15].

6.1 Introduction: starting equation

There is known the generalised Klein–Fock Gordon wave equation for a scalar charged particle with additional electromagnetic characteristic, polarisability [1]. In the presence of the external Coulomb field, this equation after separation of the variables, gives the following radial equation

$$\Big[\frac{d^2}{dR^2} + \frac{2}{R}\frac{d}{dR} + \frac{1}{\hbar^2 c^2}\Big(E + \frac{e^2}{R}\Big)^2 - \frac{m^2c^2}{\hbar^2}$$
$$-\frac{j(j+1)}{R^2} + \sigma(\frac{e^2}{\hbar c})^2 \frac{1}{R^4}\frac{\hbar}{m^2c^2}\Big] f(R) = 0, \tag{6.1}$$

where the quantum number of the angular momentum takes the values $j = 0, 1, 2, 3, ...$; the dimensionless parameter σ is responsible for additional electromagnetic characteristics of the particle associated with its polarisability. In dimensionless units:

$$\frac{E}{mc^2} = \epsilon, \qquad \frac{\hbar}{mc} = \lambda, \qquad \frac{e}{\hbar c} = \frac{1}{137} \equiv \alpha, \qquad r = \frac{R}{\lambda}$$

eq. (6.1) reads

$$\left(\frac{d^2}{dr^2} + \frac{2}{r}\frac{d}{dr} + (\epsilon + \frac{\alpha}{r})^2 - 1 - \frac{j(j+1)}{r^2} + \sigma\frac{\alpha^2}{r^4}\right) f = 0 \,. \tag{6.2}$$

This equation is related to the known class of Heun type. In the present chapter, we will perform analytical and numerical study of the problem.

DOI: 10.1201/9781003472377-6

6.2 Formal exact solutions

Let us re-write eq. (6.2) as an equation with polynomial coefficients

$$r^4\, f'' + 2r^3\, f' + \left[\alpha^2\sigma + r^2\left(\alpha^2 - j^2 - j\right) + r^4\left(\epsilon^2 - 1\right) + 2\alpha r^3\epsilon\right] f = 0. \tag{6.3}$$

Note that standard requirements for solutions describing bound quantum-mechanical states are

$$f(r) \to 0 \;\text{ at }\; r \to 0\,, \;\text{ and }\; f(r) \to 0 \;\text{ at }\; r \to \infty\,. \tag{6.4}$$

Let us construct solutions of eq. (6.3) as functions obeying the additional 1st-order constraint

$$f'(r) = \frac{\gamma_2 r^2 + \gamma_1 r + \gamma_0}{r^2}\, f(r), \tag{6.5}$$

where γ_2, γ_1, and γ_0 are some numerical coefficients. After simple calculations, we arrive at the set of constraints on parameters

$$\begin{aligned} &\alpha^2\sigma + \gamma_0^2 = 0,\; 2\gamma_0\gamma_1 = 0,\\ &\alpha^2 + \gamma_1^2 + \gamma_1 + 2\gamma_0\gamma_2 - j^2 - j = 0,\\ &2\alpha\epsilon + 2\gamma_1\gamma_2 + 2\gamma_2 = 0,\\ &\gamma_2^2 + \epsilon^2 - 1 = 0. \end{aligned} \tag{6.6}$$

In general, the system (6.6) permits six different solutions. When $\sigma \neq 0$ only two solutions are possible:

$$\begin{aligned} &\gamma_2 = -\frac{\alpha}{\sqrt{\alpha^2+1}}, \qquad \gamma_1 = 0, \qquad \gamma_0 = \frac{\sqrt{\alpha^2+1}\,(\alpha^2 - j^2 - j)}{2\alpha},\\ &\sigma = \frac{\left(\alpha^2+1\right)\left(\alpha^2 - j^2 - j\right)^2}{4\alpha^4}, \qquad \epsilon = \frac{1}{\sqrt{\alpha^2+1}}; \end{aligned} \tag{6.7}$$

$$\begin{aligned} &\gamma_2 = \frac{\alpha}{\sqrt{\alpha^2+1}}, \qquad \gamma_1 = 0, \qquad \gamma_0 = \frac{\sqrt{\alpha^2+1}\,(-\alpha^2 + j^2 + j)}{2\alpha},\\ &\sigma = -\frac{\left(\alpha^2+1\right)\left(-\alpha^2 + j^2 + j\right)^2}{4\alpha^4}, \qquad \epsilon = -\frac{1}{\sqrt{\alpha^2+1}}. \end{aligned} \tag{6.8}$$

Because the energy values do not depend on the number j and the parameter σ must have a different value depending on j, we can conclude that such solutions are of small physical interest in the context of quantum mechanics. Nevertheless, let us complete this line of consideration. In the case (6.7), eq. (6.5) results in

$$f = C_1 f_1(r), \qquad f_1(r) = \exp\frac{j(1+j) + (j + j^2 - 1 - 2r^2)\alpha^2 - \alpha^4}{2r\alpha\sqrt{1+\alpha^2}}, \tag{6.9}$$

where C_1 stands for an arbitrary constant. Relation (6.9) provides us with the one-parametric set of solutions of eq. (6.3)

$$\begin{aligned} r^4 f'' + 2r^3 f' + \Big[&\frac{\left(\alpha^2+1\right)\left(\alpha^2 - j^2 - j\right)^2}{4\alpha^2} + r^2\left(\alpha^2 - j^2 - j\right)\\ &+\Big(\frac{1}{\alpha^2+1} - 1\Big) r^4 + \frac{2\alpha r^3}{\sqrt{\alpha^2+1}}\Big] f = 0. \end{aligned} \tag{6.10}$$

A second particular solution can be easily found (see, for instance, in [2])

$$f_2(r) = \exp\left(\frac{j(j+1) - \alpha^4 + \alpha^2\left(j^2 + j - 2r^2 - 1\right)}{2\alpha\sqrt{\alpha^2+1}r}\right)$$
$$\times \int \frac{\exp\left(-\frac{-\alpha^4+\alpha^2\left(j^2+j-2r^2-1\right)+j(j+1)}{\alpha\sqrt{\alpha^2+1}r}\right)}{r^2} dr. \quad (6.11)$$

General solution of eq. (6.10) reads

$$f(r) = C_1 f_1(r) + C_2 f_2(r), \quad (6.12)$$

where C_1 and C_2 are arbitrary constants. To investigate the behaviour of the solution at the boundary points, we plot three particular solutions determined by eq. (6.12) (see Fig. 6.1).

6.3 Zero polarisability, numerical simulation

Let us consider the case when $\sigma = 0$. The main equation takes the form

$$r^2 f''(r) + 2rf'(r) + f(r)\left[\left(\alpha^2 - j^2 - j\right) + r^2\left(\epsilon^2 - 1\right) + 2\alpha r\epsilon\right] = 0. \quad (6.13)$$

We will specify a solutions which tend to zero when $r \to 0$; it can be expressed through confluent hypergeometric functions as follows:

$$f(r) = Ce^{-r\sqrt{1-\epsilon^2}} r^{\sqrt{(j+1/2)^2-\alpha^2}-1/2}$$
$$\times U\left(\frac{1}{2} - \frac{\alpha\epsilon}{\sqrt{1-\epsilon^2}} + \sqrt{(j+1/2)^2 - \alpha^2}, 2\sqrt{(j+1/2)^2 - \alpha^2} + 1, 2r\sqrt{1-\epsilon^2}\right).$$

To have polynomial solutions, we must require

$$\frac{1}{2} - \frac{\alpha\epsilon}{\sqrt{1-\epsilon^2}} + \sqrt{(j+1/2)^2 - \alpha^2} = -n, \; n = 0, 1, 2, \ldots ; \quad (6.14)$$

this leads to the known energy spectrum

$$\epsilon = \frac{1}{\sqrt{1 + \alpha^2/N^2}}, \qquad N = \frac{1}{2} + n + \sqrt{(j+1/2)^2 - \alpha^2}.$$

If $n = 0$, then from eq. (6.14) it follows

$$\epsilon = \frac{\sqrt{2j^4 + 4j^3 - 2\left(\alpha^2 - 1\right)j^2 - 2\alpha^2 j + \alpha^2\left(\sqrt{(2j+1)^2 - 4\alpha^2} + 1\right)}}{\sqrt{2}\sqrt{\alpha^2 + j^2(j+1)^2}}; \quad (6.15)$$

at $j = 1, \ldots, 10$, we obtain ten energy levels

$$j = 1 : \epsilon = 0.9999933400, \; j = 2 : \epsilon = 0.9999970400, \; j = 3 : \epsilon = 0.9999983350,$$

$$j = 4 : \epsilon = 0.9999989344, \; j = 5 : \epsilon = 0.9999992600, \; j = 6 : \epsilon = 0.9999994563,$$

$$j=7:\epsilon=0.9999995838,\ \ j=8:\epsilon=0.9999996711,\ \ j=9:\epsilon=0.9999997336,$$
$$j=10:\epsilon=0.9999997798.$$

Note that all values for ϵ belong to vicinity of 1 (from the left) and they become closer to 1 as j increases.

Taking into account the value ϵ $(n=0,\ j=1)$, we get expressions for two asymptotical terms

$$f(r)=e^{-\sqrt{\frac{685-\sqrt{168917}}{20571098}}\,r}\,r^{\frac{\sqrt{168917}}{274}-\frac{1}{2}}, \tag{6.16}$$

the relevant graph is given in Fig. 6.2. Evidently, this solution satisfies condition (6.4).

Now, let it be $n=1$

$$\frac{\alpha\epsilon}{\sqrt{1-\epsilon^2}}-\frac{1}{2}\sqrt{-4\alpha^2+4j^2+4j+1}-\frac{1}{2}=1; \tag{6.17}$$

resolving equation under ϵ we obtain

$$\epsilon=\frac{\sqrt{2j^4+4j^3-2\left(\alpha^2+3\right)j^2-2\left(\alpha^2+4\right)j+\alpha^2\left(3\sqrt{(2j+1)^2-4\alpha^2}+13\right)+8}}{\sqrt{2}\sqrt{9\alpha^2+\left(j^2+j-2\right)^2}}.$$

Taking for j the values $1,...,10$ we get

$$j=1:\epsilon=0.9999970400,\ \ j=2:\epsilon=0.9999983350,\ \ j=3:\epsilon=0.9999989344,$$
$$j=4:\epsilon=0.9999992600,\ \ j=5:\epsilon=0.9999994563,\ \ j=6:\epsilon=0.9999995838,$$
$$j=7:\epsilon=0.9999996711,\ \ j=8:\epsilon=0.9999997336,\ \ j=9:\epsilon=0.9999997798,$$
$$j=10:\epsilon=0.9999998150.$$

Energy ϵ becomes closer to 1 when j increases. Besides, all these values are bigger than the values for ϵ from eq. (6.15). Substituting $j=1$ and respective ϵ $(n=1,\ j=1)$, we get the asymptotic behaviour

$$f(r)=\frac{1}{411}e^{-\sqrt{\frac{1}{2}-\frac{\sqrt{168917}}{822}}\,r}\,r^{\frac{\sqrt{168917}}{274}-\frac{1}{2}}\left(\sqrt{822\left(411-\sqrt{168917}\right)}r-3\left(137+\sqrt{168917}\right)\right),$$

its graph is given in Fig. 6.3. Note that this function has only one zero. The graphs for functions at

$$j=2,\ C=1,\ \text{and}\ j=3,\ C=10^{-3}, \tag{6.18}$$

demonstrates the same behaviour, see Fig. 6.4.

Let us consider other series of levels at $n=2$:

$$\epsilon=\frac{\sqrt{2j^4+4j^3-2\left(\alpha^2+11\right)j^2-2\left(\alpha^2+12\right)j+\alpha^2\left(5\sqrt{(2j+1)^2-4\alpha^2}+37\right)+72}}{\sqrt{50\alpha^2+2\left(j^2+j-6\right)^2}}.$$

Taking $j=1,...,10$, we obtain ten energy levels

$$j=1:\epsilon=0.9999983350,\ \ j=2:\epsilon=0.9999989344,\ \ j=3:\epsilon=0.9999992600,$$
$$j=4:\epsilon=0.9999994563,\ \ j=5:\epsilon=0.9999995838,\ \ j=6:\epsilon=0.9999996711,$$

$$j=7:\epsilon=0.9999997336,\ j=8:\epsilon=0.9999997798,\ j=9:\epsilon=0.9999998150,$$
$$j=10:\epsilon=0.9999998424.$$

As j increases, the energy ϵ increases as well.

Substituting $j=1$ and the respective value ϵ $(n=2,\ j=1)$, we derive the following expression (let $C=1$)

$$f(r)=r^{\frac{\sqrt{168917}}{274}-\frac{1}{2}}e^{-\frac{\sqrt{\theta}r}{\sqrt{82290146}}}$$
$$\times\Big(\frac{975980}{137\theta}-\frac{75056\sqrt{168917}}{187690}-\frac{10\sqrt{168917}r^2}{41145073}+\frac{34r^2}{300329}$$
$$-14\sqrt{\frac{337834}{41145073}}\sqrt{\frac{1}{\theta}}r+\frac{412878}{137}\sqrt{\frac{2}{41145073}}\sqrt{\frac{1}{\theta}}r\Big), \tag{6.19}$$

where $\theta\equiv 2329-5\sqrt{168917}$; its graph is given in Fig. 6.5.

Now, let it be $n=3$:

$$\frac{\alpha\epsilon}{\sqrt{1-\epsilon^2}}-\sqrt{(j+1/2)^2-\alpha^2}-\frac{1}{2}=3. \tag{6.20}$$

In the case ϵ $(n=3,\ j=1)$, we derive an explicit expression of the complete solution (let $C=1$)

$$f=e^{-\frac{\sqrt{3973-7\theta}r}{\sqrt{514284026}}}r^{\frac{\theta}{274}-\frac{1}{2}}$$
$$\times U\Big(\frac{1}{274}\left(\theta-\sqrt{14}\sqrt{137\theta+77757}+137\right),\frac{\theta}{137}+1,\sqrt{\frac{2}{257142013}}r\sqrt{3973-7\theta}\Big),$$

$\theta\equiv\sqrt{168917}$; its graph is given in Fig. 6.6; the corresponding function has three zeros.

Behaviour of the function $f(r)$ at $n=3$, for $j=2,3,4$, and 5 is illustrated by Fig. 6.7.

It should be noted that graphs for functions $f(r)$ are very sensible under small shifts of the values ϵ. Indeed, let us take a value $\epsilon=0.999995$, that is located between $\epsilon=\sqrt{\frac{20570413+\sqrt{168917}}{20571098}}$ $(j=1)$ and $\epsilon=\sqrt{\frac{1}{822}\left(411+\sqrt{168917}\right)}$ $(j=1)$.

6.4 Numerical study at nonzero polarisability

Now we will examine eq. (6.3) for nonzero polarisability. We will apply numerical calculation and computer modelling with the use of the functional object *DifferentialRoot* [5]. Applying visualizing tools, we will compare the final results. For definiteness, let r varies in the interval $[0,\ 7000]$, and $\sigma=-10^{-3}$; the sign is substantial for having bound states (at negative σ, the bound state does not exist).

In order to fix the initial data, we use the approximate equation in the vicinity of the singular point $r=0$

$$f''(r)+\frac{2f'(r)}{r}+\frac{\alpha^2\sigma f(r)}{r^4}=0. \tag{6.21}$$

Allowing for the values $\alpha=\frac{1}{137}$, $\sigma=-10^{-3}$, and integrating eq. (6.21), we get

$$f(r)=C_1\cosh\left(\frac{1}{1370\sqrt{10}r}\right)-iC_2\sinh\left(\frac{1}{1370\sqrt{10}r}\right). \tag{6.22}$$

To obtain a solution that tends to zero near the singular point $r = 0$, we choose the arbitrary constants as $C_1 = 1$ and $C_2 = -i$, which yields

$$f = \exp\Big(-\frac{1}{1370\sqrt{10r}}\Big). \tag{6.23}$$

The graph of this solution is given in Fig. 6.8. As a start, we take the value $r_0 = 10^{-5}$. Therefore, the starting value for $f(r)$ is

$$f(10^{-5}) = 9.450969564 \cdot 10^{-11}. \tag{6.24}$$

Further, we obtain the starting value for the first derivative

$$f'(10^{-5}) = 0.0002181502914. \tag{6.25}$$

For simplicity, we will use the energy values when polarisability equals zero

$$\begin{aligned}\epsilon = \big(&\alpha^2 + 2j^4 + 4j^3 - 2j^2\left(\alpha^2 + 2n(n+1) - 1\right) \\ &+\alpha^2\sqrt{(2j+1)^2 - 4\alpha^2} - 2j\left(\alpha^2 + 2n(n+1)\right) \\ &+2\alpha^2 n\sqrt{(2j+1)^2 - 4\alpha^2} + 2n(n+1)\left(3\alpha^2 + n^2 + n\right)\big)^{1/2} \\ &\times\left(\sqrt{2}\sqrt{(j-n)^2(j+n+1)^2 + (\alpha + 2\alpha n)^2}\right)^{-1},\end{aligned} \tag{6.26}$$

where $\alpha = \frac{1}{137}$, $j \in \mathbb{N}$, $n = 0, 1, 2, \ldots$. Relation (6.26) arises as a solution of eq. (6.14). It is the first approximation, it is reasonable because little perturbation of parameters should produce a small perturbation of corresponding solutions.

When $\sigma = -10^{-3}$, $n = 0$, and $j = 1$ the energy is

$$\epsilon = \sqrt{\frac{20570413 + \sqrt{168917}}{20571098}}. \tag{6.27}$$

To numerically solve eq. (6.3) at initial conditions (6.24), (6.25), we use the Command [4]:

$$\rho1 = \{\alpha \to \frac{1}{137},\ \sigma \to -10^{-3},\ \epsilon \to \sqrt{\frac{20570413 + \sqrt{168917}}{20571098}}\};$$

$$sol1 = NDSolve[\{0 == eq[\epsilon, \sigma, j, r]/.\rho1./j \to 1,$$

$$f[10^{-5}] = N[\phi_0, 100],\ f'[10^{-5}] = N[\phi'_0, 100]\}, f, \{r, 0.0001, 7000\}]//First$$

where $0 == eq[\epsilon, \sigma, j, r]$ determines eq. (6.4). It provides us with interpolation for

$$\{f \to \text{InterpolatingFunction}[\{\{0.0001, 7000\}\}, <>]\}$$

its graph is given in Fig. 6.9 and constructed with the help of the Command

$$gr1 = Plot[Evaluate[\{f[r]/.sol1\}], \{r, 0.0001, 7000\}, PlotRange \to All,$$

$$PlotStyle \to \{Black, Thickness \to 0.005\}, AxesStyle \to Directive[13],$$

$$AxesLabel \to \{Style[r, 14], Style[\text{"}f(r)\text{"}, 14]\}]$$

In order to find a solution in symbolic form, we use the Command

$$sol2 = DSolve[\{0 == (eq[\epsilon, \sigma, j, r]/.\rho1./j \to 1, f[10^{-5}] = \phi_0,$$

$$f'[10^{-5}] = \phi'_0, f, r]//First$$

which results in

$$\{f \to \text{DifferentialRoot}[Function[y, x, \{(-75077 - 2818165349000x^2 + 1000\times$$

$$\sqrt{20571098(20570413 + \sqrt{168917})x^3 - 46922500x^4 + 68500\sqrt{168917}x^4})y[x]+$$

$$2818240426000x^3y'[x] + 1409120213000x^4y''[x] == 0,$$

$$y[\frac{1}{100000}] == \frac{108271}{1145607329178882}, y'[\frac{1}{100000}] == \frac{220588659091}{1011177467159038}\}]]\}]$$

the graph of this solution is given in Fig. 6.10 and constructed by means of the Command

$$gr2 = Plot[Evaluate[\{f[r]/.sol2\}], \{r, 0.0001, 7000\}, PlotRange \to All,$$

$$PlotStyle \to \{Gray, Dashing[\{0.03, 0.04\}], Thickness \to 0.02\}], AxesStyle \to Directive[13],$$

$$AxesLabel \to \{Style[r, 13], Style[\text{“}f\text{”}, 13]\}]$$

Let us compare curves in Figs. 6.9 and 6.10. With the use of the Command Show[gr1, gr2], we get Fig. 6.11 – these graphs coincide ideally in the interval [0,7000].

Similarly, we can examine other energy levels. For instance, let us take

$$\epsilon(n = 1,\ j = 1) = \sqrt{\frac{1}{822}(411 + \sqrt{168917})}. \tag{6.28}$$

Then, for numerical integration of eq. (6.3), we use the Command

$$\rho2 = \{\alpha \to \frac{1}{137},\ \sigma \to -10^{-3},\ \epsilon \to \sqrt{\frac{1}{822}(411 + \sqrt{168917})}\};$$

$$sol3 = NDSolve[\{0 == eq[\epsilon, \sigma, j, r]/.\rho2./j \to 1,$$

$$f[10^{-5}] = \phi_0,\ f'[10^{-5}] = \phi'_0\}, f, \{r, 0.001, 7000\}]//First$$

Here, $0 == eq[\epsilon, \sigma, j, r]$ determines eq. (6.4). In this way we produce interpolation for

$$\{f \to \text{InterpolatingFunction}[\{\{0.001, 7000\}\}, <>]\}$$

its graph is given in Fig. 6.12 and constructed with the help of the Command

$$gr3 = Plot[f[r]/.sol3, \{r, 0.0001, 7000\}, PlotRange \to All,$$

$$PlotStyle \to \{Black, Thickness \to 0.005\}, AxesStyle \to Directive[13],$$

$$AxesLabel \to \{Style[r, 13], Style[\text{“}f(r)\text{”}, 13]\}]$$

To find solution in symbolic form we will apply the Command

$$sol4 = DSolve[\{0 == (eq[\epsilon, \sigma, j, r]/.\rho2./j \to 1, f[10^{-5}] = \phi_0,$$

$$f'[10^{-5}] = \phi'_0, f, r]//First$$

which results in

$$\{f \to \text{DifferentialRoot}[Function[y, x, \{(-3 - 112611000x^2 + 1000\times$$

$$\sqrt{882(411 + \sqrt{168917})x^3 - 28153500x^4 + 68500\sqrt{168917}x^4})y[x]+$$

$$112614000x^3y'[x] + 56307000x^4y''[x] == 0,$$

$$y[\frac{1}{100000}] == E^{-\frac{10^3\sqrt{10}}{137}}, y'[\frac{1}{100000}] == \frac{10^8}{137}\sqrt{10}E^{-\frac{10^3\sqrt{10}}{137}}\}]]\}]$$

Its graph (Fig. 6.13) is produced through the Command

$$gr4 = Plot[f[r]/.sol\,4], \{r, 0.0001, 7000\}, PlotRange \to All,$$

$$PlotStyle \to \{Gray, Dashing[\{0.04, 0.05\}], Thickness \to 0.015\}], AxesStyle \to Directive[13],$$

$$AxesLabel \to \{Style[r, 13], Style[\text{“}f(r)\text{”}, 3]\}]$$

The last two curves are coincided excellently in the interval [0, 7000] (Fig. 6.14).

While changing $j = 1$ to $j = 2$, $j = 3$ and so on, the general behaviour of curves remains the same, though the amplitude increases as j becomes greater.

The present section provides the testing of possible computational methods rather than a detailed examination of the real physical problem of searching perturbations for energy levels due to $\sigma \neq 0$. Evidently, we need additional study of this problem.

6.5 Conclusions

Methods for solving the differential equation describing the wave functions of a polarisable particle in the Coulomb potential are discussed. Relations between the coefficients under which the general solutions of this equation can be found in analytical form are detailed. For the case of zero polarisability, the general solution to this equation in terms of special functions is obtained; for the first values of the parameter j, plots of the corresponding solutions are presented. For nonzero polarisability and certain specially chosen values of the parameters, solutions possessing the required physical properties are constructed with the use of numerical methods and functional objects of the type DifferentialRoot. Instructions in Mathematica are presented which permit to apply elaborated methods in studing other problems in physics and mathematics.

6.6 Figures

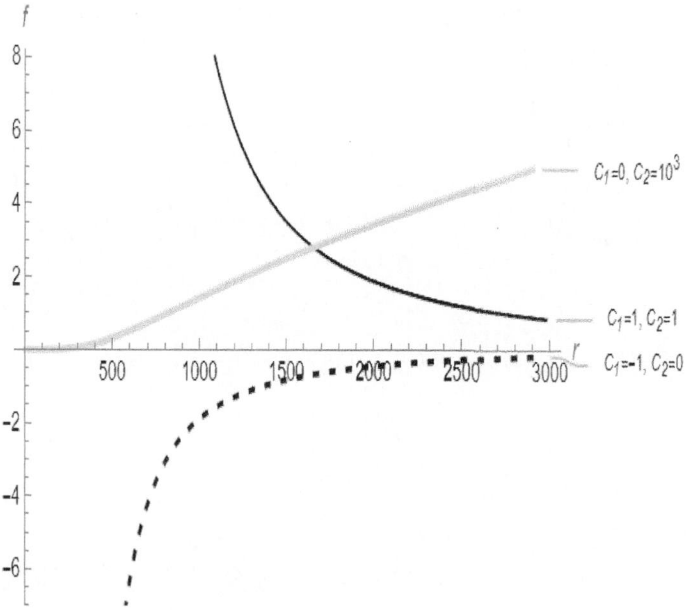

FIGURE 6.1
Graphs for solutions (6.12) $r \in [0.2, 100]$.

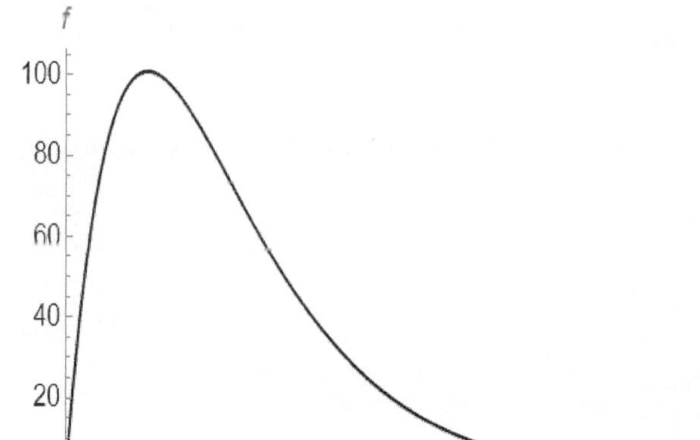

FIGURE 6.2
Plot of function (6.16) on the interval [0, 2000].

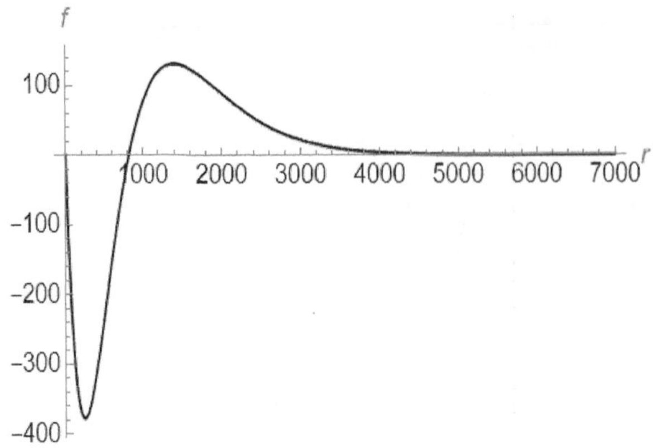

FIGURE 6.3
Plot of function (6.18) on the interval [0, 8000].

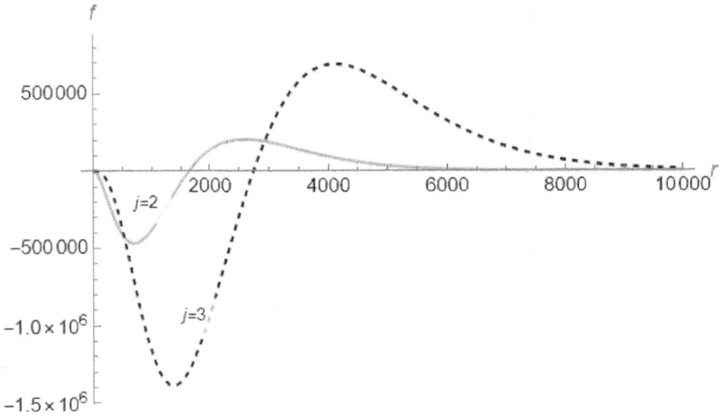

FIGURE 6.4
Plots of functions (6.14) on the interval [0, 10000] corresponding to parameters (6.18).

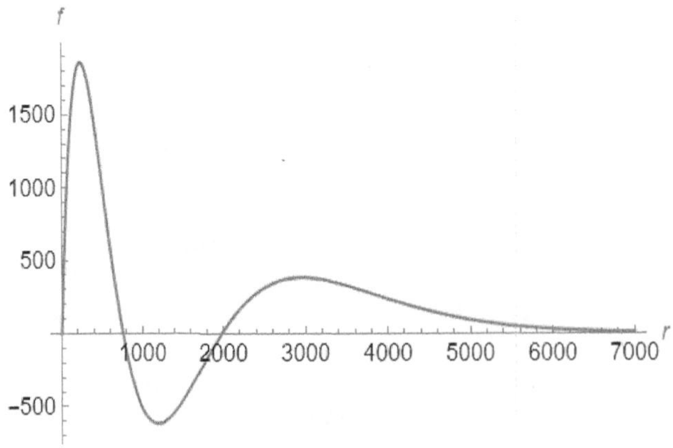

FIGURE 6.5
Plot of function (6.19) on the interval [0, 7000].

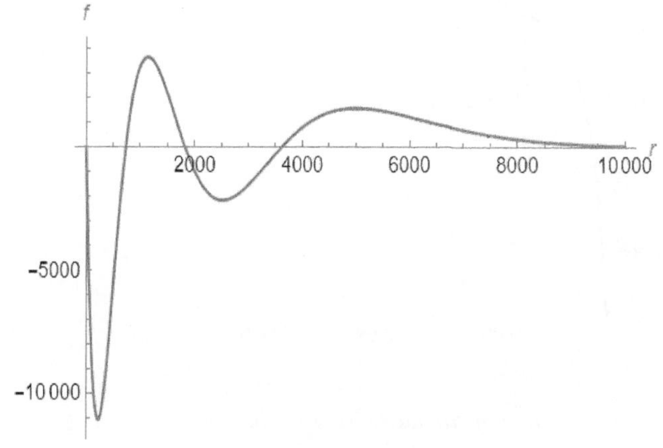

FIGURE 6.6
Plot of function (6.21) on the interval [0, 10000].

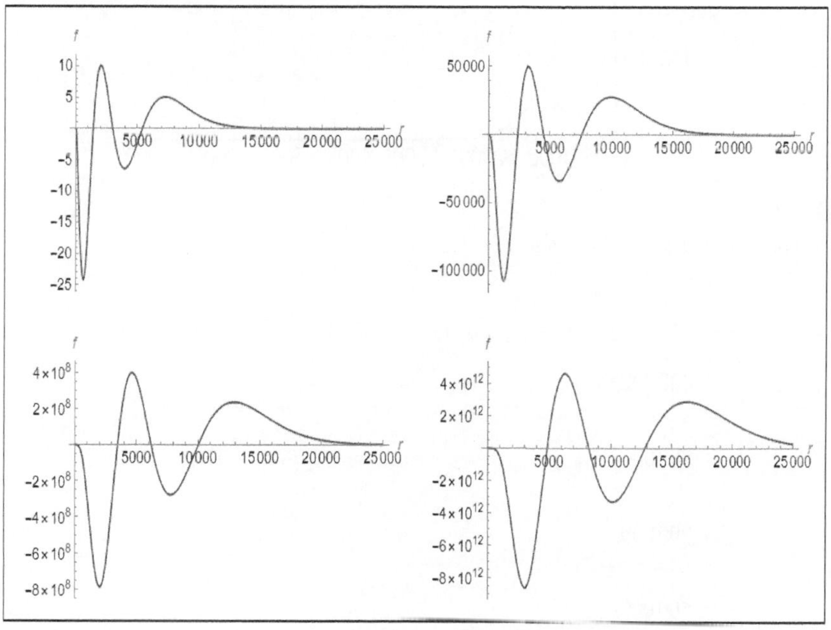

FIGURE 6.7
Plots of function (6.14), (6.20), $C = 10^{-6}$, $j = 2, 3, 4, 5$ on the interval [0, 25000].

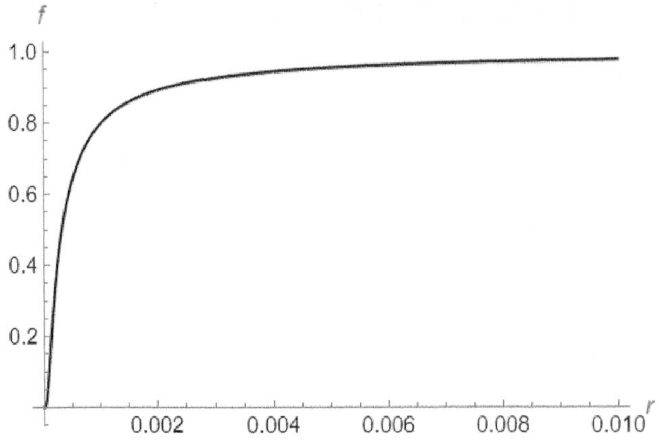

FIGURE 6.8
Plot of function (6.23) in the region close to $r = 0$.

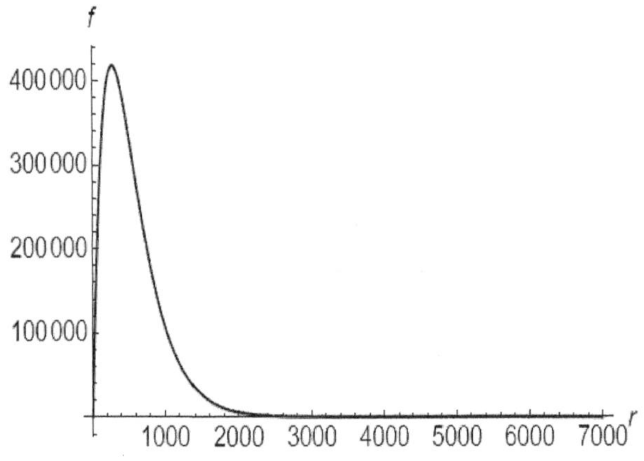

FIGURE 6.9
Plot of Interpolating Function sol1.

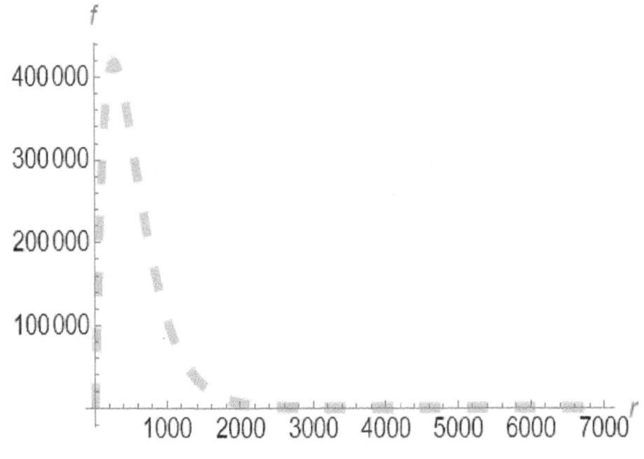

FIGURE 6.10
Plot of the function sol2.

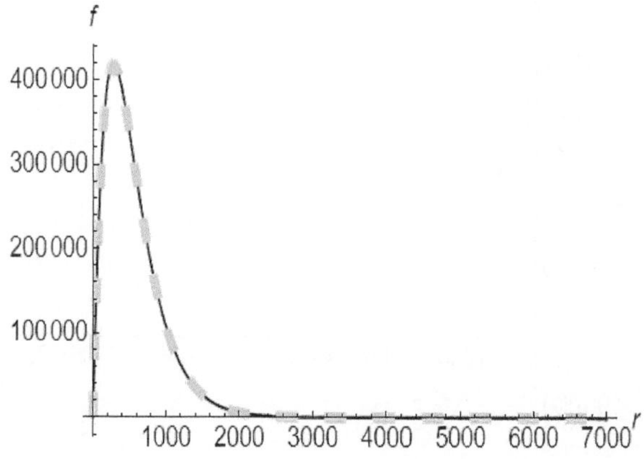

FIGURE 6.11
Comparing functions sol1 and sol2.

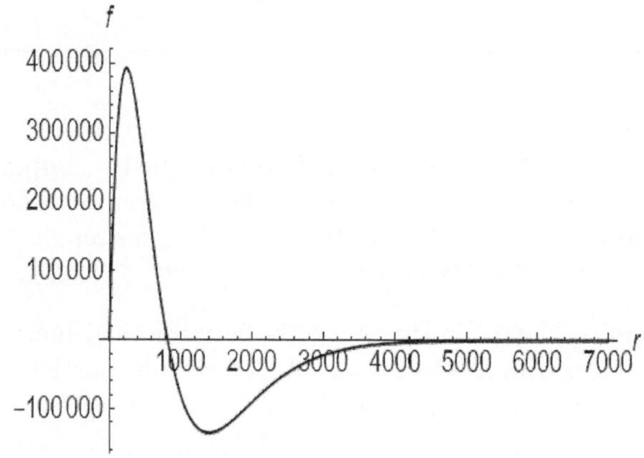

FIGURE 6.12
Plot of the Interpolating Function sol3.

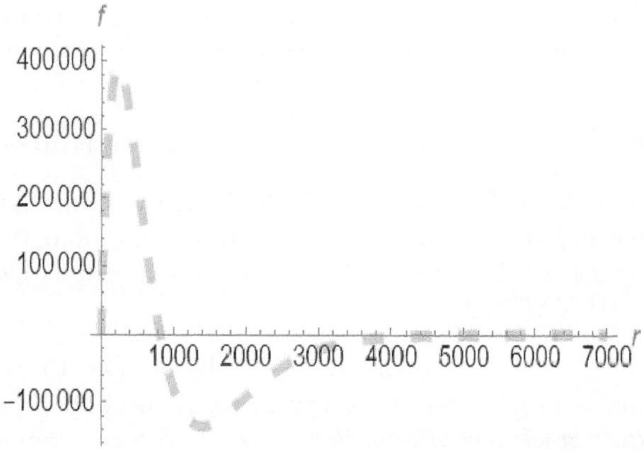

FIGURE 6.13
Plot of the function sol4.

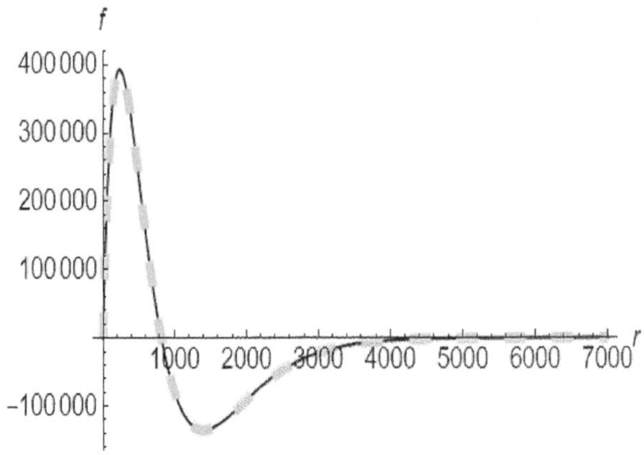

FIGURE 6.14
Comparing functions sol3 and sol4.

Bibliography

[1] N.G. Tokarevskaya, V.V. Kisel and V.M. Red'kov. On 15-component theory of a scalar particle with polarisability in Coulomb and Dirac monopole fields. In *Proceedings of the 12th International School & Conference "Foundation & Advances in Nonlinear Science", September 27-30, Minsk, Belarus*, pages 28–41, 2004.

[2] V.M. Red'kov and A.V. Chichurin. A symbolic-numerical method for solving the differential equation describing the states of polarisable particle in coulomb potential. *Programming and Computer Software*, 40(2): 86–92, 2014.

[3] A. Chichurin and V. Red'kov. Quantum mechanical scalar particle with polarisability in the Coulomb field, analytical and numerical consideration. *Studia i Materialy EWSIE*, 6: 73–89, 2013.

[4] A.A. Bogush, N.G. Tokarevskaya, V.V. Kisel and V.M. Red'kov. Yang-Mills equations based on extended 15-component description of spin 1 massless field. In *Proceedings of the X Ann. Sem. Nonlinear Phenomena in Complex Systems*, pages 89–95. Minsk. IP NASB, 2001.

[5] http://reference.wolfram.com/mathematica/ref/DifferentialRoot.html

[6] N.G. Tokarevskaya, V.M. Kisel and V.M. Red'kov. On non-relativistic approximation for charged scalar and vector particles with polarisability and Wigner's time reversion. In *Proceedings of the XI Ann. Sem. Nonlinear Phenomena in Complex Systems*, pages 382–388. Minsk, IP NASB, 2001.

[7] V.M. Red'kov, N.G. Tokarevskaya and V.V. Kisel. On 15-component theory of a charged spin-1 particle with polarisability in Coulomb and Dirac monopole fields. In *Proceedings of the 12th International School & Conference "Foundation & Advances in Nonlinear Science"*, pages 42–57. Minsk, Belarus, 2004.

[8] V.V. Kisel, N.G. Tokarevskaya and V.M. Red'kov. Scalar particle in Riemannian space, non-minimal interaction, and non-relativistic approximation. In *Proceedings of the 5th*

International Conference Bolyai-Gauss-Lobachevsky: Non-Euclidean Geometry In Modern Physics (BGL-5), pages 122–127. Minsk, Belarus, 2006.

[9] V.V. Kisel, G.G. Krylov, E.M. Ovsiyuk, M. Amirfachrian and V.M. Red'kov. Wave functions of the particle with polarisability in Coulomb field. *Nonlinear Dynamics and Applications*, 18: 168–179, 2011.

[10] E.M. Ovsiyuk, V.V. Kisel and V.M. Red'kov. *Maxwell Electrodynamics and Boson Fields in Spaces of Constant Curvature*. Nova Science Publishers Inc., NY, USA, 2014.

[11] O.V. Veko, K.V. Kazmerchuk, V.V. Kisel, E.M. Ovsiyuk and V.M. Red'kov. Quantum mechanical scalar particle with intrinsic structure in external magnetic and electric fields: influence of geometrical background. *Nonlinear Phenomena in Complex Systems*, 17(4): 464–466, 2014.

[12] V.V. Kisel, E.M. Ovsiyuk, O.V. Veko, Y.A. Voynova, V. Balan and V.M. Red'kov. *Elementary Particles with Internal Structure in External Fields.* Vol I. General Theory, Nova Science Publishers Inc., NY, USA, 2018.

[13] V.V. Kisel, E.M. Ovsiyuk, O.V. Veko, Y.A. Voynova, V. Balan and V.M. Red'kov. *Elementary Particles with Internal Structure in External Fields.* Vol II. Physical Problems, Nova Science Publishers Inc., NY, USA, 2018.

[14] E.M. Ovsiyuk, O.V. Veko, Ya.A. Voynova, A.D. Koral'kov, V.V. Kisel and V.M. Red'kov. On describing bound states for a spin 1 particle in the external Coulomb field. *Proceedings of Balkan Society of Geometers*, 25: 59–78, 2018.

[15] E.M. Ovsiyuk, A.V. Chichurin, Ya.A. Voynova and V.M. Red'kov. Spin 1 particle with polarisability in the external Coulomb field, nonrelativistic description. *Proceedings of Balkan Society of Geometers*, 26: 74–96, 2019.

7

Dirac particle in the Coulomb field in curved models

The known systems of radial equations describing relativistic hydrogen atoms on the basis of the Dirac equation in spherical Riemann spaces are investigated. The relevant 2nd-order differential equations have six regular singular points, and there solutions of Frobenius type are constructed. To produce the quantisation rule for energy values, we use the known condition separating transcendental Frobenius solutions. This provides us with energy spectra that are physically interpretable and similar to spectra arising from the scalar Klein–Fock–Gordon equation in these geometrical models. The spectra coincide with those previously found when studying the same radial equations within the semi-classical method. The convergence of the series involved is proved analytically and numerically. The squared integrability of solutions is demonstrated numerically. Visualisation of the results is given.

7.1 Introduction

Quantum mechanics had been started with the theory of the hydrogen atom, so when considering quantum mechanics in Riemannian spaces, it is natural to turn first to just this simplest system. A common quantum-mechanical hydrogen atom description is based materially on the assumption of the Euclidean character of the physical 3-space geometry. In this context, natural questions arise: what in the description is determined by this special assumption, and which changes will be entailed by allowing for other spatial geometries, for instance, Lobachevsky's H_3 or Riemann's S_3. The questions are of fundamental significance, even beyond their possible experimental testing.

For the first time, the hydrogen atom in a 3-dimensional space of constant positive curvature S_3 was considered by Schrödinger [1]. He had studied the so-called factorisation method. in quantum mechanics; in particular, the application of this technique to a discrete part of the energy spectrum for hydrogen atoms had been elaborated. An idea was to modify the basic atom system in such a way that to cover all the energy spectrum including the region $E > 0$ as well. However, the placing of the atom system inside a finite box in order to make the whole energy spectrum discrete did not seem attractive, so Schrödinger had placed the atom into the curved background of the Riemann space model S_3. Due to its compactness, this geometry may simulate the effect of the finite box.

The hydrogen atom in the Lobachevsky space H_3 was first considered by Infeld and Shild [2]. It turned out that the number of discrete levels in Lobachevsky space is always finite, and the number of levels varies and depends on the value of the curvature radius.

Thus, the models of the hydrogen atom in Euclid, Riemann, and Lobachevsky spaces significantly differ from each other, which is the result of differences in three spatial geometries: E_3, H_3, and S_3.

DOI: 10.1201/9781003472377-7

At present, we see plenty of investigations on this subject. In the first place, it concerns details of relevant wave functions, generalised spectra, and the role of spin effects [2,5,6,15, 20,22,25,26,28,32,34,36,39,40,43,47]. Also, there were discussed the role of hydrogen atoms as a probe system in the cosmology context: [4, 11, 13, 14, 16–18, 23, 24]. Much interest was given to studying the hidden symmetry in the nonrelativistic description of the hydrogen in curved geometrical models, see [7–10], and the role of generalised parabolic coordinates in this context: [12, 19, 29, 44, 46]. A number of papers on other aspects were published: scattering theory [41, 45]; path integral treatment [27, 31, 32]; classical Coulomb problem [33, 35]; quasi-classical approach [26, 42]. A comprehensive account of the matter was given in the books [48, 49].

The most difficult and still unsolved is the case of spin 1/2 particle in the Coulomb field on the background of curved models. For this system, resulting radial 2nd-order differential equation turns out to be rather complicated, it contains six regular singularities, and with the help of special transformation the task may be reduced to a 2nd-order equation with five singular points (see [48, 49]). In [26, 42], some energy spectra were found on the basis of semi-classical approximation, which seems to be quite appropriate from a physical point of view. Besides they are similar to the spectra arising when solving the Klein–Fock–Gordon equation for this system.

In the present chapter, we have studied exact solutions of the Frobenius type for these radial equations. As a quantisation rule, we apply the transcendency condition to Frobenius solutions, so producing simple algebraic equations, which provide us with physically reasonable energy spectra. In fact, they coincide with those obtained from a semiclassical study [26, 42]. It should be noted that for spherical Riemann space, exactly the same spectrum was derived as well in [37], though it was claimed mistakenly that this spectrum refers to polynomial solutions.

7.2 Hydrogen atom in the Lobachevsky space

In spherical coordinates of the space H_3, a diagonal tetrad is taken in the form

$$\begin{gathered} dS^2 = dt^2 - dr^2 - \sinh^2 r(d\theta^2 + \sin^2 d\phi^2),\ r \in (0,\infty), \\ e^{\alpha}_{(0)} = (1,0,0,0)\ ,\ e^{\alpha}_{(3)} = (0,1,0,0), \\ e^{\alpha}_{(1)} = (0,0,\frac{1}{\sinh r},0), e^{\alpha}_{(2)} = (0,0,0,\frac{1}{\sinh r \sin\theta}); \end{gathered} \tag{7.1}$$

the radial variable is dimensionless due to dividing by the curvature radius ρ. For this tetrad, the Ricci rotation coefficients are $\gamma_{ab0} = 0, \gamma_{ab3} = 0$, and

$$\gamma_{ab1} = \begin{vmatrix} 0 & 0 & 0 & 0 \\ 0 & 0 & 0 & -\coth r \\ 0 & 0 & 0 & 0 \\ 0 & +\coth r & 0 & 0 \end{vmatrix},$$

$$\gamma_{ab2} = \begin{vmatrix} 0 & 0 & 0 & 0 \\ 0 & 0 & \cot\theta \sinh^{-1} r & 0 \\ 0 & -\cot\theta\sinh^{-1} r & 0 & -\coth r \\ 0 & 0 & +\coth r & 0 \end{vmatrix}.$$

Generally covariant Dirac equation (the notation according to [48, 49] is used)

$$\left[i\gamma^c \left(e^\alpha_{(c)} \partial_\alpha + \frac{1}{2} j^{ab} \gamma_{abc} \right) - m \right] \Psi = 0$$

takes the form (let $\Psi = \sinh^{-1} r \, \psi$)

$$\left[i\gamma^0 \frac{\partial}{\partial t} + i\gamma^3 \frac{\partial}{\partial r} + \frac{1}{\sinh r} \Sigma_{\theta\phi} - m \right] \tilde{\Psi} = 0, \quad \Sigma_{\theta,\phi} = i\gamma^1 \partial_\theta + \gamma^2 \frac{i\partial_\phi + i\sigma^{12}}{\sin\theta}. \tag{7.2}$$

To diagonalise the operators $i\partial_t, \vec{J}^2, J_3$, one takes the wave function in the form [48, 49]

$$\psi = e^{-i\epsilon t} \left| \begin{array}{c} f_1(r) \, D_{-1/2} \\ f_2(r) \, D_{+1/2} \\ f_3(r) \, D_{-1/2} \\ f_4(r) \, D_{+1/2} \end{array} \right| , \tag{7.3}$$

where the Wigner functions [50] are noted as $D_\sigma = D^j_{-m.\sigma}(\phi, \theta.0)$. After separating the variables we get four radial equations (let $j + 1/2 = \nu$)

$$\begin{aligned} &\epsilon f_3 - i\frac{d}{dr} f_3 - i\frac{\nu}{\sinh r} f_4 - m f_1 = 0, \quad \epsilon f_4 + i\frac{d}{dr} f_4 + i\frac{\nu}{\sinh r} f_3 - m f_2 = 0 \, , \\ &\epsilon f_1 + i\frac{d}{dr} f_1 + i\frac{\nu}{\sinh r} f_2 - m f_3 = 0, \quad \epsilon f_2 - i\frac{d}{dr} f_2 - i\frac{\nu}{\sinh r} f_1 - m f_4 = 0 \, . \end{aligned} \tag{7.4}$$

In spherical tetrad (7.1), the space reflection operator is given by the formula

$$\hat{\Pi}_{sph} \;=\; \left| \begin{array}{cccc} 0 & 0 & 0 & -1 \\ 0 & 0 & -1 & 0 \\ 0 & -1 & 0 & 0 \\ -1 & 0 & 0 & 0 \end{array} \right| \otimes \hat{P} \, , \quad \hat{P}(\theta, \phi) = (\pi - \theta, \phi + \pi).$$

From eigenvalues equation $\hat{\Pi}_{sph} \, \Psi_{jm} \;=\; \Pi \, \Psi_{jm}$ we obtain

$$\Pi = \; \delta \, (-1)^{j+1}, \quad \delta = \pm 1 \, , \quad f_4 = \; \delta \, f_1, \quad f_3 = \; \delta \, f_2 \, . \tag{7.5}$$

This simplifies the system (7.4)

$$\left(\frac{d}{dr} + \frac{\nu}{\sinh r} \right) f + (\epsilon + \delta m) \; g = 0 \, , \quad \left(\frac{d}{dr} - \frac{\nu}{\sinh r} \right) g - (\epsilon - \delta m) \; f = 0 \, , \tag{7.6}$$

where instead of f_1 and f_2, the new variables f and g are used

$$f = \frac{f_1 + f_2}{\sqrt{2}} \, , \qquad g = \frac{f_1 - f_2}{i\sqrt{2}} \, .$$

In presence of the Coulomb field, we have equations

$$\begin{aligned} &\left(\frac{d}{dr} + \frac{\nu}{\sinh r} \right) f + \left(E + \frac{e}{\tanh r} + m \right) g = 0, \\ &\left(\frac{d}{dr} - \frac{\nu}{\sinh r} \right) g - \left(E + \frac{e}{\tanh r} - m \right) f = 0. \end{aligned} \tag{7.7}$$

After transforming the system (7.7) to the variable $\tanh \frac{r}{2} = z, z \in (0, 1)$, we obtain

$$\begin{aligned} &\frac{d}{dz} f + \frac{\nu}{z} f + \left(\frac{e}{z} + \frac{-E - e - m}{z - 1} + \frac{E - e + m}{z + 1} \right) g = 0 \, , \\ &\frac{d}{dz} g - \frac{\nu}{z} g - \left(\frac{e}{z} + \frac{-E - e + m}{z - 1} + \frac{E - e - m}{z + 1} \right) f = 0 \, . \end{aligned} \tag{7.8}$$

Whence it follows a 2nd-order equation for $f(z)$:

$$\frac{d^2 f}{dz^2} + \Big[\frac{1}{z} + \frac{1}{z-1} + \frac{1}{z+1} - 2\,\frac{ez+E+m}{ez^2+2(E+m)\,z+e}\Big]\,\frac{df}{dz}$$

$$+\Big[2\,\frac{2Ee^2-(E+m)\,\nu}{ez} + \frac{-(E+e)^2+m^2+\nu}{z-1} + \frac{(E-e)^2-m^2-\nu}{z+1}$$
$$+\frac{e^2-\nu^2}{z^2} + \frac{(E+e)^2-m^2}{(z-1)^2} + \frac{(E-e)^2-m^2}{(z+1)^2}$$
$$+\frac{2\,\nu\,[ez(E+m)+2\,(E+m)^2-e^2]}{e\,[\,ez^2+2(E+m)\,z+e\,]}\Big]\,f = 0\,. \qquad (7.9)$$

The last equation has six singular points (let $\frac{E+m}{e} = \sigma > 0$):

$$0\,, \quad \infty\,, \quad \pm 1, \quad z_{1,\,2} = -\sigma \pm \sqrt{\sigma^2-1} \quad (z_1 z_2 = 1\,, \;\; z_1 + z_2 = -2\sigma)\,. \qquad (7.10)$$

Eq. (7.9) may be re-written differently

$$\frac{d^2 f}{dz^2} + \Big[\frac{1}{z} + \frac{1}{z-1} + \frac{1}{z+1} - \frac{1}{z-z_1} - \frac{1}{z-z_2}\Big]\,\frac{df}{dz}$$

$$+\Big[\frac{4Ee-2\sigma\,\nu}{z} - \frac{(E+e)^2-m^2-\nu}{z-1} + \frac{(E-e)^2-m^2-\nu}{z+1} + \frac{e^2-\nu^2}{z^2}$$
$$+\frac{(E+e)^2-m^2}{(z-1)^2} + \frac{(E-e)^2-m^2}{(z+1)^2} + \frac{A}{z-z_1} + \frac{B}{z-z_2}\Big]\,f = 0\,, \qquad (7.11)$$

where

$$2\nu\,\frac{\sigma\, z + 2\,\sigma^2 - 1}{(z-z_1)(z-z_2)} = \frac{A}{z-z_1} + \frac{B}{z-z_2},$$

$$A = 2\nu\,\frac{\sigma z_1 + 2\sigma^2 - 1}{z_1 - z_2}, \quad B = 2\nu\,\frac{\sigma z_2 + 2\sigma^2 - 1}{z_2 - z_1}.$$

For shortness let us apply notations

$$C = (E+e)^2 - m^2, \quad D = (E-e)^2 - m^2, \quad 4Ee = C - D,$$

then eq. (7.11) reads

$$\frac{d^2 f}{dz^2} + \Big(\frac{1}{z} + \frac{1}{z-1} + \frac{1}{z+1} - \frac{1}{z-z_1} - \frac{1}{z-z_2}\Big)\,\frac{df}{dz}$$
$$+\Big(\frac{C-D-2\sigma\nu}{z} - \frac{C-\nu}{z-1} + \frac{D-\nu}{z+1} + \frac{e^2-\nu^2}{z^2}$$
$$+\frac{C}{(z-1)^2} + \frac{D}{(z+1)^2} + \frac{A}{z-z_1} + \frac{B}{z-z_2}\Big)\,f = 0. \qquad (7.12)$$

Near the points $z = 0, +1, -1, z_1, z_2$ solutions behave as

$$f \sim (z-1)^{\alpha},\; \alpha = \pm\sqrt{-C}\,; \quad f \sim (z+1)^{\beta},\; \beta = \pm\sqrt{-D}\,;$$
$$f \sim z^{M},\; M = \pm\sqrt{\nu^2 - e^2}; \quad f \sim (z-z_1)^{\gamma},\; f \sim (z-z_2)^{\gamma},\; \gamma = 0, 2\,. \qquad (7.13)$$

Let us search Frobenius type solutions in the form

$$f(z) = x^{M}(z-1)^{\alpha}(z+1)^{\beta}F(z) = \varphi(z)F(z)\,; \qquad (7.14)$$

for function $F(z)$ we derive the equation

$$\frac{d^2F}{dz^2} + \Big[\frac{2M+1}{z} + \frac{2\alpha+1}{z-1} + \frac{2\beta+1}{z+1} - \frac{1}{z-z_1} - \frac{1}{z-z_2}\Big]\frac{dF}{dz}$$

$$+\Big[\frac{M^2+e^2-\nu^2}{z^2} + \frac{\alpha^2+C}{(z-1)^2} + \frac{\beta^2+D}{(z+1)^2}$$

$$+\frac{C-D-(\alpha-\beta)\,(2M+1)-2\sigma(\nu+M)}{z} + \frac{M\,(z_1+2\,\sigma\,z_1\,z_2+z_2)}{z\,z_2\,z_1}$$

$$+\frac{M+\alpha/2+\beta/2-C+\nu+2M\alpha+\alpha\beta}{z-1} - \frac{\alpha\,(1-z_1\,z_2)}{(z-1)\,(z_1-1)\,(z_2-1)}$$

$$-\frac{M+\alpha/2+\beta/2-D+\nu+2M\beta+\alpha\beta}{z+1} + \frac{\beta\,(1-z_1\,z_2)}{(z+1)\,(z_1+1)\,(z_2+1)}$$

$$+\frac{1}{z-z_1}\Big(A-\frac{\alpha}{z_1-1}-\frac{\beta}{z_1+1}-\frac{M}{z_1}\Big)+\frac{1}{z-z_2}\Big(B-\frac{\alpha}{z_2-1}-\frac{\beta}{z_2+1}-\frac{M}{z_2}\Big)\Big]\,F=0\,.$$

Impose restrictions on parameters M, α, β:

$$\begin{aligned} M &= \pm\sqrt{\nu^2-e^2}, \\ \alpha &= \pm\sqrt{-C} = \pm\sqrt{m^2-(E+e)^2}, \\ \beta &= \pm\sqrt{-D} = \pm\sqrt{m^2-(E-e)^2}; \end{aligned} \tag{7.15}$$

it should be emphasised that bound states may correspond to the following values for parameters

$$M = +\sqrt{\nu^2-e^2}\,, \quad \alpha = +\sqrt{m^2-(E+e)^2}\,, \quad \beta = \pm\sqrt{m^2-(E-e)^2}\,. \tag{7.16}$$

With (7.16) in mind, for function $F(z)$ we obtain the equation

$$\frac{d^2F}{dz^2} + \Big[\frac{2M+1}{z} + \frac{2\alpha+1}{z-1} + \frac{2\beta+1}{z+1} - \frac{1}{z-z_1} - \frac{1}{z-z_2}\Big]\frac{dF}{dz}$$

$$+\Big[\frac{C-D-(\alpha-\beta)\,(2M+1)-2\sigma(\nu+M)}{z} + \frac{M\,(z_1+2\,\sigma\,z_1\,z_2+z_2)}{z\,z_2\,z_1}$$

$$+\frac{M+\alpha/2+\beta/2-C+\nu+2M\alpha+\alpha\beta}{z-1} - \frac{\alpha\,(1-z_1\,z_2)}{(z-1)\,(z_1-1)\,(z_2-1)}$$

$$-\frac{M+\alpha/2+\beta/2-D+\nu+2M\beta+\alpha\beta}{z+1} + \frac{\beta\,(1-z_1\,z_2)}{(z+1)\,(z_1+1)\,(z_2+1)}$$

$$+\frac{1}{z-z_1}\Big(A-\frac{\alpha}{z_1-1}-\frac{\beta}{z_1+1}-\frac{M}{z_1}\Big)+\frac{1}{z-z_2}\Big(B-\frac{\alpha}{z_2-1}-\frac{\beta}{z_2+1}-\frac{M}{z_2}\Big)\Big]\,F=0\,.$$

It is convenient to use its shortening presentation

$$\frac{d^2F}{dz^2} + \Big(\frac{P_1}{z} + \frac{P_2}{z-1} + \frac{P_3}{z+1} - \frac{1}{z-z_1} - \frac{1}{z-z_2}\Big)\frac{dF}{dz}$$

$$+\Big(\frac{Q_1}{z} + \frac{Q_2}{z-1} + \frac{Q_3}{z+1} + \frac{Q_4}{z-z_1} + \frac{Q_5}{z-z_2}\Big)F = 0\,.$$

Multiplying the last equation by $z(z-1)(z+1)(z-z_1)(z-z_2)$ we get

$$\big[z^5 + (-z_1-z_2)\,z^4 + (z_1\,z_2-1)\,z^3 + (z_1+z_2)\,z^2 - z_1\,z_2\,z\big]\,\frac{d^2F}{dz^2}$$

$$+\left[(P_1+P_2+P_3-2)\,z^4+\{(1-P_1-P_2-P_3)\,z_1+(1-P_1-P_2-P_3)\,z_2+P_2-P_3\}\,z^3\right.$$
$$+(2-P_1-P_2\,z_1+P_3\,z_2+P_2\,z_1\,z_2+P_1\,z_1\,z_2+P_3\,z_1\,z_2+P_3\,z_1-P_2\,z_2)\,z^2$$
$$\left.+(-z_1-z_2+P_1\,z_1-P_3\,z_1\,z_2+P_1\,z_2+P_2\,z_1\,z_2)\,z-P_1\,z_1\,z_2\right]\frac{dF}{dz}$$
$$+[(Q_1+Q_2+Q_3+Q_4+Q_5)z^4+\{(-Q_1-Q_2-Q_3-Q_5)z_1$$
$$+(-Q_1-Q_2-Q_3-Q_4)z_2+Q_2-Q_3\}z^3$$
$$+(Q_3\,z_1\,z_2+Q_2\,z_1\,z_2+Q_3\,z_2-Q_1-Q_4-Q_5+Q_1\,z_1\,z_2+Q_3\,z_1-Q_2\,z_2-Q_2\,z_1)\,z^2$$
$$+(Q_1\,z_2+Q_2\,z_1\,z_2+Q_5\,z_1+Q_1\,z_1-Q_3\,z_1\,z_2+Q_4\,z_2)\,z-Q_1\,z_1\,z_2]\,F=0\,.$$

Solutions $F(z)$ may be constructed in the form of power series, $F(z)=\sum_{n=0}^{\infty}d_n z^n$; after performing needed calculation, we derive 6-term recurrent relations:

$$k\geq 4,\qquad (Q_1+Q_2+Q_3+Q_4+Q_5)\,d_{k-4}$$
$$+[(k-3)(k-4)+(P_1+P_2+P_3-2)\,(k-3)$$
$$+(-Q_1-Q_2-Q_3-Q_5)\,z_1+(-Q_1-Q_2-Q_3-Q_4)\,z_2+Q_2-Q_3]\,d_{k-3}$$
$$+[(-z_1-z_2)\,(k-2)(k-3)+\{(1-P_1-P_2-P_3)\,z_1+(1-P_1-P_2-P_3)\,z_2+P_2-P_3\}\,(k-2)$$
$$+\,Q_3z_1z_2+Q_2z_1z_2+Q_3z_2-Q_1-Q_4-Q_5+Q_1z_1z_2+Q_3\,z_1-Q_2z_2-Q_2z_1]\,d_{k-2}$$
$$+[(z_1z_2-1)(k-1)(k-2)+(2-P_1-P_2z_1+P_3z_2$$
$$+P_2z_1z_2+P_1z_1z_2+P_3z_1z_2+P_3z_1-P_2z_2)(k-1)$$
$$+Q_1\,z_2+Q_2\,z_1\,z_2+Q_5\,z_1+Q_1\,z_1-Q_3\,z_1\,z_2+Q_4\,z_2]\,d_{k-1}$$
$$+\,[(z_1+z_2)\,k\,(k-1)+(-z_1-z_2+P_1\,z_1-P_3\,z_1\,z_2+P_1\,z_2+P_2\,z_1\,z_2)\,k-Q_1\,z_1\,z_2]\,d_k$$
$$+\,[-z_1\,z_2\,(k+1)\,k-P_1\,z_1\,z_2\,(k+1)]\,d_{k+1}=0\,. \tag{7.17}$$

To analyse convergence of power series, we apply Poincaré–Perron method, so divide relation (7.17) by $k^2 d_{k-4}$

$$(Q_1+Q_2+Q_3+Q_4+Q_5)$$
$$+[(k-3)(k-4)+(P_1+P_2+P_3-2)\,(k-3)$$
$$+(-Q_1-Q_2-Q_3-Q_5)\,z_1+(-Q_1-Q_2-Q_3-Q_4)\,z_2+Q_2-Q_3]\,\frac{d_{k-3}}{d_{k-4}}$$
$$+[(-z_1-z_2)(k-2)(k-3)+\{(1-P_1-P_2-P_3)z_1+(1-P_1-P_2-P_3)z_2+P_2-P_3\}(k-2)$$
$$+\,Q_3z_1z_2+Q_2z_1z_2+Q_3z_2-Q_1-Q_4-Q_5+Q_1z_1z_2+Q_3z_1-Q_2z_2-Q_2z_1]\,\frac{d_{k-2}}{d_{k-3}}\frac{d_{k-3}}{d_{k-4}}$$
$$+[(z_1z_2-1)(k-1)(k-2)+(2-P_1-P_2z_1+P_3z_2+P_2z_1z_2$$
$$+P_1z_1z_2+P_3z_1z_2+P_3z_1-P_2z_2)(k-1)$$
$$+Q_1\,z_2+Q_2\,z_1\,z_2+Q_5\,z_1+Q_1\,z_1-Q_3\,z_1\,z_2+Q_4\,z_2]\,\frac{d_{k-1}}{d_{k-2}}\frac{d_{k-2}}{d_{k-3}}\frac{d_{k-3}}{d_{k-4}}$$
$$+[(z_1+z_2)k(k-1)+(-z_1-z_2+P_1z_1-P_3z_1z_2$$
$$+P_1z_2+P_2z_1z_2)k-Q_1z_1z_2]\frac{d_k}{d_{k-1}}\frac{d_{k-1}}{d_{k-2}}\frac{d_{k-2}}{d_{k-3}}\frac{d_{k-3}}{d_{k-4}}$$
$$+\,[-z_1z_2(k+1)k-P_1\,z_1\,z_2\,(k+1)]\,\frac{d_{k+1}}{d_k}\frac{d_k}{d_{k-1}}\frac{d_{k-1}}{d_{k-2}}\frac{d_{k-2}}{d_{k-3}}\frac{d_{k-3}}{d_{k-4}}=0\,,$$

and tend $k\to\infty$. In this way, for quantity $R=\lim_{k\to\infty}(d_{k-3}/d_{k-4})$ we derive an algebraic equation with simple roots:

$$R-(z_1+z_2)\,R^2+(z_1z_2-1)\,R^3+(z_1+z_2)\,R^4-z_1z_2R^5=0\quad\Longrightarrow\quad R=0,\pm 1,\frac{1}{z_1},\frac{1}{z_2}.$$

Therefore, possible convergence radii are

$$R_{\text{conv}} = \left| \frac{1}{R} \right| = +1\,, \; +\infty\,, \; |z_1|\,, \; |z_2|. \tag{7.18}$$

Turning to recurrent formulas (7.17), we can see that coefficient at d_{k-4} vanish identically:

$$Q_1 + Q_2 + Q_3 + Q_4 + Q_5 = 0.$$

This means that (7.17) actually leads to a 5-term recurrent relation

$$S_{k-3} d_{k-3} + S_{k-2} d_{k-2} + S_{k-1} d_{k-1} + S_k d_k + S_{k+1} d_{k+1} = 0\,. \tag{7.19}$$

As a quantisation rule, let us apply transcendency condition for Frobenius type functions, this yields

$$\begin{gathered} S_{k-3} = 0, \quad k \geq 3, \qquad (k-3)(k-4) + (P_1 + P_2 + P_3 - 2)\,(k-3) \\ + (-Q_1 - Q_2 - Q_3 - Q_5)\,z_1 + (-Q_1 - Q_2 - Q_3 - Q_4)\,z_2 + Q_2 - Q_3 = 0\,, \end{gathered} \tag{7.20}$$

with the use of the above formulas for coefficients it reads

$$k^2 + (2\,M + 2\,\alpha + 2\,\beta - 6)\,k - (B - 2\,\sigma\,\nu)\,z_1 - (A - 2\,\sigma\,\nu)\,z_2$$

$$+ (2\,M + 2\,\beta - 6)\,\alpha + (2\,M - 6)\,\beta - 6\,M - C - D + 2\,\nu + 9 = 0\,.$$

Whence substituting expressions for A, B, C, D, we derive

$$\begin{gathered} k \geq 3, \;\; dk^2 + 2\,(M + \alpha + \beta - 3)\,k + 2\sigma\nu\,(z_1 + z_2) + 2\,(M + \beta - 3)\,\alpha \\ +2\,(M - 3)\,\beta + 9 - 6M + 2m^2 + 4\sigma^2\nu - 2e^2 - 2E^2 = 0. \end{gathered} \tag{7.21}$$

Now let us allow for expressions for parameters

$$\alpha = +\sqrt{m^2 - (E+e)^2}, \quad \beta = \pm\sqrt{m^2 - (E-e)^2},$$

$$M = \sqrt{\nu^2 - e^2}, \quad z_1 z_2 = 1, \quad z_1 + z_2 = -2\sigma = -2\frac{E+m}{e}\;;$$

then eq. (7.21) takes the form (two variants arise depending on the choice for β):

$$\begin{gathered} \beta = +\sqrt{m^2 - (E-e)^2}, \quad \Big(\sqrt{m^2 - (E+e)^2} + \sqrt{m^2 - (E-e)^2} \\ +k - 3 + \sqrt{\nu^2 - e^2}\Big)^2 - (\nu^2 - e^2) = 0\,; \end{gathered} \tag{7.22}$$

$$\begin{gathered} \beta = -\sqrt{m^2 - (E-e)^2}, \quad \Big(\sqrt{m^2 - (E+e)^2} - \sqrt{m^2 - (E-e)^2} \\ +k - 3 + \sqrt{\nu^2 - e^2}\Big)^2 - (\nu^2 - e^2) = 0\,. \end{gathered} \tag{7.23}$$

We will follow both variants ($\pm$ signs). Let us factorise expressions (7.22) and (7.23) in product of two terms (let $k - 3 = n, \; n = 0, 1, ...$):

$$\Big(\sqrt{m^2 - (E+e)^2} \pm \sqrt{m^2 - (E-e)^2} + n + \sqrt{\nu^2 - e^2} - \sqrt{\nu^2 - e^2}\Big)$$

$$\times \Big(\sqrt{m^2 - (E+e)^2} \pm \sqrt{m^2 - (E-e)^2} + n + \sqrt{\nu^2 - e^2} + \sqrt{\nu^2 - e^2}\Big) = 0,$$

that is

$$\left(\sqrt{m^2-(E+e)^2}\pm\sqrt{m^2-(E-e)^2}+n\right)$$
$$\times\left(\sqrt{m^2-(E+e)^2}\pm\sqrt{m^2-(E-e)^2}+n+\sqrt{\nu^2-e^2}+\sqrt{\nu^2-e^2}\right)=0\,. \qquad (7.24)$$

For upper sign (when $\beta>0$), the first multiplier is positive and cannot be equal to zero; therefore, it remains only the following equation

$$\beta>0,\quad \left(\sqrt{m^2-(E+e)^2}+\sqrt{m^2-(E-e)^2}+n+\sqrt{\nu^2-e^2}+\sqrt{\nu^2-e^2}\right)=0\,;$$

however, it does not have any physical solutions because all terms are positive.

For lower sign (when $\beta<0$), we have an equation

$$\left(\sqrt{m^2-(E+e)^2}-\sqrt{m^2-(E-e)^2}+n\right)$$
$$\times\left(\sqrt{m^2-(E+e)^2}-\sqrt{m^2-(E-e)^2}+n+\sqrt{\nu^2-e^2}+\sqrt{\nu^2-e^2}\right)=0\,.$$

There arise two possibilities:

$$\sqrt{m^2-(E+e)^2}-\sqrt{m^2-(E-e)^2}+n=0\,, \qquad (7.25)$$

and

$$\sqrt{m^2-(E+e)^2}-\sqrt{m^2-(E-e)^2}+n+\sqrt{\nu^2-e^2}+\sqrt{\nu^2-e^2}=0\,. \qquad (7.26)$$

Equation (7.25) does not contain the angular parameter $\nu=j+1/2$, and it is of no physical interest. The most promising is the variant (7.26):

$$\sqrt{m^2-E^2-e^2+2eE}-\sqrt{m^2-E^2-e^2-2eE}=n+2\sqrt{\nu^2-e^2}=2N>0\,. \qquad (7.27)$$

This equation gives

$$m^2-E^2-e^2+2eE=m^2-E^2-e^2-2eE+4N\sqrt{m^2-E^2-e^2-2eE}+4N^2,$$

that is $eE-N^2=+N\sqrt{m^2-E^2-e^2-2eE}$, and further

$$E^2(e^2+N^2)=N^2(m^2-e^2)-N^4,$$

whence we arrive at the following formula for energy spectrum

$$\frac{E}{m}=\sqrt{\frac{1-(e^2+N^2)/m^2}{1+\frac{e^2}{N^2}}}\,,\qquad N=\frac{n}{2}+\sqrt{\nu^2-e^2}\,. \qquad (7.28)$$

Expression under the square root in eq. (7.28) must be positive, this provides us with the restriction

$$\frac{e^2+N^2}{m^2}<1\,. \qquad (7.29)$$

Note that this spectrum coincides with that found in [10, 14] when studying the same problem for the Dirac equation in Lobachevsky space within the WKB-approach.

TABLE 7.1
The values of ϵ and $\epsilon = E/m$

m	E	ϵ
350	349.9926417	0.9999790
500	499.9918300	0.9999837
10^3	999.9870350	0.9999870
$5 \cdot 10^3$	999.9870350	0.9999881
10^4	9999.8814869	0.9999881
$1.5 \cdot 10^4$	14999.8223240	0.9999882
$2 \cdot 10^4$	19999.7631425	0.9999882

The next question may be posed: does there exist or not the possibility to get the energy spectrum (7.28) by imposing polynomial conditions? To this end, we should turn to the recurrent formula (7.19)

$$S_{k-3}d_{k-3} + S_{k-2}d_{k-2} + S_{k-1}d_{k-1} + S_k d_k + S_{k+1}d_{k+1} = 0$$

and for energies given by eq. (7.28) check the values of three coefficients of the power series:

$$d_{k-2} = 0, \quad d_{k-1} = 0, \quad d_k = 0\,. \tag{7.30}$$

If the equalities (7.30) are valid, then from the recurrent formula, it follows that the series becomes polynomials

$$d_{k+1} = 0, \quad d_{k+2} = 0, \quad d_{k+3} = 0, \quad \dots\,.$$

Numerical study in the next section shows that eq. (7.30) cannot be satisfied.

7.3 Numerical study

Let us fix the parameters

$$e = \frac{1}{137}, \quad m = 10^3, \quad \nu = 1\;(j = \frac{1}{2}), \quad n = 1, \quad \epsilon_{n=1} = 0.99998703496159\,; \tag{7.31}$$

recall that in this case, $k - 3 = n = 1$. For different parameters m, we have the values for energy ϵ see the Table 7.1

I. Consider the variant (see Figs. 7.1–7.3):

$$e = \frac{1}{137}, \; m = 2 \cdot 10^3, \; \nu = 1, \; n = 5, \quad E = 1999.9925881514\,; \tag{7.32}$$

for these parameters the corresponding series $F(z) = \sum_{i=0}^{12} d_i z^i$ explicitly reads (restricting ourselves by 12 terms in the variable z)

$$F(z) = 1 + 730657.434 \cdot z - 4.57148738 \cdot 10^6 z^2 + 5.274935 \cdot 10^6 z^3 - 602972.6478 \cdot z^4$$
$$+591399.332 \cdot z^5 + 586542.255 \cdot z^6 + 634277.967 \cdot z^7 + 680299.381272 \cdot z^8$$
$$+723806.99277 \cdot z^9 + 764600.05174 \cdot z^{10} + 802779.5976 \cdot z^{11} + 838564.1712 \cdot z^{12} + \dots$$

Evidently, here we have an infinite series, so conditions (7.30) cannot be valid. Therefore, exact solutions do not exist in polynomials. It is readily checked that equation $F(z) = 0$ has two roots in the physical region

$$F(z) = 0\,, \quad z_1 = 0.209647, \quad z_2 = 0.612462\,. \tag{7.33}$$

II. Consider the variant (see Figs. 7.4–7.6):

$$e = \frac{1}{137}, \; m = 2 \cdot 10^4, \; \nu = 1, \; n = 5, \quad E = 19999.956199892. \tag{7.34}$$

construct the series for $F(z)$; equation $F(z) = 0$ has two roots in physical region

$$z_1 = 0.0195201, \quad z_2 = 0.0549307\,. \tag{7.35}$$

III. Consider the variant (see Figs. 7.7–7.9):

$$e = \frac{1}{137}, \; m = 5 \cdot 10^3, \; \nu = 1, \; n = 10, \; E = 4999.99270004484. \tag{7.36}$$

construct the series for $F(z)$; eq. $F(z) = 0$ has four roots in physical region

$$z_1 = 0.0745704, \; z_2 = 0.186869, \; z_3 = 0.366653, \; z_4 = 0.65265. \tag{7.37}$$

IV. Consider the variant (see Figs. 7.10–7.12):

$$e = \frac{1}{137}, \; m = 10^4, \; \nu = 1, \; n = 10, \quad E = 9999.990800048\,; \tag{7.38}$$

construct the series for $F(z)$; eq. $F(z) = 0$ has four roots in physical region

$$z_1 = 0.0363635, \; z_2 = 0.0889571, \; z_3 = 0.169237, \; z_4 = 0.293264. \tag{7.39}$$

7.4 Hydrogen atom is spherical Riemann space

In spherical Riemann space S_3, we use the following coordinates and tetrad

$$\begin{gathered} dS^2 = dt^2 - dr^2 - \sin^2 r (d\theta^2 + \sin^2 d\phi^2), \; r \in (0, \pi), \\ e^{\alpha}_{(0)} = (1,0,0,0)\,, \qquad e^{\alpha}_{(3)} = (0,1,0,0), \\ e^{\alpha}_{(1)} = (0,0,\frac{1}{\sin r},0), e^{\alpha}_{(2)} = (0,0,0,\frac{1}{\sin r \sin\theta}). \end{gathered} \tag{7.40}$$

After separating the variables, we obtain the radial system [18, 20]:

$$\begin{aligned} &\Big(\frac{d}{dr} + \frac{\nu}{\sin r}\Big) f + \Big(E + \frac{e}{\tan r} + m\Big) g = 0, \\ &\Big(\frac{d}{dr} - \frac{\nu}{\sin r}\Big) g - \Big(E + \frac{e}{\tan r} - m\Big) f = 0; \end{aligned} \tag{7.41}$$

the radial coordinate varies in the interval $r \in [0, \pi]$. In other variable

$$z = i \tan\frac{r}{2}, \quad \cos r = \frac{1+z^2}{1-z^2}, \quad \sin r = \frac{-2iz}{1-z^2}, \quad z \in [0, +i\infty), \tag{7.42}$$

the above system takes the form

$$\frac{df}{dz}+\frac{\nu}{z}f+\Big(\frac{e}{z}+\frac{iE-e+im}{z-1}+\frac{-iE-e-im}{z+1}\Big)g=0\,,$$
$$\frac{dg}{dz}-\frac{\nu}{z}g+\Big(-\frac{e}{z}+\frac{-iE+e+im}{z-1}+\frac{iE+e-im}{z+1}\Big)f=0\,, \tag{7.43}$$

whence it follows the 2nd-order equation for $f(z)$:

$$\frac{d^2f}{dz^2}+\Big[\frac{1}{z}+\frac{1}{z-1}+\frac{1}{z+1}+2\,\frac{-ez+iE+im}{ez^2-2i\,(E+m)\,z+e}\Big]\frac{df}{dz}$$
$$+\Big[-2i\,\frac{2\,Ee^2-(E+m)\,\nu}{ez}+\frac{e^2-\nu^2}{z^2}+\frac{(E+ie)^2-m^2+\nu}{z-1}+\frac{-(E+ie)^2+m^2}{(z-1)^2}$$
$$+\frac{-(E-ie)^2+m^2-\nu}{z+1}+\frac{-(E-ie)^2+m^2}{(z+1)^2}$$
$$+\frac{2\nu\left[iez(E+m)+2(E+m)^2+e^2\right]}{e\left[-ez^2+2i\,(E+m)\,z-e\right]}\Big]f=0. \tag{7.44}$$

Equation (7.44) has six singular points (let $\frac{E+m}{e}=\sigma>0$)

$$0\,,\ \infty\,,\ \pm1,\ z_{1,\,2}=i\left(\sigma\pm\sqrt{\sigma^2+1}\right)\,; \tag{7.45}$$

physical region for the variable z is the interval $z\in[0,+i\infty)$.

Equation (7.44) may be written differently

$$\frac{d^2f}{dz^2}+\Big[\frac{1}{z}+\frac{1}{z-1}+\frac{1}{z+1}-\frac{1}{z-z_1}-\frac{1}{z-z_2}\Big]\frac{df}{dz}$$
$$+\Big[\frac{-4\,iEe+2\,i\sigma\,\nu}{z}+\frac{e^2-\nu^2}{z^2}+\frac{(E+ie)^2-m^2+\nu}{z-1}+\frac{-(E+ie)^2+m^2}{(z-1)^2}$$
$$+\frac{-(E-ie)^2+m^2-\nu}{z+1}+\frac{-(E-ie)^2+m^2}{(z+1)^2}+\frac{A}{z-z_1}+\frac{B}{z-z_2}\Big]f=0\,, \tag{7.46}$$

where

$$A=-\frac{2\nu\,\left(iz_1\,\sigma+1+2\,\sigma^2\right)}{z_1-z_2}\,,\qquad B=-\frac{2\nu\,\left(iz_2\,\sigma+1+2\,\sigma^2\right)}{z_2-z_1}$$

and

$$C=-\,(E+ie)^2+m^2,\quad D=-\,(E-ie)^2+m^2,\quad -4iEe=C-D.$$

Then eq. (7.46) takes the form

$$\frac{d^2f}{dz^2}+\Big[\frac{1}{z}+\frac{1}{z-1}+\frac{1}{z+1}-\frac{1}{z-z_1}-\frac{1}{z-z_2}\Big]\frac{df}{dz}$$
$$+\Big[\frac{C-D+2\,i\sigma\,\nu}{z}+\frac{e^2-\nu^2}{z^2}-\frac{C-\nu}{z-1}+\frac{C}{(z-1)^2}$$
$$+\frac{D-\nu}{z+1}+\frac{D}{(z+1)^2}+\frac{A}{z-z_1}+\frac{B}{z-z_2}\Big]f=0\,.$$

Frobenius solutions in vicinity of the point $z = 0$ are searched in the form

$$f(z) = z^M (z-1)^\alpha (z+1)^\beta \, g(z) = \varphi(z) g(z)\,; \tag{7.47}$$

the function $g(z)$ obeys the equation

$$\frac{d^2 g}{dz^2} + \Big[\frac{2M+1}{z} + \frac{2\alpha+1}{z-1} + \frac{2\beta+1}{z+1} - \frac{1}{z-z_1} - \frac{1}{z-z_2}\Big]\frac{dg}{dz}$$

$$+ \Big[\frac{M^2+e^2-\nu^2}{z^2} + \frac{\alpha^2+C}{(z-1)^2} + \frac{\beta^2+D}{(z+1)^2}$$

$$+ \frac{C-D-(\alpha-\beta)(2M+1)+2i\sigma(\nu+M)}{z} + \frac{M(z_1 - 2\,i\sigma\, z_1 z_2 + z_2)}{z\, z_2\, z_1}$$

$$+ \frac{M+\alpha/2+\beta/2-C+\nu+2M\alpha+\alpha\beta}{z-1} - \frac{\alpha\,(1-z_1 z_2)}{(z-1)(z_1-1)(z_2-1)}$$

$$- \frac{M+\alpha/2+\beta/2-D+\nu+2M\beta+\alpha\beta}{z+1} + \frac{\beta\,(1-z_1 z_2)}{(z+1)(z_1+1)(z_2+1)}$$

$$+ \frac{1}{z-z_1}\Big(A - \frac{\alpha}{z_1-1} - \frac{\beta}{z_1+1} - \frac{M}{z_1}\Big) + \frac{1}{z-z_2}\Big(B - \frac{\alpha}{z_2-1} - \frac{\beta}{z_2+1} - \frac{M}{z_2}\Big)\Big]\, g = 0\,.$$

Impose restrictions

$$M = \pm\sqrt{\nu^2 - e^2}\,,$$

$$\alpha = \pm\sqrt{-C} = \pm\sqrt{(E+ie)^2 - m^2} = \pm\sqrt{E^2 - m^2 - e^2 + 2ieE}\,,$$
$$\beta = \pm\sqrt{-D} = \pm\sqrt{(E-ie)^2 - m^2} = \pm\sqrt{E^2 - m^2 - e^2 - 2ieE}\,. \tag{7.48}$$

To have solutions vanishing at the point $z = 0$ ($r = 0$), we must use positive value for M: $M = +\sqrt{\nu^2 - e^2}$; near the point $z = +\infty$ ($r = \pi$) the multiplier φ before $g(z)$ behaves as follows

$$\varphi = z^M (z-1)^\alpha (z+1)^\beta \sim x^{\sqrt{\nu^2-e^2}+(\alpha+\beta)}, \quad z = ix; \tag{7.49}$$

depending on signs at α, β there exist four possibilities:

$$(-,-) \qquad \alpha+\beta = -\sqrt{E^2 - m^2 - e^2 + 2ieE} - \sqrt{E^2 - m^2 - e^2 - 2ieE} < 0\,,$$
$$(+,+) \qquad \alpha+\beta = \sqrt{E^2 - m^2 - e^2 + 2ieE} + \sqrt{E^2 - m^2 - e^2 - 2ieE} > 0\,;$$
$$(+,-) \qquad \alpha+\beta = \sqrt{E^2 - m^2 - e^2 + 2ieE} - \sqrt{E^2 - m^2 - e^2 - 2ieE} \;\; \text{imaginary}\,;$$
$$(-,+) \qquad \alpha+\beta = -\sqrt{E^2 - m^2 - e^2 + 2ieE} + \sqrt{E^2 - m^2 - e^2 - 2ieE} \;\; \text{imaginary}\,.$$

We can see that only two first variants may give multipliers tending to zero; this requires the following inequality: $M + \alpha + \beta < 0$. The inequality $M + \alpha + \beta < 0$ is definitely true for the case $(-,-)$.

Now we turn to equation for $g(z)$:

$$\frac{d^2 g}{dz^2} + \Big[\frac{2M+1}{z} + \frac{2\alpha+1}{z-1} + \frac{2\beta+1}{z+1} - \frac{1}{z-z_1} - \frac{1}{z-z_2}\Big]\frac{dg}{dz}$$

$$+ \Big[\frac{C-D-(\alpha-\beta)(2M+1)+2i\sigma(\nu+M)}{z} + \frac{M(z_1 - 2\,i\sigma\, z_1 z_2 + z_2)}{z\, z_2\, z_1}$$

$$+\frac{M+\alpha/2+\beta/2-C+\nu+2M\alpha+\alpha\beta}{z-1}-\frac{\alpha\,(1-z_1\,z_2)}{(z-1)\,(z_1-1)\,(z_2-1)}$$
$$-\frac{M+\alpha/2+\beta/2-D+\nu+2M\beta+\alpha\beta}{z+1}+\frac{\beta\,(1-z_1\,z_2)}{(z+1)\,(z_1+1)\,(z_2+1)}$$
$$+\frac{1}{z-z_1}\Big(A-\frac{\alpha}{z_1-1}-\frac{\beta}{z_1+1}-\frac{M}{z_1}\Big)+\frac{1}{z-z_2}\Big(B-\frac{\alpha}{z_2-1}-\frac{\beta}{z_2+1}-\frac{M}{z_2}\Big)\Big]\,g=0\,,$$

re-write it shorter

$$\frac{d^2g}{dz^2}+\Big(\frac{P_1}{z}+\frac{P_2}{z-1}+\frac{P_3}{z+1}-\frac{1}{z-z_1}-\frac{1}{z-z_2}\Big)\frac{dg}{dz}$$
$$+\Big(\frac{Q_1}{z}+\frac{Q_2}{z-1}+\frac{Q_3}{z+1}+\frac{Q_4}{z-z_1}+\frac{Q_5}{z-z_2}\Big)g=0\,, \tag{7.50}$$

and multiply it by $z(z-1)(z+1)(z-z_1)(z-z_2)$:

$$\left[z^5+(-z_1-z_2)\,z^4+(z_1\,z_2-1)\,z^3+(z_1+z_2)\,z^2-z_1\,z_2\,z\right]\frac{d^2g}{dz^2}$$
$$+\left[(P_1+P_2+P_3-2)\,z^4+\{(1-P_1-P_2-P_3)\,z_1+(1-P_1-P_2-P_3)\,z_2+P_2-P_3\}\,z^3\right.$$
$$+(2-P_1-P_2\,z_1+P_3\,z_2+P_2\,z_1\,z_2+P_1\,z_1\,z_2+P_3\,z_1\,z_2+P_3\,z_1-P_2\,z_2)\,z^2$$
$$\left.+(-z_1-z_2+P_1\,z_1-P_3\,z_1\,z_2+P_1\,z_2+P_2\,z_1\,z_2)\,z-P_1\,z_1\,z_2\right]\frac{dg}{dz}$$
$$+[(Q_1+Q_2+Q_3+Q_4+Q_5)z^4+\{(-Q_1-Q_2-Q_3-Q_5)z_1$$
$$+(-Q_1-Q_2-Q_3-Q_4)z_2+Q_2-Q_3\}z^3$$
$$+(Q_3\,z_1\,z_2+Q_2\,z_1\,z_2+Q_3\,z_2-Q_1-Q_4-Q_5+Q_1\,z_1\,z_2+Q_3\,z_1-Q_2\,z_2-Q_2\,z_1)\,z^2$$
$$+(Q_1\,z_2+Q_2\,z_1\,z_2+Q_5\,z_1+Q_1\,z_1-Q_3\,z_1\,z_2+Q_4\,z_2)\,z-Q_1\,z_1\,z_2]\,g=0\,.$$

Solutions for $g(z)$ are constructed as power series with 6-term recurrent relations

$$k\geq 4,\qquad (Q_1+Q_2+Q_3+Q_4+Q_5)\,d_{k-4}$$
$$+[(k-3)(k-4)+(P_1+P_2+P_3-2)\,(k-3)$$
$$+(-Q_1-Q_2-Q_3-Q_5)\,z_1+(-Q_1-Q_2-Q_3-Q_4)\,z_2+Q_2-Q_3]\,d_{k-3}$$
$$+[(-z_1-z_2)\,(k-2)(k-3)+\{(1-P_1-P_2-P_3)\,z_1$$
$$+(1-P_1-P_2-P_3)\,z_2+P_2-P_3\}\,(k-2)$$
$$+Q_3\,z_1\,z_2+Q_2\,z_1\,z_2+Q_3\,z_2-Q_1-Q_4-Q_5+Q_1\,z_1\,z_2+Q_3\,z_1-Q_2\,z_2-Q_2\,z_1]\,d_{k-2}$$
$$+[(z_1z_2-1)\,(k-1)(k-2)+(2-P_1-P_2z_1+P_3z_2$$
$$+P_2z_1z_2+P_1z_1z_2+P_3z_1z_2+P_3z_1-P_2z_2)(k-1)$$
$$+Q_1\,z_2+Q_2\,z_1\,z_2+Q_5\,z_1+Q_1\,z_1-Q_3\,z_1\,z_2+Q_4\,z_2]\,d_{k-1}$$
$$+[(z_1+z_2)\,k\,(k-1)+(-z_1-z_2+P_1\,z_1-P_3\,z_1\,z_2+P_1\,z_2+P_2\,z_1\,z_2)\,k-Q_1\,z_1\,z_2]\,d_k$$
$$+[-z_1\,z_2\,(k+1)\,k-P_1\,z_1\,z_2\,(k+1)]\,d_{k+1}=0\,. \tag{7.51}$$

Possible convergence radii are

$$R_{\rm conv}=\left|\frac{1}{R}\right|=+1\,,\ +\infty\,,\ |z_1|\,,\ |z_2|. \tag{7.52}$$

It is readily checked that the coefficient at d_{k-4} (7.51) vanishes identically, so in (7.51) we have 5-term recurrent relations

$$k \geq 4, \quad S_{k-3}d_{k-3} + S_{k-2}d_{k-2} + S_{k-1}d_{k-1} + S_k d_k + S_{k+1}d_{k+1} = 0 \,. \tag{7.53}$$

As a quantisation rule, we apply the known transcendency condition

$$k \geq 3, \quad S_{k-3} = 0, \qquad (k-3)(k-4) + (P_1 + P_2 + P_3 - 2)\,(k-3)$$
$$+(-Q_1 - Q_2 - Q_3 - Q_5)\,z_1 + (-Q_1 - Q_2 - Q_3 - Q_4)\,z_2 + Q_2 - Q_3 = 0\,, \tag{7.54}$$

which yields

$$k^2 + (2\,M + 2\,\alpha + 2\,\beta - 6)\,k - (B + 2\,i\sigma\,\nu)\,z_1 - (A + 2\,i\sigma\,\nu)\,z_2$$
$$+\,(2\,M + 2\,\beta - 6)\,\alpha + (2\,M - 6)\,\beta - 6\,M - C - D + 2\,\nu + 9 = 0\,.$$

Whence, taking into account expressions for A, B, C, D:

$$A = -\frac{2\nu\,\left(iz_1\,\sigma + 1 + 2\,\sigma^2\right)}{z_1 - z_2}\,, \quad B = -\frac{2\nu\,\left(iz_2\,\sigma + 1 + 2\,\sigma^2\right)}{z_2 - z_1}\,,$$

$$C = -\,(E + ie)^2 + m^2\,, \quad D = -\,(E - ie)^2 + m^2\,, \quad -4iEe = C - D\,,$$

we arrive at

$$k^2 + 2\,k\,(M + \alpha + \beta - 3) - 2\,i\sigma\,\nu\,\left(z_1 + z_2\right) + 2\,\left(M + \beta - 3\right)\alpha + 2\,\left(M - 3\right)\beta$$
$$+9 - 6\,M - 2\,m^2 - 4\,\nu\,\sigma^2 - 2\,e^2 + 2\,E^2 = 0\,. \tag{7.55}$$

We will follow two possibilities. The first one is

$$M = \sqrt{\nu^2 - e^2}\,, \quad \alpha = +\sqrt{(E + ie)^2 - m^2}\,, \quad \beta = +\sqrt{(E - ie)^2 - m^2}\,, \tag{7.56}$$

then eq. (7.55) takes the form

$$\left(\sqrt{(E + ie)^2 - m^2} + \sqrt{(E - ie)^2 - m^2} + k - 3 + \sqrt{\nu^2 - e^2}\right)^2 - \left(\nu^2 - e^2\right) = 0$$

or differently

$$\left(\sqrt{(E + ie)^2 - m^2} + \sqrt{(E - ie)^2 - m^2} + k - 3\right)$$
$$\times\left(\sqrt{(E + ie)^2 - m^2} + \sqrt{(E - ie)^2 - m^2} + k - 3 + 2\sqrt{\nu^2 - e^2}\right) = 0\,.$$

Here arise two equations, both of small physical interest (let $n = k - 3$):

$$\sqrt{(E + ie)^2 - m^2} + \sqrt{(E - ie)^2 - m^2} + n = 0\,, \tag{7.57}$$

$$\sqrt{(E + ie)^2 - m^2} + \sqrt{(E - ie)^2 - m^2} + n + 2\sqrt{\nu^2 - e^2} = 0\,. \tag{7.58}$$

Now consider the second variant

$$M = \sqrt{\nu^2 - e^2}\,, \quad \alpha = -\sqrt{(E + ie)^2 - m^2}\,, \quad \beta = -\sqrt{(E - ie)^2 - m^2}\,, \tag{7.59}$$

then we have transcendency condition in the form

$$\left(\sqrt{(E+ie)^2-m^2}+\sqrt{(E-ie)^2-m^2}-(k-3)-\sqrt{\nu^2-e^2}\right)^2-(\nu^2-e^2)=0$$

or differently

$$\left(\sqrt{(E+ie)^2-m^2}+\sqrt{(E-ie)^2-m^2}-(k-3)\right)$$
$$\times\left(\sqrt{(E+ie)^2-m^2}+\sqrt{(E-ie)^2-m^2}-(k-3)-2\sqrt{\nu^2-e^2}\right)=0\,.$$

So we obtain two alternative equations

$$\sqrt{(E+ie)^2-m^2}+\sqrt{(E-ie)^2-m^2}-n=0\,, \tag{7.60}$$

$$\sqrt{(E+ie)^2-m^2}+\sqrt{(E-ie)^2-m^2}-n-2\sqrt{\nu^2-e^2}=0\,. \tag{7.61}$$

Interesting is only the second one (7.61), it yields

$$\sqrt{(E+ie)^2-m^2}=2N-\sqrt{(E-ie)^2-m^2},\quad N=n/2+\sqrt{\nu^2-e^2}$$

or

$$N\sqrt{E^2-m^2-e^2-2\,iEe}=-iEe+N^2\,.$$

Further, we obtain

$$N^2(E^2-m^2-e^2-2ieE)=N^4-2ieEN^2-e^2E^2,$$

whence it follows the needed energy spectrum

$$E=m\sqrt{\frac{1+(e^2+N^2)/m^2}{1+e^2/N^2}},\quad N=\frac{n}{2}+\sqrt{\nu^2-e^2},\quad m=\frac{Mc\rho}{\hbar}\,. \tag{7.62}$$

This spectrum coincides with that found in [26, 42], when studying the same problem for the Dirac equation in the spherical Riemann space within the semi-classical approach.

Below, we will demonstrate that there is no possibility to get the energy spectrum (7.62) in polynomials.

7.5 Numerical study

It is convenient to use a dimensionless parameter $\epsilon=E/m$. Let us present the expression for ϵ in the product form $\epsilon=\epsilon_1\epsilon_2$, where ϵ_1 determines the energy spectrum in flat Minkowski space and ϵ_2 describes the influence of cirved geometry (it depends on parameter m).

Let us calculate the values for ϵ_1, ϵ_2, and ϵ at $n = \overline{1, 12}$ and $m = 1, 10^3, 10^4$:

ϵ_1	ϵ_2	$\epsilon_1\epsilon_2$	
0.9999881599,	1.802768249,	1.802746904,	$n = 1, m = 1$
0.9999933400,	2.236056064,	2.236041171,	$n = 2, m = 1$
0.9999957376,	2.692567563,	2.692556086,	$n = 3, m = 1$
0.9999970400,	3.162260811,	3.162251451,	$n = 4, m = 1$
0.9999978253,	3.640036648,	3.640028732,	$n = 5, m = 1$
0.9999983350,	4.123086242,	4.123079377,	$n = 6, m = 1$
0.9999986844,	4.609752002,	4.609745938,	$n = 7, m = 1$
0.9999989344,	5.098998615,	5.098993182,	$n = 8, m = 1$
0.9999991193,	5.590148499,	5.590143576,	$n = 9, m = 1$
0.9999992600,	6.082740632,	6.082736131,	$n = 10, m = 1$
0.9999993695,	6.576450940,	6.576446793,	$n = 11, m = 1$
0.9999994563,	7.071045207,	7.071041363,	$n = 12, m = 1$

ϵ_1	ϵ_2	$\epsilon_1\epsilon_2$	
0.9999881599,	1.000001125,	0.9999892849,	$n = 1, m = 1000$
0.9999933400,	1.000002000,	0.9999953399,	$n = 2, m = 1000$
0.9999957376,	1.000003125,	0.9999988625,	$n = 3, m = 1000$
0.9999970400,	1.000004500,	1.000001540,	$n = 4, m = 1000$
0.9999978253,	1.000006125,	1.000003950,	$n = 5, m = 1000$
0.9999983350,	1.000008000,	1.000006335,	$n = 6, m = 1000$
0.9999986844,	1.000010125,	1.000008809,	$n = 7, m = 1000$
0.9999989344,	1.000012500,	1.000011434,	$n = 8, m = 1000$
0.9999991193,	1.000015125,	1.000014244,	$n = 9, m = 1000$
0.9999992600,	1.000018000,	1.000017260,	$n = 10, m = 1000$
0.9999993695,	1.000021125,	1.000020494,	$n = 11, m = 1000$
0.9999994563,	1.000024500,	1.000023956,	$n = 12, m = 1000$

ϵ_1	ϵ_2	$\epsilon_1\epsilon_2$	
0.9999881599,	1.000000011,	0.9999881712,	$n = 1, m = 10^4$
0.9999933400,	1.000000020,	0.9999933600,	$n = 2, m = 10^4$
0.9999957376,	1.000000031,	0.9999957688,	$n = 3, m = 10^4$
0.9999970400,	1.000000045,	0.9999970850,	$n = 4, m = 10^4$
0.9999978253,	1.000000061,	0.9999978866,	$n = 5, m = 10^4$
0.9999983350,	1.000000080,	0.9999984150,	$n = 6, m = 10^4$
0.9999986844,	1.000000101,	0.9999987857,	$n = 7, m = 10^4$
0.9999989344,	1.000000125,	0.9999990594,	$n = 8, m = 10^4$
0.9999991193,	1.000000151,	0.9999992706,	$n = 9, m = 10^4$
0.9999992600,	1.000000180,	0.9999994400,	$n = 10, m = 10^4$
0.9999993695,	1.000000211,	0.9999995807,	$n = 11, m = 10^4$
0.9999994563,	1.000000245,	0.9999997013,	$n = 12, m = 10^4$

Let us find the values for the possible convergence radii $|z_{1(n)}|$ and $|z_{2(n)}|$ at $n = \overline{1, 10}$.

Let $m = 1$:

$\sigma_{(n)}$	$\lvert z_{1(n)}\rvert$	$\lvert z_{2(n)}\rvert$
383.9763259	767.9539539	0.0013022
443.3376405	886.6764088	0.0011278
505.8801838	1011.7613559	0.0009884
570.2284488	1140.4577745	0.0008768
635.6839363	1271.3686592	0.0007866
701.8618747	1403.724462	0.000712
768.5351935	1537.071037	0.000651
835.5620659	1671.124730	0.000598
902.8496699	1805.699894	0.000554
970.3348500	1940.670215	0.000515

Now let $m = 10^3$:

$\sigma_{(n)}$	$\lvert z_{1(n)}\rvert$	$\lvert z_{2(n)}\rvert$
273998.53203243916184	547997.06406670315098	$1.82482729 \cdot 10^{-6}$
273999.36157030053471	547998.72314242589120	$1.82482177 \cdot 10^{-6}$
273999.84416671429123	547999.68833525340102	$1.82481856 \cdot 10^{-6}$
274000.21096908314543	548000.42193999110697	$1.82481611 \cdot 10^{-6}$
274000.54117852697293	548001.08235887875978	$1.82481391 \cdot 10^{-6}$
274000.86787818381619	548001.73575819244411	$1.82481174 \cdot 10^{-6}$
274001.20687271021047	548002.41374724523043	$1.82480948 \cdot 10^{-6}$
274001.56648614990005	548003.13297412460718	$1.82480709 \cdot 10^{-6}$
274001.95144098015059	548003.90288378510571	$1.82480452 \cdot 10^{-6}$
274002.36457819477482	548004.72915821435140	$1.82480177 \cdot 10^{-6}$

Let $m = 10^4$:

$\sigma_{(n)}$	$\lvert z_{1(n)}\rvert$	$\lvert z_{2(n)}\rvert$
$2.7399837945238901875 \cdot 10^6$	$5.4799675890479628579 \cdot 10^6$	$1.824828 \cdot 10^{-7}$
$2.7399909031599425982 \cdot 10^6$	$5.4799818063200676787 \cdot 10^6$	$1.824824 \cdot 10^{-7}$
$2.7399942033085950102 \cdot 10^6$	$5.4799884066173725025 \cdot 10^6$	$1.824821 \cdot 10^{-7}$
$2.7399960064450303417 \cdot 10^6$	$5.4799920128902431654 \cdot 10^6$	$1.824820 \cdot 10^{-7}$
$2.7399971045818599740 \cdot 10^6$	$5.4799942091639024299 \cdot 10^6$	$1.824819 \cdot 10^{-7}$
$2.7399978285521337912 \cdot 10^6$	$5.4799956571044500644 \cdot 10^6$	$1.824819 \cdot 10^{-7}$
$2.7399983364043434328 \cdot 10^6$	$5.4799966728088693474 \cdot 10^6$	$1.824819 \cdot 10^{-7}$
$2.7399987113811094465 \cdot 10^6$	$5.4799974227624013747 \cdot 10^6$	$1.824818 \cdot 10^{-7}$
$2.7399990007096444048 \cdot 10^6$	$5.4799980014194712913 \cdot 10^6$	$1.824818 \cdot 10^{-7}$
$2.7399992328025831516 \cdot 10^6$	$5.4799984656053487851 \cdot 10^6$	$1.824818 \cdot 10^{-7}$

Because solutions near the points z_1, z_2 are given by the formulas

$$f(z) \sim (z - z_1)^a, \quad (z - z_2)^a, \quad \text{where } a = 0, 2,$$

we may state that at these points, the series has no singular behaviour. Therefore, we may assume that the convergence radius of the power series under consideration equals the unit, $R_{\text{conv}} = 1$.

7.6 Solutions in the half-spaces, $r \in (0, \pi/2)$ and $r \in (\pi/2, \pi)$

The complete spherical space consists of two-half spaces:

$$S_3^+, \quad r \in (0, \frac{\pi}{2}), \quad z = i \tan\frac{r}{2} = ix, \quad x \in (0, +1);$$

$$S_3^-, \quad r \in (\frac{\pi}{2}, \pi), \quad z = i \tan\frac{r}{2} = ix, \quad x \in (+1, +\infty).$$

Fig. 7.13, shows behaviour of the factor $|\varphi_n(x)|$ in the half-space S_3^+; Fig. 7.14 shows behaviour of the complete solution $|f_n(x)|$, $n = \bar{1}$, in the same half-space S_3^+. This figure proves the finiteness of solutions in S_3^+, they are quadratically integrable.

In order to construct solutions $f(z)$ in the half-space S_3^-. To this end, let us transform eq. (7.44)

$$\frac{d^2 f}{dz^2} + \Big[\frac{1}{z} + \frac{1}{z-1} + \frac{1}{z+1} + 2\frac{-ez + iE + im}{ez^2 - 2i\,(E+m)\,z + e}\Big]\frac{df}{dz}$$
$$+\Big[-2i\,\frac{2\,Ee^2 - (E+m)\,\nu}{ez} + \frac{e^2 - \nu^2}{z^2} + \frac{(E+ie)^2 - m^2 + \nu}{z-1} + \frac{-(E+ie)^2 + m^2}{(z-1)^2}$$
$$+\frac{-(E-ie)^2 + m^2 - \nu}{z+1} + \frac{-(E-ie)^2 + m^2}{(z+1)^2} + \frac{2\nu\left[iez(E+m) + 2(E+m)^2 + e^2\right]}{e\left[-ez^2 + 2i\,(E+m)\,z - e\right]}\Big]f = 0$$

to the new variable

$$y = \frac{1}{z} = -i\frac{1}{x} = i\bar{x}, \quad \bar{x} \in (-\infty, 0). \tag{7.63}$$

Frobenius type solutions in vicinity of the point $y = 0$ ($r = \pi$) are searched in the form

$$f(y) = y^c (y-1)^a (y+1)^b g(y) = \varphi(y) g(y). \tag{7.64}$$

To bound states there corresponds the following parameters

$$c = \sqrt{\nu^2 - e^2}, \quad a = \sqrt{-e^2 + 2ie\epsilon - m^2 + \epsilon^2}, \quad b = \sqrt{-e^2 - 2ie\epsilon - m^2 + \epsilon^2}. \tag{7.65}$$

Solutions for $g(y)$ are given by power series with 5-term recurrence relations (details are omitted). The transcendency condition gives yet another known formula (7.62) for energy levels. We have studied constructed solutions in the half-space S_3^- numerically (in the domain $\bar{x} \in (-1, 0)$). Fig. 7.15 demonstrates behaviour of complete solutions with different energies:

$$|f_n(y)|, \quad n = \bar{1}, \quad \nu = 1, \quad m = 10^4.$$

The Fig. 7.15 proves the finiteness of solutions in this half-space, they are quadratically integrable.

Let us compare the values of solutions at contiguity point $r = \pi/2$ (both functions are defined up to multipliers):

$$f_-(x \to +1 - 0) \approx 1.2 \cdot 10^{162}, \quad f_+(\bar{x} \to -1 + 0) \approx 6.2 \cdot 10^{146};$$

whence we find the relative parameter Λ:

$$f_- = \Lambda f_+, \quad \Lambda \approx 1.9 \cdot 10^{15}.$$

All graphs may be re-calculated to the initial variable r. Let us illustrate this for one case, $m = 10^4$, $\nu = 1$, $n = 1$ – see Figs. 7.16 and 7.17.

7.7 Conclusion

The known systems of radial equations describing relativistic hydrogen atoms on the basis of Dirac equation in spherical Riemann spaces are investigated. The relevant 2nd-order differential equations have six regular singular points, there solutions of Frobenius type are constructed. To produce the quantisation rule for energy values, we use the known condition separating transcendental Frobenius solutions. This provides us with energy spectra which are physically interpretable and similar to spectra arising from scalar Klein–Fock–Gordon equation in these geometrical models. The spectra coincide with those previously found when studying the same radial equations within the semi-classical method. The squared of the series involved is proved analytically and numerically. The squared integrability of solutions is demonstrated numerically. A visualisation of the results is given.

7.8 Figures for the problem in Lobachevsky space

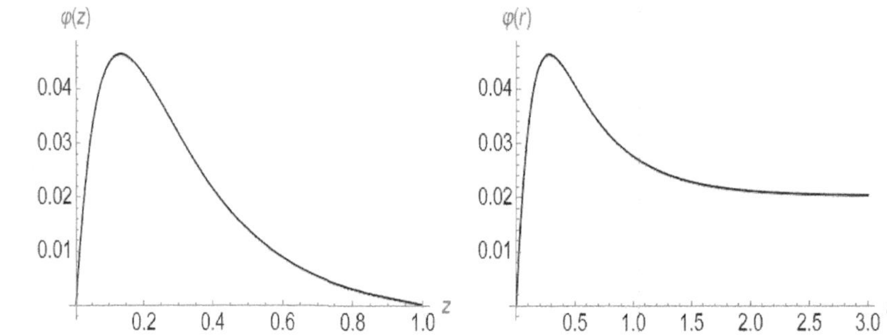

FIGURE 7.1
The graphs of the factors $\varphi(z)$ and $\varphi(r)$, at $m = 2 \cdot 10^3, n = 5$.

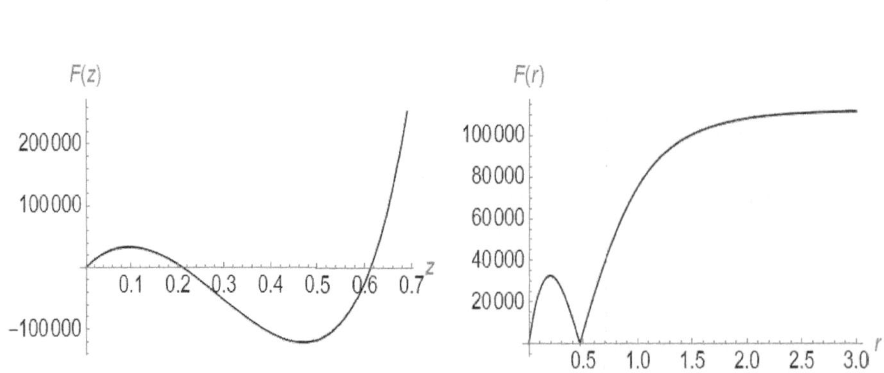

FIGURE 7.2
The graphs of the series $F(z)$ and $F(r)$, at $m = 2 \cdot 10^3, n = 5$.

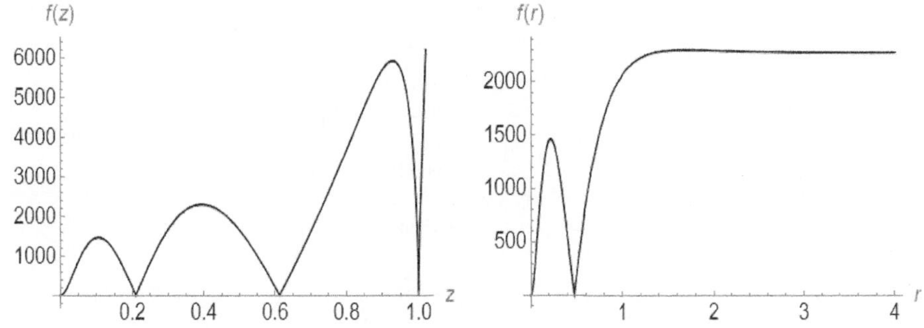

FIGURE 7.3
The graphs of the function $f(z)$ and $f(r)$, at $m = 2 \cdot 10^3, n = 5$.

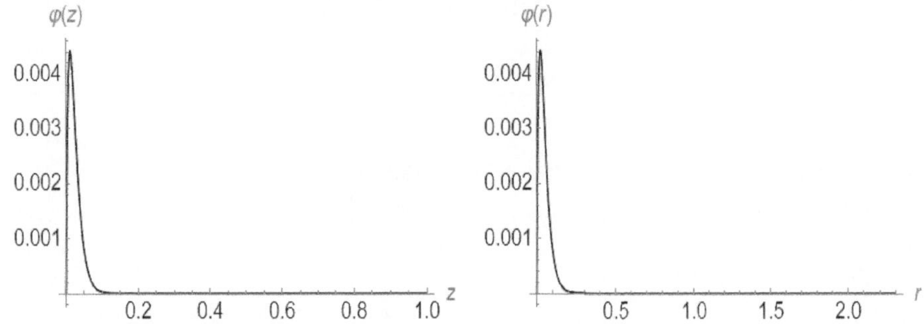

FIGURE 7.4
The graphs of the factors $\varphi(z)$ and $\varphi(r)$, at $m = 2 \cdot 10^4, n = 5$.

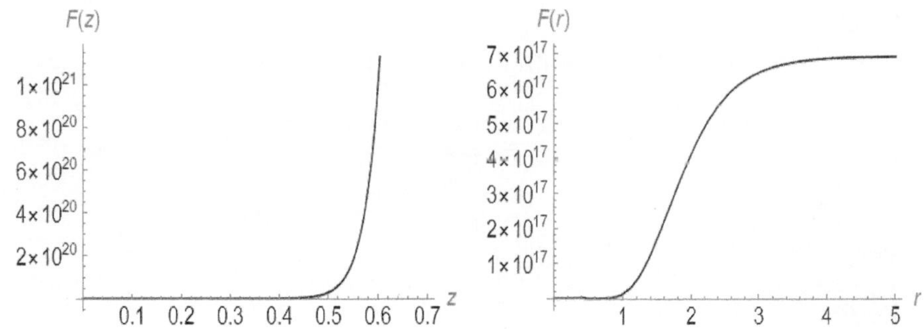

FIGURE 7.5
The graphs of the series $F(z)$ and $F(r)$, at $m = 2 \cdot 10^4, n = 5$.

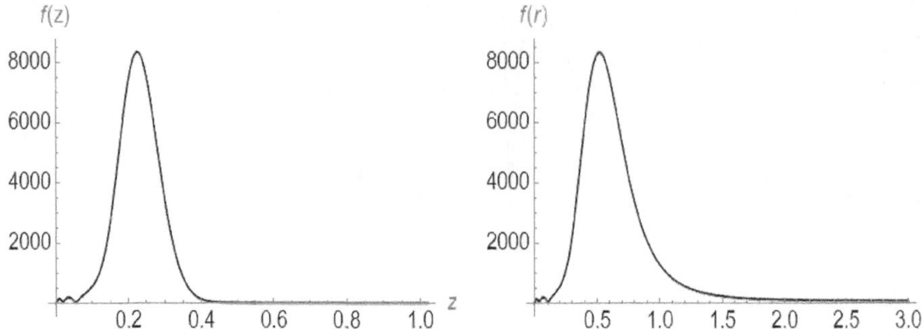

FIGURE 7.6
The graphs of the function $f(z)$ and $f(r)$, at $m = 2 \cdot 10^4, n = 5$.

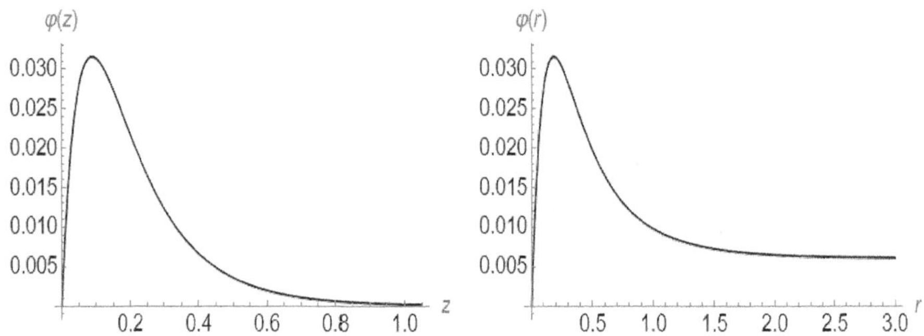

FIGURE 7.7
The graphs of the factors $\varphi(z)$ and $\varphi(r)$, at $m = 5 \cdot 10^3, n = 10$.

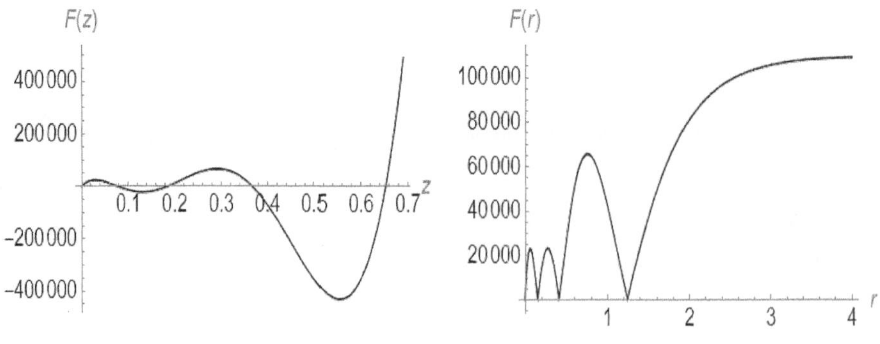

FIGURE 7.8
The graphs of the series $F(z)$ and $F(r)$, at $m = 5 \cdot 10^3, n = 10$.

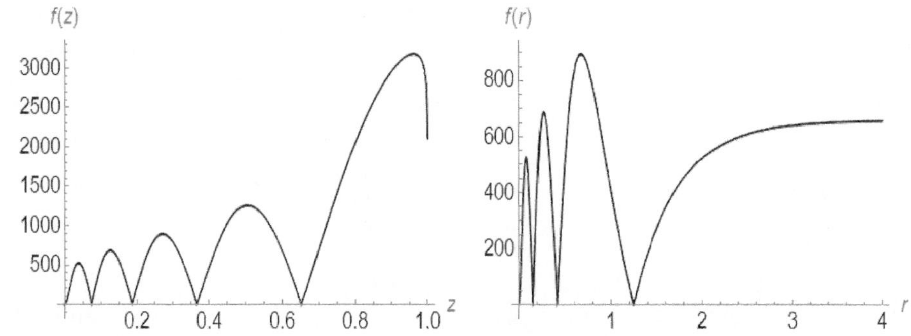

FIGURE 7.9
The graphs of the function $f(z)$ and $f(r)$, at $m = 5 \cdot 10^3, n = 10$.

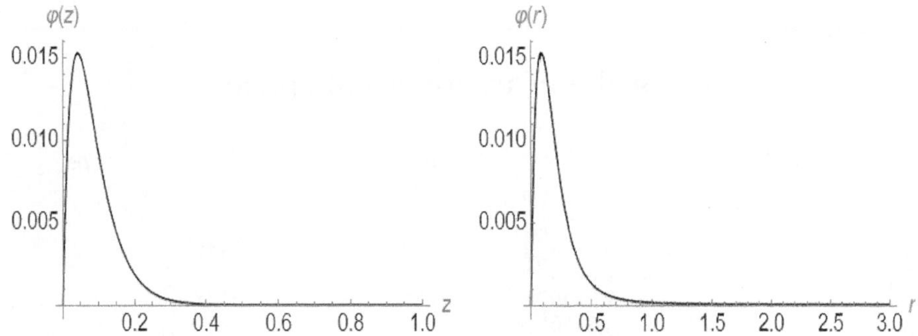

FIGURE 7.10
The graphs of the factors $\varphi(z)$ and $\varphi(r)$, at $m = 10^4, n = 10$.

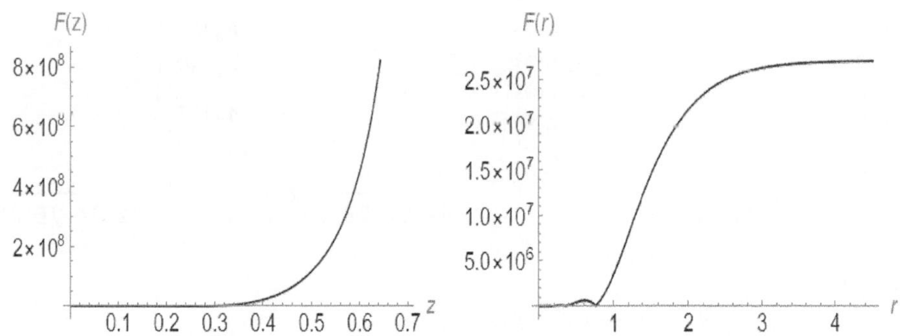

FIGURE 7.11
The graphs of the series $F(z)$ and $F(r)$, at $m = 10^4, n = 10$.

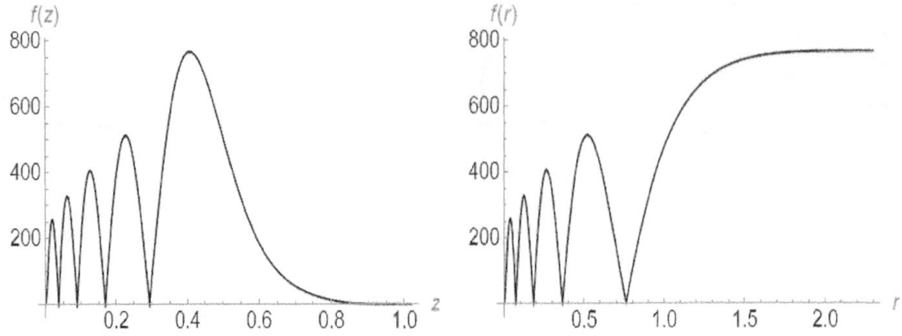

FIGURE 7.12
The graphs of the function $f(z)$ and $f(r)$, at $m = 10^4, n = 10$.

7.9 Figures for the problem in spherical space

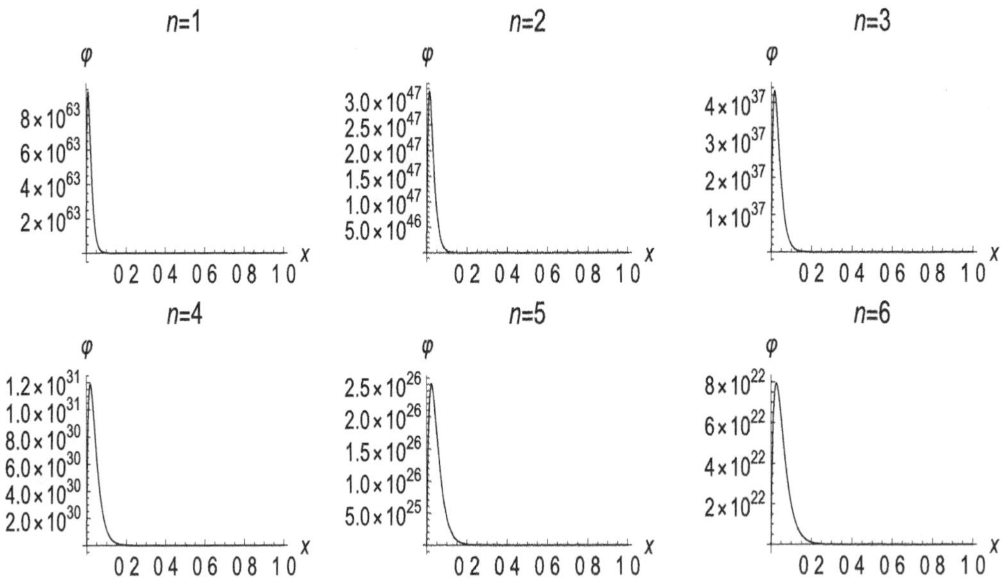

FIGURE 7.13
The graph for $\varphi_n(x)$, when $m = 10^4$, $n = \overline{1,6}$.

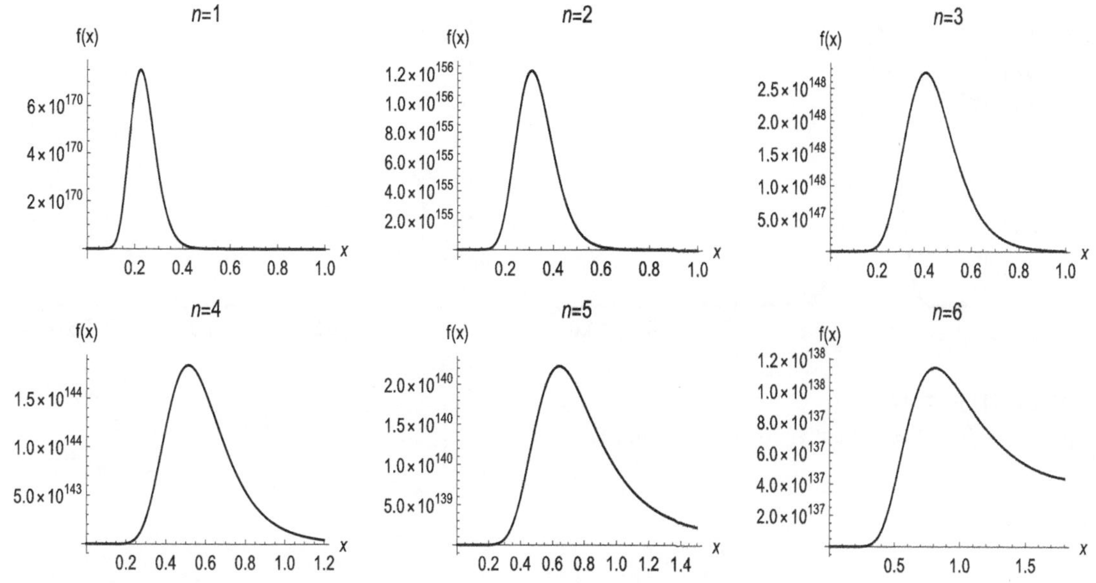

FIGURE 7.14
Graphs for $|f_n(x)|$, *when* $n = \overline{1,6}$.

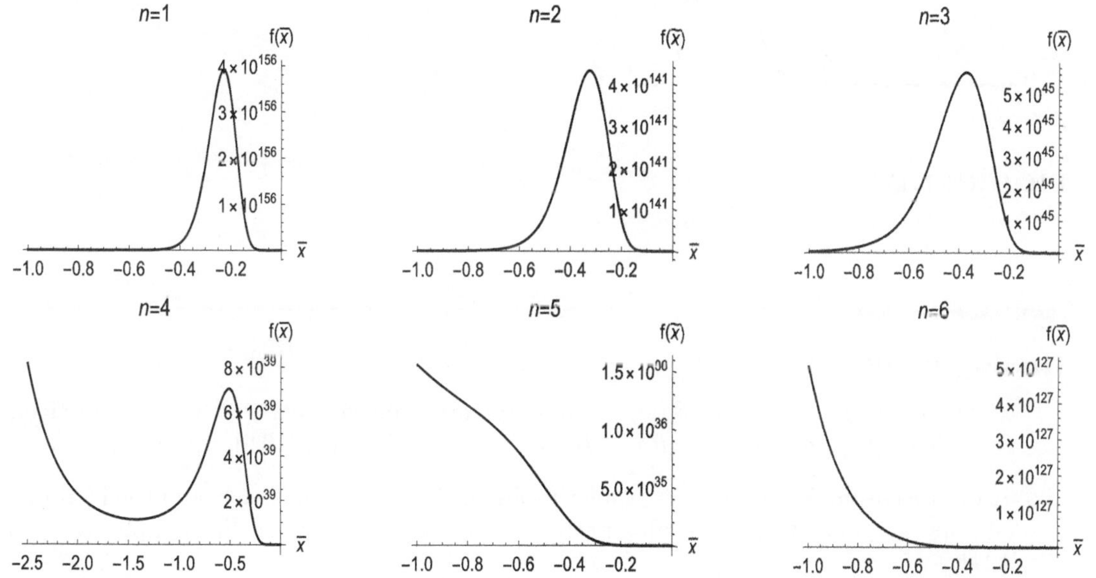

FIGURE 7.15
Graphs for $|f_n(x)|$, *when* $n = \overline{1,6}$.

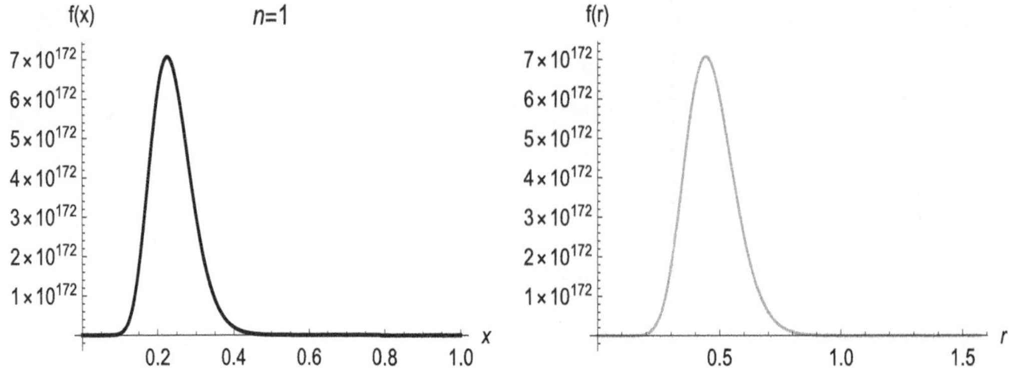

FIGURE 7.16
Graphs for $|f(r)|$ *in* $r \in (0, \pi/2)$*, when* $n = 1$.

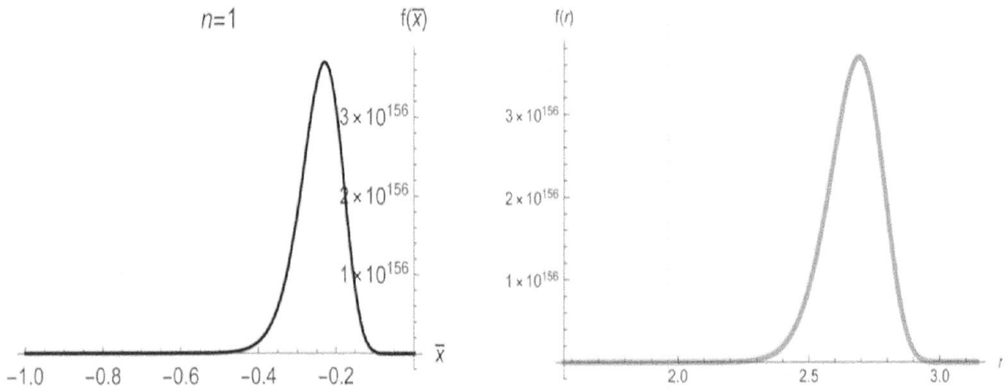

FIGURE 7.17
Graphs for $|f(r)|$ *in* $r \in (\pi/2, \pi)$*, when* $n = 1$.

Bibliography

[1] E. Schrödinger. A method of determining quantum-mechanical eigenvalues and eigenfunctions. *Proceedings of the Royal Irish Academy*, Vol. 46A, 1940, pp. 9–16.

[2] A.F. Stevenson. A note on the "Kepler problem" in a spherical space, and the factorisation method of solving eigenvalue problems. *Journals of Physical Review*, 59: 842–843, 1941.

[3] L. Infeld and A. Schild. A note on the Kepler problem in a space of constant negative curvature. *Journals of Physical Review*, 67(2): 121–122, 1945.

[4] M.E. Gertsnshtein and Yu.S. Sayasov. On spectrum for stationary states of the Dirac equation. In *Proceedings of the 2-nd Soviet Gravitational Conference "Cosmology and Quantum Mechanics, Problems of Gravitation"*, pages 182–183. Tbilisi, 1965.

[5] N. Bessis and G. Bessis. Electronic wave functions in a space of constant curvature. *Journal of Physics A*, 12(11): 1991–1997, 1979.

[6] R. Shamseddine. On the resolution of the wave equations of electron in a space of constant curvature. *Canadian Journal of Physics*, 75(11): 805–811, 1997.

[7] P.W. Higgs. Dynamical symmetries in a spherical geometry. I. *Journal of Physics A*, 12(3): 309–323, 1979.

[8] H.I. Leemon. Dynamical symmetries in a spherical geometry. II. *Journal of Physics A*, 12(14): 489–501, 1979.

[9] Yu.A. Kurochkin and V.S. Otchik. Analogue of Runge-Lenz vector and energy spectrum in Kepler problem on 3-dimensiomnal sphere. *DAN BSSR*, 23(11): 987–990, 1979.

[10] A.A. Bogush, Yu.A. Kurochkin and V.S. Otchik. On quantum-mechanical Kepler problem in Lobachevsky space. *DAN BSSR*, 24(1): 19–22, 1980.

[11] G.A. Ringwood and J.T. Devreese. The hydrogen atom: Quantum mechanics on the quotient of a conformally flat manifold. *Journal of Mathematical Physics*, 21: 1390–1392, 1980.

[12] K. Kobayshi. A derivation of the Pauli-Lenz vector and its variants. *Journal of Physics A*, 13: 425–430, 1980.

[13] L. Parker. One-electron atom in curved space-time. *Physical Review Letters*, 44: 1559–1562, 1980.

[14] L. Parker. The atom as a probe of curved space-time. *General Relativity and Gravitation*, 13: 307–311, 1981.

[15] N. Bessis, G. Bessis and R. Shamseddine. Atomic fine-structure in a space of constant curvature. *Journal of Physics A*, 15(10): 3131–3144, 1982.

[16] T. Iwai. Quantisation of the conformal Kepler problem and its application to the hydrogen atom. *Journal of Mathematical Physics*, 23: 1093–1099, 1982.

[17] J.M. Cohen and R.T. Powers. The general relativistic hydrogen-atom. *Communications in Mathematical Physics*, 86: 69–86, 1982.

[18] H. Grinberg, J. Maranon and H. Vucetich. The hydrogen atom as a projection of an homogeneous space. *Zeitschrift für Physik C. Particles and Fields*, 20: 147–149, 1983.

[19] A.A. Bogush, V.S. Otchik and V.M. Red'kov. Separation of variables in Schrödinger equation and normed wave functions for the Kepler problem in three-dimensional spaces of constant curvature. *Proceedings of the National Academy of Sciences of Belarus. Ser. fiz.-mat.*, 19(3): 56–62, 1983.

[20] N. Bessis, G. Bessis and R. Shamseddine. Space-curvature effects in atomic fine- and hyperfine-structure calculations. *Journal of Physical Review A*, 29(5): 2375–2388, 1984.

[21] N. Bessis, G. Bessis and D. Roux. Atomic fine-structure calculations in a space of constant negative curvature. *Journal of Physical Review A*, 30(2): 1094–1097, 1984.

[22] Bessis N. and Bessis G. Atomic fine and hyper-fine structure caclulations in a space of constant curvature. *Lectures Notes in Physics*, 212: 143–153, 1984.

[23] C.M. Xu and D.Y. Xu. Dirac equation and energy levels of hydrogen-like atoms in Robertson – Walker metrics. *Il Nuovo Cimento B*, 83: 162–172, 1984.

[24] V.N. Melnikov and G.N. Shikin. Hydrogen-like atom in gravitational field of the universe. *Izvestia Vuzov, Fizika*, 1: 55–59, 1985.

[25] R. Shamseddine. Structure fine et hyperfine atomique dans un espace à courbure constante. *Journal of Physics A*, 19: 717–724, 1986.

[26] V.S. Otchik and V.M. Red'kov. Quantum-mechanical Kepler problem in spaces of constant curvature. *Preprint 298, Institute of Physics, AN BSSR*, Minsk, Belarus, 1986.

[27] A.O. Barut, A. Inomata and G. Junker. Path integral treatment of the hydrogen atom in a curved space of constant curvature. *Journal of Physics A: Mathematical and General*, 20: 6271–6280, 1987.

[28] N. Bessis, G. Bessis and D. Roux. Space-curvature effects in the interaction between atom and external fields: Zeeman and Stark effects in a space of constant positive curvature. *Journal of Physical Review*, 33: 324–336, 1988.

[29] A.A. Bogush, V.S. Otchik and V.M. Red'kov. Complex parabolic coordinates and hydrogen atom on the sphere. *Deposited in VINITI 12.04.88, 2722 - B88*, Minsk,1988. (in Russian)

[30] C. Groshe. The path integral for the Kepler problem on the pseudosphere. *Annals of Physics*, 204: 208–222, 1990.

[31] A.O. Barut, A. Inomata and G. Junker. Path integral treatment of the hydrogen atom in a curved space of constant curvature. II. Hyperbolic space curvature. *Journal of Physics A: Mathematical and General*, 23: 1179–1190, 1990.

[32] N. Katayama. A note on the Kepler problem in a space of constant curvature. *Il Nuovo Cimento B*, 105(1): 113–119, 1990.

[33] N.A. Chernikov. The Kepler problem in the Lobachevsky space and its solution. *Acta Physica Polonica B*, 23: 115–122, 1992.

[34] L.G. Mardoyan, A.N. Sisakyan. The hydrogen-atom in curved space – orthogonality of the radial wave-functions with respect to the orbital angular momentum. *Soviet Journal of Nuclear Physics-USSR*, 55(9): 1366–1367, 1992.

[35] V.V. Kozlov and A.O. Harin. Kepler's problem in constant curvature spaces. *Celestial Mechanics and Dynamical Astronomy*, 54: 393–399, 1992.

[36] S.I. Vinitskii, L.G. Mardoyan, G.S. Pogosyan, A.N. Sisakyan and T.A. Strizh. Hydrogen-atom in curved space – expansion in free solutions on a 3-dimensional sphere. *Physics of Atomic Nuclei*, 56: 321–327, 1993.

[37] R. Shamseddine. On the resolution of the wave equations of electron in a space of constant curvature. *Canadian Journal of Physics*, 75: 805–811, 1997.

[38] A.A. Bogush, Yu.A. Kurochkin and V.S. Otchik. Algebra of conserved operators for the Kepler-Coulomb problem in the spaces of constant curvature. *Yadernaya fizika*, 61: 1889–1892, 1998.

[39] V.S. Otchik. On the connection between spherical and parabolic bases in the quantum mechanical Kepler problem in Lobachevsky space. *Proceeding of the National Academy of Sciences of Belarus. Phys. Math. ser.*, 35(4). 67–72, 1999.

[40] A. Nersessian and G. Pogosyan. Relation of the oscillator and Coulomb systems on spheres and pseudospheres. *Journal of Physical Review A*, 63(2): 020103(R), 2001.

[41] A.A. Bogush, Yu.A. Kurochkin and V.S. Otchik. Coulomb scattering in the Lobachevsky space. *Nonlinear Phenomena in Complex Systems*, 6(4): 894–897, 2003.

[42] V.M. Red'kov. On WKB-quantisation in Lobachevski and Riemann 3-spaces. *Nonlinear Phenomena in Complex Systems*, 6(2): 654–668, 2003.

[43] Yu.A. Kurochkin, V.S. Otchik and Dz.V. Shoukavy. MIC-Kepler scattering problem in the three-dimensional Lobachevsky space. Non-Euclidean Geometry in Modern Physics. In *Proceedings of the 5th International Conference Bolyai–Gauss–Lobachevsky: Non-Euclidean Geometry In Modern Physics (BGL-5)*, pages 116–121. Minsk, Belarus, 2006.

[44] A.A. Bogush, V.C. Otchik and V.M. Red'kov. The Runge-Lenz vector for quantum Kepler problem in the space of positive constant curvature and complex parabolic coordinates. In *Proceedings of the 5th International Conference Bolyai–Gauss–Lobachevsky: Non-Euclidean Geometry In Modern Physics (BGL-5)*, pages 135–144, arxiv:hep-th/0612178, Minsk, Belarus, 2006.

[45] Yu. Kurochkin and Dz. Shoukavy. Regge trajectories of the Coulomb potential in the space of constant negative curvature. *Journal of Mathematical Physics*, 47: Paper 022103, 2006.

[46] V.M. Red'kov and E.M. Ovsiyuk. Parabolic coordinates and the hydrogen atom in spaces H_3 and S_3. *Nonlinear Phenomena in Complex Systems*, 14(2): 1–20, 2011.

[47] E.M. Ovsiyuk. Quantum Kepler problem for spin 1/2 particle in spaces on constant curvature. I. Pauli theory. *Nonlinear Phenomena in Complex Systems*, 14(1): 14–26, 2011.

[48] V.M. Red'kov and E.M. Ovsiyuk. *Quantum Mechanics in Spaces of Constant Curvature*. Nova Science Publishers. Inc., New York, USA, 2012.

[49] E.M. Ovsiyuk. *Exactly Solvable Problems of Quantum Mechanics and Classical Field Theory in Spaces with Non-Euclidean Geometry*. National Institute for Higher Education, Minsk, 2013 (in Russian).

[50] D.A. Varshalovich, A.N. Moskalev and V.K. Xersonskiy. *Quantum Theory of Angular Momentum*. Leningrad, 1975 (in Russian).

8

Particle with spin 1 in the Coulomb field

We have studied the system of six equations which describe the quantum states of a spin 1 particle with parity $P = (-1)^j$ in the external Coulomb field. It is shown that due to the Lorentz condition, one of the radial functions must be equal to zero. Any of five remaining functions may be taken as a primary one. For such a primary function, we derive two different 2nd-order differential equations. Their Frobenius solutions are constructed, and the convergence of the involved power series is studied. As a quantisation rule, we apply so called transcendency condition to Frobenius solutions. In this way, for both equations, we have found different reasonable, from physical point of view, energy spectra.

8.1 Separation of the variables

Many years ago, a very peculiar behaviour of a spin 1 particle in the presence of the external Coulomb field was noticed by I.E. Tamm [1]. As far as we know the whole situation with this system stays much the same. In the present chapter, we examine the problem anew on the basis of the Duffin–Kemmer–Petiau formalism with the use of the tetrad generally covariant tetrad technique, it turns out to be more convenient than a common Proca tensor approach:

$$\Big\{ i\beta^c \Big[i(e^{\beta}_{(c)}\partial_\beta + \frac{1}{2} j^{ab}\gamma_{abc}(x)) \;-eA_c \Big] - M \Big\} \Psi = 0. \tag{8.1}$$

Choosing a diagonal spherical tetrad

$$\begin{aligned} dS^2 &= dt^2 - dr^2 - r^2(d\theta^2 + \sin^2\theta d\phi^2), \\ e^{\alpha}_{(0)} &= (1,0,0,0), \quad e^{\alpha}_{(3)} = (0,1,0,0), \\ e^{\alpha}_{(1)} &= (0,0,\frac{1}{r},0), \quad e^{\alpha}_{(2)} = (1,0,0,\frac{1}{r\sin\theta}), \end{aligned} \tag{8.2}$$

we reduce the above eq. (8.1) to the form

$$\Big[\, \beta^0 \Big(\epsilon + \frac{\alpha}{r}\Big) + i\Big[\beta^3\partial_r + \frac{1}{r}\,(\beta^1 j^{31} + \beta^2 j^{32})\Big] + \frac{1}{r}\Sigma_{\theta,\phi} - M \Big] \Phi(x) = 0, \tag{8.3}$$

where $\epsilon = E/(c\hbar), \alpha = e^2/(c\hbar)$, and $M = mc/\hbar$. The angular operator $\Sigma_{\theta,\phi}$ is given by the formula

$$\Sigma_{\theta,\phi} = i\beta^1\partial_\theta + \beta^2\, \frac{i\partial + ij^{12}\cos\theta}{\sin\theta}\,, \tag{8.4}$$

DOI: 10.1201/9781003472377-8

its form signifies that we have here a generalised Schrödinger–Pauli basis.

Spherical waves with (j, m) quantum numbers should be constructed within the following general substitution

$$\Psi(x) = \{ \; \Phi_0(x), \; \vec{\Phi}(x), \; \mathbf{E}(x), \mathbf{H}(x) \; \},$$

$$\Phi_0(x) = e^{-iEt/\hbar} \; \Phi_0(r) \; D_0, \qquad \vec{\Phi}(x) = e^{-iEt/\hbar} \left| \begin{array}{c} \Phi_1(r) \; D_{-1} \\ \Phi_2(r) \; D_0 \\ \Phi_3(r) \; D_{+1} \end{array} \right| ,$$
$$\mathbf{E}(x) = e^{-iEt/\hbar} \left| \begin{array}{c} E_1(r) \; D_{-1} \\ E_2(r) \; D_0 \\ E_3(r) \; D_{+1} \end{array} \right| , \quad \mathbf{H}(x) = e^{-iEt/\hbar} \left| \begin{array}{c} H_1(r) \; D_{-1} \\ H_2(r) \; D_0 \\ H_3(r) \; D_{+1} \end{array} \right| ; \tag{8.5}$$

short notation for Wigner functions is used: $D_\sigma = D^j_{-m,\sigma}(\phi, \; \theta, \; 0), \;\; \sigma = 0, \; +1, \; -1$. The quantum number j takes values $0, 1, 2, ...$. With the help of recurrent formulas

$$\nu = \sqrt{j(j+1)}, \qquad a = \sqrt{(j-1)(j+2)},$$

$$\begin{aligned} \partial_\theta \; D_{-1} &= \frac{1}{2} \; (\; a \; D_{-2} - \nu \; D_0 \;), \quad \frac{m - \cos\theta}{\sin\theta} \; D_{-1} = \frac{1}{2} \; (\; a \; D_{-2} + \nu \; D_0 \;), \\ \partial_\theta \; D_0 &= \frac{1}{2} \; (\; \nu \; D_{-1} - \nu \; D_{+1} \;), \quad \frac{m}{\sin\theta} \; D_0 = \frac{1}{2} \; (\; \nu \; D_{-1} + \nu \; D_{+1} \;), \\ \partial_\theta \; D_{+1} &= \frac{1}{2} \; (\; \nu \; D_0 - a \; D_{+2} \;), \quad \frac{m + \cos\theta}{\sin\theta} \; D_{+1} = \frac{1}{2} \; (\; \nu \; D_0 + \; a \; D_{+2} \;), \end{aligned} \tag{8.6}$$

after simple algebraic calculation, we arrive at the radial equations (for clarity, the corresponding Proca tensor equations are written down as well; the notation $\nu = \sqrt{j(j+1)/2}$ is used):

$$D^b \; \Phi_{ab} = M \; \Phi_a,$$

$$\begin{aligned} -\Big(\frac{d}{dr} + \frac{2}{r}\Big) E_2 - \frac{\nu}{r}(E_1 + E_3) &= M\Phi_0, \\ +i\Big(\epsilon + \frac{\alpha}{r}\Big) E_1 + i\Big(\frac{d}{dr} + \frac{1}{r}\Big) H_1 + i\frac{\nu}{r} H_2 &= M\Phi_1, \\ +i\Big(\epsilon + \frac{\alpha}{r}\Big) E_2 - i\frac{\nu}{r}(H_1 - H_3) &= M\Phi_2, \\ +i\Big(\epsilon + \frac{\alpha}{r}\Big) E_3 - i(\frac{d}{dr} + \frac{1}{r}) H_3 - i\frac{\nu}{r} H_2 &= M\Phi_3 \,; \end{aligned} \tag{8.7}$$

$$D_a \Phi_b - D_b \Phi_a = M \; \Phi_{ab},$$

$$\begin{aligned} -i\Big(\epsilon + \frac{\alpha}{r}\Big)\Phi_1 + \frac{\nu}{r}\Phi_0 - ME_1 &= 0, \\ -i\Big(\epsilon + \frac{\alpha}{r}\Big)\Phi_2 - \frac{d}{dr}\Phi_0 - ME_2 &= 0, \\ -i\Big(\epsilon + \frac{\alpha}{r}\Big)\Phi_3 + \frac{\nu}{r}\Phi_0 - ME_3 &= 0, \\ -i\Big(\frac{d}{dr} + \frac{1}{r}\Big)\Phi_1 - i\frac{\nu}{r}\Phi_2 - MH_1 &= 0, \\ +i\frac{\nu}{r}(\Phi_1 - \Phi_3) - MH_2 &= 0, \\ +i\Big(\frac{d}{dr} + \frac{1}{r}\Big)\Phi_3 + i\frac{\nu}{r}\Phi_2 - MH_3 &= 0. \end{aligned} \tag{8.8}$$

Concurrently with $\mathbf{J}^2, J_3$, let us diagonalise the operator of the spacial inversion $\hat{\Pi}$. After transition to spherical tetrad basis, and also to cyclic representation for DKP-matrices β^a, for this discrete operator we get

$$\hat{\Pi} = \left| \begin{array}{cccc} 1 & 0 & 0 & 0 \\ 0 & \Pi_3 & 0 & 0 \\ 0 & 0 & \Pi_3 & 0 \\ 0 & 0 & 0 & -\Pi_3 \end{array} \right| \hat{P}, \quad \Pi_3 = \left| \begin{array}{ccc} 0 & 0 & -1 \\ 0 & -1 & 0 \\ -1 & 0 & 0 \end{array} \right| . \tag{8.9}$$

Eigenvalue equation $\hat{\Pi}\Psi = P\,\Psi$ results in two classes of states different in parity:

$$P = (-1)^{j+1}, \quad \Phi_0 = 0,\ \Phi_3 = -\Phi_1,\ \Phi_2 = 0,\ E_3 = -E_1,\ E_2 = 0,\ H_3 = H_1; \tag{8.10}$$

$$P = (-1)^j, \qquad \Phi_3 = \Phi_1,\ E_3 = +E_1,\ H_3 = -H_1,\ H_2 = 0\,. \tag{8.11}$$

Correspondingly, 10 equations in (8.7)–(8.8) yield subsystems of four and six equations:

$$\begin{gathered} P = (-1)^{j+1}, \qquad +i(\epsilon + \frac{\alpha}{r})\; E_1 + i(\frac{d}{dr} + \frac{1}{r})H_1 + i\frac{\nu}{r}H_2 = M\Phi_1, \\ -i(\epsilon + \frac{\alpha}{r})\; \Phi_1 = ME_1, \;\; -i(\frac{d}{dr} + \frac{1}{r})\Phi_1 = MH_1, \;\; 2i\frac{\nu}{r}\Phi_1 = MH_2; \end{gathered} \tag{8.12}$$

whence excluding E_1, H_1, and H_2, we get a 2nd-order differential equation for Φ_1

$$\left[\frac{d^2}{dr^2} + \frac{2}{r}\frac{d}{dr} + (\epsilon + \frac{\alpha}{r})^2 - M^2 - \frac{j(j+1)}{r^2} \right] \Phi_1 = 0. \tag{8.13}$$

It coincides with that arising in the case of a scalar particle in Coulomb potential. Its solution is well known and provides us with the following energy spectrum (in usual units)

$$E = \frac{mc^2}{\sqrt{1 + \alpha^2/N^2}}, \qquad N = n + \frac{1}{2} + \sqrt{(j+1/2)^2 - \alpha^2}. \tag{8.14}$$

For states with parity $P = (-1)^j$, we have the system

$$\begin{aligned} (\frac{d}{dr} + \frac{2}{r})E_2 + 2\frac{\nu}{r}E_1 + M\Phi_0 = 0, \\ +i(\epsilon + \frac{\alpha}{r})\; E_1 + i(\frac{d}{dr} + \frac{1}{r})H_1 - M\Phi_1 = 0, \\ +i(\epsilon + \frac{\alpha}{r})E_2 - 2i\frac{\nu}{r}H_1 - M\Phi_2 = 0, \\ -i(\epsilon + \frac{\alpha}{r})\; \Phi_1 + \frac{\nu}{r}\Phi_0 - ME_1 = 0, \\ i(\epsilon + \frac{\alpha}{r})\Phi_2 + \frac{d}{dr}\Phi_0 + ME_2 = 0, \\ i(\frac{d}{dr} + \frac{1}{r})\Phi_1 + i\frac{\nu}{r}\Phi_2 + MH_1 = 0. \end{aligned} \tag{8.15}$$

8.2 The case of minimal value $j = 0$

States with the minimal value $j = 0$ should be treated separately. In this case, we should start with a more simple substitution for the wave function

$$\Phi_0(x) = e^{-i\epsilon t}\Phi_0(r), \qquad \vec{\Phi}(x) = e^{-i\epsilon t}\begin{vmatrix} 0 \\ \Phi_2(r) \\ 0 \end{vmatrix},$$
$$\vec{E}(x) = e^{-i\epsilon t}\begin{vmatrix} 0 \\ E_2(r) \\ 0 \end{vmatrix}, \qquad \vec{H}(x) = e^{-i\epsilon t}\begin{vmatrix} 0 \\ H_2(r) \\ 0 \end{vmatrix}. \tag{8.16}$$

The operator $\Sigma_{\theta,\phi}$ acts on this wave function as a zero operator, and the parity for this state equals $P = (-1)^{0+1} = -1$.

The corresponding radial system is as follows (for eliminating imaginary unit i, we use slightly different variables: $\Phi_0 = \varphi_0,\ -i\Phi_1 = \varphi_1,\ -i\Phi_2 = \varphi_2$)

$$H_2 = 0, \qquad -\Big(\frac{d}{dr} + \frac{2}{r}\Big)E_2 = M\varphi_0,$$
$$\Big(\epsilon + \frac{\alpha}{r}\Big)E_2 = M\varphi_2, \qquad \Big(\epsilon + \frac{\alpha}{r}\Big)\varphi_2 - \frac{d}{dr}\varphi_0 = ME_2. \tag{8.17}$$

Whence it follows a 2nd-order equation (let $E_2(r) = r^{-1}f(r)$)

$$\frac{d^2}{dr^2}f + \Big(\epsilon^2 - M^2 + \frac{2\alpha\epsilon}{r} - \frac{2-\alpha^2}{r^2}\Big)f = 0\,. \tag{8.18}$$

In dimensionless variables

$$x = r\epsilon = \frac{rE}{c\hbar}, \quad \frac{M^2}{\epsilon^2} = \frac{m^2c^4}{E^2} = \lambda^2,$$

it reads

$$\frac{d^2}{dx^2}f + \Big(1 - \lambda^2 + \frac{2\alpha}{x} - \frac{2-\alpha^2}{x^2}\Big)f = 0. \tag{8.19}$$

With the substitution $f(x) = x^a e^{-bx}F(x)$, for $F(x)$ we obtain

$$xF'' + (2a - 2bx)F' + \Big[\frac{a(a-1)+\alpha^2-2}{x} + (b^2+1-\lambda^2)x + (2\alpha - 2ab)\Big]F = 0.$$

Requiring

$$a = \frac{1 \pm \sqrt{9-4\alpha^2}}{2}, \quad b = \pm\sqrt{\lambda^2 - 1} = \pm\frac{\sqrt{m^2c^4 - E^2}}{E},$$

the choice of upper signs in the formulas provides us with appropriate parameters for bound states, we get

$$x\,F'' \;+\; 2(a - bx)\;F' \;+\; 2(\alpha - ab)\;F = 0.$$

In the variable $y = 2bx$, it takes the form of the confluent hypergeometric equation

$$\frac{d^2}{dy^2}F + (2a - y)\frac{d}{dt}F - \frac{ab-\alpha}{b}F = 0\,.$$

To get polynomial solutions, we must require $(ab - \alpha)/b = -n$, which leads to the quantisation formula for energies

$$E = \frac{mc^2}{\sqrt{1 + \alpha^2/(\Gamma + n)^2}}, \quad \Gamma = \frac{1 + \sqrt{9-4\alpha^2}}{2}\,. \tag{8.20}$$

8.3 Nonrelativistic approximation, energy spectra

In this section, we examine nonrelativistic approximation in the theory. First, consider states with parity $P = (-1)^{j+1}$:

$$\begin{gathered} i\Big(\epsilon + \frac{\alpha}{r}\Big)E_1 + i\Big(\frac{d}{dr} + \frac{1}{r}\Big)H_1 + i\frac{\nu}{r}H_2 = M\Phi_1, \\ -i\Big(\epsilon + \frac{\alpha}{r}\Big)\Phi_1 = ME_1, \ \ -i\Big(\frac{d}{dr} + \frac{1}{r}\Big)\Phi_1 = MH_1, \ \ 2i\frac{\nu}{r}\Phi_1 = MH_2. \end{gathered} \tag{8.21}$$

Here the H_1, H_2 represent so-called non-dynamical variables, excluding them we obtain

$$i\Big(\epsilon + \frac{\alpha}{r}\Big)E_1 + \frac{1}{M}\Big(\frac{d}{dr} + \frac{1}{r}\Big)^2\Phi_1 - \frac{2\nu^2}{Mr^2}\Phi_1 = M\Phi_1, -i\Big(\epsilon + \frac{\alpha}{r}\Big)\Phi_1 = ME_1. \tag{8.22}$$

Now we should make special substitutions, introducing the big and small constituents, $\Psi_1(r)$ and $\psi_1(r)$:

$$\Phi_1 = \Psi_1 + \psi_1, \qquad iE_1 = \Psi_1 - \psi_1; \tag{8.23}$$

correspondingly eq. (8.21) take the form

$$\begin{gathered} \Big(\epsilon + \frac{\alpha}{r}\Big)(\Psi_1 - \psi_1) + \frac{1}{M}\Big(\frac{d}{dr} + \frac{1}{r}\Big)^2(\Psi_1 + \psi_1) - \frac{2\nu^2}{Mr^2}(\Psi_1 + \psi_1) \\ = M(\Psi_1 + \psi_1), \quad \Big(\epsilon + \frac{\alpha}{r}\Big)(\Psi_1 + \psi_1) = M(\Psi_1 - \psi_1). \end{gathered} \tag{8.24}$$

Summing and subtracting these two equations, we get

$$\begin{gathered} 2\Big(\epsilon + \frac{\alpha}{r}\Big)\Psi_1 + \frac{1}{M}\Big(\frac{d}{dr} + \frac{1}{r}\Big)^2(\Psi_1 + \psi_1) - \frac{2\nu^2}{Mr^2}(\Psi_1 + \psi_1) = 2M\Psi_1, \\ -2\Big(\epsilon + \frac{\alpha}{r}\Big)\psi_1 + \frac{1}{M}\Big(\frac{d}{dr} + \frac{1}{r}\Big)^2(\Psi_1 + \psi_1) - \frac{2\nu^2}{Mr^2}(\Psi_1 + \psi_1) = 2M\psi_1. \end{gathered} \tag{8.25}$$

Now we should separate the rest energy. To this end, it suffices to make a formal change $\epsilon \Longrightarrow \epsilon + M$, which results in

$$2\Big(\epsilon + \frac{\alpha}{r}\Big)\Psi_1 + \frac{1}{M}\Big(\frac{d}{dr} + \frac{1}{r}\Big)^2(\Psi_1 + \psi_1) - \frac{2\nu^2}{Mr^2}(\Psi_1 + \psi_1) = 0,$$

$$-2\Big(\epsilon + \frac{\alpha}{r}\Big)\psi_1 + \frac{1}{M}\Big(\frac{d}{dr} + \frac{1}{r}\Big)^2(\Psi_1 + \psi_1) - \frac{2\nu^2}{Mr^2}(\Psi_1 + \psi_1) = 4M\psi_1.$$

Thus, we produce equation for the big $\Psi_1(r)$ and small $\psi_1(r)$ components:

$$\begin{gathered} 2\Big(\epsilon + \frac{\alpha}{r}\Big)\Psi_1 + \frac{1}{M}\Big(\frac{d}{dr} + \frac{1}{r}\Big)^2\Psi_1 - \frac{j(j+1)}{Mr^2}\Psi_1 = 0, \\ \Big(\frac{d}{dr} + \frac{1}{r}\Big)^2\Psi_1 - \frac{j(j+1)}{r^2}\Psi_1 = 4M^2\psi_1. \end{gathered} \tag{8.26}$$

Equation for the big component $\Psi_1(r)$ can be written as a Schrödinger-like equation:

$$\Big[\frac{d^2}{dr^2} + \frac{2}{r}\frac{d}{dr} + 2M(\epsilon + \frac{\alpha}{r}) - \frac{j(j+1)}{r^2}\Big]\Psi_1 = 0. \tag{8.27}$$

Now let us consider radial equations for states with the opposite parity $P = (-1)^j$:

$$
\begin{gathered}
-\Big(\frac{d}{dr}+\frac{2}{r}\Big)E_2 - 2\frac{\nu}{r}E_1 = M\Phi_0, \quad +i\Big(\epsilon+\frac{\alpha}{r}\Big)E_1 + i\Big(\frac{d}{dr}+\frac{1}{r}\Big)H_1 = M\Phi_1, \\
+i\Big(\epsilon+\frac{\alpha}{r}\Big)E_2 - 2i\frac{\nu}{r}H_1 = M\Phi_2, \quad -i\Big(\epsilon+\frac{\alpha}{r}\Big)\Phi_1 + \frac{\nu}{r}\Phi_0 = ME_1, \\
-i\Big(\epsilon+\frac{\alpha}{r}\Big)\Phi_2 - \frac{d}{dr}\Phi_0 = ME_2, \quad -i\Big(\frac{d}{dr}+\frac{1}{r}\Big)\Phi_1 - i\frac{\nu}{r}\Phi_2 = MH_1.
\end{gathered} \tag{8.28}
$$

Among the four dynamical functions Φ_1, Φ_2, E_1, E_2, the separation of big and small constituents is performed as follows

$$
\Phi_1 = \Psi_1 + \psi_1, \qquad \Phi_2 = \Psi_2 + \psi_2, \quad iE_1 = \Psi_1 - \psi_1, \qquad iE_2 = \Psi_2 - \psi_2. \tag{8.29}
$$

Excluding the nondynamical variables Φ_0, H_1, we obtain the system (the rest energy is separated by the formal change $\epsilon \Longrightarrow \epsilon + M$)

$$
\begin{aligned}
i\Big(\epsilon+M+\frac{\alpha}{r}\Big)E_1 + \frac{1}{M}\,(\frac{d}{dr}+\frac{1}{r})\Big[\Big(\frac{d}{dr}+\frac{1}{r}\Big)\Phi_1 + \frac{\nu}{r}\Phi_2\Big] &= M\Phi_1, \\
i\Big(\epsilon+M+\frac{\alpha}{r}\Big)E_2 - \frac{2\nu}{Mr}\Big[\Big(\frac{d}{dr}+\frac{1}{r}\Big)\Phi_1 + \frac{\nu}{r}\Phi_2\Big] &= M\Phi_2, \\
-i\Big(\epsilon+M+\frac{\alpha}{r}\Big)\Phi_1 + \frac{\nu}{Mr}\Big[-\Big(\frac{d}{dr}+\frac{2}{r}\Big)E_2 - \frac{2\nu}{r}E_1\Big] &= ME_1, \\
-i\Big(f\epsilon+M+\frac{\alpha}{r}\Big)\Phi_2 - \frac{1}{M}\frac{d}{dr}\Big[-(\frac{d}{dr}+\frac{2}{r})E_2 - \frac{2\nu}{r}E_1\Big] &= ME_2.
\end{aligned} \tag{8.30}
$$

Taking into account relations (8.29), we transform eq. (8.30) into the form

$$
\Big(\epsilon+M+\frac{\alpha}{r}\Big)(\Psi_1-\psi_1) + \frac{1}{M}\Big(\frac{d}{dr}+\frac{1}{r}\Big)^2(\Psi_1+\psi_1) + \frac{\nu}{M}\Big(\frac{d}{dr}+\frac{1}{r}\Big)\frac{1}{r}(\Psi_2+\psi_2) = M(\Psi_1+\psi_1),
$$

$$
\Big(\epsilon+M+\frac{\alpha}{r}\Big)(\Psi_2-\psi_2) - \frac{2\nu}{Mr}\Big(\frac{d}{dr}+\frac{1}{r}\Big)(\Psi_1+\psi_1) - \frac{2\nu^2}{Mr^2}(\Psi_2+\psi_2) = M(\Psi_2+\psi_2),
$$

$$
\Big(\epsilon+M+\frac{\alpha}{r}\Big)(\Psi_1+\psi_1) - \frac{\nu}{mr}\Big(\frac{d}{dr}+\frac{2}{r}\Big)(\Psi_2-\psi_2) - \frac{2\nu^2}{Mr^2}(\Psi_1-\psi_1) = M(\Psi_1-\psi_1),
$$

$$
\Big(\epsilon+M+\frac{\alpha}{r}\Big)(\Psi_2+\psi_2) + \frac{1}{M}\frac{d}{dr}\Big(\frac{d}{dr}+\frac{2}{r}\Big)(\Psi_2-\psi_2) + \frac{2\nu}{M}\frac{d}{dr}\frac{1}{r}(\Psi_1-\psi_1) = M(\Psi_2-\psi_2)\,.
$$

Whence we get

$$
\Big(\epsilon+\frac{\alpha}{r}\Big)(\Psi_1-\psi_1) + \frac{1}{M}\Big(\frac{d}{dr}+\frac{1}{r}\Big)^2(\Psi_1+\psi_1) + \frac{\nu}{M}\Big(\frac{d}{dr}+\frac{1}{r}\Big)\frac{1}{r}(\Psi_2+\psi_2) = +2M\,\psi_1\,,
$$

$$
\Big(\epsilon+\frac{\alpha}{r}\Big)(\Psi_1+\psi_1) - \frac{\nu}{Mr}\Big(\frac{d}{dr}+\frac{2}{r}\Big)(\Psi_2-\psi_2) - \frac{2\nu^2}{Mr^2}(\Psi_1-\psi_1) = -2M\psi_1;
$$

$$
\Big(\epsilon+\frac{\alpha}{r}\Big)(\Psi_2-\psi_2) - \frac{2\nu}{Mr}\Big(\frac{d}{dr}+\frac{1}{r}\Big)(\Psi_1+\psi_1) - \frac{2\nu^2}{Mr^2}(\Psi_2+\psi_2) = +2M\,\psi_2\,,
$$

$$
\Big(\epsilon+\frac{\alpha}{r}\Big)(\Psi_2+\psi_2) + \frac{1}{M}\frac{d}{dr}\Big(\frac{d}{dr}+\frac{2}{r}\Big)(\Psi_2-\psi_2) + \frac{2\nu}{M}\frac{d}{dr}\frac{1}{r}(\Psi_1-\psi_1) = -2M\,\psi_2.
$$

Summing and subtracting equations within the first pair and doing the same within the second pair, we arrive at

$$
\Big(\epsilon+\frac{\alpha}{r}\Big)(\Psi_1-\psi_1) + \frac{1}{M}\Big(\frac{d}{dr}+\frac{1}{r}\Big)^2(\Psi_1+\psi_1) + \frac{\nu}{M}\Big(\frac{d}{dr}+\frac{1}{r}\Big)\frac{1}{r}(\Psi_2+\psi_2)
$$

$$+\Big(\epsilon+\frac{\alpha}{r}\Big)(\Psi_1+\psi_1)-\frac{\nu}{Mr}\Big(\frac{d}{dr}+\frac{2}{r}\Big)(\Psi_2-\psi_2)-\frac{2\nu^2}{Mr^2}(\Psi_1-\psi_1)=0,$$

$$\Big(\epsilon+\frac{\alpha}{r}\Big)(\Psi_1-\psi_1)+\frac{1}{M}\Big(\frac{d}{dr}+\frac{1}{r}\Big)^2(\Psi_1+\psi_1)+\frac{\nu}{M}\Big(\frac{d}{dr}+\frac{1}{r}\Big)\frac{1}{r}(\Psi_2+\psi_2)$$
$$-\Big(\epsilon+\frac{\alpha}{r}\Big)(\Psi_1+\psi_1)+\frac{\nu}{Mr}\Big(\frac{d}{dr}+\frac{2}{r}\Big)(\Psi_2-\psi_2)+\frac{2\nu^2}{Mr^2}(\Psi_1-\psi_1)=+4M\,\psi_1,$$

$$\Big(\epsilon+\frac{\alpha}{r}\Big)(\Psi_2-\psi_2)-\frac{2\nu}{Mr}\Big(\frac{d}{dr}+\frac{1}{r}\Big)(\Psi_1+\psi_1)-\frac{2\nu^2}{Mr^2}(\Psi_2+\psi_2)$$
$$+\Big(\epsilon+\frac{\alpha}{r}\Big)(\Psi_2+\psi_2)+\frac{1}{M}\frac{d}{dr}\Big(\frac{d}{dr}+\frac{2}{r}\Big)(\Psi_2-\psi_2)+\frac{2\nu}{M}\frac{d}{dr}\frac{1}{r}(\Psi_1-\psi_1)=0,$$

$$(\epsilon+\frac{\alpha}{r})(\Psi_2-\psi_2)-\frac{2\nu}{Mr}(\frac{d}{dr}+\frac{1}{r})(\Psi_1+\psi_1)-\frac{2\nu^2}{Mr^2}(\Psi_2+\psi_2)$$
$$-\Big(\epsilon+\frac{\alpha}{r}\Big)(\Psi_2+\psi_2)-\frac{1}{M}\frac{d}{dr}\Big(\frac{d}{dr}+\frac{2}{r}\Big)(\Psi_2-\psi_2)-\frac{2\nu}{M}\frac{d}{dr}\frac{1}{r}(\Psi_1-\psi_1)=+4M\,\psi_2.$$

Now, taking in mind that Ψ_1,Ψ_2 are big components, and ψ_1,ψ_2 are small, we arrive at two equations for the big components, and two equations defining the small components through the big ones:

$$\frac{1}{M}\Big(\frac{d}{dr}+\frac{1}{r}\Big)^2\Psi_1+\frac{\nu}{M}\Big(\frac{d}{dr}+\frac{1}{r}\Big)\frac{1}{r}\Psi_2+\frac{\nu}{Mr}\Big(\frac{d}{dr}+\frac{2}{r}\Big)\Psi_2+\frac{2\nu^2}{Mr^2}\Psi_1=+4M\,\psi_1,$$

$$-\frac{2\nu}{Mr}\Big(\frac{d}{dr}+\frac{1}{r}\Big)\Psi_1-\frac{2\nu^2}{Mr^2}\Psi_2-\frac{1}{M}\frac{d}{dr}(\frac{d}{dr}+\frac{2}{r})\Psi_2-\frac{2\nu}{M}\frac{d}{dr}\frac{1}{r}\Psi_1=+4M\,\psi_2,$$

$$2\Big(\epsilon+\frac{\alpha}{r}\Big)\Psi_2-\frac{2\nu}{Mr}\Big(\frac{d}{dr}+\frac{1}{r}\Big)\Psi_1-\frac{2\nu^2}{Mr^2}\Psi_2+\frac{1}{M}\frac{d}{dr}\Big(\frac{d}{dr}+\frac{2}{r}\Big)\Psi_2+\frac{2\nu}{M}\frac{d}{dr}\frac{1}{r}\Psi_1=0,$$

$$2\Big(\epsilon+\frac{\alpha}{r}\Big)\Psi_1+\frac{1}{M}\Big(\frac{d}{dr}+\frac{1}{r}\Big)^2\Psi_1+\frac{\nu}{M}(\frac{d}{dr}+\frac{1}{r})\frac{1}{r}\Psi_2-\frac{\nu}{Mr}\Big(\frac{d}{dr}+\frac{2}{r}\Big)\Psi_2-\frac{2\nu^2}{Mr^2}\Psi_1=0.$$

The last two equations provide us with the nonrelativistic radial equations, which can be written as follows

$$\begin{aligned} r^2\Big[\frac{d^2}{dr^2}+\frac{2}{r}\frac{d}{dr}+2M\Big(\epsilon+\frac{\alpha}{r}\Big)-\frac{2\nu^2}{r^2}\Big]\Psi_2&=2\Psi_2+4\nu\,\Psi_1,\\ r^2\Big[\frac{d^2}{dr^2}+\frac{2}{r}\frac{d}{dr}+2M\Big(\epsilon+\frac{\alpha}{r}\Big)-\frac{2\nu^2}{r^2}\Big]\Psi_1&=2\nu\,\Psi_2. \end{aligned}\tag{8.31}$$

It is convenient to presents eq. (8.31) in the matrix form

$$\frac{1}{2}r^2\Delta\left|\begin{array}{c}\Psi_1\\ \Psi_2\end{array}\right|=\left|\begin{array}{cc}0&\nu\\ 2\nu&1\end{array}\right|\left|\begin{array}{c}\Psi_1\\ \Psi_2\end{array}\right|.\tag{8.32}$$

The right-hand part can be brought to a diagonal form

$$\left|\begin{array}{c}f_1\\ f_2\end{array}\right|=\left|\begin{array}{cc}a&c\\ d&b\end{array}\right|\left|\begin{array}{c}\Psi_1\\ \Psi_2\end{array}\right|,\qquad r^2\Delta\left|\begin{array}{c}f_1\\ f_2\end{array}\right|=\left|\begin{array}{cc}\lambda_1&0\\ 0&\lambda_2\end{array}\right|\left|\begin{array}{c}f_1\\ f_2\end{array}\right|.$$

the last task reduces to the following equation

$$\left|\begin{array}{cc}a&c\\ d&b\end{array}\right|\left|\begin{array}{cc}0&\nu\\ 2\nu&1\end{array}\right|=\left|\begin{array}{cc}\lambda_1&0\\ 0&\lambda_2\end{array}\right|\left|\begin{array}{cc}a&c\\ d&b\end{array}\right|,$$

whence it follows

$$\begin{cases} \lambda_1 \, a - 2\nu \, c = 0 \\ -\nu \, a + (\lambda_1 - 1) \, c = 0, \end{cases} \qquad \lambda_1 = \frac{1 + \sqrt{1 + 4j(j+1)}}{2} = j + 1, \qquad c = \frac{\lambda_1}{2\nu} a;$$

$$\begin{cases} \lambda_2 \, d - 2\nu \, b = 0 \\ -\nu \, d + (\lambda_2 - 1) \, b = 0, \end{cases} \qquad \lambda_2 = \frac{1 - \sqrt{1 + 4j(j+1)}}{2} = -j \, , \qquad b = \frac{\lambda_2}{2\nu} d.$$

Thus, the needed transformation reads

$$\begin{vmatrix} f_1 \\ f_2 \end{vmatrix} = \begin{vmatrix} a & \lambda_1 a/2\nu \\ d & \lambda_2 d/2\nu \end{vmatrix} \begin{vmatrix} \Psi_1 \\ \Psi_2 \end{vmatrix} . \tag{8.33}$$

In this way, we obtain two separate 2nd-order differential equations:

$$\begin{aligned} &\left[\frac{d^2}{dr^2} + \frac{2}{r}\frac{d}{dr} + 2M\left(\epsilon + \frac{\alpha}{r}\right) - \frac{2\nu^2}{r^2} - \frac{2\lambda_1}{r^2} \right] f_1 = 0, \\ &\left[\frac{d^2}{dr^2} + \frac{2}{r}\frac{d}{dr} + 2M\left(\epsilon + \frac{\alpha}{r}\right) - \frac{2\nu^2}{r^2} - \frac{2\lambda_2}{r^2} \right] f_2 = 0. \end{aligned} \tag{8.34}$$

Note two identities

$$\frac{2\nu^2}{r^2} + \frac{2\lambda_1}{r^2} = \frac{j(j+1) + 2(j+1)}{r^2} = \frac{(j+1)(j+2)}{r^2},$$

$$\frac{2\nu^2}{r^2} + \frac{2\lambda_2}{r^2} = \frac{j(j+1) - 2j}{r^2} = \frac{(j-1)j}{r^2}.$$

Here we have two problems of the same type (below the equations for f_1 and f_2, correspond to $\mu = j - 1$ and $\mu = j + 1$, respectively)

$$\left[\frac{d^2}{dr^2} + \frac{2}{r}\frac{d}{dr} + 2M(\epsilon + \frac{\alpha}{r}) - \frac{\mu(\mu+1)}{r^2} \right] f = 0. \tag{8.35}$$

Changing the variable $x = 2\sqrt{-2\epsilon M}\, r$, we obtain

$$\left[\frac{d^2}{dx^2} + \frac{2}{x}\frac{d}{dx} - \frac{1}{4} - \frac{\alpha M}{x\sqrt{-2\epsilon M}} - \frac{\mu(\mu+1)}{x^2} \right] f(x) = 0.$$

Further, introducing the substitution $f(x) = x^a e^{-bx} \, F(x)$, we arrive at

$$x\frac{d^2 F}{dx^2} + (2a + 2 - 2bx)\frac{dF}{dx}$$

$$+ \left[\frac{a(a+1) - \mu(\mu+1)}{x} - 2b - 2ab + \frac{\alpha M}{\sqrt{-2\epsilon M}} + (b^2 - \frac{1}{4})x \right] F = 0 \, .$$

When $b = +1/2$, $a = +\mu$, the last equation simplifies

$$x\frac{d^2 F}{dx^2} + (2\nu + 2 - x)\,\frac{dF}{dx} - \left[1 + \mu - \frac{\alpha M}{\sqrt{-2\epsilon M}} \right] F = 0,$$

what is the confluent hypergeometric equation for $F(A, C; \, x)$ with parameters given by

$$A = 1 + \mu - \frac{\alpha M}{\sqrt{-2\epsilon M}}, \qquad C = 2\mu + 2.$$

The quantisation condition is $A = -n$, which gives the following energy spectrum

$$1 + \mu - \frac{\alpha M}{\sqrt{-2\epsilon M}} = -n \implies \epsilon = -\frac{\alpha^2 M}{2(1 + \mu + n)^2} = -\frac{Me^4}{2\hbar^2(1 + \mu + n)^2}; \tag{8.36}$$

recall that to linearly independent solutions correspond the different values of μ: $\mu = j - 1$ and $\mu = j + 1$.

8.4 The Lorentz condition in presence of the Coulomb field

As shown above, the states with parity $P = (-1)^j$ are described by six equations

$$
\begin{gathered}
\Big(\frac{d}{dr} + \frac{2}{r}\Big) E_2 + 2\frac{\nu}{r} E_1 + m\Phi_0 = 0, \quad i\Big(\epsilon + \frac{\alpha}{r}\Big)\ E_1 + i\Big(\frac{d}{dr} + \frac{1}{r}\Big) H_1 - m\Phi_1 = 0, \\
i\Big(\epsilon + \frac{\alpha}{r}\Big) E_2 - 2i\frac{\nu}{r} H_1 - m\Phi_2 = 0, \quad -i\Big(\epsilon + \frac{\alpha}{r}\Big)\ \Phi_1 + \frac{\nu}{r}\Phi_0 - mE_1 = 0, \\
i\Big(\epsilon + \frac{\alpha}{r}\Big)\Phi_2 + \frac{d}{dr}\Phi_0 + mE_2 = 0, \quad i\Big(\frac{d}{dr} + \frac{1}{r}\Big)\Phi_1 + i\frac{\nu}{r}\Phi_2 + mH_1 = 0\,.
\end{gathered} \tag{8.37}
$$

There is known that for the spin 1 particle in the presence of electromagnetic fields there must exist a generalised Lorentz condition. In radial form for states with $P = (-1)^j$, it has the form [2]

$$
-i\ \Big(\epsilon + \frac{\alpha}{r}\Big)\ \Phi_0 - \ \Big(\frac{d}{dr} + \frac{2}{r}\Big)\ \Phi_2 - \frac{2\nu}{r}\ \Phi_1 = \frac{i\alpha}{2mr^2}\ E_2\,. \tag{8.38}
$$

Taking into account (8.38), from the system (8.37), we may derive an additional constraint on radial functions. Indeed, from eq. (8.38), let us exclude the function Φ_2 with the help of the third equation in eq. (8.37)

$$
-i\ \Big(\epsilon + \frac{\alpha}{r}\Big)\ m\Phi_0 - \ \Big(\frac{d}{dr} + \frac{2}{r}\Big)\ \Big[i(\epsilon + \frac{\alpha}{r})E_2 - \frac{2i\nu}{r}H_1\Big] - \frac{2m\nu}{r}\ \Phi_1 = \frac{i\alpha}{2r^2}\ E_2,
$$

whence it follows

$$
i\ \Big(\epsilon + \frac{\alpha}{r}\Big)\ m\Phi_0 + i\Big(\epsilon + \frac{\alpha}{r}\Big)\Big(\frac{d}{dr} + \frac{2}{r}\Big)E_2 - \frac{2i\nu}{r}\Big(\frac{d}{dr} + \frac{1}{r}\Big)H_1 + \frac{2m\nu}{r}\ \Phi_1 = \frac{i\alpha}{2r^2}\ E_2\,.
$$

Transforming the second and the third terms with the help of the first and the second equations in eq. (8.37), we obtain

$$
i\ \Big(\epsilon + \frac{\alpha}{r}\Big)\ m\Phi_0 - im\ \Big(\epsilon + \frac{\alpha}{r}\Big)\ \Phi_0 = \frac{i\alpha}{2r^2}\ E_2 \quad \Longrightarrow \quad E_2 = 0\,. \tag{8.39}
$$

Thus, in the system (8.37) we have six equations for only five unknown functions. Allowing for the constraint $E_2 = 0$, we reduce the system (8.37) to the form

$$
\begin{gathered}
1)\ \ 2\frac{\nu}{r}E_1 + m\Phi_0 = 0, \quad 2)\quad + i\Big(\epsilon + \frac{\alpha}{r}\Big)E_1 + i\Big(\frac{d}{dr} + \frac{1}{r}\Big)H_1 - m\Phi_1 = 0, \\
3) - 2i\frac{\nu}{r}H_1 - m\Phi_2 = 0, \quad 4)\quad - i\Big(\epsilon + \frac{\alpha}{r}\Big)\Phi_1 + \frac{\nu}{r}\Phi_0 - mE_1 = 0, \\
5)\ \ i\Big(\epsilon + \frac{\alpha}{r}\Big)\Phi_2 + \frac{d}{dr}\Phi_0 = 0, \quad 6)\ \ i\Big(\frac{d}{dr} + \frac{1}{r}\Big)\Phi_1 + i\frac{\nu}{r}\Phi_2 + mH_1 = 0\,.
\end{gathered} \tag{8.40}
$$

Introducing new variables $i\Phi_1 = \varphi_1$, $i\Phi_2 = \varphi_2$, we may exclude the imaginary unit from equations:

$$
\begin{gathered}
1)\ \ mE_1 = -\frac{m^2}{2\nu}r\Phi_0, \quad 2)\ \ \Big(\epsilon + \frac{\alpha}{r}\Big)mE_1 + \Big(\frac{d}{dr} + \frac{1}{r}\Big)mH_1 + m^2\varphi_1 = 0, \\
3)\ \ mH_1 = \frac{m^2}{2\nu}r\varphi_2, \quad 4)\ \ \Big(\epsilon + \frac{\alpha}{r}\Big)\varphi_1 - \frac{\nu}{r}\Phi_0 + mE_1 = 0, \\
5)\ \ \Big(\epsilon + \frac{\alpha}{r}\Big)\varphi_2 + \frac{d}{dr}\Phi_0 = 0, \quad 6)\ \ \Big(\frac{d}{dr} + \frac{1}{r}\Big)\varphi_1 + \frac{\nu}{r}\varphi_2 + mH_1 = 0\,.
\end{gathered} \tag{8.41}
$$

With the help of equations 1) and 3), we may exclude E_1, H_1, then remain four equations for the variables $\Phi_0, \varphi_1, \varphi_2$:

$$
\begin{gathered}
2) \quad -\Big(\epsilon + \frac{\alpha}{r}\Big)\frac{m^2}{2\nu} r\Phi_0 + \Big(\frac{d}{dr} + \frac{1}{r}\Big)\frac{m^2}{2\nu} r\varphi_2 + m^2\varphi_1 = 0, \\
4) \quad \Big(\epsilon + \frac{\alpha}{r}\Big)\varphi_1 - \frac{\nu}{r}\Phi_0 - \frac{m^2}{2\nu} r\Phi_0 = 0, \\
5) \quad \Big(\epsilon + \frac{\alpha}{r}\Big)\varphi_2 + \frac{d}{dr}\Phi_0 = 0, \quad 6) \quad \Big(\frac{d}{dr} + \frac{1}{r}\Big)\varphi_1 + \frac{\nu}{r}\varphi_2 + \frac{m^2}{2\nu} r\varphi_2 = 0 .
\end{gathered} \tag{8.42}
$$

It is evident that from physical point of view, there must exist two independent classes of states described within the initial system (8.37) and the system (8.42) as well.

There exist two simple possibilities to get second order differential equations for functions φ_1 and φ_2. The first is as follows: from equation 2) with the use of equation 6) we exclude the variable φ_2, and then with the help of equation 4) we exclude the variable φ_1, so we get

$$
\begin{gathered}
\frac{d^2\varphi_1}{dr^2} + \Big(\frac{4}{r} - \frac{2r}{r^2 + 2\nu^2/m^2}\Big)\frac{d\varphi_1}{dr} \\
+\Big(-m^2 + \epsilon^2 + \frac{2\,\alpha\,\epsilon}{r} - \frac{2}{r^2 + 2\,\nu^2/m^2} + \frac{-2\,\nu^2 + \alpha^2 + 2}{r^2}\Big)\varphi_1 = 0 .
\end{gathered} \tag{8.43}
$$

This equation has four singular points: three are regular points, and the point $r = \infty$ is an irregular one of rank 2. Only two singular points belong to the physical region of the variable r. Having known the function φ_1, we may find all remaining functions.

Consider the second possibility: from 4) we express Φ_0 through φ_1; then from 2) express φ_1 through φ_2 and substitute these in equation 6); in this way, we obtain the following equation for function φ_2:

$$
\begin{gathered}
\frac{d^2\varphi_2}{dr^2} + \Big[\frac{4}{r} + \frac{2\,m^2 r}{m^2 r^2 + 2\,\nu^2} + \frac{2\,(-m^2 r + \epsilon^2 r + \alpha\,\epsilon)}{m^2 r^2 - \epsilon^2 r^2 - 2\,\alpha\epsilon\, r + 2\,\nu^2 - \alpha^2}\Big]\frac{d\varphi_2}{dr} \\
+\Big[-m^2 + \epsilon^2 + \frac{2\,\alpha\,\epsilon\,\Big(2\,\nu^2 - \alpha^2 + 2\Big)}{\Big(2\,\nu^2 - \alpha^2\Big) r} + \frac{4\,m^2}{m^2 r^2 + 2\,\nu^2} + \frac{-2\,\nu^2 + \alpha^2 + 2}{r^2} \\
+\frac{-4\,m^2 r\alpha\,\epsilon + 4\,r\alpha\,\epsilon^3 - 8\,m^2\nu^2 + 4\,m^2\alpha^2 + 8\,\nu^2\epsilon^2 + 4\,\alpha^2\epsilon^2}{\Big(m^2 r^2 - \epsilon^2 r^2 - 2\,\alpha\epsilon\, r + 2\,\nu^2 - \alpha^2\Big)\Big(2\,\nu^2 - \alpha^2\Big)}\Big]\varphi_2 = 0 .
\end{gathered} \tag{8.44}
$$

Having known φ_2, we may find all remaining functions.

8.5 The study of equation for φ_1

Let us transform eq. (8.43) to dimensionless variable:

$$
x = mr = \frac{Mc}{\hbar} r = \frac{r}{\lambda}, \quad \frac{\epsilon}{m} = E, \quad 2\nu^2 = \Gamma^2;
$$

this results in

$$
\frac{d^2\varphi_1}{dx^2} + \Big(\frac{4}{x} - \frac{2x}{x^2 + \Gamma^2}\Big)\frac{d\varphi_1}{dx} + \Big(E^2 - 1 + \frac{2\alpha E}{x} - \frac{2}{x^2 + \Gamma^2} + \frac{\alpha^2 + 2 - \Gamma^2}{x^2}\Big)\varphi_1 = 0, \tag{8.45}
$$

or differently

$$\frac{d^2\varphi_1}{dx^2} + \Big(\frac{4}{x} - \frac{1}{x+i\Gamma} - \frac{1}{x-i\Gamma}\Big)\frac{d\varphi_1}{dx} + \Big(E^2 - 1 + \frac{2\alpha E}{x} - \frac{i/\Gamma}{x+i\Gamma} + \frac{i/\Gamma}{x-i\Gamma} - \frac{\mu^2}{x^2}\Big)\varphi_1 = 0, \quad (8.46)$$

where the notation $\mu^2 = \Gamma^2 - 2 - \alpha^2$ is used. The last equation has three regular points $x = 0, -i\Gamma, +i\Gamma$, and one irregular, $x = \infty$, of the rank 2. Its Frobenius solutions are constructed in the form

$$\varphi_1(x) = \varphi(x) f(x) = x^A e^{Bx} f(x).$$

For function $f(x)$ we obtain the following equation

$$f'' + \Big(\frac{2A}{x} + 2B + \frac{4}{x} - \frac{1}{x+i\Gamma} - \frac{1}{x-i\Gamma}\Big)f' + \Big(\frac{A(A-1)}{x^2} + \frac{2AB}{x} + B^2$$

$$+\frac{4}{x}\frac{A}{x} - \frac{1}{x+i\Gamma}\frac{A}{x} - \frac{1}{x-i\Gamma}\frac{A}{x} + \frac{4B}{x} - \frac{B}{x+i\Gamma} - \frac{B}{x-i\Gamma}$$

$$+E^2 - 1 + \frac{2\alpha E}{x} - \frac{i/\Gamma}{x+i\Gamma} + \frac{i/\Gamma}{x-i\Gamma} - \frac{\mu^2}{x^2}\Big)f = 0\,.$$

Allowing for restrictions $A = -3/2 \pm \sqrt{(3/2)^2 + \mu^2}$, $B = \pm\sqrt{1-E^2}$, we arrive at a simpler equation

$$\frac{d^2 f}{dx^2} + \Big(2B + \frac{4+2A}{x} - \frac{1}{x-i\Gamma} - \frac{1}{x+i\Gamma}\Big)\frac{df}{dx}$$
$$+\Big(\frac{2AB + 2E\alpha + 4B}{x} + \frac{iA - B\Gamma + i}{\Gamma\,(x-i\Gamma)} - \frac{iA + B\Gamma + i}{\Gamma\,(x+i\Gamma)}\Big)f = 0\,. \quad (8.47)$$

To get solutions suitable to describe bound states, we should take the following parameters

$$A = -\frac{3}{2} + \sqrt{(\frac{3}{2})^2 + \mu^2} > 0, \quad B = -\sqrt{1-E^2} < 0. \quad (8.48)$$

To present eq. (8.47), it is convenient to apply the shortening notations:

$$\frac{d^2 f}{dx^2} + \Big(K + \frac{L}{x} - \frac{1}{x-i\Gamma} - \frac{1}{x+i\Gamma}\Big)\frac{df}{dx} + \Big(\frac{a}{x} + \frac{b}{x-i\Gamma} - \frac{c}{x+i\Gamma}\Big)f = 0\,. \quad (8.49)$$

We construct solutions of the last equation in the form of power series, $f(x) = \sum_{n=0}^{\infty} d_n x^n$. Having performed needed calculations, we get 4-term recurrent relations for coefficients of the series:

$$[\,K(k-2) + (a+b-c)\,]\,d_{k-2}$$
$$+[\,(k-1)(k-2) + (L-2)(k-1) + i\Gamma(b+c)\,]\,d_{k-1}$$
$$+[\,K\Gamma^2 k + a\Gamma^2\,]\,d_k + [\Gamma^2(k+1)k + L\Gamma^2(k+1)\,]\,d_{k+1} = 0\,. \quad (8.50)$$

Having studied the convergence of the series by Poincaré–Perron method, we get two possible convergence radii: $R_{conv} = \Gamma, +\infty$. It is easy to show that behaviour of solutions near the singular point $\pm i\Gamma$ is given by the simple structure, $\varphi_1 \sim (x \pm i\Gamma)^D$, $D = 0, +2$. Therefore, we may expect that the series converges in all points.

Let us try to get some quantisation rules for energy levels by applying the known condition [10], which separates so-called transcendental Frobenius solutions. The needed restriction has the form (see eq. (8.50))

$$K(k-2) + (a+b-c) = 0, \quad k \geq 2\,. \quad (8.51)$$

Explicitly the last condition reads

$$\sqrt{1-E^2}\,N = E\alpha, \quad \text{where } N \equiv (k-2)+A+1\,; \tag{8.52}$$

whence it follows the formula for energy levels (let $k-2=n=0,1,2,3,...$)

$$E = \frac{1}{\sqrt{1+\alpha^2/N^2}}, \quad N = n - \frac{1}{2} + \sqrt{(\frac{3}{2})^2+\mu^2}. \tag{8.53}$$

Allowing for the equality $\mu^2 = j(j+1)-2-\alpha^2$, we obtain final formulas for energy levels

$$E = \frac{1}{\sqrt{1+\alpha^2/N^2}}, \quad N = n - \frac{1}{2} + \sqrt{j(j+1)+\frac{1}{4}-\alpha^2}, \tag{8.54}$$

where $j=1,2,3,...$; $n=0,1,2,3,...$ To illustrate the spectrum, let us find a number of energy values, $\epsilon^{(1)}_{n,j}$ at $j=1,\ n=\overline{0,20}$:

$$\begin{array}{llll}
n=0, & 0.9999733604, & & \\
n=1, & 0.9999933400, & n=11, & 0.9999998150,\\
n=2, & 0.9999970400, & n=12, & 0.9999998424,\\
n=3, & 0.9999983350, & n=13, & 0.9999998641,\\
n=4, & 0.9999989344, & n=14, & 0.9999998816,\\
n=5, & 0.9999992600, & n=15, & 0.9999998959,\\
n=6, & 0.9999994563, & n=16, & 0.9999999078,\\
n=7, & 0.9999995838, & n=17, & 0.9999999178,\\
n=8, & 0.9999996711, & n=18, & 0.9999999262,\\
n=9, & 0.9999997336, & n=19, & 0.9999999334,\\
n=10, & 0.9999997798, & n=20, & 0.9999999396.
\end{array} \tag{8.55}$$

This spectrum is illustrated by Fig. 8.1. Behaviour of the factor $\varphi(x)$ in Frobenius solutions $\varphi(x)$ at $j=1, n=1,5,\ 10,\ 20$ is illustrated by Fig. 8.2. Behaviour of the factor $\varphi(x)$ at $j=1, n=0$ is illustrated by Fig. 8.3. Also, for these parameters we calculate coefficients of the polynomial approximation $P_{k=20}$ of the series:

$$\begin{gathered}
P_{k=20} = 1 + 0.004515x + 0.1299x^2 + 0.0005247x^3 - 0.01032x^4 - 0.0000354x^5\\
+0.001825x^6 + 5.58\cdot 10^{-6}x^7 - 0.000431x^8 - 1.21\cdot 10^{-6}x^9\\
+0.0001190x^{10} + 3.11\cdot 10^{-7}x^{11} - 0.0000364x^{12} - 9.0\cdot 10^{-8}x^{13} + 0.00001194x^{14} + 2.8\cdot 10^{-8}x^{15}\\
-4.13\cdot 10^{-6}x^{16} - 9.4\cdot 10^{-9}x^{17} + 1.49\cdot 10^{-6}x^{18} + 3.3\cdot 10^{-9}x^{19} - 5.55\cdot 10^{-7}x^{20}.
\end{gathered} \tag{8.56}$$

Besides, numerical study show that polynomial $P_{k=20}$ has the following zero points, $P_{k=20}(x)=0$:

$$\begin{array}{ccc}
-2.1,\quad +2.1; & -2.03-0.65i, -2.03+0.65i; & -1.72-1.23i, -1.72+1.23i;\\
-1.23-1.66i, -1.23+1.66i; & -0.63-1.89i, -0.63+1.89i; & -1.81i, \quad +1.81i;\\
0.63-1.89i; 0.63+1.89i; & 1.23-1.67i; 1.23+1.67i; & 1.72-1.23i, 1.72+1.23i;\\
2.03-0.65i, 2.03+0.65, & &
\end{array}$$

where we can see only one zero point $x=+2.1$ in the physical region of the variable.

Behaviour of the function $f(x)$ and the complete Frobenius solution $\varphi_1(x)=\varphi(x)f(x)$ at $n=0$ is illustrated by Fig. 8.4. A numerical study of the series $f_1(x)$ for other values of n proves that the qualitative situation does not change. In particular, the polynomial approximation gives only one zero points in the physical region; they are localised in a small interval. Figure 8.5 illustrates behaviour of the complete Frobenius solutions at $n=5,10,15$, and 20.

8.6 Equation for function φ_2

Let us turn to the eq. (8.44) for function φ_2. In the variables $x = mr, \epsilon/m = E, 2\nu^2 = \Gamma^2$ it has the form

$$\frac{d^2\varphi_2}{dx^2} + \left[\frac{4}{x} + \frac{2x}{x^2+\Gamma^2} + \frac{2\alpha E - 2(1-E^2)x}{x^2(1-E^2) - 2\alpha E x + \Gamma^2 - \alpha^2}\right]\frac{d\varphi_2}{dx}$$

$$+\Big[-(1-E^2) + \frac{2\alpha E(\Gamma^2-\alpha^2+2)}{\Gamma^2-\alpha^2}\frac{1}{x} + \frac{4}{x^2+\Gamma^2} + \frac{2-(\Gamma^2-\alpha^2)}{x^2}$$

$$+\frac{1}{\Gamma^2-\alpha^2}\frac{-4\alpha E(1-E^2)x - 4\Gamma^2(1-E^2) + 4\alpha^2(1+E^2)}{x^2(1-E^2) - 2x\alpha E + \Gamma^2 - \alpha^2}\Big]\varphi_2 = 0\,. \tag{8.57}$$

The roots of the polynomial $x^2(1-E^2) - 2x\alpha E + \Gamma^2 - \alpha^2 = 0$ are

$$x_{1,2} = \frac{\alpha E \pm \sqrt{\alpha^2E^2 - (\Gamma^2-\alpha^2)(1-E^2)}}{1-E^2}. \tag{8.58}$$

Equation (8.57) may be presented differently

$$\frac{d^2\varphi_2}{dx^2} + \Big[\frac{4}{x} + \frac{1}{x+i\Gamma} + \frac{1}{x-i\Gamma} - \frac{1}{x-x_1} - \frac{1}{x-x_2}\Big]\frac{d\varphi_2}{dx}$$

$$+\Big[-(1-E^2) + \frac{2\alpha E(\Gamma^2-\alpha^2+2)}{(\Gamma^2-\alpha^2)}\frac{1}{x} + \frac{\alpha^2-\Gamma^2+2}{x^2} + \frac{2i}{\Gamma}\frac{1}{x+i\Gamma} - \frac{2i}{\Gamma}\frac{1}{x-i\Gamma}$$

$$+\frac{1}{\Gamma^2-\alpha^2}\frac{-4\alpha E(1-E^2)x - 4\Gamma^2(1-E^2) + 4\alpha^2(1+E^2)}{x^2(1-E^2) - 2x\alpha E + \Gamma^2 - \alpha^2}\Big]\varphi_2 = 0\,. \tag{8.59}$$

It has five regular singular points and one irregular one, $x = \infty$ of the rank 2:

$$0, \quad x_1, \quad x_2, \quad -i\Gamma, \quad +i\Gamma\,; \quad \infty_{[2]}\,.$$

In the neighbourhood of the points x_1, x_2, and $x = \pm i\Gamma$ solutions behave as follows

$$\varphi_2(x) = (x-x_1)^\rho, \quad \rho = 0,2; \quad \varphi_2(x) = (x-x_2)^\rho, \quad \rho = 0,2\,; \tag{8.60}$$

$$\varphi_2(x) = (x \pm i\Gamma)^\sigma, \qquad \sigma(\sigma-1) + \sigma = 0, \quad \sigma = 0. \tag{8.61}$$

Near the point $x = 0$, solutions have the simple structure $\varphi_2(x) = x^A$:

$$A(A-1) + 4A = (\Gamma^2-\alpha^2) - 2, \quad A = -\frac{3}{2} \pm \sqrt{(\frac{3}{2})^2 + \Gamma^2 - 2 - \alpha^2}\,; \tag{8.62}$$

recall that $\Gamma^2 = j(j+1) \geq 2$. In infinity, solutions behave as

$$\varphi_2(x) = e^{Bx}, \quad B = \pm\sqrt{1-E^2}\,. \tag{8.63}$$

To proceed further, we apply for eq. (8.59) a shorter form

$$\varphi_2'' + \Big(\frac{a_1}{x} + \frac{a_2}{x-i\Gamma} + \frac{a_3}{x+i\Gamma} + \frac{a_4}{x-x_1} + \frac{a_5}{x-x_2}\Big)\varphi_2'$$

$$+\Big(D + \frac{b_1}{x} + \frac{b}{x^2} + \frac{b_2}{x-i\Gamma} + \frac{b_3}{x+i\Gamma} + \frac{b_4}{x-x_1} + \frac{b_5}{x-x_2}\Big)\varphi_2 = 0. \tag{8.64}$$

Its Frobenius solution are searched in the form

$$\varphi_2(x) = x^A e^{Bx} f(x),$$

for function $f(x)$ we have the following equation

$$f'' + \Big(2B + \frac{a_1 + 2A}{x} + \frac{a_2}{r - i\Gamma} + \frac{a_3}{r + i\Gamma} + \frac{a_4}{x - x_1} + \frac{a_5}{x - x_2}\Big) f'$$

$$+ \Big((D + B^2) + \frac{a_1 A + b + A(A-1)}{x^2}$$

$$+ \frac{a_1 B - a_2 A/i\Gamma + a_3 A/i\Gamma - a_4 A/x_1 - a_5 A/x_2 + b_1 + 2AB}{x}$$

$$+ \frac{a_2 A/i\Gamma + a_2 B + b_2}{x - i\Gamma} + \frac{-a_3 A/i\Gamma + a_3 B + b_3}{x + i\Gamma}$$

$$+ \frac{a_4 A/x_1 + a_4 B + b_4}{x - x_1} + \frac{a_5 A/x_2 + a_5 B + b_5}{x - x_2}\Big) f = 0.$$

To follow for bound state solutions, we should use parameters

$$B = -\sqrt{1 - E^2} < 0, \quad A = \frac{3 + \sqrt{1 + 4(\Gamma^2 - \alpha^2)}}{2} > 0. \tag{8.65}$$

Equation for $f(x)$ may be presented shortly as

$$f'' + \Big(2C + \frac{C_1}{x} + \frac{C_2}{x - i\Gamma} + \frac{C_3}{x + i\Gamma} + \frac{C_4}{x - x_1} + \frac{C_5}{x - x_2}\Big) f'$$
$$+ \Big(\frac{D_1}{x} + \frac{D_2}{x - i\Gamma} + \frac{D_3}{x + i\Gamma} + \frac{D_4}{x - x_1} + \frac{D_5}{x - x_2}\Big) f = 0. \tag{8.66}$$

Searching solutions in the form of power series, $f = \sum_{n=0}^{\infty} d_n x^n$, we derive 6-term recurrent relations

$$[2C(n-4) + D_1 + D_2 + D_3 + D_4 + D_5]\ d_{n-4}$$
$$+[(n-3)(n-4) - 2C(x_1 + x_2)(n-3)$$
$$+C_1(n-3) + C_2(n-3) + C_3(n-3) + C_4(n-3) + C_5(n-3)$$
$$-D_1(x_1 + x_2) - D_2(x_1 + x_2 - i\Gamma) - D_3(x_1 + x_2 + i\Gamma) - D_4 x_2 - D_5 x_1]\ d_{n-3}$$
$$+[-(x_1 + x_2)(n-2)(n-3) + 2C(x_1 x_2 + \Gamma^2)(n-2)$$
$$-C_1(x_1 + x_2)(n-2) - C_2(x_1 + x_2 - i\Gamma)(n-2)$$
$$-C_3(x_1 + x_2 + i\Gamma)(n-2) - C_4 x_2(n-2) - C_5 x_1(n-2)$$
$$+D_1(x_1 x_2 + \Gamma^2) - D_2(x_1 i\Gamma + x_2 i\Gamma - x_1 x_2) + D_3(x_1 i\Gamma + x_2 i\Gamma + x_1 x_2) + D_4\Gamma^2 + D_5\Gamma^2]\ d_{n-2}$$
$$+[(x_1 x_2 + \Gamma^2)(n-1)(n-2) - 2C(x_1 + x_2)\Gamma^2(n-1) + C_1(x_1 x_2 + \Gamma^2)(n-1)$$
$$-C_2(x_1 i\Gamma + x_2 i\Gamma - x_1 x_2)(n-1) + C_3(x_1 i\Gamma + x_2 i\Gamma + x_1 x_2)(n-1) + C_4\Gamma^2(n-1) + C_5\Gamma^2(n-1)$$
$$-D_1(x_1 + x_2)\Gamma^2 + D_2 i\Gamma x_1 x_2 - D_3 i\Gamma x_1 x_2 - D_4\Gamma^2 x_2 - D_5\Gamma^2 x_1]\ d_{n-1}$$
$$+[-(x_1 + x_2)\Gamma^2 n(n-1) + 2C\Gamma^2 x_1 x_2 n - C_1(x_1 + x_2)\Gamma^2 n$$
$$+C_2 i\Gamma x_1 x_2 n - C_3 i\Gamma x_1 x_2 n - C_4\Gamma^2 x_2 n - C_5\Gamma^2 x_1 n + D_1\Gamma^2 x_1 x_2]\ d_n$$
$$+ [\Gamma^2 x_1 x_2 n(n+1) + C_1\Gamma^2 x_1 x_2(n+1)]\ d_{n+1} = 0. \tag{8.67}$$

Having used the Poincaré–Perron method, we find four possible convergence radii:

$$R_{conv} = \infty, \quad |\Gamma|, \quad |x_1|, \quad |x_2|. \tag{8.68}$$

In order to derive some quantisation rule, let us applying the known transcendency condition, here it has the form (see relation (8.67))

$$P_{k-4} = 0, \; k \geq 4, \quad 2C(k-4) + D_1 + D_2 + D_3 + D_4 + D_5 = 0, \tag{8.69}$$

where

$$2C = -2\sqrt{1-E^2},$$

$$D_1 = a_1 B - a_2 A/i\Gamma + a_3 A/i\Gamma - a_4 A/x_1 - a_5 A/x_2 + b_1 + 2AB,$$

$$D_2 = a_2 A/i\Gamma + a_2 B + b_2\,, \quad D_3 = -a_3 A/i\Gamma + a_3 B + b_3,$$

$$D_4 = a_4 A/x_1 + a_4 B + b_4, \quad D_5 = a_5 A/x_2 + a_5 B + b_5\,.$$

Taking into account expressions for involved parameters:

$$x_1 = \frac{\alpha E - \sqrt{\alpha^2 - \Gamma^2 + E^2\Gamma^2}}{1-E^2}, \quad x_2 = \frac{\alpha E + \sqrt{\alpha^2 - \Gamma^2 + E^2\Gamma^2}}{1-E^2},$$

$$a_1 = 4\,, \quad a_2 = 1\,, \quad a_3 = 1\,, \quad a_4 = -1\,, \quad a_5 = -1\,,$$

$$b_1 = \frac{2\,\alpha\,E\,(\Gamma^2 - \alpha^2 + 2)}{\Gamma^2 - \alpha^2}, \quad b_2 = \frac{-2\,i}{\Gamma}, \quad b_3 = \frac{2\,i}{\Gamma},$$

$$b_4 = -\frac{2\,(\alpha\,E + \sqrt{\alpha^2 - \Gamma^2 + E^2\Gamma^2})}{\Gamma^2 - \alpha^2}, \quad b_5 = \frac{2\,(-\alpha\,E + \sqrt{\alpha^2 - \Gamma^2 + E^2\Gamma^2})}{\Gamma^2 - \alpha^2},$$

from (8.69) we derive an algebraic equation

$$\sqrt{1-E^2}\,[\,(k-4) + (A+2)\,] = \alpha E\,, \quad k - 4 \geq 0. \tag{8.70}$$

Let us apply notation $(k-4) + (A+2) = N, k-4 = n = 0, 1, 2, ...$; then the formulas for energy levels reads

$$E = \frac{1}{\sqrt{1+\alpha^2/N^2}}, \quad N = n + \frac{1}{2} + \sqrt{\frac{1}{4} + j(j+1) - \alpha^2}. \tag{8.71}$$

It should be noted that two formulas (8.54) and (8.71) determine the same spectrum. This agrees with the fact that according to eqs. 4) and 5) in eq. (8.40), the functions φ_1 and φ_2 differ only in an elementary multiplier.

8.7 Studying the first equation for Φ_0

There exists possibility to get a 2nd-order equation for function Φ_0. To this end, from eq. 4) in eq. (8.41) one finds relationship between Φ_0 and φ_1:

$$\frac{2\nu}{r}\Big(\epsilon + \frac{\alpha}{r}\Big)\varphi_1(r) = \Big(\frac{2\nu^2}{r^2} + m^2\Big)\Phi_0(r),$$

or in dimensionless form

$$\Phi_0(x) = 2\nu \frac{Ex+\alpha}{2\nu^2+x^2}\,\varphi_1(x) = F(x)\,\varphi_1(x)\,. \tag{8.72}$$

Now taking in mind the known 2nd-order equation for φ_1: $\Delta\varphi_1 = 0$, $\varphi_1 = F^{-1}\Phi_0$, one finds 2nd-order equation for Φ_0:

$$\Delta F^{-1}\Phi_0 = 0 \quad \Longrightarrow \quad F\Delta F^{-1}\,\Phi_0 = 0\,.$$

Thus, equation for Φ_0 reads

$$\frac{Ex+\alpha}{2\nu^2+x^2}\Delta\frac{2\nu^2+x^2}{Ex+\alpha}\,\Phi_0 = 0, \tag{8.73}$$

where

$$\Delta = \frac{d^2}{dx^2} + \Big(\frac{4}{x} - \frac{2x}{x^2+\Gamma^2}\Big)\frac{d}{dx} + E^2 - 1 + \frac{2\alpha E}{x} + \frac{\alpha^2+2-\Gamma^2}{x^2} - \frac{2}{x^2+\Gamma^2}\,.$$

Equation (8.73) readily reduces to an explicit form

$$\frac{d^2\Phi_0}{dx^2} + \Big[\frac{4}{x} - \frac{2\,E}{Ex+\alpha} + \frac{2\,x}{x^2+2\,\nu^2}\Big]\frac{d\Phi_0}{dx}$$
$$+\Big[\frac{2\,E\,(\alpha^2-2)}{x\alpha} + \frac{-2\,\nu^2+\alpha^2+2}{x^2} + \frac{8\,\nu^2}{(x^2+2\,\nu^2)^2} + \frac{2\,(4\,E^2\nu^2+3\,\alpha^2)\,E^2}{(2\,E^2\nu^2+\alpha^2)\,(Ex+\alpha)\,\alpha}$$
$$+E^2 - 1 + \frac{2\,E^2}{(Ex+\alpha)^2} + \frac{-2\,E\alpha\,x+4\,E^2\nu^2+4\,\alpha^2}{(x^2+2\,\nu^2)\,(2\,E^2\nu^2+\alpha^2)}\Big]\Phi_0 = 0\,. \tag{8.74}$$

We do not need to study this equation because it must lead to the energy spectrum, which coincides with that derived from the study of equations for φ_1 and φ_2.

8.8 Second equation for function Φ_0

Now we prove that for variable $\Phi_0(x)$ there exists another 2nd-order equation, which provides us with a different energy spectrum. To derive this equation, we start with equations

$$\begin{gathered} 1)\quad E_1 = -\frac{1}{2\nu}x\Phi_0, \quad 2)\quad \Big(E+\frac{\alpha}{x}\Big)E_1 + \Big(\frac{d}{dx}+\frac{1}{x}\Big)H_1 + \varphi_1 = 0, \\ 3)\quad \frac{2\nu}{x}H_1 = \varphi_2, \quad 4)\quad -\Big(E+\frac{\alpha}{x}\Big)\varphi_1 + \frac{\nu}{x}\Phi_0 = E_1, \\ 5)\quad \Big(E+\frac{\alpha}{x}\Big)\varphi_2 + \frac{d}{dx}\Phi_0 = 0, \quad 6)\quad \Big(\frac{d}{dx}+\frac{1}{x}\Big)\varphi_1 + \frac{\nu}{x}\varphi_2 + H_1 = 0\,. \end{gathered} \tag{8.75}$$

With the help of eqs. 3) and 4), we exclude the variables φ_2 and E_1:

$$\begin{gathered} 1)\quad -\frac{2\nu}{x}\Big(E+\frac{\alpha}{x}\Big)\varphi_1 + \Big(\frac{2\nu^2}{x^2}+1\Big)\Phi_0 = 0, \\ 2)\quad \Big(\frac{d}{dx}+\frac{1}{x}\Big)H_1 + \Big(E+\frac{\alpha}{x}\Big)\frac{\nu}{x}\Phi_0 + \Big[1 - \Big(E+\frac{\alpha}{x}\Big)^2\Big]\varphi_1 = 0, \\ 5)\quad \frac{d}{dx}\Phi_0 + \frac{2\nu}{x}\Big(E+\frac{\alpha}{x}\Big)H_1 = 0, \quad 6)\quad \Big(\frac{d}{dx}+\frac{1}{x}\Big)\varphi_1 + \Big(1+\frac{2\nu^2}{x^2}\Big)H_1 = 0\,. \end{gathered} \tag{8.76}$$

Act by operator $\frac{d}{dx}$ on eq. 5) in eq. (8.76):

$$\frac{d^2}{dx^2}\Phi_0 - \frac{2\nu}{x^2}\Big(E+\frac{\alpha}{x}\Big)H_1 - \frac{2\nu}{x}\frac{\alpha}{x^2}H_1 + \frac{2\nu}{x}\Big(E+\frac{\alpha}{x}\Big)\frac{d}{dx}H_1 = 0.$$

Then with the help of eq. 2) in eq. (8.76):

$$\frac{d}{dx}H_1 = -\Big(\frac{1}{x}H_1 + \Big(E+\frac{\alpha}{x}\Big)\frac{\nu}{x}\Phi_0 + \Big[1-\Big(E+\frac{\alpha}{x}\Big)^2\Big]\varphi_1\Big)$$

we get

$$\Big[\frac{d^2}{dx^2} - \frac{2\nu^2}{x^2}\Big(E+\frac{\alpha}{x}\Big)^2\Big]\Phi_0$$

$$-\Big[\frac{2\nu}{x^2}\Big(E+\frac{\alpha}{x}\Big) + \frac{2\nu}{x}\frac{\alpha}{x^2} + \frac{2\nu}{x^2}\Big(E+\frac{\alpha}{x}\Big)\Big]H_1 - \frac{2\nu}{x}\Big(E+\frac{\alpha}{x}\Big)\Big[1-(E+\frac{\alpha}{x})^2\Big]\varphi_1 = 0.$$

In order to exclude the function φ_1, let us use equation 1) from eq. (8.76):

$$\frac{2\nu}{x}\Big(E+\frac{\alpha}{x}\Big)\varphi_1 = \Big(1+\frac{2\nu^2}{x^2}\Big)\Phi_0;$$

this yields

$$\Big[\frac{d^2}{dx^2} + \Big(E+\frac{\alpha}{x}\Big)^2 - 1 - \frac{2\nu^2}{x^2}\Big]\Phi_0 - \frac{4\nu}{x^2}\Big(E+\frac{\alpha}{x}\Big)H_1 - \frac{2\nu}{x}\frac{\alpha}{x^2}H_1 = 0. \tag{8.77}$$

Now with the use of eq. 5) from eq. (8.76):

$$\frac{d}{dx}\Phi_0 + \frac{2\nu}{x}\Big(E+\frac{\alpha}{x}\Big)H_1 = 0;$$

we obtain

$$\Big[\frac{d^2}{dx^2} + \Big(E+\frac{\alpha}{x}\Big)^2 - 1 - \frac{2\nu^2}{x^2} + \frac{2}{x}\frac{d}{dx}\Big]\Phi_0 - \frac{2\nu}{x}\frac{\alpha}{x^2}H_1 = 0.$$

Finally, taking into account eq. 5) from eq. (8.76):

$$-\frac{x}{2\nu}\frac{1}{(E+\frac{\alpha}{x})}\frac{d}{dx}\Phi_0 = H_1,$$

we arrive at the needed 2nd-order equation for function Φ_0:

$$\Big[\frac{d^2}{dx^2} + \Big(\frac{3}{x} - \frac{E}{Ex+\alpha}\Big)\frac{d}{dx} + E^2 - 1 + \frac{2E\alpha}{x} + \frac{\alpha^2-2\nu^2}{x^2}\Big]\Phi_0 = 0. \tag{8.78}$$

In the variable z, $x = -\frac{\alpha}{E}z, z \in (-\infty, 0)$ it takes the form

$$\frac{d^2\Phi_0}{dz^2} + \Big(\frac{3}{z} - \frac{1}{z-1}\Big)\frac{d\Phi_0}{dz} + \Big(\alpha^2 - \frac{\alpha^2}{E^2} - \frac{2\alpha^2}{z} - \frac{2\nu^2-\alpha^2}{z^2}\Big)\Phi_0 = 0. \tag{8.79}$$

Below we will apply the following notations

$$\gamma^2 = 2\nu^2 - \alpha^2 = j(j+1) - \alpha^2 > 0, \quad -\Lambda^2 = -(-\alpha^2 + \frac{\alpha^2}{E^2}) = -\alpha^2\frac{1-E^2}{E^2} < 0; \tag{8.80}$$

then eq. (8.79) reads

$$\frac{d^2\Phi_0}{dz^2} + \Big(\frac{3}{z} - \frac{1}{z-1}\Big)\frac{d\Phi_0}{dz} + \Big(-\Lambda^2 - \frac{2\alpha^2}{z} - \frac{\gamma^2}{z^2}\Big)\Phi_0 = 0. \tag{8.81}$$

It has two regular singular points $z = 0, 1$ and one irregular point $z = \infty$ of the rank 2. Near the point $z = 0$, the equation becomes simpler, and its solutions behave as

$$\frac{d^2\Phi_0}{dz^2} + \frac{3}{z}\frac{d\Phi_0}{dz} - \frac{\gamma^2}{z^2}\Phi_0 = 0, \quad \Phi_0 \sim z^A, \quad A = -1 + \sqrt{1+\gamma^2} > 0. \tag{8.82}$$

To bound states, there correspond positive values for A. Find asymptotical behaviour of solutions as $z \to -\infty$:

$$\frac{d^2\Phi_0}{dz^2} + \frac{2}{z}\frac{d\Phi_0}{dz} - \Lambda^2\Phi_0 = 0, \quad \Phi_0 = e^{\pm\sqrt{\Lambda^2}\,z} = e^{\mp\sqrt{M^2c^4-E^2}\,r/\hbar c}. \tag{8.83}$$

Frobenius solutions for eq. (8.81) are searched in the form $\Phi_0(z) = z^A e^{Bz}\, f(z)$; where $f(z)$ obeys the equation

$$f'' + \Big(2B + \frac{2A+3}{z} - \frac{1}{z-1}\Big)f'$$
$$+\Big((B^2-\Lambda^2) + \frac{A^2+2A-\gamma^2}{z^2} + \frac{2AB+A+3B-2\alpha^2}{z} - \frac{A+B}{z-1}\Big)f = 0.$$

Below we use the following parameters, $A = -1+\sqrt{1+\gamma^2}$, $B = +\sqrt{\Lambda^2}$, the previous equation becomes simpler

$$f'' + \Big(2B + \frac{2A+3}{z} - \frac{1}{z-1}\Big)f'$$
$$+\Big(\frac{2AB+A+3B-2\alpha^2}{z} - \frac{A+B}{z-1}\Big)f = 0. \tag{8.84}$$

It may be identified with confluent Heun equation [10,11]:

$$f'' + \Big(-t + \frac{c}{z} + \frac{d}{z-1}\Big)f' + \frac{\lambda - ta\,z}{z(z-1)}f = 0, \tag{8.85}$$

its parameters are determined by the formulas

$$t = -2B, \qquad c = 2A+3, \quad d = -1,$$
$$-\lambda = 2AB + 3B + A - 2\alpha^2, \quad -ta = 2BA + 2B - 2\alpha^2. \tag{8.86}$$

For parameter a we readily obtain expression

$$a = A + 1 - \frac{\alpha^2}{B} = +\sqrt{1+\gamma^2} - \frac{\alpha^2}{\Lambda}. \tag{8.87}$$

Solutions for $f(z)$ may be constructed in the form of a power series $f(z) = \sum_{k=0}^{\infty} d_k z^k$ with 3-term recurrent relations;

$$c\,d_1 + \lambda\,d_0 = 0, \quad k = 1,2,3,..., \qquad t\,(k-1+a)\,d_{k-1}$$
$$-[\,k(k-1+t+d+c)+\lambda\,]\,d_k + (k+1)\,(k+c)\,d_{k+1} = 0. \tag{8.88}$$

The recurrent relation (8.88) may be re-written differently

$$k = 1,2,... \qquad P_k d_k - (Q_k+\lambda)\,d_{k+1} + R_k\,d_{k+2} = 0,$$
$$P_k = t\,(k-1+a), \quad Q_k = k(k-1+t+d+c), \quad R_k = (k+1)\,(k+c), \tag{8.89}$$

it is equivalent to

$$\frac{1}{k^2}P_k - \frac{1}{k^2}(Q_k+\lambda)\,\frac{d_{k+1}}{d_k} + \frac{1}{k^2}R_k\,\frac{d_{k+2}}{d_{k+1}}\frac{d_{k+1}}{d_k} = 0;$$

whence in the limit $k \to \infty$ it follows the simple algebraic equation

$$\lim_{k\to\infty} \frac{d_{k+1}}{d_k} = \lim_{k\to\infty} \frac{d_{k+2}}{d_{k+1}} = r, \quad -r + r^2 = 0 \quad \Longrightarrow \quad r = 0, 1.$$

According to Poincaré–Perron method, we conclude that convergence radius is $R_{conv} = 1, \infty$. Behaviour of solutions near the point $z = 1$ is determined by the formula $\Phi_0 \sim (z-1)^\rho, \rho = 0, 2$.

Imposing restriction $P_n = 0$, we get the class of the transcendental confluent Heun functions

$$P_k = 0 \quad -a = k - 1 = n, \quad n = 0, 1, 2, 3.... \tag{8.90}$$

The constraint (8.90) gives the following quantisation rule

$$n + \sqrt{1 + \gamma^2} = \frac{\alpha^2}{\Lambda} \equiv \alpha\sqrt{\frac{E^2}{1 - E^2}};$$

whence it follows the formula for energies

$$E = \frac{1}{\sqrt{1 + \alpha^2/N^2}}, \quad N = n + \sqrt{j(j+1) + 1 - \alpha^2}. \tag{8.91}$$

It should be emphasised that the last formula does not coincides with the previously derived formula for energies (8.71):

$$E = \frac{1}{\sqrt{1 + \alpha^2/N^2}} \quad N = n + \frac{1}{2} + \sqrt{j(j+1) + \frac{1}{4} - \alpha^2} \tag{8.92}$$

This means that the 2nd-order equation for Φ_0 describes the class of bound states different from the class of states determined by the 2nd-order equation for φ_1 (and the equation for φ_2). This statement is valid even if both of these classes do not represent genuine spectra. As said above, by general physical grounds, we should expect existence of two series of bound states with parity $P = (-1)^j$, determined by six eqs. (8.37).

We may note that two variables $\Phi_0(x)$ and $\varphi_1(x)$ relate to each other by a simple multiplier

$$\Phi_0(x) = 2\nu \, \frac{\alpha + Ex}{2\nu^2 + x^2} \, \varphi_1(x). \tag{8.93}$$

If one substitutes the last expression for Φ_0 into eq. (8.78) for Φ_0, the one will derive for $\varphi_1(x)$ an equation different from the above known eq. (8.45). Indeed, starting with the relationships

$$\Phi_0(x) = f(x)\varphi_1(x), \qquad f(x) = \frac{xE + \alpha}{2\nu^2 + x^2},$$

and from eq. (8.78) for Φ_0

$$\left[\frac{d^2}{dx^2} + \left(\frac{3}{x} - \frac{E}{Ex + \alpha}\right)\frac{d}{dx} + E^2 - 1 + \frac{2E\alpha}{x} + \frac{\alpha^2 - 2\nu^2}{r^2}\right]\Phi_0 = 0 \tag{8.94}$$

we obtain the following equation for φ_1:

$$\varphi_1'' + \left(\frac{3}{x} - \frac{E}{Ex + \alpha} + \frac{2E}{Ex + \alpha} - \frac{4x}{2\nu^2 + x^2}\right)\varphi_1'$$

$$+\Big[\Big(\frac{E}{Ex+\alpha}-\frac{2x}{2\nu^2+x^2}\Big)^2-\frac{E^2}{(Ex+\alpha)^2}-\frac{2}{2\nu^2+x^2}+\frac{4x^2}{(2\nu^2+x^2)^2}$$
$$+(\frac{3}{x}-\frac{E}{Ex+\alpha})(\frac{E}{Ex+\alpha}-\frac{2x}{2\nu^2+x^2})+E^2-1+\frac{2E\alpha}{x}+\frac{\alpha^2-2\nu^2}{x^2}\Big]\varphi_1=0. \qquad (8.95)$$

This equation for φ_1 does not coincides with previously studied one (8.46). This means that classes of solutions determined by eqs. (8.46) and (8.95) for variable $\varphi_1(x)$ are different.

Let us perform numerical study of the corresponding energy spectrum $\epsilon^{(2)}_{n,j}$. The formula (8.91) gives the following values for energies at $j=1,\ n=\overline{0,20}$:

$$\begin{array}{ll} n=0, 0.999991120, & \\ n=1, 0.999996430, & n=11, 0.999999835, \\ n=2, 0.999998087, & n=12, 0.999999858, \\ n=3, 0.999998810, & n=13, 0.999999877, \\ n=4, 0.999999189, & n=14, 0.999999892, \\ n=5, 0.999999412, & n=15, 0.999999904, \\ n=6, 0.999999554, & n=16, 0.999999915, \\ n=7, 0.999999650, & n=17, 0.999999924, \\ n=8, 0.999999718, & n=18, 0.999999931, \\ n=9, 0.999999768, & n=19, 0.999999938, \\ n=10, 0.999999806, & n=20, 0.999999943. \end{array} \qquad (8.96)$$

This set may be illustrated by Fig. 8.6.

It is helpful to compare two series of energy levels:

$$\Delta_n=\epsilon_{n,j=1}-\epsilon^{(2)}_{n,j=1}; \qquad (8.97)$$

we readily obtain the values for Δ_n:

$$\begin{array}{lllll} -3.091\cdot 10^{-6}, & -1.047\cdot 10^{-6}, & -4.75\cdot 10^{-7}, & -2.55\cdot 10^{-7}, & -1.52\cdot 10^{-7}, \\ -9.8\cdot 10^{-8}, & -6.7\cdot 10^{-8}, & -4.8\cdot 10^{-8}, & -3.5\cdot 10^{-8}, & -2.7\cdot 10^{-8}, \\ -2.1\cdot 10^{-8}, & -1.6\cdot 10^{-8}, & -1.3\cdot 10^{-8}, & -1.1\cdot 10^{-8}, & -9.\cdot 10^{-9}, \\ -7.\cdot 10^{-9}, & -6.\cdot 10^{-9}, & -5.\cdot 10^{-9}, & -5.\cdot 10^{-9}, & -4.\cdot 10^{-9} \end{array}$$

which is illustrated by Fig. 8.9. For completeness, let us detail the third spectrum $\epsilon^{(3)}_{n,j}$, related to the states with parity $P=(-1)^{j+1}$ and described in terms of hypergeometric functions

$$E=\frac{1}{\sqrt{1+\alpha^2/N^2}},\qquad N=n+\frac{1}{2}+\sqrt{(j+\frac{1}{2})^2-\alpha^2}; \qquad (8.98)$$

this formula gives the following values for energy at $j=1,\ n=\overline{0,20}$:

$$\begin{array}{lll} n=0, & 0.999993340, & \\ n=1, & 0.999997040, & n=11, 0.999999842, \\ n=2, & 0.999998335, & n=12, 0.999999864, \\ n=3, & 0.999998934, & n=13, 0.999999881, \\ n=4, & 0.999999260, & n=14, 0.999999895, \\ n=5, & 0.999999456, & n=15, 0.999999907, \\ n=6, & 0.999999583, & n=16, 0.999999917, \\ n=7, & 0.999999671, & n=17, 0.999999926, \\ n=8, & 0.999999733, & n=18, 0.999999933, \\ n=9, & 0.999999779, & n=19, 0.999999939, \\ n=10, & 0.999999815, & n=20, 0.999999945. \end{array} \qquad (8.99)$$

which is illustrated in Fig. 8.10. Let us compare the series of energy levels:

$$\tilde{\Delta}_n = \epsilon^{(1)}_{n,j=1} - \epsilon^{(3)}_{n,j=1};$$

we readily get for $\tilde{\Delta}_n$:

$$\begin{array}{lllll} -3.700\cdot 10^{-6}, & -1.295\cdot 10^{-6}, & -5.99\cdot 10^{-7}, & -3.26\cdot 10^{-7}, & -1.96\cdot 10^{-7}, \\ -1.27\cdot 10^{-7}, & -8.7\cdot 10^{-8}, & -6.28\cdot 10^{-8}, & -4.6\cdot 10^{-8}, & -3.5\cdot 10^{-8}, \\ -2.7\cdot 10^{-8}, & -2.2\cdot 10^{-8}, & -1.8\cdot 10^{-8}, & -1.4\cdot 10^{-8}, & -1.2\cdot 10^{-8}, \\ -1.\cdot 10^{-8}, & -8.\cdot 10^{-9}, & -7.\cdot 10^{-9}, & -6.\cdot 10^{-9}, & -5.\cdot 10^{-9}, \end{array}$$

which is illustrated in Fig. 8.11.

8.9 Fourth-order equations, the first method

Let us turn again to six equations (it is convenient to numerate them):

$$\begin{array}{ll} 1) & +i\Big(\epsilon+\dfrac{\alpha}{r}\Big)E_2 - 2i\dfrac{\nu}{r}H_1 - M\Phi_2 = 0, \\ 2) & -i\left(\epsilon+\dfrac{\alpha}{r}\right)\Phi_1 + \dfrac{\nu}{r}\Phi_0 - ME_1 = 0, \\ 3) & \Big(\dfrac{d}{dr}+\dfrac{2}{r}\Big)E_2 + 2\dfrac{\nu}{r}E_1 + M\Phi_0 = 0, \\ 4) & +i\Big(\epsilon+\dfrac{\alpha}{r}\Big)E_1 + i\Big(\dfrac{d}{dr}+\dfrac{1}{r}\Big)H_1 - M\Phi_1 = 0, \\ 5) & i\Big(\epsilon+\dfrac{\alpha}{r}\Big)\Phi_2 + \dfrac{d}{dr}\Phi_0 + ME_2 = 0, \\ 6) & i\Big(\dfrac{d}{dr}+\dfrac{1}{r}\Big)\Phi_1 + i\dfrac{\nu}{r}\Phi_2 + MH_1 = 0. \end{array} \tag{8.100}$$

Equations 2) and 4):

$$-i\left(\epsilon+\frac{\alpha}{r}\right)\Phi_1 + \frac{\nu}{r}\Phi_0 - ME_1 = 0, \quad +i\left(\epsilon+\frac{\alpha}{r}\right)E_1 + i\Big(\frac{d}{dr}+\frac{1}{r}\Big)H_1 - M\Phi_1 = 0$$

permit us to express Φ_1 and E_1 through Φ_0 and H_1:

$$\begin{aligned} E_1 &= \frac{(\epsilon r+\alpha)}{(M^2-\epsilon^2)r^2 - 2\alpha\epsilon r - \alpha^2}\left(r\frac{dH_1}{dr}+H_1\right) + \frac{M\nu r}{\Big(M^2-\epsilon^2\Big)r^2 - 2\alpha\epsilon r - \alpha^2}\Phi_0, \\ \Phi_1 &= \frac{iMr}{(M^2-\epsilon^2)r^2 - 2\alpha\epsilon r - \alpha^2}\left(r\frac{dH_1}{dr}+H_1\right) + \frac{i\nu(\epsilon r+\alpha)}{(M^2-\epsilon^2)r^2 - 2\alpha\epsilon r - \alpha^2}\Phi_0 . \end{aligned} \tag{8.101}$$

Similarly, eqs. 1) and 5):

$$+i(\epsilon+\frac{\alpha}{r})E_2 - 2i\frac{\nu}{r}H_1 - M\Phi_2 = 0, \quad i\left(\epsilon+\frac{\alpha}{r}\right)\Phi_2 + \frac{d}{dr}\Phi_0 + ME_2 = 0$$

permit to express Φ_2 and E_2 through Φ_0 and H_1:

$$\begin{aligned} E_2 &= -\frac{r^2M}{(M^2-\epsilon^2)r^2 - 2\alpha\epsilon r - \alpha^2}\frac{d\Phi_0}{dr} - \frac{2\nu(\epsilon r+\alpha)}{(M^2-\epsilon^2)r^2 - 2\alpha\epsilon r - \alpha^2}H_1 , \\ \Phi_2 &= -\frac{ir(\epsilon r+\alpha)}{M^2r^2 - \epsilon^2r^2 - 2\alpha\epsilon r - \alpha^2}\frac{d\Phi_0}{dr} - \frac{2irM\nu}{M^2r^2 - \epsilon^2r^2 - 2\alpha\epsilon r - \alpha^2}H_1 . \end{aligned} \tag{8.102}$$

If in eqs. 3) and 6):

$$\left(\frac{d}{dr}+\frac{2}{r}\right)E_2+2\frac{\nu}{r}E_1+M\Phi_0=0,\quad i\left(\frac{d}{dr}+\frac{1}{r}\right)\Phi_1+i\frac{\nu}{r}\Phi_2+MH_1=0$$

one excludes with the help of derived relations (8.101) and (8.102) the variables Φ_1, E_1, Φ_2, and E_2, then one obtains equations which contain only the variables Φ_0 and H_1:

$$\begin{gathered}
-r^2M\,\frac{d^2\Phi_0}{dr^2}+\Big[-2\,Mr+2\,\frac{\alpha\,\epsilon\,M}{M^2-\epsilon^2}\\
+\frac{M\alpha^2}{(Mr-\epsilon\,r-\alpha)\,(M-\epsilon)}+\frac{M\alpha^2}{(Mr+\epsilon\,r+\alpha)\,(M+\epsilon)}\Big]\frac{d\Phi_0}{dr}\\
+M\left[\left(M^2-\epsilon^2\right)r^2-2\,\alpha\,\epsilon\,r+2\,\nu^2-\alpha^2\right]\Phi_0+\\
\Big[-\frac{2\,\nu\,M\alpha}{(\epsilon-M)\,r+\alpha}-\frac{2\,\alpha\,\nu}{r}+\frac{2\,\nu\,M\alpha}{(M+\epsilon)\,r+\alpha}\Big]H_1=0\,,
\end{gathered}\tag{8.103}$$

$$\begin{gathered}
-r^2M\,\frac{d^2H_1}{dr^2}+\Big[-2\,Mr+2\,\frac{\alpha\,\epsilon\,M}{M^2-\epsilon^2}\\
+\frac{M\alpha^2}{(Mr-\epsilon\,r-\alpha)\,(M-\epsilon)}+\frac{M\alpha^2}{(Mr+\epsilon\,r+\alpha)\,(M+\epsilon)}\Big]\frac{dH_1}{dr}\\
+\Big[M\left(\left(M^2-\epsilon^2\right)r^2-2\,\alpha\,\epsilon\,r+2\,\nu^2-\alpha^2\right)+\frac{M\alpha}{(M-\epsilon)\,r-\alpha}-\frac{M\alpha}{(M+\epsilon)\,r+\alpha}\Big]H_1\\
+\Big[-\frac{\nu\,M\alpha}{(\epsilon-M)\,r+\alpha}-\frac{\alpha\,\nu}{r}+\frac{\nu\,M\alpha}{(M+\epsilon)\,r+\alpha}\Big]\Phi_0=0\,.
\end{gathered}\tag{8.104}$$

Whence it follows a 4th-order equation for Φ_0:

$$\begin{gathered}
\frac{d^4\Phi_0}{dr^4}+\Big[\frac{-4\,M^2r-4\,\epsilon^2r-4\,\alpha\,\epsilon}{M^2r^2+\epsilon^2r^2+2\,\alpha\,\epsilon\,r+\alpha^2}+\frac{14}{r}\Big]\frac{d^3\Phi_0}{dr^3}\\
+\Big[-2\,M^2+2\,\epsilon^2+\frac{2\,\alpha^2-4\,\nu^2+52}{r^2}+\frac{\left(-28\,M^2+44\,\epsilon^2\right)\alpha+36\,r\epsilon\,\left(M^2+\epsilon^2\right)}{\alpha\,\left(\left(M^2+\epsilon^2\right)r^2+2\,\alpha\,\epsilon\,r+\alpha^2\right)}\\
-8\frac{M^2\alpha^2}{\left(\left(M^2+\epsilon^2\right)r^2+2\alpha\,\epsilon\,r+\alpha^2\right)^2}+\frac{4\alpha^2\epsilon-36\epsilon}{\alpha r}\Big]\frac{d^2\Phi_0}{dr^2}\\
+\Big[4\,\frac{\epsilon\,\left(5\,\alpha^2+2\,\nu^2-16\right)}{\alpha\,r^2}+\frac{10\,\alpha^2-20\,\nu^2+48}{r^3}\\
+\frac{1}{\left(M^2r^2+\epsilon^2r^2+2\,\alpha\,\epsilon\,r+\alpha^2\right)\alpha^2}\\
\times[8\,M^4\alpha^2r-8\,M^4\nu^2r+8\,M^2\alpha^2\epsilon^2r+8\,\epsilon^4\nu^2r+8\,M^2\alpha^3\epsilon-24\,M^2\alpha\,\epsilon\,\nu^2+8\,\alpha\,\epsilon^3\nu^2+\\
+60\,M^4r-40\,M^2\epsilon^2r-100\,\epsilon^4r+184\,M^2\alpha\,\epsilon-136\,\alpha\,\epsilon^3]\\
+\frac{24\,M^4r+16\,M^2\epsilon^2r-8\,\epsilon^4r+56\,M^2\alpha\,\epsilon-8\,\alpha\,\epsilon^3}{\left(M^2r^2+\epsilon^2r^2+2\,\alpha\,\epsilon\,r+\alpha^2\right)^2}\\
+\frac{-18\,M^2\alpha^2+8\,M^2\nu^2+10\,\alpha^2\epsilon^2-8\,\epsilon^2\nu^2-60\,M^2+100\,\epsilon^2}{r\alpha^2}\Big]\frac{d\Phi_0}{dr}+\;Big[M^4-2\,M^2\epsilon^2+\epsilon^4
\end{gathered}$$

$$+\frac{-4\left(\alpha^2+8\right)(\epsilon-M)(M+\epsilon)\nu^2+\left(-2\alpha^4-44\alpha^2\right)M^2+6\epsilon^2\alpha^2\left(\alpha^2+2\right)}{r^2\alpha^2}$$

$$+\frac{4\nu^4+\left(-4\alpha^2-20\right)\nu^2+\alpha^4+6\alpha^2}{r^4}-2\frac{\alpha^2\nu^2}{r^6M^2}+4\frac{\epsilon\left(\alpha^4-2\alpha^2\nu^2+5\alpha^2+6\nu^2\right)}{r^3\alpha}$$

$$+\frac{1}{\alpha^3((M^2+\epsilon^2)r^2+2\alpha\epsilon r+\alpha^2)}$$

$$\times[+\left(32M^4-48M^2\epsilon^2\right)\alpha^3-40\epsilon M^2r\left(M^2+\epsilon^2\right)\alpha^2$$

$$-48\nu^2\left(\epsilon^4-5M^2\epsilon^2+2/3M^4\right)\alpha-40\epsilon r\nu^2\left(-3M^2+\epsilon^2\right)\left(M^2+\epsilon^2\right)]$$

$$+32\frac{\left(1/2M^2\alpha^3+3/2\left(\epsilon^2-1/3M^2\right)\nu^2\alpha+\epsilon r\nu^2\left(M^2+\epsilon^2\right)\right)M^2}{\alpha\left(\left(M^2+\epsilon^2\right)r^2+2\alpha\epsilon r+\alpha^2\right)^2}$$

$$4\frac{\epsilon\left(M^2\alpha^4-\alpha^4\epsilon^2-10M^2\alpha^2+30M^2\nu^2-10\epsilon^2\nu^2\right)}{r\alpha^3}\Big]\Phi_0=0; \qquad (8.105)$$

two singular points $0,\infty$ belong to physical region of the variable r:

$$-\frac{\alpha}{\epsilon+iM}, \quad -\frac{\alpha}{\epsilon-iM}, \quad 0, \quad \infty.$$

Similarly, we derive a 4th-order equation for H_1:

$$\frac{d^4H_1}{dr^4}+\left[\frac{-4M^2r-4\epsilon^2r-4\alpha\epsilon}{M^2r^2+\epsilon^2r^2+2\alpha\epsilon r+\alpha^2}+\frac{14}{r}\right]\frac{d^3H_1}{dr^3}$$

$$+\Big[-2M^2+2\epsilon^2-\frac{8M^2\alpha^2}{\left(\left(M^2+\epsilon^2\right)r^2+2\alpha\epsilon r+\alpha^2\right)^2}+\frac{2\alpha^2-4\nu^2+52}{r^2}$$

$$+\frac{\left(-28M^2+44\epsilon^2\right)\alpha+36r\epsilon\left(M^2+\epsilon^2\right)}{\alpha\left(\left(M^2+\epsilon^2\right)r^2+2\alpha\epsilon r+\alpha^2\right)}+\frac{4\alpha^2\epsilon-36\epsilon}{\alpha r}\Big]\frac{d^2H_1}{dr^2}$$

$$+\Big[\frac{24M^4r+16M^2\epsilon^2r-8\epsilon^4r+56M^2\alpha\epsilon-8\alpha\epsilon^3}{\left(M^2r^2+\epsilon^2r^2+2\alpha\epsilon r+\alpha^2\right)^2}$$

$$+4\frac{\epsilon\left(5\alpha^2+2\nu^2-18\right)}{\alpha r^2}+\frac{10\alpha^2-20\nu^2+52}{r^3}$$

$$+\frac{1}{\left(M^2r^2+\epsilon^2r^2+2\alpha\epsilon r+\alpha^2\right)\alpha^2}$$

$$\times[8M^4\alpha^2r-8M^4\nu^2r+8M^2\alpha^2\epsilon^2r+8\epsilon^4\nu^2r+8M^2\alpha^3\epsilon-24M^2\alpha\epsilon\nu^2+8\alpha\epsilon^3\nu^2$$

$$+64M^4r-48M^2\epsilon^2r-112\epsilon^4r+200M^2\alpha\epsilon-152\alpha\epsilon^3]$$

$$+\frac{-18M^2\alpha^2+8M^2\nu^2+10\alpha^2\epsilon^2-8\epsilon^2\nu^2-64M^2+112\epsilon^2}{r\alpha^2}\Big]\frac{dH_1}{dr}$$

$$+\Big[M^4-2M^2\epsilon^2+\epsilon^4+\frac{1}{\alpha\left(M^2r^2+(\epsilon r+\alpha)^2\right)^2}$$

$$[((32\nu^2-24)r\epsilon-16\nu^2\alpha+16\alpha^3+8\alpha)M^4$$

$$+32\epsilon^2\left(r(\nu^2-1/2)\epsilon+\alpha\left(3/2\nu^2-3/2\right)\right)M^2+8r\epsilon^5+8\alpha\epsilon^4]$$

$$+\frac{-4\left(\alpha^2+8\right)(\epsilon-M)(M+\epsilon)\nu^2+\left(-2\alpha^4-44\alpha^2-16\right)M^2+6\epsilon^2\left(\alpha^4+2\alpha^2+8\right)}{r^2\alpha^2}$$

$$+4\,\frac{\epsilon\,(\alpha^4-2\,\alpha^2\nu^2+5\,\alpha^2+6\,\nu^2-6)}{r^3\alpha}-2\,\frac{\alpha^2\nu^2}{r^6M^2}+\frac{1}{(M^2r^2+\epsilon^2r^2+2\,\alpha\,\epsilon\,r+\alpha^2)\,\alpha^3}$$
$$\times[(-40\,\nu^2+80)\,r\epsilon^5+(-48\,\nu^2+112)\,\alpha\,\epsilon^4+80\,(\nu^2-1/2\,\alpha^2)\,rM^2\epsilon^3$$
$$+240\,(\nu^2-1/5\,\alpha^2-4/5)\,\alpha\,M^2\epsilon^2+120\,(\nu^2-1/3\,\alpha^2-2/3)\,rM^4\epsilon$$
$$-32\,\alpha\,(\nu^2-\alpha^2-1/2)\,M^4]+\frac{4\,\nu^4+(-4\,\alpha^2-20)\,\nu^2+\alpha^4+6\,\alpha^2+8}{r^4}$$
$$+40\,\frac{\epsilon\,((-1/10\,\alpha^4-3\,\nu^2+\alpha^2+2)\,M^2+\epsilon^2\,(1/10\,\alpha^4+\nu^2-2))}{r\alpha^3}\Big]H_1=0\,, \qquad (8.106)$$

it has the same structure of singular points:

$$-\frac{\alpha}{\epsilon+iM},\quad -\frac{\alpha}{\epsilon-iM},\quad 0,\quad \infty.$$

8.10 Fourth-order equations, the second method

Equations 1) and 6) in (8.100):

$$+i(\epsilon+\frac{\alpha}{r})E_2-2i\frac{\nu}{r}H_1-M\Phi_2=0,\quad i\Big(\frac{d}{dr}+\frac{1}{r}\Big)\Phi_1+i\frac{\nu}{r}\Phi_2+MH_1=0$$

permit to express the functions H_1, Φ_2 through the functions E_2, Φ_1:

$$H_1=-\frac{iMr}{M^2r^2+2\,\nu^2}\Big(r\,\frac{d\Phi_1}{dr}+\Phi_1\Big)+\frac{(\epsilon\,r+\alpha)\nu}{M^2r^2+2\,\nu^2}\,E_2\,,$$
$$\Phi_2=-\frac{2\,\nu}{M^2r^2+2\,\nu^2}\Big(r\,\frac{d\Phi_1}{dr}+\Phi_1\Big)+\frac{i(\epsilon\,r+\alpha)Mr}{M^2r^2+2\,\nu^2}\,E_2. \qquad (8.107)$$

Similarly, eqs. 2) and 3) in eq. (8.100):

$$-i\left(\epsilon+\frac{\alpha}{r}\right)\,\Phi_1+\frac{\nu}{r}\Phi_0-ME_1=0,\quad \Big(\frac{d}{dr}+\frac{2}{r}\Big)E_2+2\frac{\nu}{r}E_1+M\Phi_0=0,$$

permit to express the functions Φ_0, E_1 through E_2, Φ_1:

$$E_1=-\frac{\nu}{M^2r^2+2\,\nu^2}\Big(r\,\frac{dE_2}{dr}+2\,E_2\Big)-\frac{i\,(\epsilon\,r+\alpha)\,Mr}{M^2r^2+2\,\nu^2}\,\Phi_1\,,$$
$$\Phi_0=-\frac{Mr}{M^2r^2+2\,\nu^2}\Big(r\,\frac{dE_2}{dr}+2\,E_2\Big)+\frac{2\,i\,(\epsilon\,r+\alpha)\,\nu}{M^2r^2+2\,\nu^2}\,\Phi_1\,. \qquad (8.108)$$

Excluding from eqs. 4) and 5)

$$+i\left(\epsilon+\frac{\alpha}{r}\right)\,E_1+i\Big(\frac{d}{dr}+\frac{1}{r}\Big)H_1-M\Phi_1=0,\quad i\left(\epsilon+\frac{\alpha}{r}\right)\Phi_2+\frac{d}{dr}\Phi_0+ME_2=0,$$

the variables H_1, Φ_2 and Φ_0, E_1, we obtain two equations which contain only the variables E_2 and Φ_1:

$$Mr^2\,\frac{d^2\Phi_1}{dr^2}+\Big(2\,Mr+\frac{4\,M\nu^2r}{M^2r^2+2\,\nu^2}\Big)\frac{d\Phi_1}{dr}$$
$$+\Big(-M^3r^2+M\epsilon^2r^2+2\,M\alpha\,\epsilon\,r+M\alpha^2-2\,M\nu^2+\frac{4\,M\nu^2}{M^2r^2+2\,\nu^2}\Big)\Phi_1$$

$$+\Big(-2\,i\nu\,\epsilon-\frac{2\,i\nu\,(M^2\alpha\,r-2\,\epsilon\,\nu^2)}{M^2r^2+2\,\nu^2}-\frac{i\nu\,\alpha}{r}\Big)E_2=0\,, \tag{8.109}$$

$$M r^2\,\frac{d^2E_2}{dr^2}+\Big(2\,Mr+\frac{4\,M\nu^2 r}{M^2r^2+2\,\nu^2}\Big)\frac{dE_2}{dr}$$
$$+\Big(-M^3r^2+M\epsilon^2r^2+2\,M\alpha\,\epsilon\,r+M\alpha^2-2\,M\nu^2-2\,M+\frac{8\,M\nu^2}{M^2r^2+2\,\nu^2}\Big)E_2$$

$$+\Big(4\,i\nu\,\epsilon+\frac{4\,i\nu\,(M^2\alpha\,r-2\,\epsilon\,\nu^2)}{M^2r^2+2\,\nu^2}+\frac{2\,i\nu\,\alpha}{r}\Big)\Phi_1=0\,. \tag{8.110}$$

We readily find a 4th-order equation for Φ_1:

$$\frac{d^4\Phi_1}{dr^4}+\Big[-\frac{12\,M^2r\,(\epsilon\,r+\alpha)}{2\,M^2\epsilon\,r^3+3\,M^2\alpha\,r^2+2\,\alpha\,\nu^2}+\frac{14}{r}\Big]\frac{d^3\Phi_1}{dr^3}$$
$$+\Big[-2\,M^2+2\,\epsilon^2+\frac{18\,M^2\alpha\,(M^2\alpha\,r^2-4\,\epsilon\,\nu^2 r-2\,\alpha\,\nu^2)}{(2\,M^2\epsilon\,r^3+3\,M^2\alpha\,r^2+2\,\alpha\,\nu^2)^2}$$
$$-\frac{6\,M^2\,(14\,\epsilon\,r+17\,\alpha)}{2\,M^2\epsilon\,r^3+3\,M^2\alpha\,r^2+2\,\alpha\,\nu^2}+\frac{2\,\alpha^2-4\,\nu^2+54}{r^2}+\frac{4\,\alpha\,\epsilon}{r}\Big]\frac{d^2\Phi_1}{dr^2}$$
$$+\Big[\frac{72\,M^2\alpha\,(3\,M^2\epsilon\,r^2+5\,M^2\alpha\,r-4\,\epsilon\,\nu^2)}{(2\,M^2\epsilon\,r^3+3\,M^2\alpha\,r^2+2\,\alpha\,\nu^2)^2}$$
$$-\frac{6\,(-2\epsilon\nu^4+(3\epsilon\alpha^2+(2M^2r+3\epsilon^2r)\,\alpha+\epsilon\,(M^2r^2+r^2\epsilon^2+10))\,\nu^2-(\epsilon r+3/2\alpha)\,r\,(\alpha^2+23)\,M^2)\,M^2}{\nu^2\,(r^2\,(\epsilon r+3/2\alpha)\,M^2+\alpha\nu^2)}$$
$$+\frac{10\,\alpha^2-20\,\nu^2+60}{r^3}+\frac{24\,\alpha\,\epsilon}{r^2}+\frac{(-6\,\alpha^2-2\,\nu^2-138)\,M^2+14\,\epsilon^2\nu^2}{\nu^2 r}\Bigg]\frac{d\Phi_1}{dr}$$
$$+\Big[M^4-2\,M^2\epsilon^2+\epsilon^4-\frac{18\,M^2}{(r^2\,(\epsilon\,r+3/2\,\alpha)\,M^2+\alpha\,\nu^2)^2}\,(1/4\,M^4\alpha^2r^2$$
$$+(-\alpha^4-4\,r\alpha^3\epsilon+(-9/4\,r^2\epsilon^2+3/2\,\nu^2-3)\,\alpha^2$$
$$+\,3\,r\epsilon\,(\nu^2-5/3)\,\alpha+(2\,\nu^2-2)\,\epsilon^2r^2)\,M^2+\epsilon^2\nu^2\alpha\,(\epsilon\,r+5/2\,\alpha))$$

$$-\frac{2\alpha^2\nu^2}{r^6M^2}-\frac{18M^2}{\nu^2\,(r^2\,(\epsilon r+3/2\alpha)\,M^2+\alpha\nu^2)\,\alpha}((-3/2\,\alpha^4-7r\alpha^3\epsilon+(-4r^2\epsilon^2+\frac{11\nu^2}{6}-9/2)\alpha^2$$
$$+5\,r\,(\nu^2-7/5)\,\epsilon\,\alpha+8/3\,\epsilon^2r^2\,(\nu-1)\,(\nu+1))\,M^2+\epsilon^2\nu^2\alpha\,(\epsilon\,r+5/2\,\alpha))$$
$$+\frac{4\,\nu^4+(-4\,\alpha^2-16)\,\nu^2+\alpha^4+4\,\alpha^2+12}{r^4}+\frac{4\,\alpha\,\epsilon\,(\alpha^2-2\,\nu^2+6)}{r^3}$$
$$+\frac{(4\,\nu^4+(-2\,\alpha^2+14)\,\nu^2-18\,\alpha^2-54)\,M^2-4\,(\nu^2-3/2\,\alpha^2-11/2)\,\nu^2\epsilon^2}{\nu^2r^2}$$

$$+\frac{4\,(((-\alpha^2+12)\,\nu^2-18\,\alpha^2-12)\,M^2+\alpha^2\epsilon^2\nu^2)\,\epsilon}{r\nu^2\alpha}\Big]\Phi_1=0\,, \tag{8.111}$$

and a 4th-order equation for function E_2:

$$\frac{d^4E_2}{dr^4}+\Big[\frac{14}{r}-\frac{12\,M^2r\,(r\epsilon+\alpha)}{2\,M^2r^3\epsilon+3\,M^2r^2\alpha+2\,\nu^2\alpha}\Big]\frac{d^3E_2}{dr^3}$$

$$+\Big[-2M^2+2\epsilon^2+4\frac{\epsilon\alpha}{r}+\frac{18M^2\alpha\left(M^2r^2\alpha-4\nu^2r\epsilon-2\nu^2\alpha\right)}{\left(2M^2r^3\epsilon+3M^2r^2\alpha+2\nu^2\alpha\right)^2}$$

$$+\frac{-4\nu^2+2\alpha^2+54}{r^2}+\frac{(-84r\epsilon-102\alpha)M^2}{(2\epsilon r^3+3\alpha r^2)M^2+2\nu^2\alpha}\Big]\frac{d^2E_2}{dr^2}$$

$$+\Big[\frac{(-2\nu^2-6\alpha^2-144)M^2+14\nu^2\epsilon^2}{\nu^2r}+\frac{72M^2\alpha\left(3M^2r^2\epsilon+5M^2\alpha r-4\epsilon\nu^2\right)}{\left(2M^2r^3\epsilon+3M^2r^2\alpha+2\nu^2\alpha\right)^2}$$

$$+\frac{-20\nu^2+10\alpha^2+60}{r^3}+\frac{24\epsilon\alpha}{r^2}-\frac{12M^2}{2\nu^2M^2r^3\epsilon+3M^2r^2\nu^2\alpha+2\nu^4\alpha}$$

$$\times(\epsilon((\nu^2-\alpha^2-24)M^2+\nu^2\epsilon^2)r^2+2((\nu^2-3/4\alpha^2-18)M^2+3/2\nu^2\epsilon^2)\alpha r$$

$$-2\epsilon(\nu^2-3/2\alpha^2-5)\nu^2)\Big]\frac{dE_2}{dr}$$

$$+\Big[M^4-2M^2\epsilon^2+\epsilon^4-\frac{2\nu^2\alpha^2}{r^6M^2}-\frac{4\epsilon\left(\left(\left(\alpha^2-12\right)\nu^2+18\alpha^2+12\right)M^2-\nu^2\alpha^2\epsilon^2\right)}{r\nu^2\alpha}$$

$$-\frac{9}{2}\frac{M^2}{\left(r^2\left(r\epsilon+3/2\alpha\right)M^2+\nu^2\alpha\right)^2}\left(M^4r^2\alpha^2+\left(-4\alpha^4-16r\alpha^3\epsilon+\left(-9r^2\epsilon^2+6\nu^2-16\right)\alpha^2\right.\right.$$

$$\left.\left.+12r\epsilon\left(\nu^2-2\right)\alpha+\left(8\nu^2-8\right)\epsilon^2r^2\right)M^2+4\alpha\epsilon^2\nu^2\left(r\epsilon+5/2\alpha\right)\right)-\frac{8\epsilon\alpha\left(\nu^2-1/2\alpha^2-3\right)}{r^3}$$

$$+\frac{\left(4M^2-4\epsilon^2\right)\nu^4+\left(\left(-2\alpha^2+14\right)M^2+6\alpha^2\epsilon^2+22\epsilon^2\right)\nu^2-18M^2\left(\alpha^2+4\right)}{\nu^2r^2}$$

$$-\frac{48M^2}{\alpha\left(r^2\left(r\epsilon+3/2\alpha\right)M^2+\nu^2\alpha\right)\nu^2}\left(\left(-\frac{9}{16}\alpha^4-\frac{21}{8}r\alpha^3\epsilon+\left(-3/2r^2\epsilon^2+\frac{11}{16}\nu^2-9/4\right)\alpha^2\right.\right.$$

$$\left.\left.+\frac{15}{8}\left(\nu^2-8/5\right)\epsilon r\alpha+r^2\epsilon^2\left(\nu-1\right)\left(\nu+1\right)\right)M^2+3/8\alpha\epsilon^2\nu^2\left(r\epsilon+5/2\alpha\right)\right)$$

$$+\frac{4\nu^4+\left(-4\alpha^2-16\right)\nu^2+\alpha^4+4\alpha^2+12}{r^4}\Big]E_2=0=0. \tag{8.112}$$

These equations have identical singular points:

$$\frac{1}{2}\frac{\sqrt[3]{-\alpha\left(M^2\alpha^2-2\sqrt{2}\nu\sqrt{M^2\alpha^2+2\epsilon^2\nu^2}\epsilon+4\epsilon^2\nu^2\right)M}}{M\epsilon}$$

$$+\frac{1}{2}\frac{M\alpha^2}{\epsilon\sqrt[3]{-\alpha\left(M^2\alpha^2-2\sqrt{2}\nu\sqrt{M^2\alpha^2+2\epsilon^2\nu^2}\epsilon+4\epsilon^2\nu^2\right)M}}-\frac{1}{2}\frac{\alpha}{\epsilon},$$

$$-\frac{1}{4}\frac{\sqrt[3]{-\alpha\left(M^2\alpha^2-2\sqrt{2}\nu\sqrt{M^2\alpha^2+2\epsilon^2\nu^2}\epsilon+4\epsilon^2\nu^2\right)M}}{M\epsilon}$$

$$-\frac{1}{4}\frac{M\alpha^2}{\epsilon\sqrt[3]{-\alpha\left(M^2\alpha^2-2\sqrt{2}\nu\sqrt{M^2\alpha^2+2\epsilon^2\nu^2}\epsilon+4\epsilon^2\nu^2\right)M}}-\frac{1}{2}\frac{\alpha}{\epsilon}$$

$$-\frac{i}{2}\sqrt{3}\left(\frac{1}{2}\frac{\sqrt[3]{-\alpha\left(M^2\alpha^2-2\sqrt{2}\nu\sqrt{M^2\alpha^2+2\epsilon^2\nu^2}\epsilon+4\epsilon^2\nu^2\right)M}}{M\epsilon}\right.$$

$$
\left. -\frac{1}{2}\frac{M\alpha^2}{\epsilon\sqrt[3]{-\alpha\left(M^2\alpha^2-2\sqrt{2}\nu\sqrt{M^2\alpha^2+2\epsilon^2\nu^2}\epsilon+4\epsilon^2\nu^2\right)M}}\right),
$$

$$
-\frac{1}{4}\frac{\sqrt[3]{-\alpha\left(M^2\alpha^2-2\sqrt{2}\nu\sqrt{M^2\alpha^2+2\epsilon^2\nu^2}\epsilon+4\epsilon^2\nu^2\right)M}}{M\epsilon}
$$

$$
-\frac{1}{4}\frac{M\alpha^2}{\epsilon\sqrt[3]{-\alpha\left(M^2\alpha^2-2\sqrt{2}\nu\sqrt{M^2\alpha^2+2\epsilon^2\nu^2}\epsilon+4\epsilon^2\nu^2\right)M}}-\frac{1}{2}\frac{\alpha}{\epsilon}
$$

$$
+\frac{i}{2}\sqrt{3}\left(\frac{1}{2}\frac{\sqrt[3]{-\alpha\left(M^2\alpha^2-2\sqrt{2}\nu\sqrt{M^2\alpha^2+2\epsilon^2\nu^2}\epsilon+4\epsilon^2\nu^2\right)M}}{M\epsilon}\right.
$$

$$
\left.-\frac{1}{2}\frac{M\alpha^2}{\epsilon\sqrt[3]{-\alpha\left(M^2\alpha^2-2\sqrt{2}\nu\sqrt{M^2\alpha^2+2\epsilon^2\nu^2}\epsilon+4\epsilon^2\nu^2\right)M}}\right), \tag{8.113}
$$

and the points $r = 0,\ \infty$.

8.11 Further study of six equations

Let us turn back to the system of six equations for states with parity $P = (-1)^j$:

$$
\begin{aligned}
+i\Big(\epsilon+\frac{\alpha}{r}\Big)E_2-2i\frac{\nu}{r}H_1-M\Phi_2&=0,\\
-i\Big(\epsilon+\frac{\alpha}{r}\Big)\Phi_1+\frac{\nu}{r}\Phi_0-ME_1&=0,\\
\Big(\frac{d}{dr}+\frac{2}{r}\Big)E_2+2\frac{\nu}{r}E_1+M\Phi_0&=0,\\
+i\Big(\epsilon+\frac{\alpha}{r}\Big)E_1+i\Big(\frac{d}{dr}+\frac{1}{r}\Big)H_1-M\Phi_1&=0,\\
i\Big(\epsilon+\frac{\alpha}{r}\Big)\Phi_2+\frac{d}{dr}\Phi_0+ME_2&=0,\\
i\Big(\frac{d}{dr}+\frac{1}{r}\Big)\Phi_1+i\frac{\nu}{r}\Phi_2+MH_1&=0.
\end{aligned} \tag{8.114}
$$

The physical dimensions of the involved quantities are

$$
M=\frac{mc}{\hbar}=\frac{1}{\lambda},\quad [M]=\frac{1}{L},\quad \epsilon=\frac{E}{\hbar c},\quad [\epsilon]=\frac{1}{L},\quad \alpha=\frac{e^2}{\hbar c}=\frac{1}{137}.
$$

It is convenient to apply the dimensionless form of equations by taking the Compton wave length λ as a unit for the length and the rest energy of a particle as a unit for energy. In this way we obtain

$$
rM \Longrightarrow x,\quad \epsilon/M = E/mc^2 \Longrightarrow \epsilon,
$$

$$
\begin{aligned}
i\Big(\epsilon+\frac{\alpha}{x}\Big)E_2-2i\frac{\nu}{x}H_1-\Phi_2=0,\\
-i\Big(\epsilon+\frac{\alpha}{x}\Big)\Phi_1+\frac{\nu}{x}\Phi_0-E_1=0,\\
\Big(\frac{d}{dx}+\frac{2}{x}\Big)E_2+2\frac{\nu}{x}E_1+\Phi_0=0,\\
+i\Big(\epsilon+\frac{\alpha}{x}\Big)E_1+i\Big(\frac{d}{dx}+\frac{1}{r}\Big)H_1-\Phi_1=0,\\
i\Big(\epsilon+\frac{\alpha}{x}\Big)\Phi_2+\frac{d}{dx}\Phi_0+E_2=0,\\
i\Big(\frac{d}{dx}+\frac{1}{x}\Big)\Phi_1+i\frac{\nu}{x}\Phi_2+H_1=0\,.
\end{aligned}
\tag{8.115}
$$

With the substitutions

$$
\Phi_1=\frac{1}{x}\varphi_1,\qquad E_2=\frac{1}{x^2}e_2,\quad H_1=\frac{1}{x}h_1,
$$

the system reduces to a more simple form

$$
\begin{aligned}
\Phi_2=i\Big(\epsilon+\frac{\alpha}{x}\Big)\frac{1}{x^2}e_2-2i\frac{\nu}{x^2}h_1,\quad E_1=-i\Big(\epsilon+\frac{\alpha}{x}\Big)\frac{1}{x}\varphi_1+\frac{\nu}{x}\Phi_0\,;\\
\frac{d}{dx}e_2=-2\nu xE_1-x^2\Phi_0,\quad \frac{d}{dx}h_1=-\Big(x\epsilon+\alpha\Big)\,E_1-i\varphi_1,\\
\frac{d}{dx}\Phi_0=-i\Big(\epsilon+\frac{\alpha}{x}\Big)\Phi_2-\frac{1}{x^2}e_2,\quad \frac{d}{dx}\varphi_1=-\nu\Phi_2+ih_1\,.
\end{aligned}
\tag{8.116}
$$

With the help of two first (non-differential) equations, we exclude the functions Φ_2, E_1:

$$
\begin{aligned}
\frac{d}{dx}e_2=2i\nu\Big(\epsilon+\frac{\alpha}{x}\Big)\varphi_1-(2\nu^2+x^2)\Phi_0,\\
\frac{d}{dx}h_1=+i\Big[\Big(\epsilon+\frac{\alpha}{x}\Big)^2-1\Big]\varphi_1-\nu\Big(\epsilon+\frac{\alpha}{x}\Big)\Phi_0\ ;
\end{aligned}
\tag{8.117}
$$

$$
\begin{aligned}
\frac{d}{dx}\varphi_1=-\frac{i\nu}{x^2}\Big(\epsilon+\frac{\alpha}{x}\Big)e_2+i\Big(\frac{2\nu^2}{x^2}+1\Big)h_1,\\
\frac{d}{dx}\Phi_0=\frac{1}{x^2}\Big[\Big(\epsilon+\frac{\alpha}{x}\Big)^2-1\Big]e_2-\frac{2\nu}{x^2}\Big(\epsilon+\frac{\alpha}{x}\Big)h_1.
\end{aligned}
\tag{8.118}
$$

We use the following notations:

$$
\begin{aligned}
a=2i\nu\frac{\epsilon x+\alpha}{x},\quad c=-(2\nu^2+x^2),\quad d=i\frac{(\epsilon x+\alpha)^2-x^2}{x^2},\quad b=-\frac{\nu(\epsilon x+\alpha)}{x},\\
A=-i\frac{\nu(\epsilon x+\alpha)}{x^3},\quad C=+i\frac{(2\nu^2+x^2)}{x^2},\quad D=\frac{(\epsilon x+\alpha)^2-x^2}{x^4},\quad B=-\frac{2\nu(\epsilon x+\alpha)}{x^3},\\
ab-cd=i\,p(x),\quad AB-CD=-i\frac{p(x)}{x^4},\quad p(x)=(\epsilon^2-1)x^2+2\alpha\epsilon x-(2\nu^2-\alpha^2)\,;
\end{aligned}
$$

also we re-designate the functions:

$$
e_2=f_1,\quad h_1=f_2,\quad \varphi_1=f_3,\quad \Phi_0=f_4\,. \tag{8.119}
$$

Then the system under consideration reads

$$
\begin{aligned}
\frac{d}{dx}f_1=af_3+cf_4,\quad \frac{d}{dx}f_2=df_3+bf_4\,;\\
\frac{d}{dx}f_3=Af_1+Cf_2,\quad \frac{d}{dx}f_4=Df_1+Bf_2\,.
\end{aligned}
\tag{8.120}
$$

Note that two excluded functions (see eq. (8.74)) are determined in these notations as follows

$$\Phi_2 = i\left(\epsilon + \frac{\alpha}{x}\right)\frac{1}{x^2}f_1 - \frac{2i\nu}{x^2}f_2, \quad E_1 = -i\left(\epsilon + \frac{\alpha}{x}\right)\frac{1}{x}f_3 + \frac{\nu}{x}f_4\,. \tag{8.121}$$

Above, with the use of the Lorentz condition, it was derived the simple constraint $E_2 = 0$ $(f_1 = 0)$. Let us apply the same constraint again, and also extend such an approach by imposing similar constraints on other functions.

I. Let $f_1(x) = 0$, then from the system (8.120) by taking into account the restriction $f_1 = 0$, we obtain equations

$$af_3 + cf_4 = 0, \quad df_3 + bf_4 = \frac{d}{dx}f_2\,; \quad \frac{d}{dx}f_3 = Cf_2, \quad \frac{d}{dx}f_4 = Bf_2\,; \tag{8.122}$$

thereby we examine projection of the solution – curve $\{f_1(x), ..., f_4(x)\}$ – on the plane $\{0, f_2, f_3, f_3\}$ in 4-space

Considering two first equations in eq. (8.122) as a linear system with respect to f_3 and f_4, we get

$$f_3 = \frac{-c}{ab-cd}\frac{d}{dx}f_2, \quad f_4 = \frac{a}{ab-cd}\frac{d}{dx}f_2\,. \tag{8.123}$$

Substituting these formulas into two remaining equations in eq. (8.122), we obtain two different 2nd-order equations for the variable f_2:

$$\frac{d}{dr}\frac{-c}{ab-cd}\frac{d}{dr}f_2^I = Cf_2^I \implies \Big[\frac{d}{dx}\frac{2\nu^2+x^2}{p(x)}\frac{d}{dx} + \frac{2\nu^2+x^2}{x^2}\Big]f_2^I = 0\,; \tag{8.124}$$

$$\frac{d}{dr}\frac{a}{ab-cd}\frac{d}{dr}f_2^{II} = Bf_2^{II} \implies \Big[\frac{d}{dx}\frac{\epsilon x+\alpha}{x\,p(x)}\frac{d}{dx} + \frac{(\epsilon x+\alpha)}{x^3}\Big]f_2^{II} = 0\,. \tag{8.125}$$

Thus, the system (8.122), describing the projection of the whole solution $\{f_1, ..., f_4\}$ on the plane $f_1 = 0$, may be solved on the basis of two main functions f_2^I and f_2^{II} – these obey different 2nd-order equations, which lead to the different non-zero remaining functions f_3 and f_4.

In other words, the projection of the whole solution – the curve $\{f_i(x)\}$ on the plane $f_1 = 0$ consists of the parts (branches), related respectively to the functions f_2^I and f_2^{II}. In fact, the concept of projection is determined by definition, which allows us to get additional information about the needed entire solutions $\{f_i(x)\}$.

II. Let $f_2(x) = 0$, then we get the equations

$$af_3 + cf_4 = \frac{d}{dx}f_1, \quad 0 = df_3 + bf_4, \quad \frac{d}{dx}f_3 = Af_1, \quad \frac{d}{dx}f_4 = Df_1, \tag{8.126}$$

which lead to

$$f_3 = \frac{b}{ab-cd}\frac{d}{dx}f_1, \quad f_4 = \frac{-d}{ab-cd}\frac{d}{dx}f_1, \quad \frac{d}{dx}f_3 = Af_1, \quad \frac{d}{dx}f_4 = Df_1. \tag{8.127}$$

Therefore, we derive two equations for $f_1(x)$:

$$\frac{d}{dx}\frac{b}{ab-cd}\frac{d}{dx}f_1^I = Af_1^I \implies \left[\frac{d}{dx}\frac{(\epsilon x+\alpha)}{xp(x)}\frac{d}{dx} + \frac{(\epsilon x+\alpha)}{x^3}\right]f_1^I = 0; \tag{8.128}$$

$$\frac{d}{dx}\frac{-d}{ab-cd}\frac{d}{dx}f_1^{II} = Df_1^{II} \Longrightarrow \left[\frac{d}{dx}\frac{(\epsilon x+\alpha)^2-x^2}{x^2p(x)}\frac{d}{dx}+\frac{(\epsilon x+\alpha)^2-x^2}{x^4}\right]f_1^{II}=0. \qquad (8.129)$$

III. Let $f_3(x)=0$, then we get equations

$$\frac{d}{dx}f_1 = cf_4, \quad \frac{d}{dx}f_2 = bf_4, \quad Af_1+Cf_2=0, \quad Df_1+Bf_2=\frac{d}{dx}f_4;$$

which result in

$$f_1 = \frac{-C}{AB-CD}\frac{d}{dx}f_4,\ f_2 = \frac{A}{AB-CD}\frac{d}{dx}f_4,\ \frac{d}{dx}f_1 = cf_4,\ \frac{d}{dx}f_2 = bf_4, \qquad (8.130)$$

so we obtain two equations for f_4:

$$\frac{d}{dx}\frac{-C}{AB-CD}\frac{d}{dx}f_4^I = cf_4^I \Longrightarrow$$
$$\left[\frac{d}{dx}\frac{(2\nu^2+x^2)x^2}{p(x)}\frac{d}{dx}+(2\nu^2+x^2)\right]f_4^I=0; \qquad (8.131)$$

$$\frac{d}{dx}\frac{A}{AB-CD}\frac{d}{dx}f_4^{II} = bf_4^{II} \Longrightarrow$$
$$\left[\frac{d}{dx}\frac{(\epsilon x+\alpha)x}{p(x)}\frac{d}{dx}+\frac{(\epsilon x+\alpha}{x}\right]f_4^{II}=0. \qquad (8.132)$$

IV. Let $f_4(x)=0$, then we have the equations

$$Af_1+Cf_2=\frac{d}{dr}f_3,\ Df_1+Bf_2=0,\ \frac{d}{dx}f_1=af_3,\ \frac{d}{dr}f_2=df_3;$$

these yield

$$f_1 = \frac{B}{AB-CD}\frac{d}{dx}f_3,\ f_2 = \frac{-D}{AB-CD}\frac{d}{dx}f_3,\ \frac{d}{dx}f_1 = af_3,\ \frac{d}{dx}f_2 = df_3, \qquad (8.133)$$

so we derive two equations for $f_3(x)$:

$$\frac{d}{dx}\frac{B}{AB-CD}\frac{d}{dx}f_3^I = af_3^I \Longrightarrow$$
$$\left[\frac{d}{dx}\frac{2\nu(\epsilon x+\alpha)x}{p(x)}\frac{d}{dx}+\frac{2\nu(\epsilon x+\alpha}{x}\right]f_3^I=0; \qquad (8.134)$$

$$\frac{d}{dx}\frac{-D}{AB-CD}\frac{d}{dx}f_3^{II} = d\ f_3^{II} \Longrightarrow$$
$$\left[\frac{d}{dx}\frac{(\epsilon x+\alpha)^2-x^2}{p(x)}+\frac{(\epsilon x+\alpha)^2-x^2}{x^2}\right]f_3^{II}=0. \qquad (8.135)$$

Let us write down the explicit form of all the derived 2nd-order differential equations and fix their singular points. Recall that

$$p(x) = (\epsilon^2-1)x^2+2\epsilon\alpha x-(2\nu^2-\alpha^2) \equiv (\epsilon^2-1)(x-x_1)(x-x_2),$$

$$x_{1,2} = \frac{\epsilon \pm \sqrt{2\nu^2\epsilon^2-(2\nu^2-\alpha^2)}}{1-\epsilon^2};$$

these roots are complex-valued in the case of bound states: $0<\epsilon<1$.

For the case of projection $f_1 = 0$, we have the equation

$$\Big[\frac{d^2}{dx^2} + \Big(\frac{2x}{x^2+2\nu^2} - \frac{p'}{p}\Big)\frac{d}{dx} + \frac{p}{x^2}\Big] f_2^I = 0, \qquad (8.136)$$

with the singular points $x_1,\ x_2,\ x_{3,4} = \pm i\sqrt{2\nu^2},\ 0,\ \infty_{[2]}$; and the equation

$$\Big[\frac{d^2}{dx^2} + \Big(\frac{\epsilon}{\epsilon x+\alpha} - \frac{1}{x} - \frac{p'}{p}\Big)\frac{d}{dx} + \frac{p}{x^2}\Big] f_2^{II} = 0, \qquad (8.137)$$

with the singular points $x_1,\ x_2,\ x_7 = -\frac{\alpha}{\epsilon},\ 0,\ \infty_{[2]}$.

For the case of projection $f_2 = 0$, we have the equation

$$\Big[\frac{d^2}{dx^2} + \Big(\frac{\epsilon}{\epsilon x+\alpha} - \frac{1}{x} - \frac{p'}{p}\Big)\frac{d}{dx} + \frac{p}{x^2}\Big] f_1^I = 0, \qquad (8.138)$$

with singular points $x_1,\ x_2, x_7 = -\frac{\alpha}{\epsilon},\ 0,\ \infty_{[2]}$; and the equation

$$\Big[\frac{d^2}{dx^2} + \Big(\frac{2(\epsilon x+\alpha)\epsilon - 2x}{(\epsilon x+\alpha)^2 - x^2} - \frac{2}{x} - \frac{p'}{p}\Big)\frac{d}{dx} + \frac{p}{x^2}\Big] f_1^{II} = 0, \qquad (8.139)$$

$$(\epsilon x+\alpha)^2 - x^2 = 0 \quad \Longrightarrow \quad x_{5,6} = -\frac{\alpha}{\epsilon+1}, \frac{\alpha}{1-\epsilon},$$

and the singular points $x_1, x_2,\ x_5, x_6,\ 0,\ \infty_{[2]}$.

For projection $f_3 = 0$, we have equation

$$\Big[\frac{d^2}{dx^2} + \Big(\frac{2x}{2\nu^2+x^2} + \frac{2}{x} - \frac{p'}{p}\Big)\frac{d}{dx} - \frac{p}{x^2}\Big] f_4^I = 0, \qquad (8.140)$$

with singular points $x_1, x_2,\ x_{3,4} = \pm i\sqrt{2\nu^2},\ 0,\ \infty_{[2]}$; and equation

$$\Big[\frac{d^2}{dx^2} + \Big(\frac{\epsilon}{\epsilon x+\alpha} + \frac{1}{x} - \frac{p'}{p}\Big)\frac{d}{dx} + \frac{p}{x^2}\Big] f_4^{II} = 0, \qquad (8.141)$$

with singular points $x_1,\ x_2,\ x_7 = -\frac{\alpha}{\epsilon},\ 0,\ \infty_{[2]}$.

For projection $f_4 = 0$, we have equation

$$\Big[\frac{d^2}{dx^2} + \Big(\frac{\epsilon}{\epsilon x+\alpha} + \frac{1}{x} - \frac{p'}{p}\Big)\frac{d}{dx} + \frac{p}{x^2}\Big] f_3^I = 0, \qquad (8.142)$$

with singular points $x_1,\ x_2,\ x_5 = -\frac{\alpha}{\epsilon},\ 0,\ \infty_{[2]}$; and equation

$$\Big[\{\frac{d^2}{dx^2} + \Big(\frac{2(\epsilon x+\alpha)\epsilon - 2x}{(\epsilon x+\alpha)^2 - x^2} - \frac{p'}{p}\Big)\frac{d}{dx} + \frac{p}{x^2}\Big] f_3^{II} = 0, \qquad (8.143)$$

with singular points $x_1, x_2,\ x_5, x_6,\ 0,\ \infty_{[2]}$.

In short, results may be presented as follows:

$$\begin{array}{lll} I, & E_2 = 0 \Leftrightarrow f_1 = 0, & f_2^I, f_2^{II} \Leftrightarrow H_1^I, H_1^{II}; \\ II, & H_1 = 0 \Leftrightarrow f_2 = 0, & f_1^I, f_1^{II} \Leftrightarrow E_1^I, E_1^{II}; \\ III, & \varphi_3 = 0 \Leftrightarrow f_3 = 0, & f_4^I, f_4^{II} \Leftrightarrow \Phi_0^I, \Phi_0^{II}; \\ IV, & \Phi_0 = 0 \Leftrightarrow f_4 = 0, & f_3^I, f_3^{II} \Leftrightarrow \varphi_1^I, \varphi_1^{II}. \end{array} \qquad (8.144)$$

8.12 The 4th-order differential equations

We start again with the system

$$\frac{d}{dx}f_1 = af_3 + cf_4, \qquad \frac{d}{dx}f_2 = df_3 + bf_4,$$
$$\frac{d}{dx}f_3 = Af_1 + Cf_2, \qquad \frac{d}{dx}f_4 = Df_1 + Bf_2. \tag{8.145}$$

This is equivalent to the following

$$f_1 = \frac{Bf_3' - Cf_4'}{AB - CD}, \quad f_2 = \frac{-Df_3' + Af_4'}{AB - CD},$$
$$f_3 = \frac{bf_1' - cf_2'}{ab - cd}, \quad f_4 = \frac{-df_1' + af_2'}{ab - cd}. \tag{8.146}$$

First, we exclude the functions f_3 and f_4:

$$f_1 = \frac{B}{(AB - CD)}\frac{d}{dx}\frac{bf_1' - cf_2'}{ab - cd} - \frac{C}{(AB - CD)}\frac{d}{dx}\frac{-df_1' + af_2'}{ab - cd},$$

$$f_2 = -\frac{D}{(AB - CD)}\frac{d}{dx}\frac{bf_1' - cf_2'}{ab - cd} + \frac{A}{(AB - CD)}\frac{d}{dx}\frac{-df_1' + af_2'}{ab - cd}.$$

By taking into account expressions for $a(x), ..., D(x)$, the last equations get the following form

$$\left[K_2(x)\frac{d^2}{dx^2} + K_1(x)\frac{d}{dx} + K_0(x)\right]f_1 = \frac{df_2}{dx},$$
$$\left[L_2(x)\frac{d^2}{dx^2} + L_1(x)\frac{d}{dx} + L_0(x)\right]f_2 = \frac{df_1}{dx}, \tag{8.147}$$

where the notations are used

$$K_2(x) = \frac{1}{2}\frac{-x^5\epsilon^2 - 2\,x^4\alpha\,\epsilon + 2\,\nu^2x^3 + x^5 - x^3\alpha^2}{x\,(2\,\epsilon\,x^3 + 3\,\alpha\,x^2 + 2\,\alpha\,\nu^2)\,\nu},$$

$$K_1(x) = \frac{1}{2}\frac{\epsilon^2 - 1}{\epsilon\,\nu} + \frac{1}{2}\frac{\alpha\,(3\,x^2 - x^2\epsilon^2 - \epsilon\,x\alpha + 2\,\nu^2)}{\epsilon\,\nu\,(2\,\epsilon\,x^3 + 3\,\alpha\,x^2 + 2\,\alpha\,\nu^2)} + \frac{1}{2}\frac{\alpha}{x\nu},$$

$$K_0(x) = -\frac{1}{4}\frac{\left((\epsilon^2 - 1)\,x^2 + 2\,\epsilon\,x\alpha - 2\,\nu^2 + \alpha^2\right)^2}{\nu\,(\epsilon\,x^3 + 3/2\,\alpha\,x^2 + \alpha\,\nu^2)},$$

$$L_2(x) = \frac{(x^5\epsilon^2 + 2\,x^4\alpha\,\epsilon - x^5 + x^3\alpha^2 - 2\,\nu^2x^3)\,x}{(x^2 + x^2\epsilon^2 + 2\,\epsilon\,x\alpha + \alpha^2)\,\nu\,\alpha},$$

$$L_1(x) = \frac{(2\,\epsilon\,x\alpha\,\nu^2 + 2\,x^3\epsilon\,\alpha + 2\,x^2\alpha^2 + 2\,\nu^2\alpha^2)\,x}{(x^2 + x^2\epsilon^2 + 2\,\epsilon\,x\alpha + \alpha^2)\,\nu\,\alpha},$$

$$L_0(x) = \frac{\left((\epsilon^2 - 1)\,x^2 + 2\,\epsilon\,x\alpha - 2\,\nu^2 + \alpha^2\right)^2x^2}{\nu\,(x^2 + x^2\epsilon^2 + 2\,\epsilon\,x\alpha + \alpha^2)\,\alpha}.$$

Let us exclude the function f_2 from the eq. (8.147):

$$f_2(x) = \int\left(K_2(x)\frac{d^2}{dx^2} + K_1(x)\frac{d}{dx} + K_0(x)\right)f_1,$$

$$\Big(L_2\frac{d}{dx}+L_1\Big)\Big(K_2\frac{d^2}{dx^2}+K_1\frac{d}{dx}+K_0\Big)f_1+L_0\int dx\Big(K_2\frac{d^2}{dx^2}+K_1\frac{d}{dx}+K_0\Big)f_1=0.$$

The second relation should be divided by $L_0(x)$ and the result be differentiated; in this way, we obtain a 4th-order equation for $f_1(x)$:

$$\Big\{\frac{d}{dx}\Big(\frac{L_2}{L_0}\frac{d}{dx}+\frac{L_1}{L_0}\Big)\Big(K_2\frac{d^2}{dx^2}+K_1\frac{d}{dx}+K_0\Big)+\Big(K_2\frac{d^2}{dx^2}+K_1\frac{d}{dx}+K_0\Big)\Big\}f_1(x)=0\,.$$

Similarly, we obtain a 4th-order equation for f_2:

$$\left\{\frac{d}{dx}\left(\frac{K_2}{K_0}\frac{d}{dx}+\frac{K_1}{K_0}\right)\left(L_2\frac{d^2}{dx^2}+L_1\frac{d}{dx}+L_0\right)+\left(L_2\frac{d^2}{dx^2}+L_1\frac{d}{dx}+L_0\right)\right\}f_2(x)=0.$$

Now, we shall exclude the functions f_1 and f_2 from the equations

$$f_1=\frac{Bf_3'-Cf_4'}{AB-CD},\quad f_2=\frac{-Df_3'+Af_4'}{AB-CD},\quad f_3=\frac{bf_1'-cf_2'}{ab-cd},\quad f_4=\frac{-df_1'+af_2'}{ab-cd};$$

this results in

$$f_3=\frac{b}{ab-cd}\frac{d}{dx}\frac{Bf_3'-Cf_4'}{AB-CD}-\frac{c}{ab-cd}\frac{d}{dx}\frac{-Df_3'+Af_4'}{AB-CD},$$

$$f_4=-\frac{d}{ab-cd}\frac{d}{dx}\frac{Bf_3'-Cf_4'}{AB-CD}+\frac{a}{ab-cd}\frac{d}{dx}\frac{-Df_3'+Af_4'}{AB-CD}.$$

Taking into account the expressions for $a(x), ..., D(x)$, we reduce the last equations to the form

$$\begin{gathered}\Big[P_2(x)\frac{d^2}{dx^2}+P_1(x)\frac{d}{dx}+P_0(x)\Big]f_3=\frac{df_4}{dx},\\ \Big[Q_2(x)\frac{d^2}{dx^2}+Q_1(x)\frac{d}{dx}+Q_0(x)\Big]f_4=\frac{df_3}{dx},\end{gathered}\tag{8.148}$$

where the following notations are used

$$P_2(x)=\frac{ix^2\left(2\,\nu^2-\epsilon^2x^2-2\,\epsilon\,x\alpha-\alpha^2+x^2\right)}{\nu\,\left(2\,x^3\epsilon+2\,\nu^2\alpha+3\,x^2\alpha\right)},$$

$$P_1(x)=\frac{2\,i\nu\,\left(\epsilon\,x\alpha+\alpha^2+2\,x^2\right)}{x\,(2\,x^3\epsilon+2\,\nu^2\alpha+3\,x^2\alpha)},$$

$$P_0(x)=\frac{-i\left(\left(\epsilon^2-1\right)x^2+2\,\epsilon\,x\alpha-2\,\nu^2+\alpha^2\right)^2}{2\,\nu\,\epsilon\,x^3+3\,\nu\,\alpha\,x^2+2\,\nu^3\alpha},$$

$$Q_2(x)=\frac{1}{2}\frac{ix^4\left(2\,\nu^2-\epsilon^2x^2-2\,\epsilon\,x\alpha-\alpha^2+x^2\right)}{\nu\,\alpha\,\left(2\,\epsilon\,x\alpha+x^2+\epsilon^2x^2+\alpha^2\right)},$$

$$Q_1(x)=\frac{ix\left(2\,\nu^2x^2-\nu^2\alpha^2-\nu^2\epsilon\,x\alpha-x^4\epsilon^2-2\,x^2\alpha^2+x^4-3\,x^3\epsilon\,\alpha\right)}{\nu\,\alpha\,\left(2\,\epsilon\,x\alpha+x^2+\epsilon^2x^2+\alpha^2\right)},$$

$$Q_0(x)=\frac{-1/2\,i\left(\left(\epsilon^2-1\right)x^2+2\,\epsilon\,x\alpha-2\,\nu^2+\alpha^2\right)^2x^2}{\nu\,\alpha\,\left(x^2+\epsilon^2x^2+2\,\epsilon\,x\alpha+\alpha^2\right)}.$$

Acting in accordance with the above method, we may derive the 4th-order equations for the functions f_3, f_4. It turns out that equations for f_1 and f_3 have the same set of singular points (three regular and two irregular):

$$(2\epsilon x^3+3\alpha x^2+2\nu^2\alpha)=2\epsilon(x-x_1)(x-x_2)(x-x_3),\ x=0_{[2]},\ x=\infty_{[2]}\,;\tag{8.149}$$

$$
\begin{aligned}
& f_1'''' + \Big[- \frac{12x\,(\epsilon x + \alpha)}{2\epsilon x^3 + 2\nu^2\alpha + 3\alpha x^2} + \frac{6}{x} \Big] f_1''' \\
& + \Big[-2 + 2\epsilon^2 - \frac{18\alpha\,(2\nu^2\alpha + 4\epsilon\nu^2 x - \alpha x^2)}{(2\epsilon x^3 + 2\nu^2\alpha + 3\alpha x^2)^2} + \frac{6 + 2\alpha^2 - 4\nu^2}{x^2} + \frac{-30\alpha - 12\epsilon x}{2\epsilon x^3 + 2\nu^2\alpha + 3\alpha x^2} + \frac{4\epsilon\alpha}{x} \Big] f_1'' \\
& + \Big[\frac{72\alpha x\,(\epsilon x + \alpha)}{(2\epsilon x^3 + 2\nu^2\alpha + 3\alpha x^2)^2} + \frac{8\epsilon\alpha}{x^2} + \frac{-4\nu^2 + 2\alpha^2}{x^3} + \frac{6\nu^2 - 6\alpha^2 - 12 + 6\epsilon^2\nu^2}{x\nu^2} \\
& + \frac{24\nu^4\epsilon - 36\epsilon\alpha^2\nu^2 - 24x\nu^2\alpha + 18\alpha^3 x - 36\alpha x\epsilon^2\nu^2 + 36\alpha x - 12x^2\nu^2\epsilon^3 + 24x^2\epsilon - 12x^2\epsilon\nu^2 + 12x^2\epsilon\alpha^2}{(2\epsilon x^3 + 2\nu^2\alpha + 3\alpha x^2)\,\nu^2} \Big] f_1' \\
& + \Big[1 - 2\epsilon^2 + \epsilon^4 + \frac{-6\alpha^2 + 6\epsilon^2\nu^2 + 6\epsilon^2\alpha^2\nu^2 + 6\nu^2 - 4\epsilon^2\nu^4 + 4\nu^4 - 2\alpha^2\nu^2}{x^2\nu^2} - \frac{4\epsilon\alpha\,(2\nu^2 - \alpha^2)}{x^3} \\
& + \frac{-18\,\epsilon^2\alpha^2\nu^2 + 18\alpha^4 - 18\alpha^2\nu^2 - 84\alpha\epsilon x\nu^2 + 120\alpha^3\epsilon x - 12\alpha\epsilon^3 x\nu^2 - 48x^2\epsilon^2\nu^2 + 72x^2\epsilon^2\alpha^2}{(2\epsilon x^3 + 2\nu^2\alpha + 3\alpha x^2)\,\nu^2\alpha} \\
& + \frac{72\alpha^4 - 180\epsilon^2\alpha^2\nu^2 - 108\alpha^2\nu^2 - 72\alpha\epsilon^3 x\nu^2 - 216\alpha\epsilon x\nu^2 + 288\alpha^3\epsilon x - 144x^2\epsilon^2\nu^2 + 162x^2\epsilon^2\alpha^2 - 18x^2\alpha^2}{(2\epsilon x^3 + 2,\nu^2\alpha + 3\alpha x^2)^2} \\
& + \frac{-4\alpha^2 + \alpha^4 + 4\nu^4 - 4\alpha^2\nu^2}{x^4} - \frac{2\alpha^2\nu^2}{x^6} + \frac{4\epsilon\,(-9\alpha^2 - \alpha^2\nu^2 + 6\nu^2 + \epsilon^2\alpha^2\nu^2)}{x\nu^2\alpha} \Big] f_1 = 0\,, \qquad (8.150)
\end{aligned}
$$

$$
\begin{aligned}
& f_3'''' + \Big[- \frac{12x\,(\epsilon x + \alpha)}{2x^3\epsilon + 3\alpha x^2 + 2\nu^2\alpha} + \frac{10}{x} \Big] f_3''' \\
& + \Big[2\epsilon^2 - 2 + \frac{2\alpha^2 - 4\nu^2 + 24}{x^2} + \frac{-66\alpha - 48\epsilon x}{2x^3\epsilon + 3\alpha x^2 + 2\nu^2\alpha} - \frac{18\alpha\,(2\nu^2\alpha + 4\,x\nu^2\epsilon - \alpha x^2)}{(2x^3\epsilon + 3\alpha x^2 + 2\nu^2\alpha)^2} + \frac{4\alpha\epsilon}{x} \Big] f_3'' \\
& + \Big[\frac{16\alpha\epsilon}{x^2} + \frac{12 + 6\alpha^2 - 12\nu^2}{x^3} - \frac{72\alpha\,(2\epsilon\nu^2 - 3\alpha x - 2\epsilon x^2)}{(2x^3\epsilon + 3\alpha x^2 + 2\nu^2\alpha)^2} \\
& + \frac{24\epsilon\nu^4 - 36\epsilon\alpha^2\nu^2 - 24\epsilon\nu^2 + 18\alpha^3 x - 24\alpha x\nu^2 - 36\alpha x\epsilon^2\nu^2 + 162\alpha x - 12x^2\nu^2\epsilon^3 + 108\epsilon x^2 + 12x^2\epsilon\alpha^2 - 12x^2\epsilon\nu^2}{(2x^3\epsilon + 3\alpha x^2 + 2\nu^2\alpha)\,\nu^2} \\
& + \frac{-54 - 6\alpha^2 + 2\nu^2 + 10\epsilon^2\nu^2}{x\nu^2} \Big] f_3' \\
& + \Big[-2\epsilon^2 + \epsilon^4 + 1 + \frac{-12\alpha^2 + 12\epsilon^2\nu^2 + 6\epsilon^2\alpha^2\nu^2 + 4\nu^4 - 2\alpha^2\nu^2 - 4\nu^4\epsilon^2 + 12\nu^2}{x^2\nu^2} \\
& + \frac{-54\epsilon^2\alpha^2\nu^2 - 42\alpha^2\nu^2 + 36\alpha^4 - 132\epsilon x\alpha\,\nu^2 + 186\alpha^3\epsilon x - 24\epsilon^3 x\alpha\nu^2 - 72x^2\epsilon^2\nu^2 + 108\alpha^2\epsilon^2 x^2}{(2x^3\epsilon + 3\alpha x^2 + 2\nu^2\alpha)\,\nu^2\alpha} \\
& - \frac{4\alpha\epsilon\,(-2 - \alpha^2 + 2\nu^2)}{x^3} - \frac{2\alpha^2\nu^2}{x^6} \\
& + \frac{72\alpha^4 - 180\epsilon^2\alpha^2\nu^2 - 108\alpha^2\nu^2 - 72\epsilon^3 x\alpha\nu^2 - 216\epsilon x\alpha\nu^2 + 288\alpha^3\epsilon x + 162\alpha^2\epsilon^2 x^2 - 18\alpha^2 x^2 - 144x^2\epsilon^2\nu^2}{(2x^3\epsilon + 3\alpha x^2 + 2\nu^2\alpha)^2} \\
& + \frac{2\epsilon\,(-27\alpha^2 - 2\alpha^2\nu^2 + 18\nu^2 + 2\epsilon^2\alpha^2\nu^2)}{\alpha x\nu^2} + \frac{-2\alpha^2 - 4\alpha^2\nu^2 - 4\nu^2 + \alpha^4 + 4\nu^4}{x^4} \Big] f_3 = 0. \qquad (8.151)
\end{aligned}
$$

The equations for f_2 and f_4 have the same set of singular points

$$
(1+\epsilon^2)x^2 + 2\epsilon\alpha x + \alpha^2 = (1+\epsilon^2)(x - x_5)(x - x_6), \quad x = 0, \quad x = \infty; \qquad (8.152)
$$

$$
f_2'''' + \Big[\frac{-4\epsilon\alpha - 4\,x\epsilon^2 - 4x}{2\epsilon x\alpha + x^2 + \alpha^2 + \epsilon^2 x^2} + \frac{10}{x} \Big] f_2''' + \Big[-2 + 2\epsilon^2 + \frac{22 - 4\nu^2 + 2\alpha^2}{x^2}
$$

$$
\begin{aligned}
&-\frac{8\alpha^2}{\left(2\epsilon x\alpha+x^2+\alpha^2+\epsilon^2x^2\right)^2}+\frac{32\epsilon^2\alpha-16\alpha+24x\epsilon^3+24\epsilon x}{\left(2\epsilon x\alpha+x^2+\alpha^2+\epsilon^2x^2\right)\alpha}+4\frac{\epsilon\left(-6+\alpha^2\right)}{\alpha x}\Bigg] f_2''\\
&+\Bigg[\frac{4\epsilon\left(2\nu^2-6+3\alpha^2\right)}{\alpha x^2}+\frac{24\epsilon\alpha-8\epsilon^3\alpha-8x\epsilon^4+8x}{\left(2\epsilon x\alpha+x^2+\alpha^2+\epsilon^2x^2\right)^2}\\
&+\frac{-72\epsilon^3\alpha+8\nu^2\epsilon^3\alpha+56\epsilon\alpha+8\epsilon\alpha^3-24\nu^2\epsilon\alpha-48x\epsilon^4+8x\nu^2\epsilon^4-32x\epsilon^2+8\epsilon^2\alpha^2x-8x\nu^2+16x+8\alpha^2x}{\left(2\epsilon x\alpha+x^2+\alpha^2+\epsilon^2x^2\right)\alpha^2}\\
&+\frac{8\nu^2-8\nu^2\epsilon^2-16+48\epsilon^2-14\alpha^2+6\epsilon^2\alpha^2}{\alpha^2x}+\frac{-12\nu^2+8+6\alpha^2}{x^3}\Bigg] f_2'+\\
&+\Bigg[\epsilon^4-2\epsilon^2+1+\frac{24\nu^2+6\epsilon^2\alpha^2+6\epsilon^2\alpha^4-24\nu^2\epsilon^2-4\nu^2\epsilon^2\alpha^2-30\alpha^2-2\alpha^4+4\alpha^2\nu^2}{x^2\alpha^2}\\
&+\frac{16\alpha^3-16\alpha\nu^2+48\alpha\nu^2\epsilon^2+32\epsilon x\nu^2+32\epsilon^3x\nu^2}{\left(2\epsilon x\alpha+x^2+\alpha^2+\epsilon^2x^2\right)^2\alpha}-\frac{2\alpha^2\nu^2}{x^6}\\
&+\frac{-40\alpha\nu^2\epsilon^4-40\alpha^3\epsilon^2+192\alpha\nu^2\epsilon^2+24\alpha^3-24\alpha\nu^2-32x\nu^2\epsilon^5+64\epsilon^3x\nu^2-32x\epsilon^3\alpha^2-32\epsilon\alpha^2x+96\epsilon x\nu^2}{\left(2\epsilon x\alpha+x^2+\alpha^2+\epsilon^2x^2\right)\alpha^3}+\\
&+\frac{4\epsilon\left(-24\nu^2+\epsilon^2\alpha^4+8\nu^2\epsilon^2+8\alpha^2-\alpha^4\right)}{\alpha^3x}+\frac{-8\nu^2-4\alpha^2\nu^2+\alpha^4+4\nu^4}{x^4}\\
&-\frac{4\epsilon\left(2\alpha^2\nu^2-\alpha^4-4\nu^2-2\alpha^2\right)}{\alpha x^3}\Bigg] f_2=0\,,
\end{aligned}
\tag{8.153}
$$

$$
\begin{aligned}
&\frac{d^4f_4}{dx^4}+\left[\frac{-4\epsilon\alpha-4\,x\epsilon^2-4x}{x^2+\alpha^2+2\epsilon x\alpha+\epsilon^2x^2}+14x^{-1}\right]\frac{d^3f_4}{dx^3}\\
&+\Bigg[2\epsilon^2-2-\frac{8\alpha^2}{\left(x^2+\alpha^2+2\epsilon x\alpha+\epsilon^2x^2\right)^2}+\frac{-4\nu^2+2\alpha^2+52}{x^2}\\
&+\frac{44\alpha\epsilon^2-28\alpha+36x\epsilon^3+36\epsilon x}{\left(x^2+\alpha^2+2\epsilon x\alpha+\epsilon^2x^2\right)\alpha}+4\frac{\epsilon\left(\alpha^2-9\right)}{\alpha x}\Bigg]\frac{d^2f_4}{dx^2}\\
&+\Bigg[\frac{-8\epsilon^3\alpha+56\epsilon\alpha-8x\epsilon^4+16x\epsilon^2+24x}{\left(x^2+\alpha^2+2\epsilon x\alpha+\epsilon^2x^2\right)^2}+\frac{4\epsilon\left(-16+5\alpha^2+2\nu^2\right)}{\alpha x^2}+\frac{48+10\alpha^2-20\nu^2}{x^3}\\
&+\frac{8\nu^2\epsilon^3\alpha-136\epsilon^3\alpha-24\epsilon\alpha\nu^2+8\alpha^3\epsilon+184\epsilon\alpha-100x\epsilon^4+8x\nu^2\epsilon^4-40x\epsilon^2+8\alpha^2x\epsilon^2+60x+8\alpha^2x-8x\nu^2}{\left(x^2+\alpha^2+2\epsilon x\alpha+\epsilon^2x^2\right)\alpha^2}\\
&+\frac{100\epsilon^2-60+10\alpha^2\epsilon^2-18\alpha^2-8\epsilon^2\nu^2+8\nu^2}{\alpha^2x}\Bigg]\frac{df_4}{dx}\\
&+\Bigg[\epsilon^4-2\epsilon^2+1+\frac{-20\nu^2+\alpha^4+6\alpha^2+4\nu^4-4\alpha^2\nu^2}{x^4}\\
&+\frac{48\alpha\epsilon^2\nu^2+16\alpha^3-16\,\alpha\nu^2+32\epsilon^3x\nu^2+32\epsilon\,x\nu^2}{\left(x^2+\alpha^2+2\epsilon x\alpha+\epsilon^2x^2\right)^2\alpha}\\
&+\frac{32\nu^2-44\alpha^2+12\alpha^2\epsilon^2+4\alpha^2\nu^2-2\alpha^4-4\alpha^2\epsilon^2\nu^2-32\epsilon^2\nu^2+6\epsilon^2\alpha^4}{x^2\alpha^2}\\
&-\frac{4\epsilon\left(-\alpha^4-5\alpha^2-6\nu^2+2\alpha^2\nu^2\right)}{\alpha x^3}-\frac{2\alpha^2\nu^2}{x^6}\\
&+\frac{-48\alpha\nu^2\epsilon^4-48\alpha^3\epsilon^2+240\alpha\epsilon^2\nu^2+32\alpha^3-32\alpha\nu^2-40x\epsilon^5\nu^2-40x\epsilon^3\alpha^2+80,\epsilon^3x\nu^2-40\epsilon x\alpha^2+120\epsilon x\nu^2}{\left(x^2+\alpha^2+2\epsilon x\alpha+\epsilon^2x^2\right)\alpha^3}\\
&+\frac{4\,\epsilon\,\left(-30\,\nu^2+10\,\alpha^2-\alpha^4+10\,\epsilon^2\nu^2+\epsilon^2\alpha^4\right)}{\alpha^3x}\Bigg] f_4=0\,.
\end{aligned}
\tag{8.154}
$$

Any of four functions f_1, f_2, f_3, and f_4 may be taken as a main one, and then all remaining functions can be straightforwardly calculated.

For instance, let the function f_1 be the main one. Then we should apply eqs. (8.146) and (8.147):

$$f_1 = \frac{Bf_3' - Cf_4'}{AB - CD}, \quad f_2 = \frac{-Df_3' + Af_4'}{AB - CD}, \quad f_3 = \frac{bf_1' - cf_2'}{ab - cd}, \quad f_4 = \frac{-df_1' + af_2'}{ab - cd},$$
$$(K_2\frac{d^2}{dx^2} + K_1\frac{d}{dx} + K_0)f_1 = \frac{df_2}{dx}, \quad (L_2\frac{d^2}{dx^2} + L_1\frac{d}{dx} + L_0)f_2 = \frac{df_1}{dx}. \tag{8.155}$$

From the fifth equation, we find f_2; then from the third and fourth, we express f_3 and f_4.

If f_2 is chosen as a main function, then from the sixth equation we express f_1, and further, from the third and fourth equations we obtain the functions f_3 and f_4.

If the main function is f_3, the we use the equations

$$f_1 = \frac{Bf_3' - Cf_4'}{AB - CD}, \quad f_2 = \frac{-Df_3' + Af_4'}{AB - CD}, \quad f_3 = \frac{bf_1' - cf_2'}{ab - cd}, \quad f_4 = \frac{-df_1' + af_2'}{ab - cd},$$
$$(P_2\frac{d^2}{dx^2} + P_1\frac{d}{dx} + P_0)f_3 = \frac{df_4}{dx}, \quad (Q_2\frac{d^2}{dx^2} + Q_1\frac{d}{dx} + Q_0)f_4 = \frac{df_3}{dx}\ ; \tag{8.156}$$

then from the firth equation we get f_4, and then, from the first and second equations we obtain the expressions for f_1 and f_2. If the main function is f_4, we get f_3 from the sixth equation, and after that from the first and second equations we find f_1 and f_2.

From four independent solutions of any 4th-order equations, only two solutions may be referred as independent series of bound states.

8.13 Conclusion

We have studied the system of six equations that describe the quantum states of a spin 1 particle with parity $P = (-1)^j$ in an external Coulomb field. It is shown that, due to the Lorentz condition, one of the radial functions must be equal to zero. Any of the five remaining functions may be taken as a primary one. For such a primary function, we derive two different 2nd-order differential equations. Their Frobenius solutions are constructed, and the convergence of the involved power series is studied. As a quantisation rule, we apply so called transcendency condition to Frobenius solutions. In this way, for both equations, we have found different reasonable, from physical point of view, energy spectra.

8.14 Figures

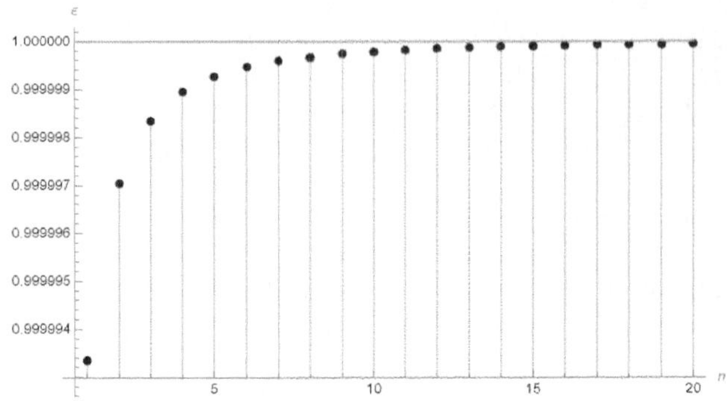

FIGURE 8.1
Energy levels; $j = 1, \;\; n = \overline{1, 20}$.

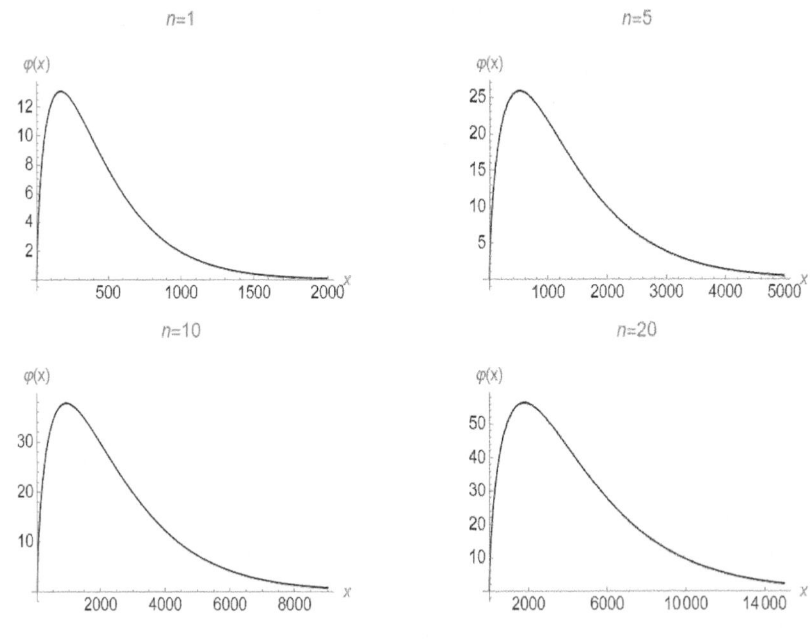

FIGURE 8.2
Graphs for factor $\varphi(x)$; $n = 1, 5, 10, 20$.

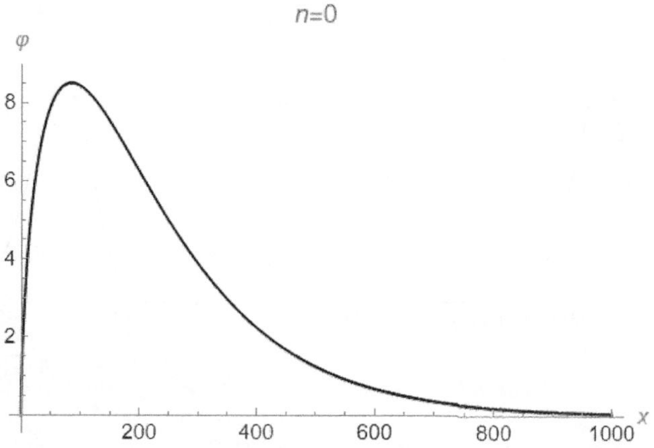

FIGURE 8.3
Graph for factor $\varphi(x); n = 0$.

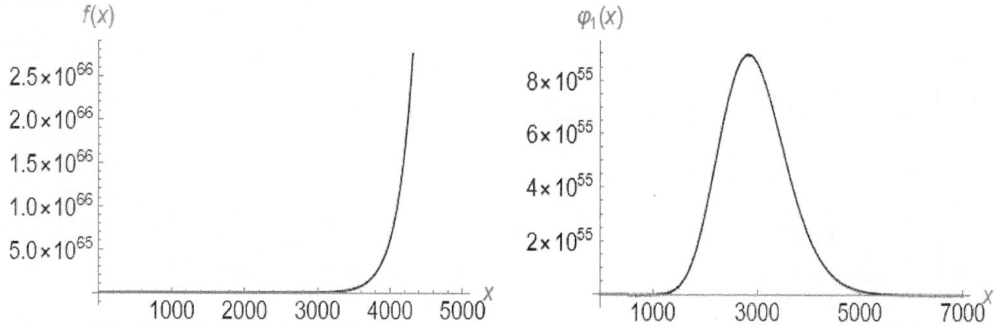

FIGURE 8.4
Frobenius solution $\varphi_1(x); n = 0,\ j = 1$.

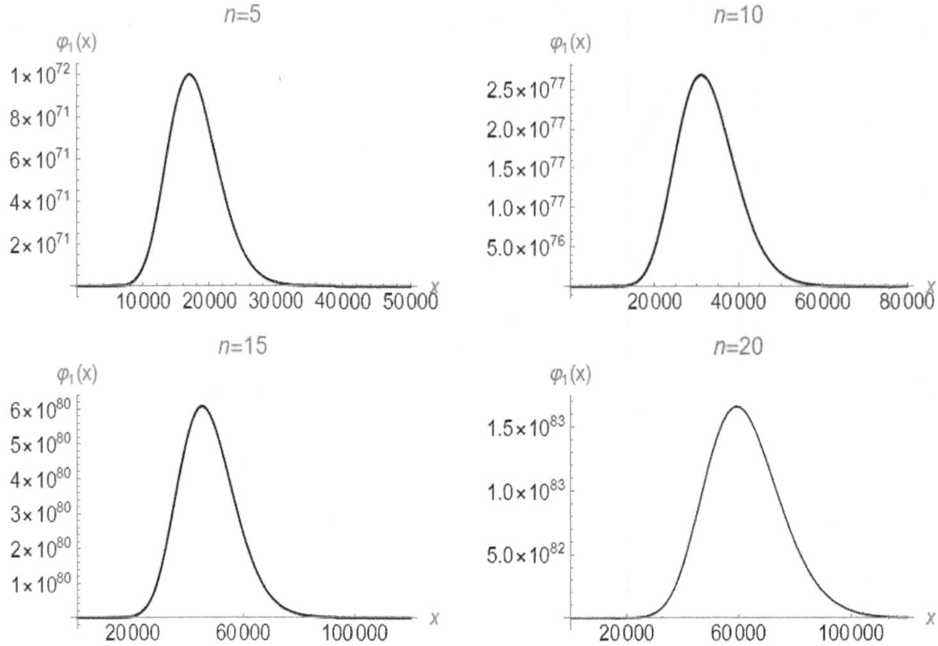

FIGURE 8.5
Frobenius solution $\varphi_1(x); n = 5, 10, 15, 20, j = 1$.

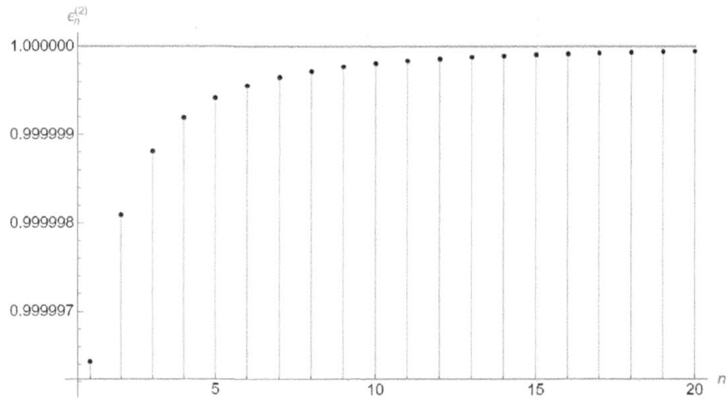

FIGURE 8.6
Energy levels $\epsilon^{(2)}_{n,j=1}$; $n = \overline{0, 20}$.

Figure 8.7 illustrates behaviour of factors φ in Frobenius solution at $j = 1$, $n = 1, 5, 10, 20$. The Fig. 8.8 illustrates behaviour of the complete Frobenius solutions at $j = 1$ and $n = 1, 5, 10, 20$.

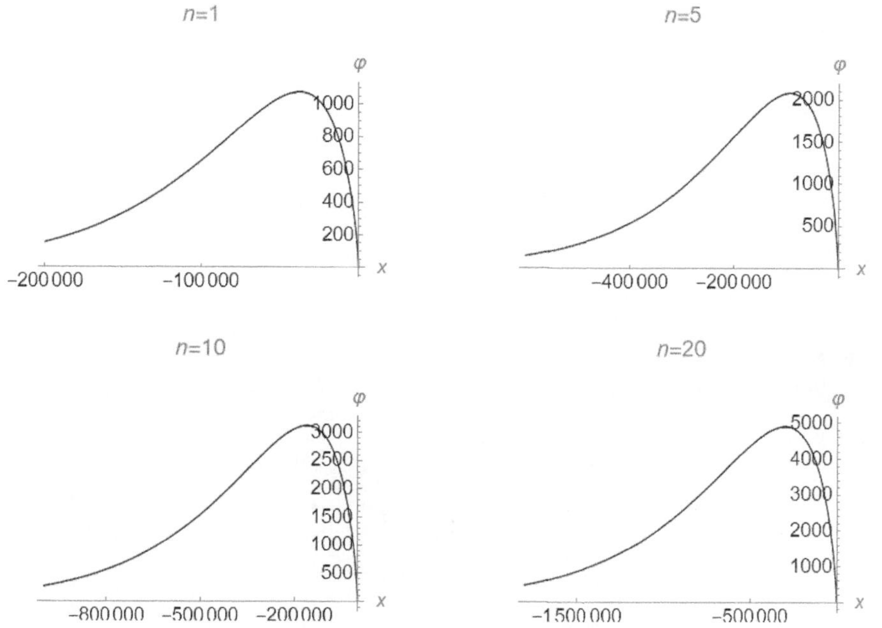

FIGURE 8.7
Graphs for factors $\varphi(x)$, $n = 1, 5, 10, 20$ in solution $\Phi_0(x)$.

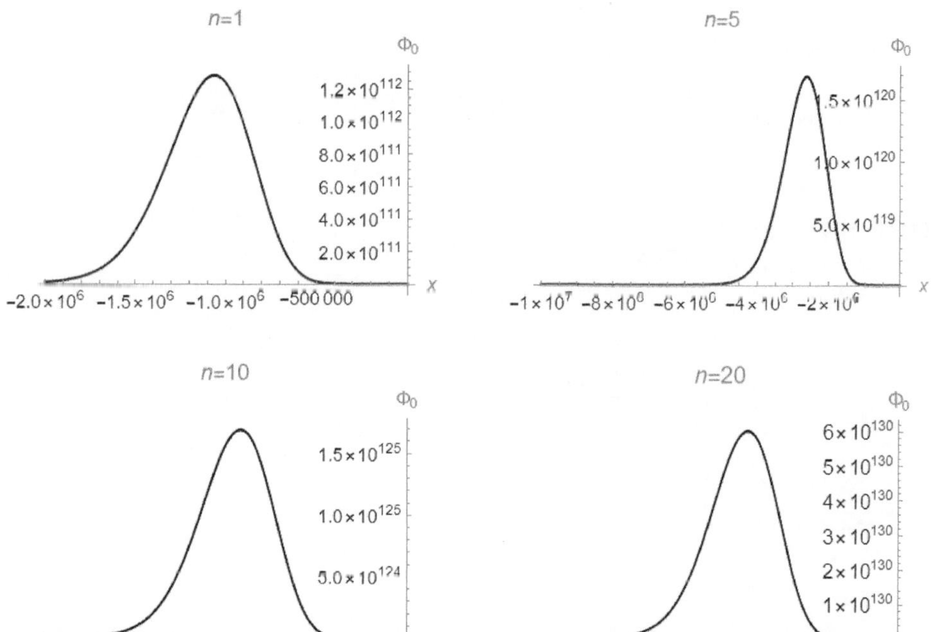

FIGURE 8.8
Graphs for complete solutions $\Phi_0(x)$, $n = 1, 5, 10, 20$.

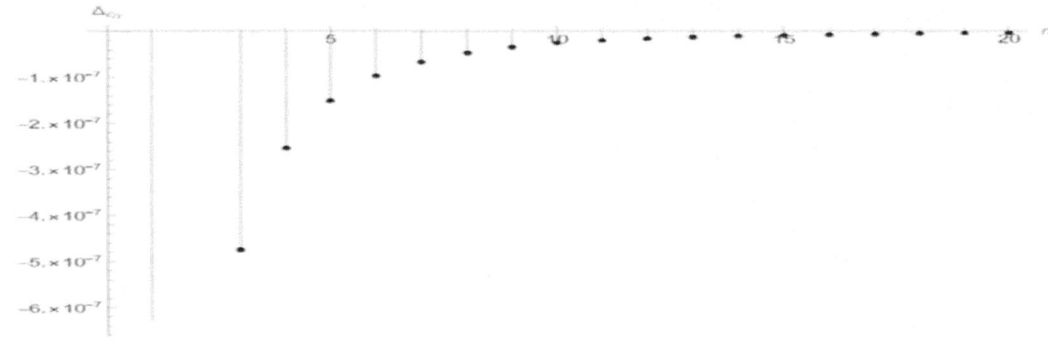

FIGURE 8.9
The graph for Δ_{ϵ_n}, $n = \overline{1, 20}$.

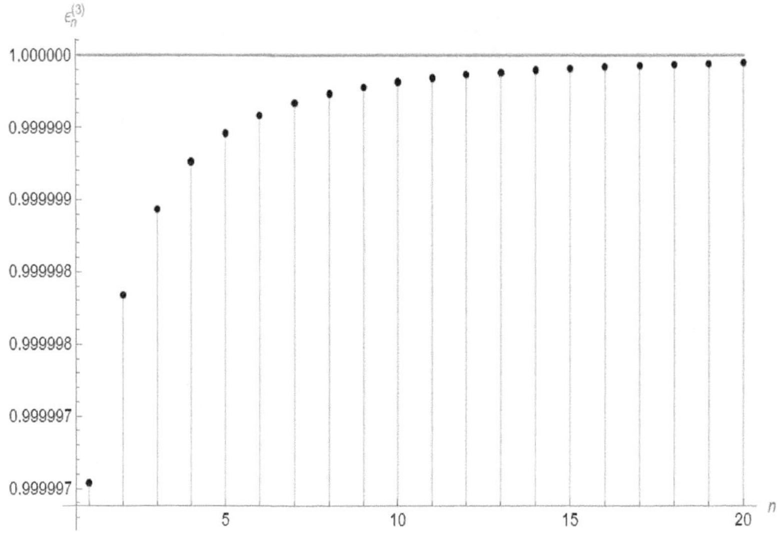

FIGURE 8.10
Energy levels $\epsilon^{(3)}_{n,j=1}$; $n = \overline{0, 20}$.

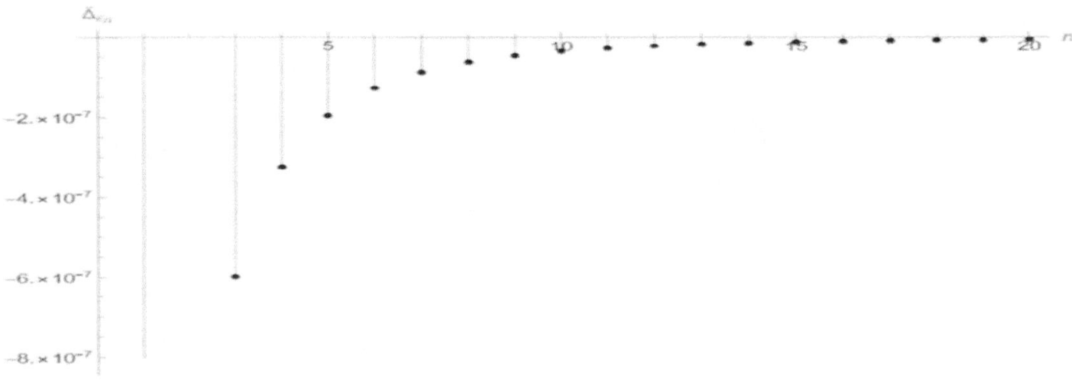

FIGURE 8.11
The graph for $\tilde{\Delta}_{\epsilon_n}$, $n = \overline{1, 20}$.

Bibliography

[1] I.E. Tamm. Motion of mesons in electromagnetic fields. *DAN USSR*, 29: 551–554, 1940.

[2] V.V. Kisel, E.M. Ovsiyuk and V.M. Red'kov. On the wave functions and energy spectrum for a spin 1 particle in external Coulomb field. *Nonlinear Phenomena in Complex Systems*, 13(4): 352–367, 2010.

[3] V.V. Kisel, V.M. Red'kov and E.M. Ovsiyuk. Wave functions and enetgy spectrum for spin 1 particle in external Coulomb field. *DAN of Belarus*, 55(1): 50–55, 2011.

[4] E. Ovsiyuk, O. Veko, K. Kazmerchuk, V. Kisel and V. Red'kov. Quantum mechanics of a spin 1 particle in the magnetic monopole potential, in spaces of Euclid, Lobachevsky, and Riemann: nonrelativistic approximation. *Ukrainian Journal of Physics*, 58(11): 1073–1083, 2013.

[5] O.V. Veko, K.V. Kazmerchuk, E.M. Ovsiyuk, V.V. Kisel, A.M. Ishkhanyan and V.M. Red'kov. Spin 1 particle in the magnetic monopole potential for Minkowski and Lobachevsky spaces: nonrelativistic approximation. *Nonlinear Phenomena in Complex Systems*, 18(2): 243–258, 2015.

[6] M. Neagu, O. Florea, O. Veko, E. Ovsiyuk and V. Red'kov. From quantum dynamics of spin 1 particle in Coulomb field to jet geometric-physical objects. *Applied Sciences*, 16: 73–99, 2014.

[7] N.G. Krylova, E.M. Ovsiyuk, V. Balan and V.M. Red'kov. Geometrization for a Quantum-Mechanical Problem of a Vector Particle in External Coulomb Field. *Nonlinear Phenomena in Complex Systems*, 21(4): 309–325, 2018.

[8] E.M. Ovsiyuk, O.V. Veko, Ya.A. Voynova, A.D. Koral'kov, V.V. Kisel and V.M. Red'kov. On desctribing the boundstates for spin 1 particle in external Coulomb field. *Problems of Physics, Informatics and Technics*, 35(2): 21–33, 2018.

[9] E.M. Ovsiyuk, O.V. Veko, Ya.A. Voynova, A.D. Koral'kov, V.V. Kisel and V.M. Red'kov. On describing bound states for a spin 1 particle in the external Coulomb field. *Proceedings of Balkan Society of Geometers*, 25: 59–78, 2018.

[10] A. Ronveaux. *Heun's Differential Equation.* Oxford University Press, Oxford, 1995.

[11] S.Yu. Slavyanov and W. Lay. *Special Functions. A Unified Theory Based on Singularities.* Oxford University Press, Oxford, 2000.

9

Geometrical modelling of the media in electrodynamics

It is known that vacuum Maxwell equations being considered on the background of any pseudo-Riemannian space-time may be interpreted as Maxwell equations in Minkowski space but specified in some effective medium, in which constitutive relations are determined by the metric of the curved space-time. In that context, we will consider space-time models with an event horizon. All of them have the same metric structure, we restrict ourselves to the spherically symmetric case and consider de Sitter, anti de Sitter, and Schwarzschild models. Also, we will study hyperbolic Lobachevsky and spherical Riemann models; parameterised coordinates with spherical and cylindric symmetry. We will prove that in all examined cases, effective tensors and of electric permittivity (ϵ_{ij}) and magnetic permeability (ϵ_{ij}) obey the same condition: $\epsilon_{ij}(x)\,\mu_{jk}(x) = \delta_{ik}$. Simplicity of expressions for these tensors $\epsilon_{ij}(x)$ and $\mu_{jk}(x)$ is misleading; for each curved space-time model, we are to solve Maxwell equations separately and anew. We will construct these solutions explicitly, applying Maxwell equations in spinor form.

9.1 Geometry and modelling the constitutive relations

It is known that vacuum Maxwell equations being considered on the background of any pseudo-Riemannian space-time may be interpreted as Maxwell equations in Minkowski space but specified in some effective medium, whose constitutive relations are determined by the metric of the curved space-time [1–45]. A detailed treatment of such a possibility for quasi-Cartesian coordinates was given in [46].

Let us start with Maxwell equation in the medium when using some curvilinear coordinates (x^σ) and respective metric $G_{\alpha\beta}(x)$. So we have

$$\partial_\alpha F_{\beta\gamma} + \partial_\beta F_{\gamma\alpha} + \partial_\gamma F_{\alpha\beta} = 0\,, \qquad \frac{1}{\sqrt{-G}}\,\partial_\beta \sqrt{-G}\,H^{\alpha\beta} = J^\alpha\,, \tag{9.1}$$

where $G = \det[G_{\alpha\beta}(x)]$ stands for a determinant of the metric tensor. We are to use two electromagnetic tensors, $F_{\alpha\beta}(x)$ and $H^{\alpha\beta}(x)$, related to each other by means of some constitutive equations.

Let us assume that a certain Riemannian space-time model is parameterised by formally similar coordinates (x^σ) with respective metric $g_{\alpha\beta}(x)$. Maxwell vacuum equations in that space-time have the form

$$\partial_\alpha f_{\beta\gamma} + \partial_\beta f_{\gamma\alpha} + \partial_\gamma f_{\alpha\beta} = 0\,, \qquad \frac{1}{\sqrt{-g}}\,\partial_\beta \sqrt{-g}\,h^{\beta\alpha} = j^\alpha\,, \tag{9.2}$$

$$h_{\alpha\beta}(x) = \epsilon_0 f_{\alpha\beta}(x)\,, \quad h^{\alpha\beta}(x) = \epsilon_0 g^{\alpha\rho}(x) g^{\beta\sigma}(x)\, f_{\rho\sigma}(x). \tag{9.3}$$

DOI: 10.1201/9781003472377-9

we have specified vacuum constitutive relations; note that $g = \det[g_{\alpha\beta}(x)]$. The second equation in eq. (9.2) may be re-written as

$$\frac{\sqrt{-G}}{\sqrt{-g}} \frac{1}{\sqrt{-G}} \partial_\beta \sqrt{-G} \frac{\sqrt{-g}}{\sqrt{-G}} h^{\beta\alpha} = j^\alpha . \tag{9.4}$$

Let us define new variables

$$\begin{gathered} F_{\alpha\beta}(x) = f_{\alpha\beta}(x), H^{\beta\alpha}(x) = \frac{\sqrt{-g}}{\sqrt{-G}} h^{\beta\alpha}(x) , \\ J^\alpha(x) = \frac{\sqrt{-g(x)}}{\sqrt{-G(x)}} j^\alpha(x) , \end{gathered} \tag{9.5}$$

then eq. (9.2) may be understood as Maxwell equations of the form (9.1) in flat space-time:

$$\partial_\alpha F_{\beta\gamma} + \partial_\beta F_{\gamma\alpha} + \partial_\gamma F_{\alpha\beta} = 0 , \quad \frac{1}{\sqrt{-G}} \partial_\beta \sqrt{-G} \, H^{\alpha\beta} = J^\alpha ; \tag{9.6}$$

and relationship between two electromagnetic tensors is governed by the formula

$$H^{\alpha\beta}(x) = \epsilon_0 \frac{\sqrt{-g(x)}}{\sqrt{-G(x)}} g^{\alpha\rho}(x) g^{\beta\sigma}(x) \, F_{\rho\sigma}(x). \tag{9.7}$$

In that context, let us consider space-time models with event horizon (let the external currents vanish); all of them have a metric of the same structure (we restrict ourselves to spherically symmetric case; for definiteness; we take in mind de Sitter, anti de Sitter, and Schwarzschild models)

$$\begin{gathered} dS^2 = \varphi dt^2 - r^2 d\theta^2 - r^2 \sin^2\theta d\phi^2 - \varphi^{-1} dr^2, \\ \text{Minkowski space } dS_0^2 = dt^2 - r^2 d\theta^2 - r^2 \sin^2\theta d\phi^2 - dr^2 . \end{gathered} \tag{9.8}$$

Because the metric determinant of the models is the same, $G(x) = g(x)$, effective constitutive relations (9.8) become simpler:

$$H^{\alpha\beta}(x) = \epsilon_0 \, g^{\alpha\rho}(x) g^{\beta\sigma}(x) \, F_{\rho\sigma}(x). \tag{9.9}$$

Taking into account the explicit form of the metric tensor (numerating coordinates as follows $x^\alpha = (t, \theta, \phi, r)$):

$$g^{\beta\alpha} = \left| \begin{array}{cccc} 1/\varphi & 0 & 0 & 0 \\ 0 & -1/r^2 & 0 & 0 \\ 0 & 0 & -1/r^2 \sin^2\theta & 0 \\ 0 & 0 & 0 & -\varphi \end{array} \right| , \tag{9.10}$$

we get constitutive relations modified by Riemannian metric

$$\begin{gathered} H^{0j} = \epsilon_0 g^{00} g^{jj} F_{0j} , \\ H^{01} = \frac{1}{\varphi} \left(-\frac{\epsilon_0}{r^2} , \quad F_{01} \right) , H^{02} = \frac{1}{\varphi} \left(-\frac{\epsilon_0}{r^2 \sin^2\theta} F_{02} \right) , \quad H^{03} = -\epsilon_0 F_{03} , \\ H^{ij} = \epsilon_0 g^{ii} g^{jj} F_{ij} , \\ H^{23} = \varphi \left(\frac{\epsilon_0}{r^2 \sin^2\theta} F_{23} \right) , \quad H^{31} = \varphi \left(\frac{\epsilon_0}{r^2} F_{31} \right) , \quad H^{12} = \left(\frac{\epsilon_0}{r^4 \sin^4\theta} F_{12} \right) . \end{gathered}$$

The last formulas may be re-written differently

$$D^{\theta} = \frac{1}{\varphi}\epsilon_0 \frac{E_\theta}{r^2}, \quad D^{\phi} = \frac{1}{\varphi}\epsilon_0 \frac{E_\phi}{r^2 \sin^2\theta}, \quad D^r = \epsilon_0 E_r\,;$$
$$H^{\theta} = \varphi \frac{1}{\mu_0}\frac{B_\theta}{r^2\sin^2\theta}, \quad H^{\phi} = \varphi \frac{1}{\mu_0}\frac{B_\phi}{r^2}, \quad H^r = \frac{1}{\mu_0}\frac{B_r}{r^4\sin^4\theta}; \tag{9.11}$$

were we have used identity $c^2 = 1/\epsilon_0\mu_0$ and definitions for two tensors

$$(F_{ab}) = \left|\begin{array}{cccc} 0 & -E^1 & -E^2 & -E^3 \\ E^1 & 0 & -cB^3 & +cB^2 \\ E^2 & +cB^3 & 0 & -cB^1 \\ E^3 & -cB^2 & +cB^1 & 0 \end{array}\right|, (H^{ab}) = \left|\begin{array}{cccc} 0 & -D^1 & -D^2 & -D^3 \\ D^1 & 0 & -H^3/c & +H^2/c \\ D^2 & +H^3/c & 0 & -H^1/c \\ D^3 & -H^2/c & +H^1/c & 0 \end{array}\right|.$$

The constitutive relations (9.11) can be presented with the help of electric permittivity and magnetic permeability tensors:

$$D^i(x) = \epsilon_0\epsilon_{ij}(x)E^{(j)}(x)\,, \quad H^i(x) = \frac{1}{\mu_0}\mu_{ij}(x)B^{(j)}(x)\,,$$
$$[\epsilon_{ij}(r)] = \left|\begin{array}{ccc} \varphi^{-1}(r) & 0 & 0 \\ 0 & \varphi^{-1}(r) & 0 \\ 0 & 0 & 1 \end{array}\right|, \quad [\mu_{ij}(r)] = \left|\begin{array}{ccc} \varphi(r) & 0 & 0 \\ 0 & \varphi(r) & 0 \\ 0 & 0 & 1 \end{array}\right|, \tag{9.12}$$

where

$$E^{(j)} = \frac{E_j}{h_j}, \quad B^{(1)} = \frac{B_1}{h_2 h_3}, \quad B^{(2)} = \frac{B_3}{h_3 h_1}, \quad B^{(3)} = \frac{B_3}{h_1 h_2}, \tag{9.13}$$

and h_j are determined by Minkowski metric (9.8) in spherical coordinates:

$$dS_0^2 = dt^2 - h_1 d\theta^2 - h_2 d\phi^2 - h_3 dr^2. \tag{9.14}$$

The simplicity of relations (9.12) is misleading; in fact, for each curved space-time model, we are to solve Maxwell equations separately and anew. Let us specify the effective constitutive equations for four models:

Minkowski ($r \in (0, +\infty)$),

$$\epsilon = \left|\begin{array}{ccc} 1 & 0 & 0 \\ 0 & 1 & 0 \\ 0 & 0 & 1 \end{array}\right|, \quad \mu = \left|\begin{array}{ccc} 1 & 0 & 0 \\ 0 & 1 & 0 \\ 0 & 0 & 1 \end{array}\right|;$$

de Sitter ($r \in (0, +1)$),

$$\epsilon = \left|\begin{array}{ccc} (1-r^2)^{-1} & 0 & 0 \\ 0 & (1-r^2)^{-1} & 0 \\ 0 & 0 & 1 \end{array}\right|, \quad \mu = \left|\begin{array}{ccc} (1-r^2) & 0 & 0 \\ 0 & (1-r^2) & 0 \\ 0 & 0 & 1 \end{array}\right|;$$

anti de Sitter ($r \in (0, +\infty)$),

$$\epsilon = \left|\begin{array}{ccc} (1+r^2)^{-1}(r) & 0 & 0 \\ 0 & (1+r^2)^{-1} & 0 \\ 0 & 0 & 1 \end{array}\right|, \quad \mu = \left|\begin{array}{ccc} (1+r^2) & 0 & 0 \\ 0 & (1+r^2) & 0 \\ 0 & 0 & 1 \end{array}\right|;$$

Schwarzschild ($r \in (1, +\infty)$),

$$\epsilon = \left|\begin{array}{ccc} (1-1/r)^{-1} & 0 & 0 \\ 0 & (1-1/r)^{-1} & 0 \\ 0 & 0 & 1 \end{array}\right|, \quad \mu = \left|\begin{array}{ccc} (1-1/r) & 0 & 0 \\ 0 & (1-1/r) & 0 \\ 0 & 0 & 1 \end{array}\right|.$$

Direct comparison is possible only for Minkowski and anti de Sitter models, due to the same region for radial coordinate.

That interpretation is possible for other space-time models as well. Let us discuss the hyperbolic Lobachevsky and spherical Riemann models being compared with the Minkowski model:

$$\begin{aligned} dS_0^2 &= dt^2 - r^2 d\theta^2 - r^2 \sin^2\theta d\phi^2 - dr^2, \\ dS^2 &= dt^2 - \sinh^2 r\, d\theta^2 - \sinh^2 r\, \sin^2\theta\, d\phi^2 - dr^2, \\ dS^2 &= dt^2 - \sin^2 r\, d\theta^2 - \sin^2 r \sin^2\theta\, d\phi^2 - dr^2. \end{aligned} \tag{9.15}$$

For Lobachevsky space we have

$$H^{0j} = \frac{\sinh^2 r}{r^2}\epsilon_0 g^{jj} F_{0j}\,, \quad H^{03} = -\frac{\sinh^2 r}{r^2}\epsilon_0 F_{03},$$

$$H^{01} = -\frac{\sinh^2 r}{r^2}\frac{\epsilon_0}{\sinh^2 r}F_{01}, H^{02} = -\frac{\sinh^2 r}{r^2}\frac{\epsilon_0}{\sinh^2 r \sin^2\theta}F_{02},$$

$$H^{ij} = \frac{\sinh^2 r}{r^2}\,\epsilon_0 g^{ii} g^{jj} F_{ij}\,, \quad H^{12} = \frac{\sinh^2 r}{r^2}\frac{\epsilon_0}{\sinh^4 r \sin^4\theta}F_{12},$$

$$H^{23} = \frac{\sinh^2 r}{r^2}\frac{\epsilon_0}{\sinh^2 r \sin^2\theta}F_{23}, \quad H^{31} = \frac{\sinh^2 r}{r^2}\frac{\epsilon_0}{\sinh^2 r}F_{31},$$

or differently

$$\begin{aligned} &H^{01} = -\frac{\epsilon_0}{r^2}F_{01}\,, \quad H^{02} = -\frac{\epsilon_0}{r^2\sin^2\theta}F_{02}, \quad H^{03} = -\frac{\sinh^2 r}{r^2}\epsilon_0 F_{03}\,, \\ &H^{23} = \frac{\epsilon_0}{r^2\sin^2\theta}F_{23}\,, \quad H^{31} = \frac{\epsilon_0}{r^2}F_{31}\,, \quad H^{12} = \frac{r^2}{\sinh^2 r}\frac{\epsilon_0}{r^4\sin^4\theta}F_{12}\,, \end{aligned} \tag{9.16}$$

which may be re-written in terms of effective tensors as follows

$$\begin{aligned} &D^i(x) = \epsilon_0\epsilon_{ij}(x)E^{(j)}(x)\,, \quad H^i(x) = \frac{1}{\mu_0}\mu_{ij}(x)B^{(j)}(x)\,, \\ &[\epsilon_{ij}(r)] = \begin{vmatrix} 1 & 0 & 0 \\ 0 & 1 & 0 \\ 0 & 0 & \sinh^2 r/r^2 \end{vmatrix}, \quad [\mu_{ij}(r)] = \begin{vmatrix} 1 & 0 & 0 \\ 0 & 1 & 0 \\ 0 & 0 & r^2/\sinh^2 r \end{vmatrix}. \end{aligned} \tag{9.17}$$

For spherical model we have similar result with evident modifications

$$\sinh^2 r, r \in (0, +\infty) \quad \Longrightarrow \quad \sin^2 r, r \in (0, \pi)\,.$$

Let us consider examples with cylindric symmetry. For flat Minkowski space we have (numerate coordinates as $x^\alpha = (t, r, \phi, z)$)

$$G_{\alpha\beta} = \begin{vmatrix} 1 & 0 & 0 & 0 \\ 0 & -1 & 0 & 0 \\ 0 & 0 & -r^2 & 0 \\ 0 & 0 & 0 & -1 \end{vmatrix}; \tag{9.18}$$

in spherical Riemann space an analogous metric is

$$g_{\alpha\beta} = \begin{vmatrix} 1 & 0 & 0 & 0 \\ 0 & -1 & 0 & 0 \\ 0 & 0 & -\sin^2 r & 0 \\ 0 & 0 & 0 & -\cos^2 z \end{vmatrix}, \ r \in (0, \frac{\pi}{2}),\ z \in (-\frac{\pi}{2}, +\frac{\pi}{2}). \tag{9.19}$$

Generated by the last metric constitutive relations have the form

$$H^{\alpha\beta} = \epsilon_0 \frac{\sin r \cos z}{r} g^{\alpha\rho} g^{\beta\sigma} F_{\rho\sigma}\,, \tag{9.20}$$

that is

$$H^{0j} = \frac{\sin r \cos z}{r} \epsilon_0 g^{jj} F_{0j}\,, \qquad H^{03} = -\frac{\sin r \cos z}{r} \frac{\epsilon_0}{\cos^2 z} F_{03},$$
$$H^{01} = -\frac{\sin r \cos z}{r} \epsilon_0 F_{01}, \qquad H^{02} = -\frac{\sin r \cos z}{r} \frac{\epsilon_0}{\sin^2 r} F_{02},$$
$$H^{ij} = \frac{\sin r \cos z}{r} \epsilon_0 g^{ii} g^{jj} F_{ij}\,, \qquad H^{12} = \frac{\sin r \cos z}{r} \frac{\epsilon_0}{\sin^2 r} F_{12},$$
$$H^{23} = \frac{\sin r \cos z}{r} \frac{\epsilon_0}{\sin^2 r \cos^2 z} F_{23}\,, \; H^{31} = \frac{\sin r \cos z}{r} \frac{\epsilon_0}{\cos^2 r} F_{31}.$$

This result may be presented with the help of two effective tensors as follows

$$D^i(x) = \epsilon_0 \epsilon_{ij}(x) E^{(j)}(x), [\epsilon_{ij}(r,z)] = \begin{vmatrix} \frac{\sin r \cos z}{r} & 0 & 0 \\ 0 & \frac{r \cos z}{\sin r} & 0 \\ 0 & 0 & \frac{\sin r}{r \cos z} \end{vmatrix},$$
$$H^i(x) = \frac{1}{\mu_0} \mu_{ij}(x) B^{(j)}(x), [\mu_{ij}(r,z)] = \begin{vmatrix} \frac{r}{\sin r \cos z} & 0 & 0 \\ 0 & \frac{\sin r}{r \cos z} & 0 \\ 0 & 0 & \frac{r \cos z}{\sin r} \end{vmatrix}, \tag{9.21}$$

where referring to cylindric coordinates in Minkowski space electromagnetic components are

$$E^{(j)} = \frac{E_j}{h_j}, \quad B^{(1)} = \frac{B_1}{h_2 h_3}, \quad B^{(2)} = \frac{B_3}{h_3 h_1}, \quad B^{(3)} = \frac{B_3}{h_1 h_2}\,,$$
$$dt^2 - dr^2 - r^2 d\phi^2 - dz^2 = dt^2 - h_1 dr^2 - h_2 d\phi^2 - h_3 dz^2.$$

In the case of hyperbolic Lobachevsky space, we have similar results with evident modifications

$$D^i(x) = \epsilon_0 \epsilon_{ij}(x) E^{(j)}(x), \quad [\epsilon_{ij}(r,z)] = \begin{vmatrix} \frac{\sinh r \cosh z}{r} & 0 & 0 \\ 0 & \frac{r \cosh z}{\sinh r} & 0 \\ 0 & 0 & \frac{\sinh r}{r \cosh z} \end{vmatrix},$$
$$H^i(x) = \frac{1}{\mu_0} \mu_{ij}(x) B^{(j)}(x), \quad [\mu_{ij}(r,z)] = \begin{vmatrix} \frac{r}{\sinh r \cosh z} & 0 & 0 \\ 0 & \frac{\sinh r}{r \cosh z} & 0 \\ 0 & 0 & \frac{r \cosh z}{\sinh r} \end{vmatrix}.$$

We can see that for all examples, effective tensors (ϵ_{ij}) and (ϵ_{ij}) obey one the same condition:

$$\epsilon_{ij}(x)\, \mu_{jk}(x) = \delta_{ik}.$$

9.2 Spinor form of Maxwell equations

To introduce spinor notations, let us start with the ordinary Dirac equation

$$(i\gamma^a \partial_a - m)\Psi = 0, \gamma^a = \begin{vmatrix} 0 & \bar{\sigma}^a \\ \sigma^a & 0 \end{vmatrix}, \Psi = \begin{vmatrix} \xi^\alpha \\ \eta_{\dot{\alpha}} \end{vmatrix}, \{\alpha, \dot{\alpha}\} = 1, 2\,; \tag{9.22}$$

$\sigma^a = (I, \sigma^j)$, $\bar{\sigma}^a = (I, -\sigma^j)$. In 2-spinor form, we have two equations

$$i\sigma^a \partial_a \xi = m\eta\,, \quad i\bar{\sigma}^a \partial_a \eta = m\xi. \tag{9.23}$$

It is convenient to attach spinor indices to Pauli matrices: $\sigma^a = (\sigma^a)_{\dot{\beta}\alpha}\,,\ \bar{\sigma}^a = (\bar{\sigma}^a)^{\beta\dot{\alpha}}$, then eq. (9.23) read

$$i(\sigma^a \partial_a)_{\dot{\beta}\alpha}\, \xi^\alpha = m\eta_{\dot{\beta}}, \quad i(\bar{\sigma}^a \partial_a)^{\beta\dot{\alpha}}\, \eta_{\dot{\alpha}} = m\xi^\beta. \tag{9.24}$$

Electromagnetic tensor is equivalent to a pair of symmetrical 2-rank spinors: $F_{mn} \Leftrightarrow \{\xi^{\alpha\beta}, \eta_{\dot{\alpha}\dot{\beta}}\}$; correspondingly, eight Maxwell equations are presented as follows

$$(\sigma^a \partial_a)_{\dot{\rho}\alpha}\, \xi^{\alpha\beta} = (\sigma^b)_{\dot{\rho}\alpha} \omega^{\alpha\beta}\, J_b, \quad (\bar{\sigma}^a \partial_a)^{\rho\dot{\alpha}}\, \eta_{\dot{\alpha}\dot{\beta}} = (\bar{\sigma}^b)^{\rho\dot{\alpha}} \omega_{\dot{\alpha}\dot{\beta}}\, J_b\,; \tag{9.25}$$

the second equation is conjugate to the first. In eq. (9.25) we use spinor metrical matrices

$$(\epsilon_{\alpha\beta}) = i\sigma^2\,, \quad (\epsilon^{\alpha\beta}) = -i\sigma^2\,; \quad (\epsilon_{\dot{\alpha}\dot{\beta}}) = i\sigma^2\,, \quad (\epsilon^{\dot{\alpha}\dot{\beta}}) = -i\sigma^2. \tag{9.26}$$

To prove the equivalence of the spinor form (9.25) to the ordinary Maxwell equation in vector notations, let us apply notations without spinor indices. To this end, we take into account identities

$$\begin{gathered} (\xi^{\alpha\beta}) = \Sigma^{mn} F_{mn} \sigma^2\,, \quad (\eta_{\dot{\alpha}\dot{\beta}}) = -\bar{\Sigma}^{mn} F_{mn} \sigma^2\,, \\ \Sigma^{mn} = \frac{1}{4}(\bar{\sigma}^m \sigma^n - \bar{\sigma}^n \sigma^m)\,, \quad \bar{\Sigma}^{mn} = \frac{1}{4}(\sigma^m \bar{\sigma}^n - \sigma^n \bar{\sigma}^m). \end{gathered} \tag{9.27}$$

Then, eq. (9.25) may be re-written as

$$\sigma^a \partial_a\, \Sigma^{mn} F_{mn} = -\sigma^b J_b\,, \qquad \bar{\sigma}^a \partial_a\, \bar{\Sigma}^{mn} F_{mn} = -\bar{\sigma}^b J_b\,. \tag{9.28}$$

We are to take the following identities into account

$$\Sigma^{mn} F_{mn} = \sigma^1 (F_{01} - iF_{23}) + \sigma^2 (F_{02} - iF_{31}) + \sigma^3 (F_{03} - iF_{12})\,,$$

$$\bar{\Sigma}^{mn} F_{mn} = \sigma^1 (-F_{01} - iF_{23}) + \sigma^2 (-F_{02} - iF_{31}) + \sigma^3 (-F_{03} - iF_{12})\,;$$

with notations

$$F_{01} = -E^1\,, F_{02} = -E^2\,, F_{03} = -E^3\,, F_{23} = B^1\,, F_{31} = B^2\,, F_{12} = B^3 \tag{9.29}$$

they read

$$\begin{gathered} \Sigma^{mn} F_{mn} = -\sigma^1 (E^1 + iB^1) - \sigma^1 (E^2 + iB^2) - \sigma^1 (E^3 + iB^3) = -\sigma^j a_j, \\ \bar{\Sigma}^{mn} F_{mn} = \sigma^1 (E^1 - iB^1) + \sigma^1 (E^2 - iB^2) + \sigma^1 (E^3 - iB^3) = +\sigma^j b_j, \end{gathered} \tag{9.30}$$

and

$$(\xi^{\alpha\beta}) = \Sigma^{mn} F_{mn} \sigma^2 = \left| \begin{array}{cc} -i(a_1 - ia_2) & ia_3 \\ ia_3 & +i(a_1 + ia_2) \end{array} \right|,$$

$$(\eta_{\dot{\alpha}\dot{\beta}}) = -\bar{\Sigma}^{mn} F_{mn} \sigma^2 = \left| \begin{array}{cc} -i(b_1 - ib_2) & ib_3 \\ ib_3 & i(b_1 + ib_2) \end{array} \right|.$$

Taking into account (9.30), Maxwell eq. (9.28) may be presented in the form

$$\sigma^a \partial_a\, \sigma^j a_j = \sigma^b J_b\,, \quad \bar{\sigma}^a \partial_a\, \sigma^j b_j = -\bar{\sigma}^b J_b, \tag{9.31}$$

or

$$(\partial_0 + \sigma^l \partial_l)\,(\sigma^k a_k) = J_0 + \sigma^j J_j\,, \quad (\partial_0 - \sigma^l \partial_l)\,(\sigma^k b_k) = -J_0 + \sigma^j J_j\,. \tag{9.32}$$

From (9.32) we derive

$$\sigma^n \partial_0 a_n + (\delta_{lk} + i\omega_{nlk}\sigma^n)\partial_l a_k\,, = J_0 + \sigma^n J_n,$$

$$\sigma^n \partial_0 b_n - (\delta_{lk} + i\omega_{nlk}\sigma^n)\partial_l b_k. = -J_0 + \sigma^n J_n\,.$$

Therefore, we have four equations

$$\partial_l a_l = J_0\,, \partial_0 a_n + i\omega_{nlk}\partial_l a_k = J_n\,, \partial_l b_l = J_0\,, \partial_0 b_n - i\omega_{nlk}\partial_l b_k = J_n\,,$$

or differently

$$\begin{array}{ll} (1) & \partial_l(E^l + iB^l) = J_0\,, \\ (2) & \partial_0(E^l + iB^l) + i\omega_{nlk}\partial_l(E^k + iB^k) = J_n\,, \\ (1') & \partial_l(E^l - iB^l) = J_0\,, \\ (2') & \partial_0(E^l - iB^l) - i\omega_{nlk}\partial_l(E^k - iB^l) = J_n. \end{array}$$

Summing and subtracting equations within each pair, we obtain

$$1+1', \quad \partial_l E^l = J_0\,, \quad 1-1', \quad \partial_l B^l = 0\,,$$

$$2+2', \quad \partial_0 E^n - \omega_{nlk}\partial_l B^k = J_k\,, \quad 2-2', \quad \partial_0 B^n + \omega_{nlk}\partial_l E^k = 0\,;$$

they may be identified with Maxwell equations in vector form

$$\operatorname{div}\mathbf{E} = J^0, \quad \operatorname{div}\mathbf{B} = 0\,, \quad \operatorname{rot}\mathbf{B} = \partial_0\mathbf{E} + \mathbf{J}\,, \quad \operatorname{rot}\mathbf{E} = -\partial_0\mathbf{B}\,, \tag{9.33}$$

where

$$\mathbf{E} = (E^n)\,, \ \mathbf{B} = (B^n)\,, \quad J^0 = J_0\,, \ \mathbf{J} = (J^n) = (-J_n).$$

9.3 Separating the variables in de Sitter models

Generally, covariant Maxwell equations in spinor form can be found by the same method which is used for generalising the Dirac equation

$$i\sigma^\alpha(x)\,[\partial_\alpha + \Sigma_\alpha(x)]\,\xi(x) = m\,\eta(x)\,, \quad i\bar{\sigma}^\alpha(x)\,\left[\partial_\alpha + \bar{\Sigma}_\alpha(x)\right]\eta(x) = m\,\xi(x). \tag{9.34}$$

So, the Maxwell equations in spinor form are to be generalised as follows [46]

$$\begin{aligned} i\sigma^\alpha(x)\,[\partial_\alpha + \Sigma_\alpha(x)\otimes I + I\otimes\Sigma_\alpha(x)]\,\xi(x) &= \sigma^\beta(x)(-i\sigma^2)J_\beta(x)\,, \\ i\bar{\sigma}^\alpha(x)\,\left[\partial_\alpha + \bar{\Sigma}_\alpha(x)\otimes I + I\otimes\bar{\Sigma}_\alpha(x)\right]\eta(x) &= \sigma^\beta(x)(+i\sigma^2)J_\beta(x)\,, \end{aligned} \tag{9.35}$$

where (see in [46]):

$$\begin{aligned} &\sigma^\alpha(x) = \sigma^a e^\alpha_{(a)}(x)\,, \quad \bar{\sigma}^\alpha(x) = \bar{\sigma}^a e^\alpha_{(a)}(x)\,, \\ &\Sigma_\alpha(x) = \frac{1}{2}\Sigma^{ab} e^\beta_{(a)}\nabla_\alpha(e_{(b)\beta}), \quad \bar{\Sigma}_\alpha(x) = \frac{1}{2}\bar{\Sigma}^{ab} e^\beta_{(x)}\nabla_\alpha(e_{(b)\beta})\,, \\ &\Sigma^{ab} = \frac{1}{4}(\,\bar{\sigma}^a\sigma^b - \bar{\sigma}^b\,\sigma^a\,), \quad \bar{\Sigma}^{ab} = \frac{1}{4}(\sigma^a\bar{\sigma}^b - \sigma^b\bar{\sigma}^a). \end{aligned} \tag{9.36}$$

As in Minkowski space, the second equation is conjugate to the first, so it suffices to study only the first one (further, we follow the case without current)

$$\sigma^{\alpha}(x)\,[\partial_{\alpha}+\Sigma_{\alpha}(x)\otimes I+I\otimes\Sigma_{\alpha}(x)]\,\xi(x)=0. \tag{9.37}$$

In eq. (9.37), the quantity $\xi(x)$ stands for a symmetric 2-rank spinor, it can be treated as a 2×-matrix function. Equation (9.37) may be presented with the help of Ricci rotation coefficients $\gamma_{abc}(x)$ as follows

$$\Big[\sigma^{c}e^{\alpha}_{(c)}(x)\partial_{\alpha}+\sigma^{c}(\frac{1}{2}\Sigma^{ab}\otimes I+I\otimes\frac{1}{2}\Sigma^{ab})\gamma_{abc}(x)\Big]\,\xi(x)=0\,;$$
$$\gamma_{abc}=-e_{(a)\beta;\alpha}e^{\beta}_{(b)}e^{\alpha}_{(c)}\,,\quad \Sigma^{ab}=\frac{1}{4}(\bar{\sigma}^{a}\sigma^{b}-\bar{\sigma}^{b}\sigma^{a})\,. \tag{9.38}$$

Let us specify eq. (9.37) in static de Sitter coordinates:

$$dS^{2}=(1-r^{2}/\rho^{2})c^{2}dt^{2}-r^{2}(d\theta^{2}+\sin^{2}\theta d\phi^{2})-\frac{dr^{2}}{1-r^{2}/\rho^{2}}\,, \tag{9.39}$$

where ρ is the curvature radius; below we will apply dimensionless coordinates $ct/\rho\Longrightarrow t,\ r/\rho\Longrightarrow r$. We use diagonal spherical tetrad

$$x^{\alpha}=(t,\ \theta,\ \phi,\ r)\,,\quad \varphi=1-r^{2}\,,\quad \varphi'=\frac{d\varphi}{dr},$$
$$e_{(0)\alpha}=(\frac{1}{\sqrt{\varphi}},\ 0,\ 0,\ 0)\,,\quad e_{(3)\alpha}=(0,\ 0,\ 0,\ \sqrt{\varphi})\,,$$
$$e_{(1)\alpha}=(0,\ \frac{1}{r},\ 0,\ 0)\,,\quad e_{(2)\alpha}=(0,0,\ \frac{1}{r\sin\theta},\ 0). \tag{9.40}$$

Local Pauli matrices are

$$\sigma^{\alpha}(x)=(\frac{1}{\sqrt{\varphi}},-\frac{\sigma^{1}}{r},-\frac{\sigma^{2}}{r\sin\theta},-\sqrt{\varphi}\sigma^{3}),\quad \sigma^{\alpha}(x)=(\frac{1}{\sqrt{\varphi}},\frac{\sigma^{1}}{r},\frac{\sigma^{2}}{r\sin\theta},\sqrt{\varphi}\sigma^{3}).$$

Taking in mind general formulas

$$\sigma^{ab}=\frac{1}{4}\left|\begin{array}{cc}\bar{\sigma}^{a}\sigma^{b}-\bar{\sigma}^{b}\sigma^{a} & 0\\ 0 & \sigma^{a}\bar{\sigma}^{b}-\sigma^{b}\bar{\sigma}^{a}\end{array}\right|=\left|\begin{array}{cc}\Sigma^{ab} & 0\\ 0 & \bar{\Sigma}^{ab}\end{array}\right|,$$

$$\Gamma_{\alpha}(x)=\frac{1}{2}\left|\begin{array}{cc}\Sigma^{ab}e^{\beta}_{(a)}\nabla_{\alpha}(\,e^{\alpha}_{(b);\beta} & 0\\ 0 & \bar{\Sigma}^{ab}e^{\beta}_{(a)}\nabla_{\alpha}(e^{\alpha}_{(b);\beta}\end{array}\right|=\left|\begin{array}{cc}\Sigma_{\alpha}(x) & 0\\ 0 & \bar{\Sigma}_{\alpha}(x)\end{array}\right|$$

and the known expressions [47] for bispinor connection in de Sitter space, we find spinor connections

$$\Gamma_{t}=\left|\begin{array}{cc}\Sigma_{t} & 0\\ 0 & \bar{\Sigma}_{t}\end{array}\right|=\frac{\varphi'}{2}\,\sigma^{03}=\frac{\varphi'}{2}\left|\begin{array}{cc}\sigma^{3}/2 & 0\\ 0 & -\sigma^{3}/2\end{array}\right|,\quad \Gamma_{r}=0\,,$$

$$\Gamma_{\theta}=\left|\begin{array}{cc}\Sigma_{\theta} & 0\\ 0 & \bar{\Sigma}_{\theta}\end{array}\right|=\sqrt{\varphi}\,\sigma^{31}=\sqrt{\varphi}\left|\begin{array}{cc}-i\sigma^{2}/2 & 0\\ 0 & -i\sigma^{2}/2\end{array}\right|, \tag{9.41}$$

$$\Gamma_{\phi}=\left|\begin{array}{cc}\Sigma_{\phi} & 0\\ 0 & \bar{\Sigma}_{\phi}\end{array}\right|=\sqrt{\varphi}\sin\theta\left|\begin{array}{cc}i\sigma^{1}/2 & 0\\ 0 & i\sigma^{1}/2\end{array}\right|+\cos\theta\left|\begin{array}{cc}-i\sigma^{3}/2 & 0\\ 0 & -i\sigma^{3}/2\end{array}\right|.$$

Therefore, eq. (9.37) takes the form

$$\Big\{ \frac{1}{\sqrt{\varphi}} \Big[\partial_t + \frac{\varphi'}{4}(\sigma^3 \otimes I + I \otimes \sigma^3) \Big] + \sigma^3 \sqrt{\varphi} \partial_r + \frac{\sigma^1}{r} \Big[\partial_\theta - i\frac{\sqrt{\varphi}}{2}(\sigma^2 \otimes I + I \otimes \sigma^2) \Big]$$
$$+ \frac{\sigma^2}{r \sin\theta} \Big[\partial_\phi + i\frac{\sqrt{\varphi}}{2} \sin\theta (\sigma^1 \otimes I + I \otimes \sigma^1) - i\frac{\cos\theta}{2}(\sigma^3 \otimes I + I \otimes \sigma^3) \Big] \Big\} \xi = 0.$$

It is convenient to re-group the terms differently

$$\Big\{ \frac{\partial}{\partial t} + \varphi[\sigma^3 \frac{\partial}{\partial r} + \frac{i}{r}(-\sigma^1 \frac{\sigma^2 \otimes I + I \otimes \sigma^2}{2} + \sigma^2 \frac{\sigma^1 \otimes I + I \otimes \sigma^1}{2}) + \frac{\varphi'}{2\varphi} \frac{\sigma^3 \otimes I + I \otimes \sigma^3}{2}]$$

$$+ \frac{\sqrt{\varphi}}{r} \Big[\sigma^1 \partial_\theta F - i\sigma^2 \frac{i\partial_\phi + \cos\theta(\sigma^3 \otimes I + I \otimes \sigma^3)/2}{\sin\theta} \Big] \Big\} \xi = 0. \tag{9.42}$$

Let us search for solutions with spherical symmetry, by diagonalizing operators of the total angular momentum. In this tetrad basis, it has Schrödinger-like structure

$$J_1 = l_1 + \frac{\cos\phi}{\sin\theta} S_3, \quad J_2 = l_2 + \frac{\sin\phi}{\sin\theta} S_3, \quad J_3 = l_3 = -i\frac{\partial}{\partial \phi},$$
$$S_3 = ij^{12} = \frac{1}{2}(\sigma^3 \otimes I + I \otimes \sigma^3), \quad \sigma^3 = \begin{vmatrix} 1 & 0 \\ 0 & -1 \end{vmatrix}. \tag{9.43}$$

We are to find eigenstates of the operator S_3:

$$\frac{1}{2}(\sigma^3 \otimes I + i \otimes \sigma^3) \begin{vmatrix} a & b \\ b & c \end{vmatrix} = \sigma \begin{vmatrix} a & b \\ b & c \end{vmatrix}. \tag{9.44}$$

Explicitly, this equation reads

$$\begin{vmatrix} a & b \\ -b & -c \end{vmatrix} + \begin{vmatrix} a & -b \\ b & -c \end{vmatrix} = 2\sigma \begin{vmatrix} a & b \\ b & c \end{vmatrix},$$

whence it follows the linear system with three different solutions

$$\left\{ \begin{array}{l} a = \sigma a, \\ 0 = \sigma b, \\ c = -\sigma c, \end{array} \right. \quad \Longrightarrow \quad \begin{array}{ll} \sigma = +1, & a = 1,\ b = 0,\ c = 0; \\ \sigma = 0, & a = 0,\ b = 1,\ c = 0; \\ \sigma = -1, & a = 0,\ b = 0,\ c = 1. \end{array} \tag{9.45}$$

So we have three eigenvalues $\sigma = -1, 0, +1$ and corresponding three eigenstates:

$$\sigma = +1, \begin{vmatrix} 1 & 0 \\ 0 & 0 \end{vmatrix}; \quad \sigma = 0, \begin{vmatrix} 0 & 1 \\ 1 & 0 \end{vmatrix}; \quad \sigma = -1, \begin{vmatrix} 0 & 0 \\ 0 & 1 \end{vmatrix}. \tag{9.46}$$

In accordance with general theory [47], we construct solutions obeying two equations

$$(J_1^2 + J_2^2 + J_3^2)\, \xi(x) = j(j+1)\, \xi(x), \quad J_3\, \xi(x) = m\, \xi(x),$$

and $\xi(x)$ should have the form (we apply the known Wigner functions [47])

$$\xi(x) = e^{-i\omega t} \begin{vmatrix} f(r)\, D_{-1} & h(r)\, D_0 \\ h(r)\, D_0 & g(r)\, D_{+1} \end{vmatrix}, \quad D_\sigma = D^j_{-m,-\sigma}(\phi, \theta, 0); \tag{9.47}$$

where f, g, h stand for unknown radial functions; indices j,m at Wigner functions are omitted. The substitution (9.45) is correct only for the following j,m:

$$j = 1,2,3,...; \quad m = -j, -j+1, ..., j-1, j\ .$$

For states with $j=0$, the initial substitution is different

$$j = 0, \qquad \xi(x) = e^{-i\omega t} \left| \begin{array}{cc} 0 & h(r) \\ h(r) & 0 \end{array} \right| . \tag{9.48}$$

Now we should separate the variables in eq. (9.42). First, we find (the factor $e^{-i\omega t}$ is omitted for brevity)

$$\xi(x) = \left| \begin{array}{cc} f(r)\,D_{-1} & h(r)\,D_0 \\ h(r)\,D_0 & g(r)\,D_{+1} \end{array} \right| ,$$

$$\frac{\partial}{\partial t}\xi = \left| \begin{array}{cc} -i\omega f\,D_{-1} & -i\omega h\,D_0 \\ -i\omega h\,D_0 & -i\omega g\,D_{+1} \end{array} \right| , \qquad \sigma^3 \frac{\partial}{\partial r}\xi = \left| \begin{array}{cc} f'D_{-1} & h'D_0 \\ -h'D_0 & -g'D_{+1} \end{array} \right| ,$$

$$\frac{i}{r}\left[-\sigma^1 \frac{\sigma^2 \otimes I + I \otimes \sigma^2}{2} + \sigma^2 \frac{\sigma^1 \otimes I + I \otimes \sigma^1}{2}) \right] \xi = \frac{1}{r} \left| \begin{array}{cc} f(r)D_{-1} & 2h(r)D_0 \\ -2h(r)D_0 & -g(r)D_{+1} \end{array} \right| ,$$

$$\frac{\sigma^3 \otimes I + I \otimes \sigma^3}{2}\xi = \left| \begin{array}{cc} f(r)\,D_{-1} & 0 \\ 0 & -g(r)\,D_{+1} \end{array} \right| .$$

Now, we find the action of angular operator $\Sigma_{\theta,\phi}$ on spinor $\xi(x)$ (taking in mind that $i\partial_\phi D^j_{-m,\sigma} = -m\ D^j_{-m,\sigma}$):

$$\Sigma_{\theta,\phi}\xi(x) = \sigma^1 \partial_\theta \xi - i\sigma^2 \frac{-m + \cos\theta(\sigma^3 \otimes I + I \otimes \sigma^3)/2}{\sin\theta}\ \xi$$

$$= \left| \begin{array}{cc} h\ \partial_\theta D_0 & g\ \partial_\theta D_{+1} \\ f\ \partial_0 D_{-1} & h\ \partial_\theta D_0 \end{array} \right| + \frac{1}{\sin\theta} \left| \begin{array}{cc} h\ mD_0 & g\ (m+\cos\theta)D_{+1} \\ f\ (-m+\cos\theta)D_{-1} & -h\ mD_0 \end{array} \right| ,$$

or

$$\Sigma_{\theta,\phi}\xi(x) = \left| \begin{array}{cc} h\ (\partial_\theta + m\sin^{-1})\ D_0 & g\ [\partial_\theta + (m+\cos\theta)\sin^{-1}\theta] D_{+1} \\ f\ [\partial_\theta + (-m+\cos\theta)\sin^{-1}\theta]\ D_{-1} & h\ (\partial_\theta - m\sin^{-1})\ D_0 \end{array} \right| .$$

With the use of the known recurrent formulas for Wigner functions [17]

$$\partial_\theta D_{-1} = \frac{1}{2}(b\ D_{-2} - a\ D_0)\,, (-m+\cos\theta)\sin^{-1}\theta\ D_{-1} = \frac{1}{2}(-b\ D_{-2} - a\ D_0)\,,$$

$$\partial_\theta D_{+1} = \frac{1}{2}(a\ D_0 - b\ D_{+2})\,, (m+\cos\theta)\sin^{-1}\theta\ D_{+1} = \frac{1}{2}(a\ D_0 + b\ D_{+2})\,,$$

$$\partial_\theta D_0 = \frac{1}{2}(a\ D_{-1} - a\ D_{+1})\,, m\sin^{-1}\theta\ D_0 = \frac{1}{2}(a\ D_{-1} + a\ D_{+1})\,,$$

$$a = \sqrt{j(j+1)}\,, \quad b = \sqrt{(j-1)(j+1)}\,,$$

we derive

$$\Sigma_{\theta,\phi}\xi = a \left| \begin{array}{cc} h\ D_{-1} & g\ D_0 \\ -f\ D_0 & -h\ D_{+1} \end{array} \right| . \tag{9.49}$$

Therefore, eq. (9.42) takes the form

$$\begin{vmatrix} -i\omega f\, D_{-1} & -i\omega h\, D_0 \\ -i\omega h\, D_0 & -i\omega g\, D_{+1} \end{vmatrix} + \varphi \left\{ \begin{vmatrix} f' D_{-1} & h' D_0 \\ -h' D_0 & -g' D_{+1} \end{vmatrix} \right.$$

$$\left. + \frac{1}{r} \begin{vmatrix} f\, D_{-1} & 2h\, D_0 \\ -2h\, D_0 & -g\, D_{+1} \end{vmatrix} + \frac{\varphi'}{2\varphi} \begin{vmatrix} f\, D_{-1} & 0 \\ 0 & -g\, D_{+1} \end{vmatrix} \right\} + \frac{\sqrt{\varphi}}{r} a \begin{vmatrix} h\, D_{-1} & g\, D_0 \\ -f\, D_0 & -h\, D_{+1} \end{vmatrix} = 0,$$

whence it follows the system of four radial equations:

$$\begin{gathered} -i\omega f + \varphi(\frac{d}{dr} + \frac{1}{r} + \frac{\varphi'}{2\varphi})f + a\frac{\sqrt{\varphi}}{r}h = 0\,, \quad +i\omega g + \varphi(\frac{d}{dr} + \frac{1}{r} + \frac{\varphi'}{2\varphi})g + a\frac{\sqrt{\varphi}}{r}h = 0\,, \\ -i\omega h + \varphi(\frac{d}{dr} + \frac{2}{r})h + a\frac{\sqrt{\varphi}}{r}g = 0\,, \quad +i\omega h + \varphi(\frac{d}{dr} + \frac{2}{r})h + a\frac{\sqrt{\varphi}}{r}f = 0\,; \end{gathered} \tag{9.50}$$

remembering that $a = \sqrt{j(j+1)}$. Equations for the case $j = 0$ follow from eq. (9.50) by setting $f = 0, g = 0$, and $a = 0$:

$$0 = 0, \quad 0 = 0\,, -i\omega h + \varphi(\frac{d}{dr} + \frac{2}{r})h = 0\,, +i\omega h + \varphi(\frac{d}{dr} + \frac{2}{r})h = 0. \tag{9.51}$$

There exists only one and trivial solution: $h(r) = 0$, which means that Maxwell equations do not have solutions with $j = 0$.

Turning to (9.50), let us sum and subtract eqs. 3 and 4, this leads to

$$2\varphi(\frac{d}{dr} + \frac{2}{r})h + a\frac{\sqrt{\varphi}}{r}(f+g) = 0, \quad h = \frac{ia}{2\omega}\frac{\sqrt{\varphi}}{r}(f-g). \tag{9.52}$$

It is readily checked that the first equation in (9.52) turns out to be an identity $0 = 0$ by substituting from eqs. 3 and 4 the variables $f(r)$ and $g(r)$ expressed through $h(r)$. This means that we have only three independent equations

$$\begin{gathered} h = \frac{ia}{2\omega}\frac{\sqrt{\varphi}}{r}(f-g)\,, \\ -i\omega f + \varphi(\frac{d}{dr} + \frac{1}{r} + \frac{\varphi'}{2\varphi})f + a\frac{\sqrt{\varphi}}{r}h = 0\,, \\ +i\omega g + \varphi(\frac{d}{dr} + \frac{1}{r} + \frac{\varphi'}{2\varphi})g + a\frac{\sqrt{\varphi}}{r}h = 0. \end{gathered} \tag{9.53}$$

Excluding the variable h, we get

$$\begin{gathered} (\frac{d}{dr} + \frac{1}{r} + \frac{\varphi'}{2\varphi} - \frac{i\omega}{\varphi})f + \frac{ia^2}{2\omega r^2}(f-g) = 0\,, \\ (\frac{d}{dr} + \frac{1}{r} + \frac{\varphi'}{2\varphi} + \frac{i\omega}{\varphi})g + \frac{ia^2}{2\omega r^2}(f-g) = 0. \end{gathered} \tag{9.54}$$

Let us sum and subtract equations in eq. (9.54), in the same time introducing new variables, $f + g = F$, $f - g = G$, this results in

$$(\frac{d}{dr} + \frac{1}{r} + \frac{\varphi'}{2\varphi})F - \frac{i\omega}{\varphi}G + \frac{ia^2}{\omega r^2}G = 0, \quad (\frac{d}{dr} + \frac{1}{r} + \frac{\varphi'}{2\varphi})G - \frac{i\omega}{\varphi}F = 0. \tag{9.55}$$

The system (9.55) is simplified by substitutions $F = (r\sqrt{\varphi})^{-1}\bar{F}$, $G = (r\sqrt{\varphi})^{-1}\bar{G}$; so we obtain

$$i\omega\frac{d}{dr}\bar{F} + (\frac{\omega^2}{\varphi} - \frac{a^2}{r^2})\bar{G} = 0\,, \quad \varphi\frac{d}{dr}\bar{G} = i\omega\bar{F}\,, \tag{9.56}$$

whence it follows a 2nd-order equation for main function $\bar{G}$:

$$\left(\frac{d^2}{dr^2} + \frac{\varphi'}{\varphi}\frac{d}{dr} + \frac{\omega^2}{\varphi^2} - \frac{j(j+1)}{r^2\varphi}\right)\bar{G} = 0, \quad \bar{F}(r) = \frac{\varphi(r)}{i\omega}\frac{d}{dr}\bar{G}(r). \tag{9.57}$$

9.4 Solutions in Minkowski space

Let us briefly consider the simplest variant of eq. (9.57) for Minkowski space:

$$\left(\frac{d^2}{dr^2} + \omega^2 - \frac{j(j+1)}{r^2}\right)\bar{G} = 0. \tag{9.58}$$

We have an equation with regular point $r = 0$ and irregular point $r = \infty$ of the rank 2, so it belongs to confluent hypergeometric type. Possible asymptotic behaviour for solutions is as follows

$$r \to 0,\ \bar{G} \sim r^{j+1},\ r^{-j}, \quad r \to \infty,\ \bar{G} \sim e^{-i\omega r}, e^{+i\omega r}. \tag{9.59}$$

With the help of substitution

$$\bar{G} = r^a r^{br} g(r), \quad a = j+1,\ -j, \quad b = \pm i\omega\,, \tag{9.60}$$

we get the following equation (in the variable $x = -2br$)

$$x\,\frac{d^2 g}{dx^2} + (2a - x)\,\frac{dg}{dx} - a\,g = 0\,, \tag{9.61}$$

which is identified with confluent hypergeometric equation

$$x\,\frac{d^2 F}{dx^2} + (c - x)\,\frac{dF}{dx} - a\,F = 0, \quad c = 2a. \tag{9.62}$$

Let us fix parameters a and b: $a = j+1$, $b = i\omega$, $x = -2br = -2i\omega r$, then the regular in the point $r = 0$ solution is (see notations in [48])

$$\bar{G}_1(x) = x^{j+1} e^{-x/2} \Phi(a, c; x)\,, \quad a = j+1\,, c = 2a. \tag{9.63}$$

Taking in mind the known Kummer identity $\Phi(a, c; x) = e^x \Phi(c - a, c; -x)$, we readily prove that solution is real in all points:

$$\bar{G}_1(x) = x^{j+1} e^{-x/2} \Phi(j+1, 2j+2; x)$$
$$= (-1)^{j+1} (x^*)^{j+1} e^{-x^*/2} \Phi(j+1, 2j+2; x^*), \quad x^* = -x.$$

Because the second parameter $c = 2(j+1)$ takes on integer values, the singular near the point $r = 0$ is given by the function (see in [48]):

$$g(x) = x^a e^{-x/2} \Psi(a, c; x)\,,\ a = -j\,, c = -2j\,; \quad g(x \to 0) = x^{-j}\,\frac{\Gamma(1+2j)}{\Gamma(j+1)}. \tag{9.64}$$

The description becomes more symmetric after transforming the main eq. (9.58) to Bessel form:

$$\bar{G}(r) = \sqrt{r}\, g(r)\,,\ z = \omega r,\ \frac{d^2 g}{dz^2} + \frac{1}{z}\frac{dg}{dz} + \left(1 - \frac{p^2}{z^2}\right) g = 0,\ p = j + 1/2. \tag{9.65}$$

Two independent solutions $J_p(z)$ and $J_{-p}(z)$ are referred to confluent hypergeometric functions:

$$J_{\pm p}(z) = (\frac{z}{2})^{\pm p} \frac{e^{iz}}{\Gamma(1 \pm p)} \Phi(\pm p + 1/2, \pm 2p + 1; -2iz) ; \tag{9.66}$$

also there are known relations

$$\begin{aligned} J_{\pm p}(z) &= (\frac{z}{2})^{\pm p} \sum_{n=0}^{\infty} \frac{1}{n!\Gamma(n+1 \pm p)} (\frac{iz}{2})^{2n} , \\ J_{\pm p}(|z| \to \infty) &= \sqrt{\frac{2}{\pi z}} \cos[z - (\frac{1}{2} \pm p)\frac{\pi}{2}] . \end{aligned} \tag{9.67}$$

9.5 Solutions in de Sitter space

Let us study the main radial equation for $\bar{G}(r)$ in de Sitter model:

$$\left(\frac{d^2}{dr^2} + \frac{\varphi'}{\varphi} \frac{d}{dr} + \frac{\omega^2}{\varphi^2} - \frac{j(j+1)}{r^2 \varphi} \right) \bar{G} = 0 , \tag{9.68}$$

explicitly it reads

$$\left(\frac{d^2}{dr^2} - \frac{2r}{1-r^2} \frac{d}{dr} + \frac{\omega^2}{(1-r^2)^2} - \frac{j(j+1)}{r^2(1-r^2)} \right) \bar{G} = 0. \tag{9.69}$$

In the variable $z = r^2$, we have

$$\begin{aligned} &\Big[\frac{d^2}{dz^2} + (\frac{1}{z-1} + \frac{1}{2z}) \frac{d}{dz} \\ &+ \frac{\omega^2}{4(-1+z)^2} - \frac{\omega^2}{4(-1+z)} + \frac{\omega^2}{4z} + \frac{j+j^2}{4(z-1)} - \frac{j+j^2}{4z^2} - \frac{j+j^2}{4z} \Big] \bar{G} = 0. \end{aligned} \tag{9.70}$$

Here, we have an equation of the hypergeometric type with tree regular points. Behaviour of solutions in the vicinity of singular points is

$$z \to 0, \;\; \bar{G} = z^a , a = \frac{j+1}{2}, -\frac{j}{2} ; \quad z \to 1, \;\; \bar{G} = (1-z)^b, \quad b = \pm \frac{i\omega}{2} .$$

Searching complete solutions in the form $\bar{G} = z^a (1-z)^b H(z)$, after performing the needed calculation we arrive at

$$\begin{aligned} &4z(1-z)H'' + [8a(1-z) - 8bz + 2(1-3z)] \, H' \\ &+ \big[[4a(a-1) + 2a - j(j+1)] \frac{1}{z} + [4b(b-1) + 4b + \omega^2] \frac{1}{1-z} \\ &-4a(a-1) - 8ab - 4b(b-1) - 6a - 6b \big] H = 0. \end{aligned}$$

Equating coefficients at z^{-1} and $(z-1)^{-1}$ to zero, we find yet known restrictions on parameters a and b, and obtain more simple equation

$$z(1-z)H'' + [2a + \frac{1}{2} - (2a + 2b + 3/2)z]H' - (a+b)(a+b+1/2)H = 0, \tag{9.71}$$

which is identified with hypergeometric equation with parameters

$$\alpha = a + b\,, \quad \beta = a + b + \frac{1}{2}\,, \quad \gamma = 2a + \frac{1}{2}. \tag{9.72}$$

In order to find asymptotic behaviour of basic solutions $F(z) = u_1(z)$ at $z \to 1$, we should apply the Kummer relation [48]

$$\begin{aligned} u_1(z) &= \frac{\Gamma(\gamma)\Gamma(\gamma-\alpha-\beta)}{\Gamma(\gamma-\alpha)\Gamma(\gamma-\beta)} u_2 + \frac{\Gamma(\gamma)\Gamma(\alpha+\beta-\gamma)}{\Gamma(\alpha)\Gamma(\beta)} u_6\,, \\ u_1 &= H(\alpha,\beta;\gamma;z)\,, \quad u_2 = H(\alpha,\beta;\alpha+\beta+1-\gamma;1-z)\,, \\ u_6 &= (1-z)^{\gamma-\alpha-\beta} H(\gamma-\alpha,\gamma-\beta;\gamma+1-\alpha-\beta;1-z). \end{aligned} \tag{9.73}$$

When $z \to 1$, relation (9.73) gives

$$F_1(z \to 1) = \frac{\Gamma(\gamma)\Gamma(\gamma-\alpha-\beta)}{\Gamma(\gamma-\alpha)\Gamma(\gamma-\beta)} + \frac{\Gamma(\gamma)\Gamma(\alpha+\beta-\gamma)}{\Gamma(\alpha)\Gamma(\beta)}(1-z)^{\gamma-\alpha-\beta}.$$

Therefore, the complete solution $\bar{G}$ at $z \to 1$ behaves as follows (taking in mind $\gamma - \alpha - \beta = -2b$, and $b = \pm i\,\omega/2$)

$$\bar{G}_1(z \to 1) = \Gamma(\gamma)\Big[\frac{\Gamma(\gamma-\alpha-\beta)}{\Gamma(\gamma-\alpha)\Gamma(\gamma-\beta)}(1-z)^b + \frac{\Gamma(\alpha+\beta-\gamma)}{\Gamma(\alpha)\Gamma(\beta)}(1-z)^{-b}\Big].$$

Due to identities

$$\begin{aligned} (\gamma-\alpha-\beta) = -2b, \quad \alpha+\beta-\gamma = +2b = (\gamma-\alpha-\beta)^*; \\ (\gamma-\alpha) = a + \frac{1}{2} - b = \beta^*, \quad (\gamma-\beta) = a - b = \alpha^*, \end{aligned}$$

we may conclude that the function $\bar{G}(z \to 1)$ is real. It is readily proved that the complete function $\bar{G}_1(z)$ is real in the whole region of the variable z. To this end, we apply the Kummer identity

$$u_1 = F(\alpha,\beta;\gamma;z) = (1-z)^{\gamma-\alpha-\beta} F(\gamma-\alpha,\gamma-\beta;\gamma;z)\,,$$

which provides us with two apparently different representations for the same function

$$z^a(1-z)^b F(\alpha,\beta;\gamma;z) = z^a(1-z)^b(1-z)^{\gamma-\alpha-\beta} F(\gamma-\alpha,\gamma-\beta;\gamma;z)\,,$$

which can be re-written as follows (remembering that a is real)

$$\bar{G}_1(z) = z^a(1-z)^b F(\alpha,\beta;\gamma;z) = z^a(1-z)^{b^*} F(\beta^*,\alpha^*;\gamma;z)\,, \tag{9.74}$$

so $\bar{G}_1(z) = [\bar{G}_1(z)]^*$.

Now, it is convenient to fix parameters, $a = (j+1)/2\,,\ b = +i\omega/2$; this choice corresponds to regular at $z = 0$ solution. Singular solution refers to the function $u_5(z)$:

$$\begin{aligned} u_5(z) &= z^{1-\gamma} F(\alpha+1-\gamma,\beta+1-\gamma;2-\gamma;z)\,; \\ \bar{G}_5(z) &= z^a z^{1-\gamma} = z^{(j+1)/2} z^{-j-1/2} = z^{-j/2}. \end{aligned} \tag{9.75}$$

For u_5, there are two possible representations

$$\begin{aligned} u_5(z) &= z^{1-\gamma} F(\alpha+1-\gamma,\beta+1-\gamma;2-\gamma;z) \\ &= z^{1-\gamma}(1-z)^{\gamma-\alpha-\beta} F(1-\alpha,1-\beta;2-\gamma;z)\,; \end{aligned}$$

so we have two representations for the complete solutions

$$\bar{G}_5(z) = z^{-j/2}(1-z)^{i\omega/2}F(-j/2+i\omega/2, 1/2-j/2+i\omega/2; 1/2-j; z)$$
$$= z^{-j/2}(1-z)^{-i\omega/2}F(1/2-j/2-i\omega/2, -j/2-i\omega/2; 1/2-j; z)\,,$$

so that $\bar{G}_5(z) = [\bar{G}_5(z)]^*$. In order to construct complex and conjugate solutions with the given behaviour at $z \to 1$:

$$u_2 \sim (1-z)^b = (1-z)^{+i\omega/2}\,, \quad u_6 \sim (1-z)^{-b} = (1-z)^{-i\omega/2}\,,$$

we have to apply Kummer solutions $u_2(z)$ and $u_6(z)$.

In order to clarify additionally the physical meaning of the arising mathematical task, we turn back to eq. (9.68)

$$\varphi\Big(\frac{d}{dr}\varphi\frac{d}{dr} + \omega^2 - \frac{j(j+1)}{r^2}\varphi\Big)\bar{G} = 0, \tag{9.76}$$

and transform it to a new variable r_*:

$$\varphi\frac{d}{dr} = \frac{d}{dr_*} \quad \Longrightarrow \quad dr_* = \frac{dr}{\varphi(r)} = \frac{dr}{1-r^2}\,;$$
$$r_* = \frac{1}{2}\ln\frac{1+r}{1-r}; \quad r \to 0, r_* \to 0\,; \quad r \to 1, r_* \to +\infty. \tag{9.77}$$

Correspondingly, eq. (9.76) reads

$$[\frac{d^2}{dr_*^2} + \omega^2 - \frac{j(j+1)}{r^2}(1-r^2)]\bar{G} = 0, \quad r \to 0,\ [\frac{d^2}{dr_*^2} - \frac{j(j+1)}{r^2}]\bar{G} = 0;$$
$$r \to +1, \quad [\frac{d^2}{dr_*^2} + \omega^2]\bar{G} = 0, \quad \bar{G} = e^{\pm i\omega r_*} = \cos\omega r_* \pm i\sin\omega r_*.$$

Near the horizon (at $r \to 1$), solutions behave as massless harmonic waves. Equation may be treated as Schrödinger-like equation with an effective potential

$$[\frac{d^2}{dr_*^2} + \omega^2 - U(r_*)]\bar{G} = 0\,, \quad U(r) = \frac{j(j+1)}{r^2}(1-r^2)\,; \tag{9.78}$$

we should take in mind relations

$$r = \frac{e^{2r_*}-1}{e^{2r_*}+1}, \quad U(r_*) = (j+1)(\frac{1}{r^2}-1) = \frac{4j(j+1)e^{2r_*}}{(e^{2r_*}-1)^2}. \tag{9.79}$$

9.6 Solutions in anti de Sitter space

The study from previous section may be extended to anti de Sitter space-time:

$$dS^2 = \varphi dt^2 - r^2(d\theta^2 + \sin^2\theta d\phi^2) - \frac{dr^2}{\varphi}, \quad \varphi = 1 + r^2,\ r \in (0, +\infty). \tag{9.80}$$

We do not need to repeat the most of the above calculation, and may start with eq. (9.57)

$$\Big(\frac{d^2}{dr^2} + \frac{\varphi'}{\varphi}\frac{d}{dr} + \frac{\omega^2}{\varphi^2} - \frac{j(j+1)}{r^2\varphi}\Big)\bar{G} = 0\,, \tag{9.81}$$

now it reads

$$\left(\frac{d^2}{dr^2} + \frac{2r}{1+r^2}\frac{d}{dr} + \frac{\omega^2}{(1+r^2)^2} - \frac{j(j+1)}{r^2(1+r^2)} \right) \bar{G} = 0. \tag{9.82}$$

Transforming it to a new variable $r^2 = y,\ y \in (0, +\infty)$, we get

$$\left(\frac{d^2}{dy^2} + P^2 \right) \bar{G} = 0, \quad P^2(y) = \frac{[\omega^2 - j(j+1)]\, y - j(j+1)}{4y^2(1+y)^2}. \tag{9.83}$$

We may interpret it as an equation of Schrödinger type with an effective linear momentum $P^2(y)$; its behaviour at singular points is described by the formulas

$$\begin{array}{l} y \to 0, \quad P^2 \sim \dfrac{-j(j+1)}{4y^2} \to -\infty \, ; \\ y \to \infty, \quad P^2 \sim \dfrac{\omega^2 - j(j+1)}{4y^3} = \begin{cases} +0, & \omega^2 - j(j+1) > 0 \quad (A), \\ -0, & \omega^2 - j(j+1) < 0 \quad (B). \end{cases} \end{array} \tag{9.84}$$

In the quantum-mechanical context, we can easily interpret only the case (A), when $\omega^2 > j(j+1)$; the situation (B) is anomalous, for instance, a corresponding classical particle cannot be moving with such parameters.

Let us transform eq. (9.83) to a new variable, $y = -z, z = -r^2, z \in (-\infty, 0)$:

$$\left(\frac{d^2}{dz^2} + \frac{1-3z}{2z(1-z)}\frac{d}{dz} - \frac{\omega^2}{4z(1-z)^2} - \frac{j(j+1)}{4z^2(1-z)} \right) \bar{G} = 0. \tag{9.85}$$

Applying the substitution $\bar{G} = z^a\,(1-z)^b\, H(z)$; we derive an equation for $H(z)$ (see result (9.71) with the change ω^2 to $-\omega^2$)

$$\begin{array}{c} 4z(1-z)H'' + [8a(1-z) - 8bz + 2(1-3z)]\, H' \\ +\big\{ [4a(a-1) + 2a - j(j+1)]\dfrac{1}{z} + [4b(b-1) + 4b - \omega^2]\dfrac{1}{1-z} \\ -4a(a-1) - 8ab - 4b(b-1) - 6a - 6b \big\} H = 0. \end{array} \tag{9.86}$$

Impose evident restrictions $a = (j+1)/2, -j/2\,,\ b = \pm\omega/2$; then fix parameters as follows

$$a = \frac{j+1}{2}\,, \ b = -\frac{\omega}{2} < 0, \quad \bar{G}(z) = z^{(j+1)/2}(1-z)^{-\omega/2}H(z). \tag{9.87}$$

All possible functions $H(z)$ must be solutions of hypergeometric equation

$$\begin{array}{c} z(1-z)F'' + [\gamma - (\alpha+\beta+1)\, z\, F' - \alpha\beta\, F = 0\,, \\ \alpha = \dfrac{j+1-\omega}{2}\,, \ \beta = \alpha + \dfrac{1}{2}\,, \ \gamma = 2a + \dfrac{1}{2} = j + 3/2. \end{array} \tag{9.88}$$

Taking $F(z)$ as the Kummer solution $u_1(z)$ (see notations in [48])

$$u_1(z) = F(\alpha, \beta, \gamma; z) = F(\frac{j+1-\omega}{2}, \frac{j+2-\omega}{2}, +\frac{3}{2}; z)\,, \tag{9.89}$$

we get situation, when it is possible to obtain solutions in polynomials, $\alpha = -n,\ n = 0, 1, 2, \ldots$:

$$\omega = 2n + j + 1\,, \quad u_1(z) = F(-n, -n + \frac{1}{2}, j + 3/2; z). \tag{9.90}$$

The corresponding complete solution is given by the formula

$$\bar{G}_1(z) = z^{(j+1)/2}(1-z)^{-n-(j+1)/2}\left(1 + c_1 z + ... + c_m z^n\right);$$

at $z \to -\infty$ we have

$$\bar{G}_1(z \to -\infty) = z^{(j+1)/2}(-z)^{-n-(j+1)/2}\left(1 + c_1 z + ... + c_n z^n\right) \to \text{const.}$$

Thus, we have constructed solutions $\bar{G}_1(z)$ in quasi-polynomial form, finite at two singular points, $r = 0$ and $r = \infty$; the quantization of parameter ω is $\omega = 2n + j + 1, \quad n = 0, 1, 2, ...$

Let us study the case of singular solutions, when $\bar{G} \sim z^{-j/2}$. To get it, we should use another Kummer solution

$$\begin{aligned} u_5(z) &= z^{1-\gamma} F(\alpha + 1 - \gamma, \beta + 1 - \gamma, 2 - \gamma; z) \\ &= z^{-j-1/2} F\left(\frac{-j-\omega}{2}, \frac{-j-\omega+1}{2}, -j + \frac{1}{2}; z\right); \end{aligned} \tag{9.91}$$

the respective complete solution is

$$\bar{G}_5(z) = z^{-j/2} z^{-\omega/2} F\left(\frac{-j-\omega}{2}, \frac{-j-\omega+1}{2}, -j + \frac{1}{2}; z\right). \tag{9.92}$$

In fact, also we can apply some quantization condition

$$\frac{-j+1-\omega}{2} = -n', \quad \omega = 2n' - j + 1,$$

$$\bar{G}_5(z) = z^{-j/2} z^{-n+j/2} F\left(-n - \frac{1}{2}, -n, -j + \frac{1}{2}; z\right).$$

The structure of this spectrum is substantially different from the previous one; in particular, at each j, there are a number of negative values for ω. To find behaviour of that solution at infinity, we apply the following Kummer formula

$$u_5(z) = \frac{\Gamma(2-\gamma)\Gamma(\beta-\alpha)}{\Gamma(1-\alpha)\Gamma(\beta+1-\gamma)} e^{i\pi(1-\gamma)}\, u_3(z) + \frac{\Gamma(2-\gamma)\Gamma(\alpha-\beta)}{\Gamma(1-\beta)\Gamma(\alpha+1-\gamma)} e^{i\pi(1-\gamma)} u_4(z);$$

at $z \to -\infty$ it gives

$$u_5(z \to -\infty) = \frac{\Gamma(2-\gamma)\Gamma(\beta-\alpha)}{\Gamma(1-\alpha)\Gamma(\beta+1-\gamma)} e^{i\pi(1-\gamma)}\,(-z)^{-\alpha}$$

$$+ \frac{\Gamma(2-\gamma)\Gamma(\alpha-\beta)}{\Gamma(1-\beta)\Gamma(\alpha+1-\gamma)} e^{i\pi(1-\gamma)}(-z)^{-\beta};$$

so the corresponding complete solution is

$$\bar{G}_5(z \to -\infty) = \frac{\Gamma(2-\gamma)\Gamma(\beta-\alpha)}{\Gamma(1-\alpha)\Gamma(\beta+1-\gamma)} e^{i\pi(1-\gamma)}\, z^{-j/2} z^{-n+j/2}(-z)^{+n}$$

$$+ \frac{\Gamma(2-\gamma)\Gamma(\alpha-\beta)}{\Gamma(1-\beta)\Gamma(\alpha+1-\gamma)} e^{i\pi(1-\gamma)} z^{-j/2} z^{-n+j/2}(-z)^{+n-1/2},$$

whence ignoring the second term we arrive at

$$\bar{G}_5(z \to -\infty) = \frac{\Gamma(2-\gamma)\Gamma(\beta-\alpha)}{\Gamma(1-\alpha)\Gamma(\beta+1-\gamma)} e^{i\pi(1-\gamma)}. \tag{9.93}$$

Thus, solution $u_5(z)$ leads to complete solution $\bar{G}_5(z)$ with quasi-polynomial structure, which is singular at $z = 0$ and regular at infinity $z = -\infty$; the corresponding quantization rule is $\omega = 2n' - j$. This type of solutions is hardly of physical interest.

In order to clarify physical sense of arising problem, let us turn back to eq. (9.81), written in the form

$$\varphi\Big(\frac{d}{dr}\varphi\frac{d}{dr} + \omega^2 - \frac{j(j+1)}{r^2}\varphi\Big)\bar{G} = 0\,; \tag{9.94}$$

and transform it to a new variable

$$\begin{gathered}\varphi\frac{d}{dr} = \frac{d}{dr_*} \quad \Longrightarrow \quad dr_* = \frac{dr}{\varphi(r)} = \frac{dr}{1+r^2}\,;\\ r_* = \arctan r, \quad \tan r_* = r\,;\; r \to 0, r_* \to 0\,; r \to +\infty, r_* \to +\frac{\pi}{2}.\end{gathered} \tag{9.95}$$

Equation (9.94) in this variable reads

$$\Big[\frac{d^2}{dr_*^2} + \omega^2 - j(j+1)(1 + \frac{1}{\tan^2 r_*})\Big]\bar{G} = 0,\; r_* \in (0, \frac{\pi}{2}). \tag{9.96}$$

Here we have Schrödinger-type equation

$$\Big[\frac{d^2}{dr_*^2} + \omega^2 - U(r_*)\Big]\bar{G} = 0, \quad U = j(j+1)(1 + \frac{1}{\tan^2 r_*})\,. \tag{9.97}$$

This problem is easily interpretable in quantum mechanics if the following inequality is valid

$$\omega^2 > j(j+1) \iff \Omega^2 > \frac{c^2}{\rho^2}j(j+1). \tag{9.98}$$

In this point we should recall that related to solutions $\bar{G}_1(z)$ spectrum for ω satisfies this requirement

$$\omega^2 - j(j+1) = 4n^2 + (4n+1)(j+1) > 0. \tag{9.99}$$

Let us consider from this point of view the spectrum related to $\bar{G}_5(z)$:

$$\omega = k - j\,, \quad \text{where} \quad k = (2n' + 1) \in \{1, 3, 5, ...\}. \tag{9.100}$$

From (9.100) it follows $\omega^2 - j(j+1) = k^2 - 2kj - j$. Taking in mind the roots

$$k_1 = j - j\sqrt{1 + 1/j}\,,\; -1 < k_1 < 0\,; \qquad k_2 = j + j\sqrt{1 + 1/j}\,,\; k_2 > 2j\,,$$

we conclude

$$\begin{gathered}\omega^2 - j(j+1) < 0, \;\text{ when } \quad (2n'+1) < j + j\sqrt{1+1/j}\,,\\ \omega^2 - j(j+1) > 0\,, \;\text{ when } \quad (2n'+1) > j + j\sqrt{1+1/j}\,;\end{gathered} \tag{9.101}$$

solutions of the type $\bar{G}_5(z)$ are relevant to the situation badly interpretable from physical point of view.

9.7 Maxwell equations in Schwarzschild metric

We may start with eq. (9.50), specifying it to Schwarzschild space-time with $\varphi = 1 - \frac{1}{r}$. The main equation formally is the same

$$\Big(\frac{d^2}{dr^2} + \frac{\varphi'}{\varphi}\frac{d}{dr} + \frac{\omega^2}{\varphi^2} - \frac{j(j+1)}{r^2\varphi}\Big)\bar{G} = 0\,, \tag{9.102}$$

but explicitly it reads

$$\Big(\frac{d^2}{dr^2} + \frac{1}{r(r-1)}\frac{d}{dr} + \frac{\omega^2 r^2}{(r-1)^2} - \frac{j(j+1)}{r^2}\frac{r}{r-1}\Big)\bar{G} = 0. \tag{9.103}$$

Here we have an equation with three singular points, the points $r = 0, 1$ are regular, and the point $r = \infty$ is irregular in rank 2; this is the class of confluent Heun equation. Equation (9.103) becomes more understandable after transforming into other variable:

$$\begin{gathered}\Big(\varphi\frac{d}{dr}\varphi\frac{d}{dr} + \omega^2 - \frac{j(j+1)}{r^2}\varphi\Big)\bar{G} = 0\,, \quad dr_* = \frac{dr}{\varphi} = dr(1 + \frac{1}{r-1})\,,\\ r_* = r + \ln(r-1), \quad r \to \infty, \quad r_* \to +\infty; \quad r \to 1+0, \quad r_* \to -\infty\,,\\ \Big(\frac{d^2}{dr_*^2} + \omega^2 - U(r_*)\Big)\bar{G} = 0\,, \quad U(r_*) = \omega^2 - \frac{j(j+1)}{r^2}\varphi.\end{gathered} \tag{9.104}$$

Let us specify behaviour of the effective potential at two infinities

$$\begin{aligned} U(r_* \to +\infty) &= \frac{j(j+1)}{r^2}\frac{r-1}{r} = +0\,,\\ U(r_* \to -\infty) &= \frac{j(j+1)}{r^2}\frac{r-1}{r} = +0\,; \end{aligned} \tag{9.105}$$

this means that here we have an effective potential of barrier type, tending to zero both at $r \to 1$ ($r_* \to -\infty$) and infinity $r \to \infty$ ($r_* \to +\infty$).

Now we are to construct formal solutions of eq. (9.103):

$$\begin{aligned} &\Big[\frac{d^2}{dr^2} + (\frac{1}{r-1} - \frac{1}{r})\frac{d}{dr}\\ &+\omega^2(1 + \frac{2}{r-1} + \frac{1}{(r-1)^2}) + j(j+1)(\frac{1}{r} - \frac{1}{r-1})\Big]\bar{G} = -0. \end{aligned} \tag{9.106}$$

In (non-physical) singular point $r = 0$ solutions behave as

$$G'' - \frac{1}{r}G' + \frac{j(j+1)}{r}G = 0\,, \quad G \sim r^c, \quad c = 0,\; c = 2. \tag{9.107}$$

Near the point $r = 1$ we have

$$\Big[\frac{d^2}{dr^2} + \frac{1}{r-1}\frac{d}{dr} + \frac{\omega^2}{(r-1)^2}\Big]\bar{G} = -0\,, \quad G \sim (r-1)^a, \quad a = \pm i\omega. \tag{9.108}$$

To find asymptotic at infinity, we transform equation to the variable $x = 1/r$:

$$\Big[\frac{d^2}{dx^2} + (\frac{2}{x} - \frac{1}{1-x})\frac{d}{dx} + \frac{\omega^2}{x^4(1-x)^2} - \frac{j(j+1)}{x^2(1-x)}\Big]G = 0.$$

It becomes simpler near the point $x = 0$:

$$\Big[\frac{d^2}{dx^2} + \frac{2}{x}\frac{d}{dx} + \frac{\omega^2}{x^4} - \frac{j(j+1)}{x^2}\Big] G = 0, \qquad G \sim x^A e^{B/x},$$

further we derive

$$-\frac{AB}{x^3} - \frac{B(A-2)}{x^3} - \frac{2B}{x^3} + \frac{B^2}{x^4} + \frac{\omega^2}{x^4} = 0,$$

whence it follows

$$\begin{gathered} B^2 + \omega^2 \quad \Longrightarrow \quad B = \pm i\omega, \\ AB + B(A-2) + 2B = 0, \quad 2AB = 0, \quad A = 0. \end{gathered} \tag{9.109}$$

Therefore, general solution of eq. (9.103) may be searched in the form $\bar{G} = r^c (r-1)^a e^{br}\, g(r)$. After needed calculation we arrive at

$$g'' + \Big(\frac{2c}{r} + \frac{2a}{r-1} + 2b + \frac{1}{r(r-1)}\Big) g'$$

$$+\Big[\frac{c(c-1)}{r^2} + \frac{a(a-1)}{(r-1)^2} + \frac{2ca}{r-1} - \frac{2ca}{r} + \frac{2cb}{r} + \frac{2ab}{r-1} + b^2$$

$$+\frac{c}{r-1} - \frac{c}{r^2} - \frac{c}{r} + \frac{a}{r} + \frac{a}{(r-1)^2} - \frac{a}{r-1} + \frac{b}{r-1} - \frac{b}{r}$$

$$+\omega^2 + \frac{2\omega^2}{r-1} + \frac{\omega^2}{(r-1)^2} + \frac{j(j+1)}{r} - \frac{j(j+1)}{r-1}\Big] g = 0.$$

Imposing evident restrictions, we get eight variants of parameters a, b, c:

$$a = \pm i\omega, \quad c = 0, 2, \quad b = \pm i\omega.$$

So resulting equation becomes simpler

$$\begin{aligned} & g'' + \Big(\frac{2a+1}{r-1} + \frac{2c-1}{r} + 2b\Big) g' \\ & + \Big[\frac{2ca + 2ab + c - a + b - j(j+1) + 2\omega^2}{r-1} \\ & + \frac{2ca + 2cb - c + a - b + j(j+1)}{r}\Big] g = 0. \end{aligned} \tag{9.110}$$

Because the physical region of radial variable is the interval $r \in (1, +\infty)$, the most interesting would be a series in the variable $x = r - 1$. Transformed to this variable x, equation reads (the prime designates derivative d/dx):

$$g'' + (p + \frac{p_1}{x} + \frac{p_2}{x+1}) g' + (\frac{q_1}{x} + \frac{q_2}{x+1}) g = 0, \quad x \in (0, +\infty). \tag{9.111}$$

Its solutions may be constructed as a power series, $g(x) = \sum_{k=0}^{\infty} c_k x^k$. After performing the needed calculations, we derive recurrent formulas

$$\begin{gathered} n = 0, \quad q_1 c_0 + p_1 c_1 = 0; \\ n = 1, \quad (q_1 + q_2) c_0 + (p + p_1 + p_2 + q_1) c_1 + (2 + 2p_1) c_2 = 0; \\ n = 2, 3, \ldots, \qquad [p(n-1) + q_1 + q_2]\, c_{n-1} \\ +[n(n-1) + (p + p_1 + p_2) n + q_1]\, c_n + [(n+1)n + p_1(n+1)]\, c_{n+1} = 0. \end{gathered}$$

Possible convergence radii are found by Poincaré–Perron method: dividing the last relation by $n^2 c_{n-1}$

$$\frac{1}{n^2}[p(n-1)+q_1+q_2]$$
$$+\frac{1}{n^2}[n(n-1)+(p+p_1+p_2)n+q_1]\,\frac{c_n}{c_{n-1}}+\frac{1}{n^2}[(n+1)n+p_1(n+1)]\,\frac{c_{n+1}}{c_n}\frac{c_n}{c_{n-1}}=0$$

and tending $n \to \infty$, we obtain algebraic equation which determines possible convergence radii:

$$\lim_{n\to\infty}\frac{c_n}{c_{n-1}}=R\,,\quad R+R^2=0\,,\quad R_{conv}=\frac{1}{|R|}=1,\infty\,.$$

Recall that complete solutions have the structure

$$\bar{G}(r)=r^c(r-1)^a e^{br}g(r) \quad \Longrightarrow \quad \bar{G}(x)=(1+x)^c x^a e^{b(1+x)}g(x)\,,$$
$$c=0,2\,;\quad a=-i\omega,+i\omega\,;\quad b=-i\omega,+i\omega\,;\quad x\in(0,+\infty)\,; \tag{9.112}$$

below we list eight types of solutions (they are collected in pairs of conjugate ones)

$$\begin{array}{llll}
c=0, & a=+i\omega, & b=+i\omega, & \bar{G}_1=x^{+i\omega}e^{+i\omega(1+x)}g_1(x)\,,\\
c=0, & a=-i\omega, & b=-i\omega, & \bar{G}_1^*=x^{-i\omega}e^{-i\omega(1+x)}g_1^*(x)\,;\\
c=0, & a=+i\omega, & b=-i\omega, & \bar{G}_2=x^{+i\omega}e^{-i\omega(1+x)}g_2(x)\,,\\
c=0, & a=-i\omega, & b=+i\omega, & \bar{G}_2^*=x^{-i\omega}e^{+i\omega(1+x)}g_2^*(x)\,;
\end{array} \tag{9.113}$$

$$\begin{array}{llll}
c=2, & a=+i\omega, & b=+i\omega, & \bar{G}_3=(1+x)^2x^{+i\omega}e^{+i\omega(1+x)}g_3(x)\,,\\
c=2, & a=+i\omega, & b=-i\omega, & \bar{G}_4=(1+x)^2x^{+i\omega}e^{-i\omega(1+x)}g_4(x)\,;\\
c=2, & a=-i\omega, & b=+i\omega, & \bar{G}_4^*=(1+x)^2x^{-i\omega}e^{+i\omega(1+x)}g_4^*(x)\,,\\
c=2, & a=-i\omega, & b=-i\omega. & \bar{G}_3^*=(1+x)^2x^{-i\omega}e^{-i\omega(1+x)}g_x^3(x).
\end{array} \tag{9.114}$$

9.8 Solutions in spherical Riemann space

Now we consider Maxwell equations in spherical Riemann model:

$$dS^2=dt^2-dr^2-\sin^2 r d\theta^2-\sin^2\theta d\phi^2\,,$$
$$x^\alpha=(t,r,\theta,\phi)\,,\quad g_{\alpha\beta}=\left|\begin{array}{cccc}1&0&0&0\\0&-1&0&0\\0&0&-\sin^2 r&0\\0&0&0&-\sin^2 r\sin^2\theta\end{array}\right|. \tag{9.115}$$

We use the following tetrad

$$e^\alpha_{(0)}=(1,0,0,0)\,,\quad e^\alpha_{(3)}=(0,1,0,0)\,,$$
$$e^\alpha_{(1)}=(0,0,\frac{1}{\sin r},0)\,,\quad e^\alpha_{(2)}=(1,0,0,\frac{1}{\sin r\sin\theta})\,; \tag{9.116}$$

by changing the numeration for coordinates $x^\alpha=(t,r,\theta,\phi)\Longrightarrow x^\alpha=(t,\theta,\phi,r)$ the tetrad

(9.116) becomes diagonal. Ricci rotation coefficients equal

$$\gamma_{ab0}=0, \quad \gamma_{ab1}=\begin{vmatrix} 0 & 0 & 0 & 0 \\ 0 & 0 & 0 & -\frac{1}{\tan r} \\ 0 & 0 & 0 & 0 \\ 0 & +\frac{1}{\tan r} & 0 & 0 \end{vmatrix},$$

$$\gamma_{ab2}=\begin{vmatrix} 0 & 0 & 0 & 0 \\ 0 & 0 & +\frac{1}{\tan\theta\sin r} & 0 \\ 0 & -\frac{1}{\tan\theta\sin r} & 0 & -\frac{1}{\tan r} \\ 0 & 0 & +\frac{1}{\tan r} & 0 \end{vmatrix}, \quad \gamma_{ab3}=0\,.$$

Starting with the general spinor equation

$$\left[\sigma^c e^{\alpha}_{(c)}(x)\partial_\alpha + \sigma^c(\frac{1}{2}\Sigma^{ab}\otimes I + I\otimes\frac{1}{2}\Sigma^{ab})\gamma_{abc}(x)\right]\xi(x)=0\,, \tag{9.117}$$

we arrive at

$$\Big[\partial_t + \Big\{\sigma^3\partial_r + \frac{i}{\tan r}\Big(-\sigma^1\frac{\sigma^2\otimes I + I\otimes\sigma^2}{2} + \sigma^2\frac{\sigma^1\otimes I + I\otimes\sigma^1}{2}\Big)\Big\}$$
$$+\frac{1}{\sin r}\Big\{\sigma^1\partial_\theta - i\sigma^2\frac{i\partial_\varphi + \cos\theta(\sigma^3\otimes I + I\otimes\sigma^3)/2}{\sin\theta}\Big\}\Big]\xi=0.$$

Comparing it with eq. (9.42), we can write down radial equations by formal changes in the system (9.50):

$$\begin{aligned} &-i\omega f + (\frac{d}{dr}+\frac{1}{\tan r})f + \frac{a}{\sin r}h=0\,, \quad i\omega g + (\frac{d}{dr}+\frac{1}{\tan r})g + \frac{a}{\sin r}h=0\,, \\ &-i\omega h + (\frac{d}{dr}+\frac{2}{\tan r})h + \frac{a}{\sin r}g=0\,, \quad i\omega h + (\frac{d}{dr}+\frac{2}{\tan r})h + \frac{a}{\sin r}f=0. \end{aligned} \tag{9.118}$$

Summing and subtracting third and fourth equations, we derive

$$2(\frac{d}{dr}+\frac{2}{\tan r})h + \frac{a}{\sin r}(f+g)=0\,, \quad 2i\omega h + \frac{a}{\sin r}(f-g)=0. \tag{9.119}$$

It is readily checked that the first equation in (9.118) is the result of combining three remaining ones. Therefore, we have only three independent equations

$$\begin{aligned} &h=-\frac{a}{2i\omega\sin r}(f-g)\,, \\ &-i\omega f + (\frac{d}{dr}+\frac{1}{\tan r})f + \frac{a}{\sin r}h=0\,, \quad i\omega g + (\frac{d}{dr}+\frac{1}{\tan r})g + \frac{a}{\sin r}h=0. \end{aligned} \tag{9.120}$$

Excluding the variable $h(r)$ we obtain

$$\begin{aligned} &(\frac{d}{dr}+\frac{1}{\tan r}-i\omega)f + \frac{ia^2}{2\omega\sin^2 r}(f-g)=0\,, \\ &(\frac{d}{dr}+\frac{1}{\tan r}+i\omega)g + \frac{ia^2}{2\omega\sin^2 r}(f-g)=0. \end{aligned} \tag{9.121}$$

Summing and subtracting these two equations, and using new variables, $f+g=F$, $f-g=G$, we arrive at the system

$$(\frac{d}{dr}+\frac{1}{\tan r})F - i\omega G + \frac{ia^2}{\omega\sin^2 r}G=0, \quad (\frac{d}{dr}+\frac{1}{\tan r})G - i\omega F=0. \tag{9.122}$$

System (9.122) may be simplified by separating multipliers, $F = \sin^{-1} r\, \bar{F}$, $G = \sin^{-1} r\, \bar{G}$, in this way we get

$$\frac{d}{dr} i\omega \bar{F} + (\omega^2 - \frac{a^2}{\sin^2 r})\bar{G} = 0\,, \quad \frac{d}{dr}\bar{G} = i\omega \bar{F}\,, \tag{9.123}$$

whence it follows an equation for main function

$$\Big(\frac{d^2}{dr^2} + \omega^2 - \frac{a^2}{\sin^2 r}\Big)\bar{G} = 0. \tag{9.124}$$

In the new variable, $y = \frac{1-\cos r}{2}$, the last equation reads

$$\Big[y(1-y)\frac{d^2}{dy^2} + (\frac{1}{2} - y)\frac{d}{dy} + \omega^2 - \frac{a^2}{4y(1-y)}\Big]\bar{G} = 0. \tag{9.125}$$

Its solution are searched in the form $\bar{G} = y^A\,(1-y)^B\,g(y)$; this results in

$$y(1-y)g'' + [2A + 1/2 - (2A + 2B + 1)y]\,g'$$

$$+\frac{1}{y}\,[A(A-1) + \frac{1}{2}A - \frac{a^2}{4}]\,g + \frac{1}{1-y}\,[B(B-1) + \frac{1}{2}B - \frac{a^2}{4}]\,g$$

$$+[\,\omega^2 - 2AB - A(A-1) - B(B-1) - A - B\,]\,g = 0.$$

Equating coefficients at y^{-1} and $(1-y)^{-1}$ to zero, we get

$$A = (j+1)/2,\ -j/2; \quad B = (j+1)/2,\ -j/2.$$

The above equation for $g(y)$ is simplified and recognised as hypergeometric equation with parameters

$$\gamma = 2A + 1/2, \quad \alpha = A + B - \omega\,, \quad \beta = A + B + \omega. \tag{9.126}$$

Let us fix parameters A and B: $A = (j+1)/2\,,\ B = (j+1)/2$, so obtaining

$$\gamma = j + 3/2\,, \quad \alpha = j + 1 - \omega\,, \quad \beta = j + 1 + \omega. \tag{9.127}$$

We get polynomials imposing evident restriction

$$\alpha = -n, \quad n = 1, 2, 3, \ldots\,, \quad \omega = n + j + 1\,; \tag{9.128}$$

corresponding complete solution has the structure

$$\bar{G}(y) = y^{(j+1)/2}\,(1-y)^{(j+1)/2}\,F(-n,\,n + 2j + 2,\,j + 3/2;\,y)\,, \tag{9.129}$$

it equals to zero at the points $y \to 0,\ y \to 1\ \ (r \to 0,\ r \to \pi)$.

9.9 Solutions in Lobachevsky space

The main radial equation reads

$$\Big(\frac{d^2}{dr^2} + \omega^2 - \frac{a^2}{\sinh^2 r}\Big)\bar{G} = 0\,, \quad r \in (0, \infty). \tag{9.130}$$

In the new variable $y = \frac{1-\cosh r}{2}$, it takes the form

$$\Big[y(y-1)\frac{d^2}{dy^2} + (y-1/2)\frac{d}{dy} + \omega^2 - \frac{a^2}{4y(y-1)}\Big]\bar{G} = 0. \tag{9.131}$$

Formally, this equation differs from that used in the previous section only in the sign at ω^2. Substitution for $\bar{G}(y)$ is the same

$$\bar{G} = y^A\,(1-y)^B\,g(y), \quad A = \frac{j+1}{2}, -\frac{j}{2}; \quad B = \frac{j+1}{2}, -\frac{j}{2}; \tag{9.132}$$

for $g(y)$ we get an equation of hypergeometric type

$$y(1-y)g'' + \Big[\,2A + \frac{1}{2} - (2A+2B+1)y\Big]\,g' - \big[(A+B)^2 + \omega^2\big]\,g = 0$$

with parameters

$$\gamma = 2A + 1/2\,, \quad \alpha = A + B - i\omega\,, \quad \beta = A + B + i\omega. \tag{9.133}$$

Let us fix parameters as follows (negative B ensures the term $(1-y)^B$ tending to zero when $y \to \infty$):

$$A = \frac{j+1}{2}\,, \; B = -\frac{j}{2}\,; \quad \gamma = j + 3/2\,, \; \alpha = 1/2 - i\omega\,, \; \beta = 1/2 + i\omega\,, \tag{9.134}$$

thus we have constructed the needed solution

$$\bar{G}_1(y) = y^{(j+1)/2}\,(1-y)^{-j/2}\,u_1(y), \quad u_1(y) = F(\alpha,\beta,\gamma;y), \tag{9.135}$$

it tends to zero at the point $y = 0$ ($r = 0$). The singular point $y = 1$ does not belong to a physical region. To find behaviour of this solution in infinity, we should apply the Kummer formula

$$u_1(y) = \frac{\Gamma(\gamma)\Gamma(\beta-\alpha)}{\Gamma(\gamma-\alpha)\Gamma(\beta)}\,u_3(y) + \frac{\Gamma(\gamma)\Gamma(\alpha-\beta)}{\Gamma(\gamma-\beta)\Gamma(\alpha)}\,u_4(y)\,, \tag{9.136}$$

where

$$u_3(y) = (-y)^{-\alpha}F(\alpha, \alpha+1-\gamma, \alpha+1-\beta; \frac{1}{y})$$

$$= (-y)^{(-1/2+i\omega)}F(1/2-i\omega, -i\omega-j, 1-2i\omega; \frac{1}{y})\,,$$

$$u_4(y) = (-y)^{-\beta}F(\beta+1-\gamma, \beta, \beta+1-\alpha; \frac{1}{y})$$

$$= (-y)^{(-1/2-i\omega)}F(i\omega-j, (1/2+i\omega), 1+2i\omega; \frac{1}{y}).$$

As $y \to +\infty$, the last formula gives

$$u_1(y) = \frac{\Gamma(\gamma)\Gamma(\beta-\alpha)}{\Gamma(\gamma-\alpha)\Gamma(\beta)}\,(-y)^{(-1/2+i\omega)} + \frac{\Gamma(\gamma)\Gamma(\alpha-\beta)}{\Gamma(\gamma-\beta)\Gamma(\alpha)}\,(-y)^{(-1/2-i\omega)}.$$

Therefore, the complete solution behaves as follows

$$\bar{G}_1\;(y \to \infty)$$

$$= (-1)^{-(j+1)/2} \left\{ \frac{\Gamma(\gamma)\Gamma(\beta-\alpha)}{\Gamma(\gamma-\alpha)\Gamma(\beta)} (-y)^{i\omega} + \frac{\Gamma(\gamma)\Gamma(\alpha-\beta)}{\Gamma(\gamma-\beta)\Gamma(\alpha)} (-y)^{-i\omega} \right\}. \tag{9.137}$$

Taking in mind identities

$$\beta - \alpha = 2i\omega\,, \ \alpha - \beta = -2i\omega\,, \ \gamma - \alpha = j + i\omega + 1\,, \ \gamma - \beta = j - i\omega + 1,$$

we conclude that $\bar{G}_1$ $(y \to \infty)$ is real up to simple phase factor. In initial variable r, asymptotic (9.137) is determined by the formulas

$$\begin{aligned} (-y)^{i\omega} &\approx (\frac{1}{4})^{i\omega} e^{i\omega r}\,, \\ \bar{G}_1\,(r \to \infty) &= M e^{i\omega r} + M^* e^{-i\omega r}, \\ M &= \frac{\Gamma(\gamma)\Gamma(\beta-\alpha)}{\Gamma(\gamma-\alpha)\Gamma(\beta)} (\frac{1}{4})^{i\omega}. \end{aligned} \tag{9.138}$$

It is readily proved that when using Kummer solutions u_3 and u_4, their corresponding complete solutions $\bar{G}_3$ and $\bar{G}_4$ are conjugate to each other and have the asymptotics

$$u_3 \sim \text{const } e^{i\omega r}\,, \qquad u_4 \sim \text{const } e^{-i\omega r}.$$

9.10 Cylindric solutions in spherical space

Let us consider spinor Maxwell equations in cylindric coordinates of the spherical Riemann model, it is specified by the formulas

$$\begin{aligned} dS^2 = dt^2 - dr^2 - \sin^2 r\, d\phi^2 - \cos^2 r\, dz^2\,, \ x^\alpha = (t, r, \phi, z)\,, \\ e^\beta_{(a)}(x) = \left| \begin{array}{cccc} 1 & 0 & 0 & 0 \\ 0 & 1 & 0 & 0 \\ 0 & 0 & \sin^{-1} r & 0 \\ 0 & 0 & 0 & \cos^{-1} r \end{array} \right|, \end{aligned} \tag{9.139}$$

these coordinates belong to: $r \in [0,\, +\pi/2]$, $\phi \in [-\pi,\, +\pi]$, $z \in [-\pi,\, +\pi]$. Ricci rotation coefficients are (we write down only non-vanishing ones)

$$\gamma_{ab0} = 0\,, \quad \gamma_{ab1} = 0\,, \quad \gamma_{122} = \frac{\cos r}{\sin r}\,, \quad \gamma_{313} = \frac{\sin r}{\cos r}\,. \tag{9.140}$$

Starting with general spinor form of Maxwell equations

$$\begin{aligned} \left[\sigma^c e^\alpha_{(c)}(x)\partial_\alpha + \sigma^c(\frac{1}{2}\Sigma^{ab} \otimes I + I \otimes \frac{1}{2}\Sigma^{ab})\gamma_{abc}(x) \right] \xi(x) = 0\,, \\ \Sigma^{0j} = \frac{1}{2}\sigma^j, \quad \Sigma^{12} = -\frac{i}{2}\sigma^3, \quad \Sigma^{23} = -\frac{i}{2}\sigma^1, \quad \Sigma^{31} = -\frac{i}{2}\sigma^2, \end{aligned} \tag{9.141}$$

we obtain

$$\begin{aligned} [\, \partial_t + \sigma^1 \partial_r + \frac{\sigma^2}{\sin r}\partial_\phi + \frac{\sigma^3}{\cos r}\partial_z \\ +\sigma^2(\Sigma^{12} \otimes I + I \otimes \Sigma^{12})\gamma_{122} + \sigma^3(\Sigma^{31} \otimes I + I \otimes \Sigma^{31})\gamma_{313}]\xi(x) = 0\,, \end{aligned}$$

that is

$$\Big[\partial_t + \sigma^1 \partial_r - \frac{i\sigma^2}{2}(\sigma^3 \otimes I + I \otimes \sigma^3)\frac{\cos r}{\sin r}$$
$$- \frac{i\sigma^3}{2}(\sigma^2 \otimes I + I \otimes \sigma^2)\frac{\sin r}{\cos r} + \frac{\sigma^2}{\sin r}\partial_\phi + \frac{\sigma^3}{\cos r}\partial_z\Big]\,\xi = 0. \tag{9.142}$$

The structure of this equation assumes the following substitution for electromagnetic spinor

$$\xi(t,r,\phi,z) = e^{-i\omega t}\, e^{im\phi}\, e^{ikz} \begin{vmatrix} f(r) & h(r) \\ h(r) & g(r) \end{vmatrix}, \tag{9.143}$$

so we derive

$$\Big\{ -i\omega I + \sigma^1 \frac{d}{dr} - \frac{i\cos r}{2\sin r}\sigma^2(\sigma^3 \otimes I + I \otimes \sigma^3) - \frac{i\sin r}{2\cos r}\sigma^3(\sigma^2 \otimes I + I \otimes \sigma^2)$$
$$+ \frac{im}{\sin r}\sigma^2 + \frac{ik}{\cos r}\sigma^3 \Big\} \begin{vmatrix} f(r) & h(r) \\ h(r) & g(r) \end{vmatrix} = 0\,, \tag{9.144}$$

and further we find the system of four equations:

$$\begin{aligned}
&(\frac{d}{dr} + \frac{m}{\sin r} - \frac{\sin r}{\cos r})h + (-i\omega + \frac{ik}{\cos r})f = 0\,,\\
&(\frac{d}{dr} - \frac{m}{\sin r} - \frac{\sin r}{\cos r})h + (-i\omega - \frac{ik}{\cos r})g = 0\,,\\
&(\frac{d}{dr} + \frac{m}{\sin r} + \frac{\cos r}{\sin r} - \frac{1}{2}\frac{\sin r}{\cos r})g + \frac{1}{2}\frac{\sin r}{\cos r}f + (-i\omega + \frac{ik}{\cos r})h = 0,\\
&(\frac{d}{dr} - \frac{m}{\sin r} + \frac{\cos r}{\sin r} - \frac{1}{2}\frac{\sin r}{\cos r})f + \frac{1}{2}\frac{\sin r}{\cos r}g + (-i\omega - \frac{ik}{\cos r})h = 0.
\end{aligned} \tag{9.145}$$

Summing and subtracting equations in each pair, we obtain (let it be $F = f + g\,, G = f - g$)

$$\begin{aligned}
&\frac{ik}{\cos r}F - i\omega G + \frac{2m}{\sin r}h = 0\,,\\
&\frac{ik}{\cos r}G - i\omega F + 2(\frac{d}{dr} - \frac{\sin r}{\cos r})h = 0\,,\\
&-\frac{2ik}{\cos r}h - \frac{m}{\sin r}F + (\frac{d}{dr} + \frac{\cos r}{\sin r} - \frac{\sin r}{\cos r})G = 0\,,\\
&-2i\omega\, h - \frac{m}{\sin r}G + (\frac{d}{dr} + \frac{\cos r}{\sin r})F = 0.
\end{aligned} \tag{9.146}$$

Let us express from first, second, and fourth equations the variables ωG, ωF, and $2i\omega h$ and substitute them into the third equation; this results in the identity $0 \equiv 0$. Therefore, only three equations in eq. (9.146) are independent:

$$\begin{aligned}
&\frac{ik}{\cos r}F - i\omega G + \frac{2m}{\sin r}h = 0\,,\\
&\frac{ik}{\cos r}G - i\omega F + 2(\frac{d}{dr} - \frac{\sin r}{\cos r})h = 0\,,\\
&-2i\omega\, h - \frac{m}{\sin r}G + (\frac{d}{dr} + \frac{\cos r}{\sin r})F = 0.
\end{aligned} \tag{9.147}$$

Taking into account identities

$$F = \frac{1}{\sin r}\bar{F}\,,\ (\frac{d}{dr} + \frac{\cos r}{\sin r})F = \frac{1}{\sin r}\frac{d\bar{F}}{dr}\,; \quad h = \frac{1}{\cos r}\bar{h}\,,\ (\frac{d}{dr} - \frac{\sin r}{\cos r})h = \frac{1}{\cos r}\frac{d\bar{h}}{dr}\,,$$

we may simplify eq. (9.147):

$$\begin{aligned} &\frac{ik}{\cos r}\,\frac{1}{\sin r}\bar{F} - i\omega G + \frac{2m}{\sin r}\,\frac{1}{\cos r}\bar{h} = 0, \\ &\frac{ik}{\cos r}G - i\omega\frac{1}{\sin r}\bar{F} + \frac{2}{\cos r}\,\frac{d\bar{h}}{dr} = 0, \\ &-2i\omega\,\frac{1}{\cos r}\bar{h} - \frac{m}{\sin r}G + \frac{1}{\sin r}\,\frac{d\bar{F}}{dr} = 0. \end{aligned} \tag{9.148}$$

Let it be $2i\bar{h} = \bar{H}$. The last system is presented as follows

$$\omega G = \frac{k\bar{F} - m\bar{H}}{\cos r \sin r}\,,$$

$$\frac{k}{\cos r}\,\omega G - \frac{\omega^2}{\sin r}\bar{F} - \frac{\omega}{\cos r}\,\frac{d\bar{H}}{dr} = 0\,, \qquad -\omega^2\,\frac{1}{\cos r}\,\bar{H} - \frac{m}{\sin r}\,\omega\, G + \frac{\omega}{\sin r}\,\frac{d\bar{F}}{dr} = 0.$$

Excluding the function G, we derive

$$\begin{aligned} &\frac{1}{\cos r}(\omega\frac{d}{dr} + \frac{km}{\cos r\sin r})\bar{H} + \frac{1}{\sin r}(\omega^2 - \frac{k^2}{\cos^2 r})\bar{F} = 0\,, \\ &\frac{1}{\sin r}(\omega\frac{d}{dr} - \frac{km}{\sin r\cos r})\bar{F} + \frac{1}{\cos r}(-\omega^2 + \frac{m^2}{\sin^2 r})\bar{H} = 0. \end{aligned} \tag{9.149}$$

Let us transform the system to new variable $\sin r = \sqrt{z}$, $z \in [0, 1]$, then we arrive at

$$\begin{aligned} &\Big[2\omega\frac{d}{dz} + \frac{km}{z(1-z)}\Big]\bar{H} + \frac{\omega^2 - k^2 - \omega^2 z}{z(1-z)}\;\bar{F} = 0\,, \\ &\Big[2\omega\frac{d}{dz} - \frac{km}{z(1-z)}\Big]\bar{F} + \frac{m^2 - \omega^2 z}{z(1-z)}\;\bar{H} = 0. \end{aligned} \tag{9.150}$$

Note that from eq. (9.150) straightforwardly follow two differential equations with four singular points:

$$\bar{H}, \quad z = 0, 1, \infty, (1 - \frac{k^2}{\omega^2})\,; \qquad \bar{F}, \quad z = 0, 1, \infty, \frac{m^2}{\omega^2}\,.$$

There exists possibility to reduce the problem to equations with three singular points. Indeed, let us define new variables, $\bar{H} = V + W$, $\bar{F} = V - W$, then the system (9.150) reads

$$\Big[2\omega\frac{d}{dz} + \frac{km}{z(1-z)}\Big](V + W) + \frac{\omega^2 - k^2 - \omega^2 z}{z(1-z)}\;(V - W) = 0\,,$$

$$\Big[2\omega\frac{d}{dz} - \frac{km}{z(1-z)}\Big](V - W) + \frac{m^2 - \omega^2 z}{z(1-z)}\;(V + W) = 0.$$

Summing and subtracting these equations we get

$$\begin{aligned} &\Big[4\omega\frac{d}{dz} + \frac{\omega^2 - k^2 + m^2 - 2\omega^2 z}{z(1-z)}\Big]V - \frac{\omega^2 - (k+m)^2}{z(1-z)}\;W = 0\,, \\ &\Big[4\omega\frac{d}{dz}W - \frac{\omega^2 - k^2 + m^2 - 2\omega^2 z}{z(1-z)}\Big]W + \frac{\omega^2 - (k-m)^2}{z(1-z)}\;V = 0. \end{aligned} \tag{9.151}$$

We readily derive a 2nd-order equation for $W(z)$:

$$z(z-1)W'' + (2z-1)W' + \Big[-\frac{\omega(\omega+2)}{4} - \frac{k^2}{4(z-1)} + \frac{m^2}{4z}\Big]W = 0. \tag{9.152}$$

Near the points $z = 0, 1$, solutions behave as

$$z \to 0, W = z^A, A = \pm\frac{|m|}{2}; \quad z \to 1, W = (z-1)^B, \ B = \pm\frac{|k|}{2}. \tag{9.153}$$

In all region of z, solutions are searched in the form $W(z) = z^A(z-1)^B\bar{W}(z)$. After needed calculation we arrive at

$$(z-1)z\bar{W}'' + [2A(z-1) + 2Bz + (2z-1)]\,\bar{W}'$$
$$+\Big((A+B)(A+B+1) - \frac{\omega(\omega+2)}{4} - \frac{k^2}{4(z-1)} + \frac{B^2}{z-1} + \frac{m^2}{4z} - \frac{A^2}{z}\Big)\bar{W} = 0.$$

Imposing yet known restrictions (9.153), we obtain

$$\begin{aligned} &z(1-z)\bar{W}'' + [2A + 1 - (2A + 2B + 2)z)]\bar{W}' \\ &-[(A+B)(A+B+1) - \frac{1}{4}\omega(\omega+2)]\,\bar{W} = 0, \end{aligned} \tag{9.154}$$

which is identified with the equation of hypergeometric type

$$z(1-z)\,\frac{d^2F}{dz^2} + [\gamma - (\alpha+\beta+1)\gamma]\,\frac{dF}{dz} - \alpha\beta\,F = 0\,.$$

Let us fix parameters A and B so that solutions be finite at the points $z = 0,\ 1$:

$$\begin{aligned} A = +\frac{|m|}{2}\,, \quad B = +\frac{|k|}{2}, \quad \gamma = |m| + 1\,, \\ \alpha = \frac{|k| + |m| - \omega}{2}\,, \quad \beta = \frac{|k| + m| + \omega}{2} + 1\,, \end{aligned} \tag{9.155}$$

and accept the standard requirement for polynomials:

$$\begin{aligned} \alpha = -n, \quad \omega = 2n + |k| + |m|\,, \quad \beta = n + 1 + |m| + |k|, \\ n = 0, 1, 2, ..., \quad W(z) = z^{|m|/2}(z-1)^{|k|/2}F(\alpha, \beta, \gamma; z). \end{aligned} \tag{9.156}$$

Now, let us turn to equation for the second function $V(z)$. There exists symmetry between two equations (9.151): the system is invariant under the formal changes

$$V \Longleftrightarrow W, \quad \omega \Longleftrightarrow -\omega, \quad m \Longleftrightarrow -m. \tag{9.157}$$

Therefore, from the 2nd-order equation (9.152) for $W(z)$, without any calculation we obtain a respective equation for $V(z)$:

$$z(z-1)V'' + (2z-1)W' + [-\frac{\omega(\omega-2)}{4} - \frac{k^2}{4(z-1)} + \frac{m^2}{4z}]V = 0. \tag{9.158}$$

We are to apply the same substitution $V(z) = z^A(z-1)^B\bar{V}(z)$. After the needed calculation we get an equation for $\bar{V}(z)$:

$$(z-1)z\bar{V}'' + [2A(z-1) + 2Bz + (2z-1)]\,\bar{V}'$$
$$+[\,(A+B)^2 + A + B - \frac{1}{4}\omega(\omega-2) - \frac{k^2}{4(z-1)} + \frac{B^2}{z-1} + \frac{m^2}{4z} - \frac{A^2}{z}\,]\,\bar{V} = 0.$$

Imposing evident restrictions on A and B, we arrive at an equation of hypergeometric type

$$z(1-z)\bar{V}'' + [2A+1-(2A+2B+2)z)]\,\bar{V}' \\ -[\,(A+B)(A+B+1) - \frac{1}{4}\omega(\omega-2)\,]\,\bar{V} = 0 \tag{9.159}$$

with parameters

$$A = +\frac{|m|}{2}\,, \quad B = +\frac{|k|}{2}, \quad \gamma' = |m|+1\,, \\ \alpha' = \frac{|k|+|m|+\omega}{2}\,, \quad \beta' = \frac{|k|+m|-\omega}{2}+1. \tag{9.160}$$

Further, applying polynomial condition, we find needed solutions

$$\beta' = -n', \quad \omega = 2(n'+1)+|k|+|m|\,, \quad \alpha' = n'+1+|m|+|k|, \\ n' = 0,1,2,..., \quad V(z) = z^{|m|/2}(z-1)^{|k|/2}F(\alpha',\beta',\gamma';z). \tag{9.161}$$

A relative coefficient between two functions, $\bar{W}(z)$ and $\bar{V}(z)$, may be found with the use of 1st-order relations, related these function.

In a similar way, we could study the spinor Maxwell equations in hyperbolic Lobachevsky space, being parameterised by cylindric coordinates according to the formulas:

$$dS^2 = dt^2 - dr^2 - \sinh^2 r\, d\phi^2 - \cosh^2 r\, dz^2\,, \quad x^\alpha = (t,r,\phi,z)\,, \\ e^{\beta}_{(a)}(x) = \left|\begin{array}{llll} 1 & 0 & 0 & 0 \\ 0 & 1 & 0 & 0 \\ 0 & 0 & \sinh^{-1} r & 0 \\ 0 & 0 & 0 & \cosh^{-1} r \end{array}\right|, \tag{9.162}$$

where $r \in [0,\ +\infty)$, $\phi \in [-\pi,\ +\pi]$, and $z \in (-\infty,\ +\infty)$. The treatment would be similar and it does not require new ideas.

9.11 Conclusions

The vacuum Maxwell equations being considered on the background of any pseudo-Riemannian space-time may be interpreted as Maxwell equations in Minkowski space but specified in some effective medium, in which constitutive relations are determined by the metric of the curved space-time. In that context, we will consider space-time models with an event horizon. All of them have a metric of one the same structure, we restrict ourselves to the spherically symmetric cases, and consider de Sitter, anti de Sitter, and Schwarzschild models. Also we have studied hyperbolic Lobachevsky and spherical Riemann models, parameterised coordinates with spherical and cylindric symmetry. We will prove that in all examined cases, effective tensors and of electric permittivity (ϵ_{ij}) and magnetic permeability (ϵ_{ij}) obey one with the same condition: $\epsilon_{ij}(x)\,\mu_{jk}(x) = \delta_{ik}$. Simplicity of expressions for these tensors $\epsilon_{ij}(x)$ and $\mu_{jk}(x)$ is misleading; for each curved space-time model, we are to solve Maxwell equations separately and anew. We have constructed these solutions explicitly, applying Maxwell's equations in spinor form.

Bibliography

[1] W. Gordon. Zur Lichtfortpanzungnach der Relativitätstheorie *Annalen der Physik. (Leipzig)*, 72: 421–456, 1923.

[2] I.E. Tamm. Electrodinamics of anisotropic media in Special relativity. *G R F Kh O.*, 56(2-3): 248–262, 1924.

[3] I.E. Tamm. Crystal optics in Special relativity and geometry of quadratic forms. *G R F Kh O.*, 54(3-4): 1, 1925.

[4] L.I. Mandelstam, I.E. Tamm. Elektrodynamik der anisotropen Medien und der speziallen Relativitatstheorie. *Mathematische Annalen*, 95: 154–160, 1925.

[5] L.D. Landau, E.M. Lifsjitz. *Theorenical Physics. Vol. 2. Field Theory.* Sience, Moscow, 1973.

[6] A. Lichnerowicz. *Théories relativistes de la gravitation et de l'électromagnétisme.* Masson et Cie., Paris, 1955.

[7] V. Novacu. *Introducere in Electrodinamica.* Ed. Acad. Rep. Romine Popul, Romine, 1955.

[8] N.L. Balazs. Effect of a gravitational field, due to a rotating body, on the plane of polarization of an electromagnetic wave. *Journals of Physical Review*, 110: 236–239, 1958.

[9] Pham Man Quan. Sur les équations de l'électromagnétisme dans la materie. *Comptes rendus de l'Académie des Sciences*, 242: 465–467, 1956.

[10] Pham Man Quan. Projections des géodésiques de longuer nulle et rayons électromagnétiques dans un milieu en movement permanent. *Comptes rendus de l'Académie des Sciences*, 242: 857, 1956.

[11] G.V. Skrotskii. The influence of gravitation on the propagation of light. *Soviet Physics. Doklady*, 2: 226–229, 1957.

[12] L.M. Tomil'chik and F.I. Fedorov. Magnetic anisotropy as metrical property of space. *Crystalography*, 4(4): 498–504, 1959.

[13] J. Plebanski. Electromagnetic waves in gravitational fields. *Journal of Physical Review*, 118: 1396–1408, 1960.

[14] E. Post. *Formal structure of Electrodynamics. General Covariance and Electromagnetics.* North-Holland, Amsterdam, 1962.

[15] M.A. Tonnelat. Sur la thórie du photon dans un espace de Riemann. *Annals of Physics, New York*, 15: 144, 1941.

[16] J.R. Ellis. *Maxwell's Equations and Theories of Maxwell Form*: Ph.D. thesis. University of London, 1964.

[17] T.H. O'Del. *The Electrodynamics of Magneto-electric Media.* North-Holland, Amsterdam, 1970.

[18] F. De Felice. On the gravitational field acting as an optical medium. *General Relativity and Gravitation*, 2: 347-357, 1971.

[19] B.M. Bolotovskiy and S.N. Stolarov. Contempostate of electrodynamica of moving bodies. *Uspexi Fizicheskix nauk*, 114(12): 569–608, 1974.

[20] B.M. Bolotovskiy and S.N. Stolarov. *Contempostate of Electrodynamica of Moving Bodies (Unlimited Media).* Eistein collection. 1974. Eds. V.L. Ginzburg. G.I. Naan, U.B. Frankfurt. Moscow: Sience, 1976. P. 179–275.

[21] G. Venuri. A geometrical formulation of electrodynamics. *Nuovo Cimento A.*, 65(1): 64–76, 1981.

[22] W. Schleich, M.O. Scully. General relativity and modern optics. In *New Trends in Atomic Physics, Les Houches, Session XXXVIII, 1982. Eds.: G. Grynberg, R. Stora.*, North-Holland, Amsterdam, 1984.

[23] A.B. Berezin, E.A. Tolkachev and F.I. Fedorov. Dual-invariant constitutive equations for gyrotropic media in rest. *Doklady AN BSSR*, 29(7): 595–597, 1985.

[24] A.B. Berezin et al. Quaternion constitutive equations for moving gyrotropic media. *Journal of Applied Spectroscopy*, 47(1): 113–118, 1987.

[25] V.N. Barykin, E.A. Tolkachev and L.M. Tomil'chik. On symetry aspects for choosing constitutive relations in microscopic electrodynamics of moving media. *Vesti AN BSSR. Ser. Phys.-mat*, 18(2): 96–98, 1982.

[26] S. Antoci. The electromagnetic properties of material media and Einstein's unified field theory. *Progress of Theoretical Physics*, 87: 1343–1357, 1992.

[27] S. Antoci and L. Mihich. A forgotten argument by Gordon uniquely selects Abraham's tensor as the energy-momentum tensor for the electromagnetic field in homogeneous, isotropic matter. *Nuovo Cimento B.*, 112: 991–1001, 1997.

[28] P. Hillion. Spinor electromagnetism in isotropic chiral media. *Advances in Applied Clifford Algebras*, 3: 107–120, 1993.

[29] P. Hillion. Constitutive relations and Clifford algebra in electromagnetism. *Advances in Applied Clifford Algebras*, 5: 141–158, 1995.

[30] W.S. Weiglhofer. On a medium constraint arising directly from Maxwell's equations. *Journal of Physics A.*, 27: 871–874, 1994.

[31] A. Lakhtakia and W.S. Weiglhofer. Lorentz covariance, Occam's razor, and a constraint on linear constitutive relations. *Physics Letters A.*, 213: 107–111, 1996.

[32] A. Lakhtakia, T.G. Mackay and S. Setiawan. Global and local perspectives of gravitationally assisted negative-phase-velocity propagation of electromagnetic waves in vacuum. *Physics Letters A.*, 336: 89–96, 2005.

[33] U. Leonhardt and P. Piwnicki. Optics of nonuniformly moving media. *Physics Letters A.*, 60: 4301–4312, 1999.

[34] U. Leonhardt. Space-time geometry of quantum dielectrics. *Physics Letters A.*, 62: 012111, 2000.

[35] Y.N. Obukhov and F.W. Hehl. Spacetime metric from linear electrodynamics. *Physics Letters B.*, 458: 466–470, 1999.

[36] V.A. De Lorenci, R. Klippert and Y.N. Obukhov. On optical black holes in moving dielectrics. *Physics Letters D.*, 68: Paper 061502, 2003.

[37] F.W. Hehl and Yu.N. Obukhov. Linear media in classical electrodynamics and the Post constraint. *Physics Letters A.*, 334: 249–259, 2005.

[38] M. Novello and J.M. Salim. Effective electromagnetic geometry. *Physics Letters D.*, 63: Paper 083511, 2001.

[39] M. Novello and S. Perez Bergliaffa. Effective Geometry. *AIP Conference Proceedings*, 668: 288–300, 2003.

[40] M. Novello, S. Perez Bergliaffa and J. Salim. Analog black holes in flowing dielectrics. *Classical and Quantum Gravity*, 20: 859–872, 2003.

[41] O.L. De Lange and R.E. Raab. Post's constrain for electromagnetic constitutive relations. *Journal of Optics A.*, 3: 23–26, 2001.

[42] R.E. Raab and O.L. de Lange. Symmetry constrains for electromagnetic constitutive relations. *Journal of Optics A.*, 3: 446–451, 2001.

[43] K.K. Nandi et al. Analog of the Fizeau effect in an effective optical medium. *Journals of Physical Review, D.*, 67: Paper 025002, 2003.

[44] P. Boonserm et al. Effective refractive index tensor for weak field gravity. *Classical and Quantum Gravity*, 22: 1905–1915, 2005.

[45] C. Barcelro, S. Liberati and M. Visser. Analogue Gravity. *Living Reviews in Relativity*, 8: 12, 2005.

[46] V.M. Red'kov. *Partiacle fields ib Riemannian space and the Lorent Grpup.* Belarusian Science, Minsk, 2009.

[47] V.M Red'kov. *Tetrad formalism, spherical symmetry and Schrödinger basis.* Belarusian Science, Minsk, 2011.

[48] H. Bateman and A. Erdélyi. *Higher Transcendental Functions.* Vol. 1, Mc Graw-Hilll Book Company, New York, 1953.

10

P-asymmetric equation for a spin 1/2 particle in external fields

Within the theory of relativistic wave equations with extended sets of Lorentz group representations, a new P-noninvariant 20-component wave equation for spin 1/2 particle is proposed. The presence of an external electromagnetic field and Riemannian space-time background are taken into account. Due to the internal structure of the particle, additional interaction terms appears, and it relates to anomalous magnetic moment of the particle. Exact solutions of the equation in the presence of the external Coulomb fields have been constructed, and radial wave functions are expressed in terms of confluent Heun functions.

10.1 Gel'fand–Yaglom basis

The goal of the paper is to construct a new P-nonivariant wave equation for a massive spin 1/2 particle. We apply the general theory of relativistic wave equations with extended sets of representations of the Lorentz group. In general, the existence of more general wave equations than commonly used ones is well known within the so-called Gel'fand–Yaglom formalism – see references [1–52] and also books [53–57].

We start with the following set of irreducible representations (it contains 20 components)

$$T = (0,1/2) \oplus (1/2,0) \oplus (0,1/2)' \oplus (1/2,0)' \oplus (1,1/2) \oplus (1/2,1)\,, \tag{10.1}$$

where the "prime" serves to distinguish repeated representations of the Lorentz group. The matrix Γ_4 of the corresponding wave equation has the following structure (in the Gel'fand–Yaglom basis)

$$\Gamma_4 = (C^{(1/2)} \otimes I_2) \oplus (C^{(3/2)} \otimes I_4)\,, \tag{10.2}$$

where $C^{(1/2)}$ and $C^{(3/2)}$ represent spin-blocks related to spins 1/2 and 3/2. With the use of the numeration of irreducible components in eq. (10.1)

$$(0,1/2) \sim 1, (0,1/2)' \sim 2, (1,1/2) \sim 3, (1/2,0) \sim 4, (1/2,0)' \sim 5, (1/2,1) \sim 6,$$

possible structure of the blocks $C^{(1/2)}$ and $C^{(3/2)}$ is given by relations

$$C^{(\frac{1}{2})} = \left| \begin{array}{cccccc} 0 & 0 & 0 & c_{14}^{(\frac{1}{2})} & c_{15}^{(\frac{1}{2})} & c_{16}^{(\frac{1}{2})} \\ 0 & 0 & 0 & c_{24}^{(\frac{1}{2})} & c_{25}^{(\frac{1}{2})} & c_{26}^{(\frac{1}{2})} \\ 0 & 0 & 0 & c_{34}^{(\frac{1}{2})} & c_{35}^{(\frac{1}{2})} & c_{36}^{(\frac{1}{2})} \\ c_{41}^{(\frac{1}{2})} & c_{42}^{(\frac{1}{2})} & c_{43}^{(\frac{1}{2})} & 0 & 0 & 0 \\ c_{51}^{(\frac{1}{2})} & c_{52}^{(\frac{1}{2})} & c_{53}^{(\frac{1}{2})} & 0 & 0 & 0 \\ c_{61}^{(\frac{1}{2})} & c_{62}^{(\frac{1}{2})} & c_{63}^{(\frac{1}{2})} & 0 & 0 & 0 \end{array} \right|, \quad C^{(\frac{3}{2})} = \left| \begin{array}{cc} 0 & c_{36}^{(\frac{3}{2})} \\ c_{63}^{(\frac{3}{2})} & 0 \end{array} \right|. \tag{10.3}$$

DOI: 10.1201/9781003472377-10

From invariance of the wave equation under proper Lorentz group, follow the constraints

$$c_{36}^{(\frac{3}{2})} = 2c_{36}^{(\frac{1}{2})}, \quad c_{63}^{(\frac{3}{2})} = 2c_{63}^{(\frac{1}{2})}. \tag{10.4}$$

Without loss of generality, the links between repeated components may be broken, which yields

$$c_{15}^{(\frac{1}{2})} = c_{51}^{(\frac{1}{2})} = c_{24}^{(\frac{1}{2})} = c_{42}^{(\frac{1}{2})} = 0. \tag{10.5}$$

Because, we wish to construct the model of a particle with single spin 1/2, eigenvalues of the block $C^{(3/2)}$ must be equal to zero. Therefore, we set

$$c_{36}^{(\frac{3}{2})} = c_{63}^{(\frac{3}{2})} = 0, \tag{10.6}$$

whence due to eq. (10.4) it follows

$$c_{36}^{(\frac{1}{2})} = c_{63}^{(\frac{1}{2})} = 0. \tag{10.7}$$

Relations (10.5)–(10.7) assume that the linking scheme for model under consideration has the form

$$\begin{array}{ccc} 1 & - & 4 \\ | & & | \\ 6 & & 3 \\ | & & | \\ 2 & - & 5. \end{array} \tag{10.8}$$

10.2 Modified Gel'fand–Yaglom basis

Let us find the form of the matrix Γ_4 for the equation

$$(\Gamma_\mu \partial_\mu + M)\Psi = 0 \tag{10.9}$$

with the use of so-called modified Gel'fand–Yaglom basis, in which the components of the complete wave function Ψ are listed as follows

$$\begin{array}{c} \epsilon^{(1)}_{\frac{1}{2},\frac{1}{2}}, \quad \epsilon^{(1)}_{\frac{1}{2},-\frac{1}{2}}, \quad \epsilon^{(2)}_{\frac{1}{2},\frac{1}{2}}, \quad \epsilon^{(2)}_{\frac{1}{2},-\frac{1}{2}}, \quad \epsilon^{(3)}_{\frac{1}{2},\frac{1}{2}}, \quad \epsilon^{(3)}_{\frac{1}{2},-\frac{1}{2}}, \\ \epsilon^{(4)}_{\frac{1}{2},\frac{1}{2}}, \quad \epsilon^{(4)}_{\frac{1}{2},-\frac{1}{2}}, \quad \epsilon^{(6)}_{\frac{1}{2},\frac{1}{2}}, \quad \epsilon^{(6)}_{\frac{1}{2},-\frac{1}{2}}, \quad \epsilon^{(5)}_{\frac{1}{2},\frac{1}{2}}, \quad \epsilon^{(5)}_{\frac{1}{2},-\frac{1}{2}}, \\ \epsilon^{(6)}_{\frac{3}{2},\frac{3}{2}}, \quad \epsilon^{(6)}_{\frac{3}{2},-\frac{3}{2}}, \quad \epsilon^{(5)}_{\frac{3}{2},\frac{3}{2}}, \quad \epsilon^{(5)}_{\frac{3}{2},-\frac{3}{2}}, \quad \epsilon^{(6)}_{\frac{3}{2},\frac{1}{2}}, \quad \epsilon^{(6)}_{\frac{3}{2},-\frac{1}{2}}, \quad \epsilon^{(5)}_{\frac{3}{2},\frac{1}{2}}, \quad \epsilon^{(5)}_{\frac{3}{2},-\frac{1}{2}}. \end{array} \tag{10.10}$$

Correspondingly, the matrix Γ_4 is presented in the form

$$\Gamma_4 = C^{(1/2)} \oplus C^{(3/2)},$$

where spin blocks $C^{(1/2)}$ and $C^{(3/2)}$ are given by the formulas (take in mind eqs. (10.5)–(10.7)), $C^{(\frac{3}{2})} = 0_8$ and

$$C^{(\frac{1}{2})} = \begin{vmatrix} 0 & 0 & c^{(\frac{1}{2})}_{14} & 0 & 0 & 0 & 0 & 0 & 0 & 0 & c^{(\frac{1}{2})}_{16} & 0 \\ 0 & 0 & 0 & c^{(\frac{1}{2})}_{14} & 0 & 0 & 0 & 0 & 0 & 0 & 0 & c^{(\frac{1}{2})}_{16} \\ c^{(\frac{1}{2})}_{41} & 0 & 0 & 0 & 0 & 0 & 0 & 0 & c^{(\frac{1}{2})}_{43} & 0 & 0 & 0 \\ 0 & c^{(\frac{1}{2})}_{41} & 0 & 0 & 0 & 0 & 0 & 0 & 0 & c^{(\frac{1}{2})}_{43} & 0 & 0 \\ 0 & 0 & 0 & 0 & 0 & 0 & c^{(\frac{1}{2})}_{25} & 0 & 0 & 0 & c^{(\frac{1}{2})}_{26} & 0 \\ 0 & 0 & 0 & 0 & 0 & 0 & 0 & c^{(\frac{1}{2})}_{26} & 0 & 0 & 0 & c^{(\frac{1}{2})}_{26} \\ 0 & 0 & 0 & 0 & c^{(\frac{1}{2})}_{52} & 0 & 0 & 0 & c^{(\frac{1}{2})}_{53} & 0 & 0 & 0 \\ 0 & 0 & 0 & 0 & 0 & c^{(\frac{1}{2})}_{52} & 0 & 0 & 0 & c^{(\frac{1}{2})}_{53} & 0 & 0 \\ 0 & 0 & c^{(\frac{1}{2})}_{34} & 0 & 0 & 0 & c^{(\frac{1}{2})}_{35} & 0 & 0 & 0 & 0 & 0 \\ 0 & 0 & 0 & c^{(\frac{1}{2})}_{34} & 0 & 0 & 0 & c^{(\frac{1}{2})}_{35} & 0 & 0 & 0 & 0 \\ c^{(\frac{1}{2})}_{61} & 0 & 0 & 0 & c^{(\frac{1}{2})}_{62} & 0 & 0 & 0 & 0 & 0 & 0 & 0 \\ 0 & c^{(\frac{1}{2})}_{61} & 0 & 0 & 0 & c^{(\frac{1}{2})}_{62} & 0 & 0 & 0 & 0 & 0 & 0 \end{vmatrix} .$$

We are to perform some transformations on eq. (10.11). In particular, the matrix block

$$\begin{vmatrix} 0 & 0 & c^{(\frac{1}{2})}_{14} & 0 \\ 0 & 0 & 0 & c^{(\frac{1}{2})}_{14} \\ c^{(\frac{1}{2})}_{41} & 0 & 0 & 0 \\ 0 & c^{(\frac{1}{2})}_{41} & 0 & 0 \end{vmatrix}$$

may be decomposed into two parts

$$\frac{1}{2}\begin{vmatrix} 0 & 0 & (c^{(\frac{1}{2})}_{14} + c^{(\frac{1}{2})}_{41}) & 0 \\ 0 & 0 & 0 & (c^{(\frac{1}{2})}_{14} + c^{(\frac{1}{2})}_{41}) \\ (c^{(\frac{1}{2})}_{14} + c^{(\frac{1}{2})}_{41}) & 0 & 0 & 0 \\ 0 & (c^{(\frac{1}{2})}_{14} + c^{(\frac{1}{2})}_{41}) & 0 & 0 \end{vmatrix}$$

$$+\frac{1}{2}\begin{vmatrix} 0 & 0 & (c^{(\frac{1}{2})}_{14} - c^{(\frac{1}{2})}_{41}) & 0 \\ 0 & 0 & 0 & (c^{(\frac{1}{2})}_{14} - c^{(\frac{1}{2})}_{41}) \\ (c^{(\frac{1}{2})}_{41} - c^{(\frac{1}{2})}_{14}) & 0 & 0 & 0 \\ 0 & (c^{(\frac{1}{2})}_{41} - c^{(\frac{1}{2})}_{14}) & 0 & 0 \end{vmatrix}$$

$$= \frac{1}{2}\left(c^{(\frac{1}{2})}_{14} + c^{(\frac{1}{2})}_{41}\right)\gamma_4 + \frac{1}{2}\left(c^{(\frac{1}{2})}_{14} - c^{(\frac{1}{2})}_{41}\right)\gamma_5\gamma_4 \,, \qquad (10.11)$$

where the Dirac matrices are used

$$\gamma_4 = \begin{vmatrix} 0 & 0 & 1 & 0 \\ 0 & 0 & 0 & 1 \\ 1 & 0 & 0 & 0 \\ 0 & 1 & 0 & 0 \end{vmatrix}, \qquad \gamma_5 = \begin{vmatrix} 1 & 0 & 0 & 0 \\ 0 & 1 & 0 & 0 \\ 0 & 0 & -1 & 0 \\ 0 & 0 & 0 & -1 \end{vmatrix} .$$

Similarly, we get

$$\left|\begin{array}{cccc} 0 & 0 & c_{25}^{(\frac{1}{2})} & 0 \\ 0 & 0 & 0 & c_{25}^{(\frac{1}{2})} \\ c_{52}^{(\frac{1}{2})} & 0 & 0 & 0 \\ 0 & c_{52}^{(\frac{1}{2})} & 0 & 0 \end{array}\right| = \frac{1}{2}(c_{25}^{(\frac{1}{2})} + c_{52}^{(\frac{1}{2})})\gamma_4 + \frac{1}{2}(c_{25}^{(\frac{1}{2})} - c_{52}^{(\frac{1}{2})})\gamma_5\gamma_4 \,,$$

$$\left|\begin{array}{cccc} 0 & 0 & c_{16}^{(\frac{1}{2})} & 0 \\ 0 & 0 & 0 & c_{16}^{(\frac{1}{2})} \\ c_{43}^{(\frac{1}{2})} & 0 & 0 & 0 \\ 0 & c_{43}^{(\frac{1}{2})} & 0 & 0 \end{array}\right| = \frac{1}{2}(c_{16}^{(\frac{1}{2})} + c_{43}^{(\frac{1}{2})})\gamma_4 + \frac{1}{2}(c_{16}^{(\frac{1}{2})} - c_{43}^{(\frac{1}{2})})\gamma_5\gamma_4 \,,$$

$$\left|\begin{array}{cccc} 0 & 0 & c_{26}^{(\frac{1}{2})} & 0 \\ 0 & 0 & 0 & c_{26}^{(\frac{1}{2})} \\ c_{53}^{(\frac{1}{2})} & 0 & 0 & 0 \\ 0 & c_{53}^{(\frac{1}{2})} & 0 & 0 \end{array}\right| = \frac{1}{2}(c_{26}^{(\frac{1}{2})} + c_{53}^{(\frac{1}{2})})\gamma_4 + \frac{1}{2}(c_{26}^{(\frac{1}{2})} - c_{53}^{(\frac{1}{2})})\gamma_5\gamma_4 \,,$$

$$\left|\begin{array}{cccc} 0 & 0 & c_{34}^{(\frac{1}{2})} & 0 \\ 0 & 0 & 0 & c_{34}^{(\frac{1}{2})} \\ c_{61}^{(\frac{1}{2})} & 0 & 0 & 0 \\ 0 & c_{61}^{(\frac{1}{2})} & 0 & 0 \end{array}\right| = \frac{1}{2}(c_{34}^{(\frac{1}{2})} + c_{61}^{(\frac{1}{2})})\gamma_4 + \frac{1}{2}(c_{34}^{(\frac{1}{2})} - c_{61}^{(\frac{1}{2})})\gamma_5\gamma_4 \,,$$

$$\left|\begin{array}{cccc} 0 & 0 & c_{35}^{(\frac{1}{2})} & 0 \\ 0 & 0 & 0 & c_{35}^{(\frac{1}{2})} \\ c_{62}^{(\frac{1}{2})} & 0 & 0 & 0 \\ 0 & c_{62}^{(\frac{1}{2})} & 0 & 0 \end{array}\right| = \frac{1}{2}(c_{35}^{(\frac{1}{2})} + c_{62}^{(\frac{1}{2})})\gamma_4 + \frac{1}{2}(c_{35}^{(\frac{1}{2})} - c_{62}^{(\frac{1}{2})})\gamma_5\gamma_4 \,.$$

Collecting results together, we find the following decomposition for spin block $C^{(1/2)}$:

$$C^{(1/2)} = \frac{1}{2}\left|\begin{array}{ccc} (c_{14}^{(\frac{1}{2})} + c_{41}^{(\frac{1}{2})}) & 0 & (c_{16}^{(\frac{1}{2})} + c_{43}^{(\frac{1}{2})}) \\ 0 & (c_{25}^{(\frac{1}{2})} + c_{52}^{(\frac{1}{2})}) & (c_{26}^{(\frac{1}{2})} + c_{53}^{(\frac{1}{2})}) \\ (c_{34}^{(\frac{1}{2})} + c_{61}^{(\frac{1}{2})}) & (c_{35}^{(\frac{1}{2})} + c_{62}^{(\frac{1}{2})}) & 0 \end{array}\right| \otimes \gamma_4$$
$$+\frac{1}{2}\left|\begin{array}{ccc} (c_{14}^{(\frac{1}{2})} - c_{41}^{(\frac{1}{2})}) & 0 & (c_{16}^{(\frac{1}{2})} - c_{43}^{(\frac{1}{2})}) \\ 0 & (c_{25}^{(\frac{1}{2})} - c_{52}^{(\frac{1}{2})}) & (c_{26}^{(\frac{1}{2})} - c_{53}^{(\frac{1}{2})}) \\ (c_{34}^{(\frac{1}{2})} - c_{61}^{(\frac{1}{2})}) & (c_{35}^{(\frac{1}{2})} - c_{62}^{(\frac{1}{2})}) & 0 \end{array}\right| \otimes \gamma_5\gamma_4 \,. \qquad (10.12)$$

It should be emphasised that in expression (10.12) the first term corresponds to a purely P-invariant model, whereas the second term relates to a purely P-noninvariant model. In the present chapter, we restrict ourselves to the second variant.

It is convenient to employ the shortening notations

$$a_1 = \frac{1}{2}(c_{14}^{(\frac{1}{2})} + c_{41}^{(\frac{1}{2})}), \quad a_2 = \frac{1}{2}(c_{16}^{(\frac{1}{2})} + c_{43}^{(\frac{1}{2})}), \quad a_3 = \frac{1}{2}(c_{25}^{(\frac{1}{2})} + c_{52}^{(\frac{1}{2})}), \quad a_4 = \frac{1}{2}(c_{26}^{(\frac{1}{2})} + c_{53}^{(\frac{1}{2})}),$$

$$a_5 = \frac{1}{2}(c_{34}^{(\frac{1}{2})} + c_{61}^{(\frac{1}{2})}), \quad a_6 = \frac{1}{2}(c_{35}^{(\frac{1}{2})} + c_{62}^{(\frac{1}{2})}),$$

$$b_1 = \frac{1}{2}(c_{14}^{(\frac{1}{2})} - c_{41}^{(\frac{1}{2})}), \quad b_2 = \frac{1}{2}(c_{16}^{(\frac{1}{2})} - c_{43}^{(\frac{1}{2})}), \quad b_3 = \frac{1}{2}(c_{25}^{(\frac{1}{2})} - c_{52}^{(\frac{1}{2})}), \quad b_4 = \frac{1}{2}(c_{26}^{(\frac{1}{2})} - c_{53}^{(\frac{1}{2})}),$$

$$b_5 = \frac{1}{2}(c_{34}^{(\frac{1}{2})} - c_{61}^{(\frac{1}{2})}), \quad b_6 = \frac{1}{2}(c_{35}^{(\frac{1}{2})} - c_{62}^{(\frac{1}{2})}).$$

Correspondingly, the spin block $C^{(1/2)}$ reads

$$C^{(1/2)} = \begin{vmatrix} a_1 & 0 & a_2 \\ 0 & a_3 & a_4 \\ a_5 & a_6 & 0 \end{vmatrix} \otimes \gamma_4 + \begin{vmatrix} b_1 & 0 & b_2 \\ 0 & b_3 & b_4 \\ b_5 & b_5 & 0 \end{vmatrix} \otimes \gamma_5\gamma_4 . \tag{10.13}$$

For purely P-noninvariant model, it becomes simpler

$$C^{(1/2)} = \begin{vmatrix} b_1 & 0 & b_2 \\ 0 & b_3 & b_4 \\ b_5 & b_5 & 0 \end{vmatrix} \otimes \gamma_5\gamma_4 , \tag{10.14}$$

after re-designating $b_i \longrightarrow ib_i$ it takes the form

$$C^{(1/2)} = i \begin{vmatrix} b_1 & 0 & b_2 \\ 0 & b_3 & b_4 \\ b_5 & b_5 & 0 \end{vmatrix} \otimes \gamma_5\gamma_4 . \tag{10.15}$$

Because we make the model for a particle with one mass parameter, the matrix

$$\begin{vmatrix} b_1 & 0 & b_2 \\ 0 & b_3 & b_4 \\ b_5 & b_5 & 0 \end{vmatrix}$$

should have only one non-vanishing eigenvalue; let it equal +1. In accordance with this, parameters b_i obey the following restrictions

$$b_1 + b_3 = 1 , \quad b_1b_3 - b_2b_5 - b_4b_6 = 0, \quad b_2b_3b_5 + b_1b_4b_6 = 0 . \tag{10.16}$$

10.3 On Lagrangian formulation of the model

Let us examine the problem of possible Lagrangian formulation of equation under consideration. Hermitian matrix of bilinear form in the Gel'fand–Yaglom basis has the structure

$$\eta = (\eta^{(1/2)} \otimes I_2) \oplus (\eta^{(3/2)} \otimes I_4), \tag{10.17}$$

where the blocks $\eta^{(1/2)}$ and $\eta^{(3/2)}$ read

$$\eta^{(\frac{1}{2})} = \begin{vmatrix} 0 & \eta_{12}^{(\frac{1}{2})} & 0 & 0 & 0 & 0 \\ \eta_{21}^{(\frac{1}{2})} & 0 & 0 & 0 & 0 & 0 \\ 0 & 0 & 0 & \eta_{34}^{(\frac{1}{2})} & 0 & 0 \\ 0 & 0 & \eta_{43}^{(\frac{1}{2})} & 0 & 0 & 0 \\ 0 & 0 & 0 & 0 & 0 & \eta_{56}^{(\frac{1}{2})} \\ 0 & 0 & 0 & 0 & \eta_{65}^{(\frac{1}{2})} & 0 \end{vmatrix} , \eta^{(\frac{3}{2})} = \begin{vmatrix} 0 & -\eta_{56}^{(\frac{1}{2})} \\ -\eta_{65}^{(\frac{1}{2})} & 0 \end{vmatrix} . \tag{10.18}$$

Because the bilinear form is not P invariant, the usually assumed condition $\eta^{(s)}_{\tau\dot{\tau}} = \eta^{(s)}_{\dot{\tau}\tau}$ does not hold. However, due to Hermiticity of the matrix η, we have $(\eta^{(s)}_{\dot{\tau}\tau})^* = \eta^{(s)}_{\tau\dot{\tau}}$. Without loss of generality, the last constraint may be satisfied by setting

$$\eta^{(s)}_{\tau\dot{\tau}} = -\eta^{(s)}_{\dot{\tau}\tau} = \pm i\,. \tag{10.19}$$

For instance, let us set

$$\eta^{(\frac{1}{2})}_{12} = -\eta^{(\frac{1}{2})}_{21} = \eta^{(\frac{1}{2})}_{34} = -\eta^{(\frac{1}{2})}_{43} = i\,, \quad \eta^{(\frac{1}{2})}_{56} = -\eta^{(\frac{1}{2})}_{65} = i\,f\,, \quad f = \pm 1\,. \tag{10.20}$$

Then, from existence of Lagrangian formulation of the theory it follows

$$c^{(s)}_{\tau\tau'}\eta^{(s)}_{\tau'\dot{\tau}'} = (c^{(s)}_{\dot{\tau}'\dot{\tau}})^*\tau^{(s)}_{\tau\dot{\tau}} \quad \Longrightarrow \quad (C^{(s)}\eta^{(s)})^+ = C^{(s)}\eta^{(s)}\,,$$

this assumes the following restrictions

$$\begin{aligned}
&[(c^{(\frac{1}{2})}_{12}]^* = -(c^{(\frac{1}{2})}_{12}\,, [c^{(\frac{1}{2})}_{21}]^* = -c^{(\frac{1}{2})}_{21}\,, [c^{(\frac{1}{2})}_{34}]^* = -c^{(\frac{1}{2})}_{34}\,, [c^{(\frac{1}{2})}_{43}]^* = -c^{(\frac{1}{2})}_{43}\,,\\
&c^{(\frac{1}{2})}_{61} = -f(c^{(\frac{1}{2})}_{25})^*, c^{(\frac{1}{2})}_{52} = -f(c^{(\frac{1}{2})}_{16})^*, c^{(\frac{1}{2})}_{63} = -f(c^{(\frac{1}{2})}_{45})^*, c^{(\frac{1}{2})}_{54} = -f(c^{(\frac{1}{2})}_{36})^*.
\end{aligned} \tag{10.21}$$

Taking in mind still existing arbitrariness in choosing elements of the block $C^{(1/2)}$, we impose restrictions

$$c^{(\frac{1}{2})}_{12} = -c^{(\frac{1}{2})}_{21} = ib_1, c^{(\frac{1}{2})}_{34} = -c^{(\frac{1}{2})}_{43} = ib_3, c^{(\frac{1}{2})}_{16} = -c^{(\frac{1}{2})}_{25} = ib_2, dc^{(\frac{1}{2})}_{36} = -c^{(\frac{1}{2})}_{45} = ib_4,$$

where b_1, b_3 are real-valued and b_2, b_4 are complex parameters. Thus, we arrive at the following representation for matric $\Gamma^{(\frac{1}{2})}_4$:

$$\Gamma^{(\frac{1}{2})}_4 = i\left|\begin{array}{cccccc}
0 & b_1 & 0 & 0 & 0 & b_2\\
-b_1 & 0 & 0 & 0 & -b_2 & 0\\
0 & 0 & 0 & b_3 & 0 & b_4\\
0 & 0 & -b_3 & 0 & -b_4 & 0\\
0 & fb_2^* & 0 & fb_4^* & 0 & 0\\
-fb_2^* & 0 & -fb_4^* & 0 & & 0
\end{array}\right| \otimes I_2 = B \otimes i\gamma_5\gamma_4\,, \tag{10.22}$$

where

$$B = i\left|\begin{array}{ccc}
b_1 & 0 & b_2\\
0 & b_3 & b_4\\
fb_2^* & fb_4^* & 0
\end{array}\right|, \tag{10.23}$$

this somewhat repeats results of previous sections. Correspondingly, constraints on parameters b_i take the form

$$b_1 + b_3 = 1\,, \quad b_1b_3 - f|b_2|^2 - f|b_4|^2 = 0\,, \quad b_1|b_4|^2 + b_3|b_2|^2 = 0\,. \tag{10.24}$$

It is readily checked that if $f = +1$ the last system for b_i is not consistent. Therefore, only the variant $f = -1$ remains. For this case, we have

$$\eta^{(\frac{1}{2})} \otimes I_2 = \left|\begin{array}{cccccc}
0 & i & 0 & 0 & 0 & 0\\
-i & 0 & 0 & 0 & 0 & 0\\
0 & 0 & 0 & i & 0 & 0\\
0 & 0 & -i & 0 & 0 & 0\\
0 & 0 & 0 & 0 & 0 & -i\\
0 & 0 & 0 & 0 & i & 0
\end{array}\right| \otimes I_2 = \left|\begin{array}{ccc}
1 & 0 & 0\\
0 & 1 & 0\\
0 & 0 & -1
\end{array}\right| \otimes i\gamma_5\gamma_4, \tag{10.25}$$

$$\Gamma_4^{(1/2)} = B \otimes i\gamma_5\gamma_4, \quad B = \begin{vmatrix} b_1 & 0 & b_2 \\ 0 & b_3 & b_4 \\ -b_2^* & -b_4^* & 0 \end{vmatrix}, \tag{10.26}$$

and

$$b_1 + b_3 = 1, \quad b_1 b_3 + |b_2|^2 + |b_4|^2 = 0, \quad b_1|b_4|^2 + b_3|b_2|^2 = 0. \tag{10.27}$$

The minimal polynomial for matrix Γ_4 has the form $\Gamma_4^2(\Gamma_4^2 - 1) = 0$; besides, the charge definiteness condition

$$(-1)^2[(\mathrm{Sp}(\Gamma_4^3\eta))^2 - (\mathrm{Sp}(\Gamma_4^2\eta))^2] > 0$$

holds due to the identities

$$\mathrm{Sp}(\Gamma_4^2\eta) = 0, \quad \mathrm{Sp}(\Gamma_4^3\eta) = 4b_1(b_1^2 - |b_2|^2) + 4b_3(b_3^2 - |b_4|^2) = 4(b_1^2 + b_3^2) > 0.$$

10.4 Spinor form of the wave equation

Now we have to find spinor form of the above equation. Representations for wave function Ψ in modified GY-basis and in canonical basis are

$$\Psi_{GY,m} = \Big\{ \Psi^{(0,1/2)}_{1/2,1/2}, \Psi^{(0,1/2)}_{1/2,-1/2}, \Psi^{(1/2,0)}_{1/2,1/2}, \Psi^{(1/2,0)}_{1/2,-1/2}, \Psi'^{(0,1/2)}_{1/2,1/2}, \Psi'^{(0,1/2)}_{1/2,-1/2}, \Psi'^{(1/2,0)}_{1/2,1/2}, \Psi'^{(1/2,0)}_{1/2,-1/2},$$

$$\Psi^{(1,1/2)}_{1/2,1/2}, \Psi^{(1,1/2)}_{1/2,-1/2}, \Psi^{(1/2,1)}_{1/2,1/2}, \Psi^{(1/2,1)}_{1/2,-1/2}, \Psi^{(1,1/2)}_{3/2,3/2}, \Psi^{(1,1/2)}_{3/2,-3/2}, \Psi^{(1/2,1)}_{3/2,3/2},$$

$$\Psi^{(1/2,1)}_{3/2,-3/2}, \Psi^{(1,1/2)}_{3/2,1/2}, \Psi^{(1,1/2)}_{3/2,-1/2}, \Psi^{(1/2,1)}_{3/2,1/2}, \Psi^{(1/2,1)}_{3/2,-1/2} \Big\},$$

$$\Psi_{can} = \Big\{ \Psi^{(0,1/2)}_{(0,1/2)}, \Psi^{(0,1/2)}_{(0,-1/2)}, \Psi^{(1/2,0)}_{(1/2,0)}, \Psi^{(1/2,0)}_{(-1/2,0)}, \Psi'^{(0,1/2)}_{(0,1/2)}, \Psi'^{(0,1/2)}_{(0,-1/2)}, \Psi'^{(1/2,0)}_{(1/2,0)}, \Psi'^{(1/2,0)}_{(-1/2,0)},$$

$$\Psi^{(1,1/2)}_{(1,1/2)}, \Psi^{(1,1/2)}_{(0,1/2)}, \Psi^{(1,1/2)}_{(-1,1/2)}, \Psi^{(1,1/2)}_{(1,-1/2)}, \Psi^{(1,1/2)}_{(0,-1/2)}, \Psi^{(1,1/2)}_{(-1,-1/2)}, \Psi^{(1/2,1)}_{(1/2,1)}, \Psi^{(1/2,1)}_{(1/2,0)},$$

$$\Psi^{(1/2,1)}_{(1/2,-1)}, \Psi^{(1/2,1)}_{(-1/2,1)}, \Psi^{(1/2,1)}_{(-1/2,0)}, \Psi^{(1/2,1)}_{(-1/2,-1)} \Big\}$$

relate to each other through linear transformation

$$\Psi_{GY,m} = B\Psi_{can},$$

where

$$B = \begin{vmatrix}
1 & \cdot & \cdot & \cdot & \cdot & \cdot & \cdot & \cdot & \cdot & \cdot & \cdot & \cdot & \cdot & \cdot & \cdot & \cdot & \cdot & \cdot & \cdot & \cdot \\
\cdot & 1 & \cdot & \cdot & \cdot & \cdot & \cdot & \cdot & \cdot & \cdot & \cdot & \cdot & \cdot & \cdot & \cdot & \cdot & \cdot & \cdot & \cdot & \cdot \\
\cdot & \cdot & 1 & \cdot & \cdot & \cdot & \cdot & \cdot & \cdot & \cdot & \cdot & \cdot & \cdot & \cdot & \cdot & \cdot & \cdot & \cdot & \cdot & \cdot \\
\cdot & \cdot & \cdot & 1 & \cdot & \cdot & \cdot & \cdot & \cdot & \cdot & \cdot & \cdot & \cdot & \cdot & \cdot & \cdot & \cdot & \cdot & \cdot & \cdot \\
\cdot & \cdot & \cdot & \cdot & 1 & \cdot & \cdot & \cdot & \cdot & \cdot & \cdot & \cdot & \cdot & \cdot & \cdot & \cdot & \cdot & \cdot & \cdot & \cdot \\
\cdot & \cdot & \cdot & \cdot & \cdot & 1 & \cdot & \cdot & \cdot & \cdot & \cdot & \cdot & \cdot & \cdot & \cdot & \cdot & \cdot & \cdot & \cdot & \cdot \\
\cdot & \cdot & \cdot & \cdot & \cdot & \cdot & 1 & \cdot & \cdot & \cdot & \cdot & \cdot & \cdot & \cdot & \cdot & \cdot & \cdot & \cdot & \cdot & \cdot \\
\cdot & \cdot & \cdot & \cdot & \cdot & \cdot & \cdot & 1 & \cdot & \cdot & \cdot & \cdot & \cdot & \cdot & \cdot & \cdot & \cdot & \cdot & \cdot & \cdot \\
\cdot & \cdot & \cdot & \cdot & \cdot & \cdot & \cdot & \cdot & \cdot & -\sqrt{\frac{1}{3}} & \cdot & \sqrt{\frac{2}{3}} & \cdot & \cdot & \cdot & \cdot & \cdot & \cdot & \cdot & \cdot \\
\cdot & \cdot & \cdot & \cdot & \cdot & \cdot & \cdot & \cdot & \cdot & \cdot & -\sqrt{\frac{2}{3}} & \cdot & \sqrt{\frac{1}{3}} & \cdot & \cdot & \cdot & \cdot & \cdot & \cdot & \cdot \\
\cdot & \cdot & \cdot & \cdot & \cdot & \cdot & \cdot & \cdot & \cdot & \cdot & \cdot & \cdot & \cdot & \cdot & \cdot & \sqrt{\frac{1}{3}} & \cdot & -\sqrt{\frac{2}{3}} & \cdot & \cdot \\
\cdot & \cdot & \cdot & \cdot & \cdot & \cdot & \cdot & \cdot & \cdot & \cdot & \cdot & \cdot & \cdot & \cdot & \cdot & \cdot & \sqrt{\frac{2}{3}} & \cdot & -\sqrt{\frac{1}{3}} & \cdot \\
\cdot & \cdot & \cdot & \cdot & \cdot & \cdot & \cdot & \cdot & 1 & \cdot & \cdot & \cdot & \cdot & \cdot & \cdot & \cdot & \cdot & \cdot & \cdot & \cdot \\
\cdot & \cdot & \cdot & \cdot & \cdot & \cdot & \cdot & \cdot & \cdot & \cdot & \cdot & \cdot & \cdot & 1 & \cdot & \cdot & \cdot & \cdot & \cdot & \cdot \\
\cdot & \cdot & \cdot & \cdot & \cdot & \cdot & \cdot & \cdot & \cdot & \cdot & \cdot & \cdot & \cdot & \cdot & 1 & \cdot & \cdot & \cdot & \cdot & \cdot \\
\cdot & \cdot & \cdot & \cdot & \cdot & \cdot & \cdot & \cdot & \cdot & \cdot & \cdot & \cdot & \cdot & \cdot & \cdot & \cdot & \cdot & \cdot & \cdot & 1 \\
\cdot & \cdot & \cdot & \cdot & \cdot & \cdot & \cdot & \cdot & \cdot & \sqrt{\frac{2}{3}} & \cdot & \sqrt{\frac{1}{3}} & \cdot & \cdot & \cdot & \cdot & \cdot & \cdot & \cdot & \cdot \\
\cdot & \cdot & \cdot & \cdot & \cdot & \cdot & \cdot & \cdot & \cdot & \cdot & \sqrt{\frac{1}{3}} & \cdot & \sqrt{\frac{2}{3}} & \cdot & \cdot & \cdot & \cdot & \cdot & \cdot & \cdot \\
\cdot & \cdot & \cdot & \cdot & \cdot & \cdot & \cdot & \cdot & \cdot & \cdot & \cdot & \cdot & \cdot & \cdot & \cdot & \sqrt{\frac{2}{3}} & \cdot & -\sqrt{\frac{1}{3}} & \cdot & \cdot \\
\cdot & \cdot & \cdot & \cdot & \cdot & \cdot & \cdot & \cdot & \cdot & \cdot & \cdot & \cdot & \cdot & \cdot & \cdot & \cdot & \sqrt{\frac{1}{3}} & \cdot & \sqrt{\frac{2}{3}} & \cdot
\end{vmatrix},$$

and canonical and spinor wave functions relate to each other as follows

$$\Psi_{spin} = \Big\{\Psi^{\dot{1}}, \Psi^{\dot{2}}, \Psi_1, \Psi_2, \Psi'^{\dot{1}}, \Psi'^{\dot{2}}, \Psi'_1, \Psi'_2, \Psi^{\dot{1}}_{(11)}, \Psi^{\dot{1}}_{(12)}, \Psi^{\dot{1}}_{(22)}, \Psi^{\dot{2}}_{(11)}, \Psi^{\dot{2}}_{(12)}, \Psi^{\dot{2}}_{(22)},$$

$$\Psi_1^{(\dot{1}\dot{1})}, \Psi_1^{(\dot{1}\dot{2})}, \Psi_1^{(\dot{2}\dot{2})}, \Psi_2^{(\dot{1}\dot{1})}, \Psi_2^{(\dot{1}\dot{2})}, \Psi_2^{(\dot{2}\dot{2})}\Big\},$$

$$\Psi_{can} = A\Psi_{spin.}, \; B = \begin{vmatrix}
1 & \cdot & \cdot & \cdot & \cdot & \cdot & \cdot & \cdot & \cdot & \cdot & \cdot & \cdot & \cdot & \cdot & \cdot & \cdot & \cdot & \cdot & \cdot & \cdot \\
\cdot & 1 & \cdot & \cdot & \cdot & \cdot & \cdot & \cdot & \cdot & \cdot & \cdot & \cdot & \cdot & \cdot & \cdot & \cdot & \cdot & \cdot & \cdot & \cdot \\
\cdot & \cdot & 1 & \cdot & \cdot & \cdot & \cdot & \cdot & \cdot & \cdot & \cdot & \cdot & \cdot & \cdot & \cdot & \cdot & \cdot & \cdot & \cdot & \cdot \\
\cdot & \cdot & \cdot & 1 & \cdot & \cdot & \cdot & \cdot & \cdot & \cdot & \cdot & \cdot & \cdot & \cdot & \cdot & \cdot & \cdot & \cdot & \cdot & \cdot \\
\cdot & \cdot & \cdot & \cdot & 1 & \cdot & \cdot & \cdot & \cdot & \cdot & \cdot & \cdot & \cdot & \cdot & \cdot & \cdot & \cdot & \cdot & \cdot & \cdot \\
\cdot & \cdot & \cdot & \cdot & \cdot & 1 & \cdot & \cdot & \cdot & \cdot & \cdot & \cdot & \cdot & \cdot & \cdot & \cdot & \cdot & \cdot & \cdot & \cdot \\
\cdot & \cdot & \cdot & \cdot & \cdot & \cdot & 1 & \cdot & \cdot & \cdot & \cdot & \cdot & \cdot & \cdot & \cdot & \cdot & \cdot & \cdot & \cdot & \cdot \\
\cdot & \cdot & \cdot & \cdot & \cdot & \cdot & \cdot & 1 & \cdot & \cdot & \cdot & \cdot & \cdot & \cdot & \cdot & \cdot & \cdot & \cdot & \cdot & \cdot \\
\cdot & \cdot & \cdot & \cdot & \cdot & \cdot & \cdot & \cdot & 1 & \cdot & \cdot & \cdot & \cdot & \cdot & \cdot & \cdot & \cdot & \cdot & \cdot & \cdot \\
\cdot & \cdot & \cdot & \cdot & \cdot & \cdot & \cdot & \cdot & \cdot & \sqrt{2} & \cdot & \cdot & \cdot & \cdot & \cdot & \cdot & \cdot & \cdot & \cdot & \cdot \\
\cdot & \cdot & \cdot & \cdot & \cdot & \cdot & \cdot & \cdot & \cdot & \cdot & 1 & \cdot & \cdot & \cdot & \cdot & \cdot & \cdot & \cdot & \cdot & \cdot \\
\cdot & \cdot & \cdot & \cdot & \cdot & \cdot & \cdot & \cdot & \cdot & \cdot & \cdot & 1 & \cdot & \cdot & \cdot & \cdot & \cdot & \cdot & \cdot & \cdot \\
\cdot & \cdot & \cdot & \cdot & \cdot & \cdot & \cdot & \cdot & \cdot & \cdot & \cdot & \cdot & \sqrt{2} & \cdot & \cdot & \cdot & \cdot & \cdot & \cdot & \cdot \\
\cdot & \cdot & \cdot & \cdot & \cdot & \cdot & \cdot & \cdot & \cdot & \cdot & \cdot & \cdot & \cdot & 1 & \cdot & \cdot & \cdot & \cdot & \cdot & \cdot \\
\cdot & \cdot & \cdot & \cdot & \cdot & \cdot & \cdot & \cdot & \cdot & \cdot & \cdot & \cdot & \cdot & \cdot & 1 & \cdot & \cdot & \cdot & \cdot & \cdot \\
\cdot & \cdot & \cdot & \cdot & \cdot & \cdot & \cdot & \cdot & \cdot & \cdot & \cdot & \cdot & \cdot & \cdot & \cdot & \sqrt{2} & \cdot & \cdot & \cdot & \cdot \\
\cdot & \cdot & \cdot & \cdot & \cdot & \cdot & \cdot & \cdot & \cdot & \cdot & \cdot & \cdot & \cdot & \cdot & \cdot & \cdot & 1 & \cdot & \cdot & \cdot \\
\cdot & \cdot & \cdot & \cdot & \cdot & \cdot & \cdot & \cdot & \cdot & \cdot & \cdot & \cdot & \cdot & \cdot & \cdot & \cdot & \cdot & 1 & \cdot & \cdot \\
\cdot & \cdot & \cdot & \cdot & \cdot & \cdot & \cdot & \cdot & \cdot & \cdot & \cdot & \cdot & \cdot & \cdot & \cdot & \cdot & \cdot & \cdot & \sqrt{2} & \cdot \\
\cdot & \cdot & \cdot & \cdot & \cdot & \cdot & \cdot & \cdot & \cdot & \cdot & \cdot & \cdot & \cdot & \cdot & \cdot & \cdot & \cdot & \cdot & \cdot & 1
\end{vmatrix},$$

we conclude that in spinor basis, the matrix Γ_4 has the form

$$\Gamma_4^{spin.} =$$

$$= i \left| \begin{array}{cccccccccc}
0 & 0 & b_1 & 0 & 0 & 0 & 0 & 0 & 0 & 0 \\
0 & 0 & 0 & b_1 & 0 & 0 & 0 & 0 & 0 & 0 \\
-b_1 & 0 & 0 & 0 & 0 & 0 & 0 & 0 & 0 & \sqrt{\frac{2}{3}}b_2 \\
0 & -b_1 & 0 & 0 & 0 & 0 & 0 & 0 & 0 & 0 \\
0 & 0 & 0 & 0 & 0 & 0 & b_3 & 0 & 0 & 0 \\
0 & 0 & 0 & 0 & 0 & 0 & 0 & b_3 & 0 & 0 \\
0 & 0 & 0 & 0 & -b_3 & 0 & 0 & 0 & 0 & \sqrt{\frac{2}{3}}b_4 \\
0 & 0 & 0 & 0 & 0 & -b_3 & 0 & 0 & 0 & 0 \\
0 & 0 & 0 & 0 & 0 & 0 & 0 & 0 & 0 & 0 \\
0 & 0 & -\sqrt{\frac{1}{6}}fb_5^* & 0 & 0 & 0 & -\sqrt{\frac{1}{6}}b_6 & 0 & 0 & 0 \\
0 & 0 & 0 & -\sqrt{\frac{2}{3}}b_5 & 0 & 0 & 0 & -\sqrt{\frac{2}{3}}b_6 & 0 & 0 \\
0 & 0 & \sqrt{\frac{2}{3}}b_5 & 0 & 0 & 0 & \sqrt{\frac{2}{3}}b_6 & 0 & 0 & 0 \\
0 & 0 & 0 & \sqrt{\frac{1}{6}}b_5 & 0 & 0 & 0 & \sqrt{\frac{1}{6}}b_6 & 0 & 0 \\
0 & 0 & 0 & 0 & 0 & 0 & 0 & 0 & 0 & 0 \\
0 & 0 & 0 & 0 & 0 & 0 & 0 & 0 & 0 & 0 \\
-\sqrt{\frac{1}{6}}b_5 & 0 & 0 & 0 & -\sqrt{\frac{1}{6}}b_6 & 0 & 0 & 0 & 0 & 0 \\
0 & -\sqrt{\frac{2}{3}}b_5 & 0 & 0 & 0 & -\sqrt{\frac{2}{3}}b_6 & 0 & 0 & 0 & 0 \\
\sqrt{\frac{2}{3}}b_5 & 0 & 0 & 0 & \sqrt{\frac{2}{3}}b_6 & 0 & 0 & 0 & 0 & 0 \\
0 & \sqrt{\frac{1}{6}}b_5 & 0 & 0 & 0 & \sqrt{\frac{1}{6}}b_6 & 0 & 0 & 0 & 0 \\
0 & 0 & 0 & 0 & 0 & 0 & 0 & 0 & 0 & 0
\end{array} \right.$$

$$i \left| \begin{array}{cccccccccc}
0 & 0 & 0 & 0 & 0 & \sqrt{\frac{2}{3}}b_2 & 0 & -\sqrt{\frac{2}{3}}b_2 & 0 & 0 \\
0 & 0 & 0 & 0 & 0 & 0 & \sqrt{\frac{2}{3}}b_2 & 0 & -\sqrt{\frac{2}{3}}b_2 & 0 \\
0 & -\sqrt{\frac{2}{3}}b_2 & 0 & 0 & 0 & 0 & 0 & 0 & 0 & 0 \\
\sqrt{\frac{2}{3}}b_2 & 0 & \sqrt{\frac{2}{3}}b_2 & 0 & 0 & 0 & 0 & 0 & & 0 \\
0 & 0 & 0 & 0 & 0 & \sqrt{\frac{2}{3}}gb_4 & 0 & -\sqrt{\frac{2}{3}}gb_4 & 0 & 0 \\
0 & 0 & 0 & 0 & 0 & 0 & \sqrt{\frac{2}{3}}b_4 & 0 & -\sqrt{\frac{2}{3}}b_4 & 0 \\
0 & -\sqrt{\frac{2}{3}}b_4 & 0 & 0 & 0 & 0 & 0 & 0 & 0 & 0 \\
\sqrt{\frac{2}{3}}b_4 & 0 & -\sqrt{\frac{2}{3}}b_4 & 0 & 0 & 0 & 0 & 0 & 0 & 0 \\
0 & 0 & 0 & 0 & 0 & 0 & 0 & 0 & 0 & 0 \\
0 & 0 & 0 & 0 & 0 & 0 & 0 & 0 & 0 & 0 \\
0 & 0 & 0 & 0 & 0 & 0 & 0 & 0 & 0 & 0 \\
0 & 0 & 0 & 0 & 0 & 0 & 0 & 0 & 0 & 0 \\
0 & 0 & 0 & 0 & 0 & 0 & 0 & 0 & 0 & 0 \\
0 & 0 & 0 & 0 & 0 & 0 & 0 & 0 & 0 & 0 \\
0 & 0 & 0 & 0 & 0 & 0 & 0 & 0 & 0 & 0 \\
0 & 0 & 0 & 0 & 0 & 0 & 0 & 0 & 0 & 0 \\
0 & 0 & 0 & 0 & 0 & 0 & 0 & 0 & 0 & 0 \\
0 & 0 & 0 & 0 & 0 & 0 & 0 & 0 & 0 & 0 \\
0 & 0 & 0 & 0 & 0 & 0 & 0 & 0 & 0 & 0 \\
0 & 0 & 0 & 0 & 0 & 0 & 0 & 0 & 0 & 0
\end{array} \right| .$$

The matrix Γ_4^{spin} acts on the wave function Ψ_{spin} as follows

$$\Gamma_4^{spin.}\Psi_{spin.}$$

$$= i \left| \begin{array}{c} b_1 \frac{1}{i}(\sigma^4)^{\dot{a}b}\Psi_b + \sqrt{\frac{2}{3}} b_2 \frac{1}{i}(\sigma^4)^b_{\dot{c}}\Psi^{(\dot{a}\dot{c})}_b \\ -b_1 \frac{1}{i}(\sigma^4)_{a\dot{b}}\Psi^{\dot{b}} - \sqrt{\frac{2}{3}} b_2 \frac{1}{i}(\sigma^4)^c_{\dot{b}}\psi^{\dot{b}}_{(ac)} \\ b_3 \frac{1}{i}(\sigma^4)^{\dot{a}b}\Psi'_b + \sqrt{\frac{2}{3}} b_4 (\sigma^4)^b_{\dot{c}}\psi^{(\dot{a}\dot{c})}_b \\ -b_3 \frac{1}{i}(\sigma^4)_{a\dot{b}}\Psi'^{\dot{b}} - \sqrt{\frac{2}{3}} b_4 (\sigma^4)^c_{\dot{b}}\Psi^{\dot{b}}_{(ac)} \\ -\frac{1}{\sqrt{6}}\frac{1}{i} b_5((\sigma^4)^{\dot{c}}_a\Psi_b + (\sigma^4)^{\dot{c}}_b\Psi_a) - \frac{1}{\sqrt{6}}\frac{1}{i} b_6((\sigma^4)^{\dot{c}}_a\Psi'_b + (\sigma^4)^{\dot{c}}_b\Psi'_a) \\ \frac{1}{\sqrt{6}}\frac{1}{i} b_5((\sigma^4)^{\dot{a}}_c\Psi^{\dot{b}} + (\sigma^4)^{\dot{b}}_c\Psi^{\dot{a}}) + \frac{1}{\sqrt{6}}\frac{1}{i} b_6((\sigma^4)^{\dot{a}}_c\Psi'^{\dot{b}} + (\sigma^4)^{\dot{b}}_c\Psi'^{\dot{a}}) \end{array} \right| , \tag{10.28}$$

where

$$(\sigma^4)_{\dot{a}b} = \left| \begin{array}{cc} i & 0 \\ 0 & i \end{array} \right| .$$

The matrices Γ_μ^{spin} act similarly, it is enough to make formal change $\sigma^4 \mapsto \sigma^\mu$, where

$$(\sigma^1)_{\dot{a}b} = \left| \begin{array}{cc} 0 & 1 \\ 1 & 0 \end{array} \right| , \qquad (\sigma^2)_{\dot{a}b} = \left| \begin{array}{cc} 0 & -i \\ i & 0 \end{array} \right| , \qquad (\sigma^3)_{\dot{a}b} = \left| \begin{array}{cc} 1 & 0 \\ 0 & -1 \end{array} \right| .$$

Taking in mind the above relations, we find the system of spinor equations

$$i\{b_1\partial^{\dot{a}b}\Psi_b + \sqrt{\frac{2}{3}} b_2 \partial^b_{\dot{c}}\Psi^{(\dot{a}\dot{c})}_b\} + M\Psi^{\dot{a}} = 0 , \tag{10.29}$$

$$-i\{b_1\partial_{a\dot{b}}\Psi^{\dot{b}} + \sqrt{\frac{2}{3}} b_2 \partial^c_{\dot{b}}\Psi^{\dot{b}}_{(ac)}\} + M\Psi_a = 0 , \tag{10.30}$$

$$i\{b_3\partial^{\dot{a}b}\Psi'_b + \sqrt{\frac{2}{3}} b_4 \partial^b_{\dot{c}}\Psi^{(\dot{a}\dot{c})}_b\} + M\Psi'^{\dot{a}} = 0 , \tag{10.31}$$

$$-i\{b_3\partial_{a\dot{b}}\Psi'^{\dot{b}} + \sqrt{\frac{2}{3}} b_4 \partial^c_{\dot{b}}\Psi^{\dot{b}}_{(ac)}\} + M\Psi'_a = 0 , \tag{10.32}$$

$$-\frac{i}{\sqrt{6}} b_5(\partial^{\dot{c}}_a\Psi_b + \partial^{\dot{c}}_b\Psi_a) - \frac{i}{\sqrt{6}} b_6(\partial^{\dot{c}}_a\Psi'_b + \partial^{\dot{c}}_b\Psi'_a) + M\Psi^{\dot{c}}_{(ab)} = 0 , \tag{10.33}$$

$$\frac{i}{\sqrt{6}} b_5(\partial^{\dot{a}}_c\Psi^{\dot{b}} + \partial^{\dot{b}}_c\Psi^{\dot{a}}) + \frac{i}{\sqrt{6}} b_6(\partial^{\dot{a}}_c\Psi'^{\dot{b}} + \partial^{\dot{b}}_c\Psi'^{\dot{a}}) + M\Psi^{(\dot{a}\dot{b})}_c = 0 , \tag{10.34}$$

where $\partial_{a\dot{b}} = \frac{1}{i}\partial_\mu\sigma^\mu_{a\dot{b}}$.

10.5 Equations in spin-tensor form

Let us obtain spit-tensor representation of eqs. (10.29)–(10.34), taking in mind identities

$$\Psi^{(\dot{a}\dot{b})}_c = \frac{1}{2}(\sigma^{\mu\dot{a}}_c\Psi^{\dot{b}}_\mu + \sigma^{\mu\dot{b}}_c\Psi^{\dot{a}}_\mu), \quad \Psi^{\dot{c}}_{(ab)} = \frac{1}{2}(\sigma^{\mu\dot{c}}_a\Psi_{\mu b} + \sigma^{\mu\dot{c}}_b\Psi_{\mu a}) ,$$
$$\Psi_a = \sigma^\mu_{a\dot{b}}\Psi^{\dot{b}}_\mu, \, \Psi^{\dot{a}} = \sigma^{\mu\dot{a}b}\Psi_{\mu b}, \quad \Psi'_a = \Psi_{a0}, \, \Psi'^{\dot{a}} = \Psi^{\dot{a}}_0 . \tag{10.35}$$

Equation (10.32) can be presented as follows

$$i\Big\{ b_1\partial^{\dot a b}\sigma^{\mu}_{b\dot c}\Psi^{\dot c}_{\mu} + \frac{1}{\sqrt{6}} b_2 \partial^{b}_{\dot c}(\sigma^{\mu\dot a}_{b}\Psi^{\dot c}_{\mu} + \sigma^{\mu\dot c}_{b}\Psi^{\dot a}_{\mu})\Big\} + M\sigma^{\mu\dot a b}\Psi_{\mu b} = 0\,,$$

or

$$i\Big\{ b_1\partial^{\dot a b}\sigma^{\mu}_{b\dot c}\Psi^{\dot c}_{\mu} + \frac{b_2}{\sqrt{6}}[-\sigma^{\mu\dot a b}\partial_{b\dot c}\Psi^{\dot c}_{\mu} + \frac{2}{i}\partial_{\mu}\Psi^{\dot a}_{\mu}]\Big\} + M\sigma^{\mu\dot a b}\Psi_{\mu b} = 0\,. \tag{10.36}$$

Similarly, from eq. (10.30) it follows

$$-i\Big\{ b_1\partial_{a\dot b}\sigma^{\mu\dot b c}\Psi_{\mu c} + \frac{b_2}{\sqrt{6}}[-\sigma^{\mu}_{a\dot b}\partial^{\dot b c}\Psi_{\mu c} + \frac{2}{i}\partial_{\mu}\Psi_{\mu a}]\Big\} + M\sigma^{\mu}_{a\dot b}\Psi^{\dot b}_{\mu} = 0\,. \tag{10.37}$$

Joining eqs. (10.36) and (10.37), we get

$$i\begin{vmatrix} I & 0 \\ 0 & -I \end{vmatrix} \Big\{ b_1 \begin{vmatrix} \partial^{\dot a b}\sigma^{\mu}_{b\dot c} & 0 \\ 0 & \partial_{a\dot b}\sigma^{\mu\dot b c} \end{vmatrix} \begin{vmatrix} \Psi^{\dot c}_{\mu} \\ \Psi_{c\mu} \end{vmatrix} + \frac{b_2}{\sqrt{6}}[\frac{2}{i} \begin{vmatrix} \Psi^{\dot a}_{\mu} \\ \Psi_{a\mu} \end{vmatrix}$$

$$- \begin{vmatrix} \sigma^{\mu\dot a b}\partial_{b\dot c} & 0 \\ 0 & \sigma^{\mu}_{a\dot b}\partial^{\dot b c} \end{vmatrix} \begin{vmatrix} \Psi^{\dot c}_{\mu} \\ \Psi_{c\mu} \end{vmatrix}]\Big\} + M \begin{vmatrix} 0 & \sigma^{\mu\dot a b} \\ \sigma^{\mu}_{a\dot b} & 0 \end{vmatrix} \begin{vmatrix} \Psi^{\dot b}_{\mu} \\ \Psi_{b\mu} \end{vmatrix} = 0\,,$$

or

$$i\gamma_5 \Big\{ b_1\hat\partial(\gamma_{\mu}\Psi_{\mu}) - \frac{4b_2}{\sqrt{6}}[-\frac{1}{4}\hat\partial(\gamma_{\mu}\Psi_{\mu}) + (\partial_{\mu}\Psi_{\mu})]\Big\} + M(\gamma_{\mu}\Psi_{\mu}) = 0\,, \tag{10.38}$$

where

$$\Psi_{\mu} = \begin{vmatrix} \Psi^{\dot a}_{\mu} \\ \Psi_{\mu b} \end{vmatrix}, \qquad \gamma_{\mu} = \frac{1}{i}\begin{vmatrix} 0 & \sigma^{\mu\dot a b} \\ \sigma^{\mu}_{a\dot b} & 0 \end{vmatrix}.$$

Acting in similar manner, from eqs. (10.31) and (10.32) we derive

$$i\Big\{ b_3\partial^{\dot a b}\Psi_{b0} + \frac{b_4}{\sqrt{6}}[-\sigma^{\mu\dot a b}\partial_{b\dot c}\Psi^{\dot c}_{\mu} + \frac{2}{i}\partial_{\mu}\Psi^{\dot a}_{\mu}]\Big\} M\Psi^{\dot a}_{0} = 0,$$

$$-i\Big\{ b_3\partial_{a\dot b}\Psi^{\dot b}_{0} + \frac{b_4}{\sqrt{6}}[-\sigma^{\mu}_{a\dot b}\partial^{\dot b c}\Psi_{\mu c} + \frac{2}{i}\partial_{\mu}\Psi_{\mu a}]\Big\} + M\Psi_{a0} = 0\,.$$

Joining them into one equation, we produce

$$i\begin{vmatrix} I & 0 \\ 0 & -I \end{vmatrix} \Big\{ b_3 \begin{vmatrix} \partial^{\dot a b}\sigma^{\mu}_{b\dot c} & 0 \\ 0 & \partial_{a\dot b}\sigma^{\mu\dot b c} \end{vmatrix} \begin{vmatrix} \Psi^{\dot c}_{0} \\ \Psi_{c0} \end{vmatrix} + \frac{b_4}{\sqrt{6}}\Big[\frac{2}{i} \begin{vmatrix} \Psi^{\dot a}_{\mu} \\ \Psi_{a\mu} \end{vmatrix}$$

$$- \begin{vmatrix} \sigma^{\mu\dot a b}\partial_{b\dot c} & 0 \\ 0 & \sigma^{\mu}_{a\dot b}\partial^{\dot b c} \end{vmatrix} \begin{vmatrix} \Psi^{\dot c}_{\mu} \\ \Psi_{c\mu} \end{vmatrix} \Big]\Big\} + M \begin{vmatrix} \Psi^{\dot b}_{0} \\ \Psi_{b0} \end{vmatrix} = 0\,,$$

or

$$i\gamma_5\Big\{ b_3\hat\partial\Psi_0 - i\frac{4b_4}{\sqrt{6}}[(\partial_{\mu}\Psi_{\mu}) - \frac{1}{4}\hat\partial(\gamma_{\mu}\Psi_{\mu})]\Big\} + M\Psi_0 = 0\,, \tag{10.39}$$

where

$$\Psi_0 = \begin{vmatrix} \Psi^{\dot a}_{0} \\ \Psi_{0b} \end{vmatrix}. \tag{10.40}$$

Now, let us consider eq. (10.34). With eq. (10.35) in mind, we derive

$$i\frac{b_5}{\sqrt{6}}\{\partial_c^{\dot{a}}\sigma^{\mu\dot{b}k}\Psi_{\mu k} + \partial_c^{\dot{b}}\sigma^{\mu\dot{a}k}\Psi_{\mu k}\} + i\frac{b_6}{\sqrt{6}}\{\partial_c^{\dot{a}}\Psi_0^{\dot{b}} + \partial_c^{\dot{b}}\Psi_0^{\dot{a}}\} + \frac{M}{2}[\sigma_c^{\mu\dot{a}}\Psi_\mu^{\dot{b}} + \sigma_c^{\mu\dot{b}}\Psi_\mu^{\dot{a}}] = 0\,.$$

After multiplying the last by $\sigma_{\dot{b}}^{\lambda a}$, we obtain

$$i\frac{b_5}{\sqrt{6}}\{\frac{2}{i}\partial_\lambda\sigma^{\mu\dot{b}k}\Psi_{\mu k} - \partial^{\dot{b}c}\sigma_{c\dot{a}}^{\lambda}\sigma^{\mu\dot{a}k}\Psi_{\mu k}\} + i\frac{b_6}{\sqrt{6}}\{\frac{2}{i}\partial_\lambda\Psi_0^{\dot{b}} - \partial^{\dot{b}c}\sigma_{c\dot{a}}^{\lambda}\Psi_0^{\dot{a}}\}$$
$$+\frac{M}{2}\{2\Psi_\lambda^{\dot{b}} - \sigma^{\mu\dot{b}c}\sigma_{c\dot{a}}^{\lambda}\psi_\mu^{\dot{a}}\} = 0\,. \tag{10.41}$$

Similarly, from eq. (10.33), we derive

$$-i\frac{b_5}{\sqrt{6}}\{\partial_a^{\dot{c}}\sigma^{\mu}_{b\dot{k}}\Psi_\mu^{\dot{k}} + \partial_b^{\dot{c}}\sigma^{\mu}_{a\dot{k}}\Psi_\mu^{\dot{k}}\} - i\frac{b_6}{\sqrt{6}}\{\partial_a^{\dot{c}}\Psi_{b0} + \partial_b^{\dot{c}}\Psi_a0\} + \frac{M}{2}\{\sigma_a^{\mu\dot{c}}\Psi_{\mu b} + \sigma_b^{\mu\dot{c}}\Psi_{\mu a}\} = 0\,.$$

Convoluting the last with $\sigma_{\dot{a}}^{\lambda b}$ we get

$$-i\frac{b_5}{\sqrt{6}}\{\frac{2}{i}\partial_\lambda\sigma^{\mu}_{b\dot{k}}\Psi_\mu^{\dot{k}} - \partial_{b\dot{c}}\sigma^{\lambda\dot{c}a}\sigma^{\mu}_{a\dot{k}}\Psi_\mu^{\dot{k}}\} - i\frac{b_6}{\sqrt{6}}\{\frac{2}{i}\partial_\lambda\Psi_{b0} - \partial_{b\dot{c}}\sigma^{\lambda\dot{c}a}\Psi_{a0}\}$$
$$+\frac{M}{2}\{2\Psi_{\mu b} - \sigma^{\mu}_{b\dot{c}}\sigma^{\lambda\dot{c}a}\Psi_{\mu a}\} = 0\,. \tag{10.42}$$

Joining eqs. (10.41) and (10.42), we find

$$\frac{i}{\sqrt{6}}\begin{vmatrix} I & 0 \\ 0 & -I\end{vmatrix}\Big\{b_5\Big[\frac{2}{i}\partial_\lambda\begin{vmatrix} 0 & \sigma^{\mu\dot{c}n} \\ \sigma^{\mu}_{c\dot{n}} & 0\end{vmatrix}\begin{vmatrix}\Psi_\mu^{\dot{n}} \\ \Psi_{n\mu}\end{vmatrix} - \begin{vmatrix} 0 & \partial^{\dot{c}a}\sigma^{\lambda}_{a\dot{b}}\sigma^{\mu\dot{b}n} \\ \partial_{c\dot{a}}\sigma^{\lambda\dot{a}b}\sigma^{\mu}_{b\dot{n}} & 0\end{vmatrix}\begin{vmatrix}\Psi_\mu^{\dot{n}} \\ \Psi_{n\mu}\end{vmatrix}\Big]$$
$$+b_6\Big[\frac{2}{i}\partial_\lambda\begin{vmatrix}\Psi_\mu^{\dot{c}} \\ \Psi_{c\mu}\end{vmatrix} - \begin{vmatrix}\partial^{\dot{c}a}\sigma^{\lambda}_{a\dot{b}} & 0 \\ 0 & \partial_{c\dot{a}}\sigma^{\lambda\dot{a}b}\end{vmatrix}\begin{vmatrix}\Psi_\mu^{\dot{b}} \\ \Psi_{b\mu}\end{vmatrix}\Big]$$
$$+\frac{M}{2}\Big[2\begin{vmatrix}\Psi_\mu^{\dot{c}} \\ \Psi_{c\mu}\end{vmatrix} - \begin{vmatrix}\sigma^{\mu\dot{c}a}\sigma^{\lambda}_{a\dot{b}} & 0 \\ 0 & \sigma^{\mu}_{c\dot{a}}\sigma^{\lambda\dot{a}b}\end{vmatrix}\begin{vmatrix}\Psi_\mu^{\dot{b}} \\ \Psi_{b\mu}\end{vmatrix}\Big] - 0\,,$$

or

$$\frac{i}{\sqrt{6}}\Big\{b_5[2\partial_\lambda(\gamma_\mu\Psi_\mu) + \hat{\partial}\gamma_\lambda(\gamma_\mu\Psi_\mu)] + b_6[\frac{2}{i}\partial_\lambda\Psi_0 - i\hat{\partial}\gamma_\lambda\Psi_0]\Big\} + \frac{M}{2}[2\Psi_\lambda + \gamma_\mu\gamma_\lambda\Psi_\mu] = 0\,.$$

Finally, we obtain

$$\frac{2i}{\sqrt{6}}\gamma_5\Big\{b_5[\partial_\lambda(\gamma_\mu\Psi_\mu) - \frac{1}{4}\gamma_\lambda\hat{\partial}(\gamma_\mu\Psi_\mu)]$$
$$-ib_6[\partial_\lambda\Psi_0 - \frac{1}{4}\gamma_\lambda\hat{\partial}\Psi_0\Big\} + M\{\Psi_\lambda - \frac{1}{4}\gamma_\lambda(\gamma_\mu\Psi_\mu)\} = 0\,. \tag{10.43}$$

Thus, the complete spin-tensor system reads

$$i\gamma_5\Big\{b_1\hat{\partial}(\gamma_\mu\Psi_\mu) - \frac{4b_2}{\sqrt{6}}[-\frac{1}{4}\hat{\partial}(\gamma_\mu\Psi_\mu) + (\partial_\mu\Psi_\mu)]\Big\} + M(\gamma_\mu\Psi_\mu) = 0, \tag{10.44}$$

$$i\gamma_5\Big\{b_3\hat{\partial}\Psi_0 - i\frac{4b_4}{\sqrt{6}}[(\partial_\mu\Psi_\mu) - \frac{1}{4}\hat{\partial}(\gamma_\mu\Psi_\mu)]\Big\} + M\Psi_0 = 0, \tag{10.45}$$

$$\frac{2i}{\sqrt{6}}\gamma_5\Big\{b_5[\partial_\lambda(\gamma_\mu\Psi_\mu) - \frac{1}{4}\gamma_\lambda\hat{\partial}(\gamma_\mu\Psi_\mu)] - ib_6[\partial_\lambda\Psi_0 - \frac{1}{4}\gamma_\lambda\hat{\partial}\Psi_0\Big\}$$
$$+M\Big\{\Psi_\lambda - \frac{1}{4}\gamma_\lambda(\gamma_\mu\Psi_\mu)\Big\} = 0\,. \tag{10.46}$$

10.6 On reducing the system to minimal form

We multiply eq. (10.44) by b_5, eq. (10.45) by $(-ib_6)$, and sum the results. This yields

$$i\gamma_5\Big\{b_1 b_5\hat{\partial}(\gamma_\mu\Psi_\mu) - \frac{4}{\sqrt{6}}b_2 b_5\Big[(\partial_\mu\Psi_\mu) - \frac{1}{4}\hat{\partial}(\gamma_\mu\Psi_\mu)\Big]\Big\} + Mb_5(\gamma_\mu\Psi_\mu)$$

$$+i\gamma_5\Big\{ - ib_3 b_6\hat{\partial}\Psi_0 - \frac{4}{\sqrt{6}}b_4 b_6\Big[(\partial_\mu\Psi_\mu) - \frac{1}{4}\hat{\partial}(\gamma_\mu\Psi_\mu)\Big]\Big\} - Mb_6\Psi_0 = 0\,,$$

or

$$i\gamma_5\Big\{b_1 b_5\hat{\partial}(\gamma_\mu\Psi_\mu) - ib_3 b_6\hat{\partial}\Psi_0 - \frac{4}{\sqrt{6}}(b_2 b_5 + b_4 b_6)\Big[(\partial_\mu\Psi_\mu) - \frac{1}{4}\hat{\partial}(\gamma_\mu\Psi_\mu)\Big]\Big\}$$
$$+M[b_5(\gamma_\mu\Psi_\mu) - ib_6\Psi_0] = 0\,.$$

Allowing for identities

$$b_1 + b_3 = 1 \quad\Longrightarrow\quad b_1 = 1 - b_3\,, \quad b_3 = 1 - b_1\,,$$

we rewrite the previous equation as follows

$$i\gamma_5\Big\{(1-b_3)b_5\hat{\partial}(\gamma_\mu\Psi_\mu) - i(1-b_1)b_6\hat{\partial}\Psi_0 - \frac{4}{\sqrt{6}}(b_2 b_5 + b_4 b_6)\Big[(\partial_\mu\Psi_\mu) - \frac{1}{4}\hat{\partial}(\gamma_\mu\Psi_\mu)\Big]\ \Big\}$$
$$+M[b_5(\gamma_\mu\Psi_\mu) - ib_6\Psi_0] = 0\,,$$

or

$$(i\gamma_5\hat{\partial} + M)[b_5(\gamma_\mu\Psi_\mu) - ib_6\Psi_0] - i\gamma_5\hat{\partial}[b_3 b_5(\gamma_\mu\Psi_\mu) - ib_1 b_6\Psi_0]$$
$$-\frac{4}{\sqrt{6}}i\gamma_5(b_2 b_5 + b_4 b_6)[(\partial_\mu\Psi_\mu) - \frac{1}{4}\hat{\partial}(\gamma_\mu\Psi_\mu)] = 0\,.$$

Whence taking into account identity $b_2 b_5 + b_4 b_6 = b_1 b_3$, we obtain

$$(i\gamma_5\hat{\partial} + M)[b_5(\gamma_\mu\Psi_\mu) - ib_6\Psi_0] - i\gamma_5\hat{\partial}[b_3 b_5(\gamma_\mu\Psi_\mu) - ib_1 b_6\Psi_0]$$
$$-\frac{4}{\sqrt{6}}b_1 b_3 i\gamma_5[(\partial_\mu\Psi_\mu) - \frac{1}{4}\hat{\partial}(\gamma_\mu\Psi_\mu)] = 0\,. \tag{10.47}$$

Now, let us multiply eq. (10.44) by $b_3 b_5$, eq. (10.45) by $(-ib_1 b_6)$, and sum the results. This yields

$$i\gamma_5\Big\{b_1 b_3 b_5\hat{\partial}(\gamma_\mu\Psi_\mu) - \frac{4}{\sqrt{6}}b_2 b_3 b_5[(\partial_\mu\Psi_\mu) - \frac{1}{4}\hat{\partial}(\gamma_\mu\Psi_\mu)]\Big\} + Mb_3 b_5(\gamma_\mu\Psi_\mu)$$

$$+i\gamma_5\Big\{ - ib_1 b_3 b_6\hat{\partial}\Psi_0 - \frac{4}{\sqrt{6}}b_1 b_4 b_6[(\partial_\mu\Psi_\mu)$$

$$-\frac{1}{4}\hat{\partial}(\gamma_\mu\Psi_\mu)]\Big\} - iMb_1 b_6\Psi_0 = 0\,,$$

or

$$M[b_3 b_5(\gamma_\mu\Psi_\mu) - ib_1 b_6\Psi_0]$$

$$+i\gamma_5\Big\{b_1 b_3[b_5\hat{\partial}(\gamma_\mu\Psi_\mu) - ib_6\hat{\partial}\Psi_0] - \frac{4}{\sqrt{6}}(b_2 b_3 b_5 + b_1 b_4 b_6)[(\partial_\mu\Psi_\mu) - \frac{1}{4}\hat{\partial}(\gamma_\mu\Psi_\mu)]\Big\} = 0\,,$$

that is

$$M[b_3 b_5(\gamma_\mu\Psi_\mu) - ib_1 b_6\Psi_0] + b_1 b_3 i\gamma_5\hat{\partial}[b_5(\gamma_\mu\Psi_\mu) - ib_6\Psi_0] = 0\,. \tag{10.48}$$

From eq. (10.48) we derive

$$b_3 b_5(\gamma_\mu \Psi_\mu) - i b_1 b_6 \Psi_0 = -\frac{b_1 b_3}{M} i\gamma_5 \hat{\partial}[b_5(\gamma_\mu \Psi_\mu) - i b_6 \Psi_0] = 0\,, \tag{10.49}$$

therefore, eq. (10.47) can be presented differently

$$(i\gamma_5\hat{\partial} + M)[b_5(\gamma_\mu \Psi_\mu) - i b_6 \Psi_0] + \frac{b_1 b_3}{M}\hat{\partial}^2[b_3 b_5(\gamma_\mu \Psi_\mu) - i b_1 b_6 \Psi_0]$$
$$-\frac{4}{\sqrt{6}} b_1 b_3 i\gamma_5[(\partial_\mu \Psi_\mu) - \frac{1}{4}\hat{\partial}(\gamma_\mu \Psi_\mu)] = 0\,. \tag{10.50}$$

Now, act on eq. (10.46) by operator ∂_λ, this gives

$$\frac{2i}{\sqrt{6}}\gamma_5 \frac{3}{4}\hat{\partial}^2[b_5(\gamma_\mu \Psi_\mu) - i b_6 \Psi_0] + M[(\partial_\mu \Psi_\mu) - \frac{1}{4}\hat{\partial}(\gamma_\mu \Psi_\mu)] = 0\,,$$

whence it follows

$$\hat{\partial}^2[b_5(\gamma_\mu \Psi_\mu) - i b_6 \Psi_0] = \frac{4M}{\sqrt{6}} i\gamma^5[(\partial_\mu \Psi_\mu) - \frac{1}{4}\hat{\partial}(\gamma_\mu \Psi_\mu)]\,. \tag{10.51}$$

Taking into account eq. (10.51) in eq. (10.50), we obtain

$$(i\gamma_5\hat{\partial} + M)[b_5(\gamma_\mu \Psi_\mu) - i b_6 \Psi_0] + \frac{4}{\sqrt{6}} b_1 b_3 i\gamma^5[(\partial_\mu \Psi_\mu) - \frac{1}{4}\hat{\partial}(\gamma_\mu \Psi_\mu)]$$

$$-\frac{4}{\sqrt{6}} b_1 b_3 i\gamma_5[(\partial_\mu \Psi_\mu) - \frac{1}{4}\hat{\partial}(\gamma_\mu \Psi_\mu)] = 0\,,$$

where two terms cancel each other and the final result is

$$(i\gamma_5\hat{\partial} + M)[b_5(\gamma_\mu \Psi_\mu) - i b_6 \Psi_0] = 0\,. \tag{10.52}$$

Below we use the notation

$$\Phi = b_5(\gamma_\mu \Psi_\mu) - i b_6 \Psi_0\,. \tag{10.53}$$

Taking into account eq. (10.52) in eq. (10.49), we get

$$b_3 b_5(\gamma_\mu \Psi_\mu) - i b_1 b_6 \Psi_0 = b_1 b_3 b_5(\gamma_\mu \Psi_\mu) - i b_1 b_3 b_6 \Psi_0 = 0\,,$$

or

$$b_3 b_5(1 - b_1)(\gamma_\mu \Psi_\mu) - i b_1 b_6 (1 - b_3)\Psi_0 = 0\,,$$

that is

$$b_3^2 b_5(\gamma_\mu \Psi_\mu) - i b_1^2 b_6 \Psi_0 = 0\,.$$

Consider together two relations

$$b_5(\gamma_\mu \Psi_\mu) - i b_6 \Psi_0 = \Phi\,, \quad b_3^2 b_5(\gamma_\mu \Psi_\mu) - i b_1^2 b_6 \Psi_0 = 0\,.$$

They make up linear system with respect to variables Ψ_0 and $(\gamma_\mu \Psi_\mu)$, its solution is

$$(\gamma_\mu \Psi_\mu) = \frac{b_1^2}{b_5(b_1^2 - b_3^2)}\Phi\,, \quad \Psi_0 = \frac{-i b_3^2}{b_6(b_1^2 - b_3^2)}\Phi\,. \tag{10.54}$$

The main function Φ satisfies the modified Dirac-like equation

$$\{\, i\gamma_5(\gamma_\mu \partial_\nu) + M \,\}\,\Phi = 0\,. \tag{10.55}$$

10.7 The presence of electromagnetic field

In the presence of external electromagnetic field, we have to start with the system (let $D_\mu = \partial_\nu + ieA_\mu$)

$$i\gamma_5 \Big\{ b_1 \hat{D}(\gamma_\mu \Psi_\mu) - \frac{4b_2}{\sqrt{6}} [(D_\mu \Psi_\mu) - \frac{1}{4}\hat{D}(\gamma_\mu \Psi_\mu)] \Big\} + M(\gamma_\mu \Psi_\mu) = 0\,, \tag{10.56}$$

$$i\gamma_5 \Big\{ b_3 \hat{D}\Psi_0 - i\frac{4b_4}{\sqrt{6}} [(D_\mu \Psi_\mu) - \frac{1}{4}\hat{D}(\gamma_\mu \Psi_\mu)] \Big\} + M\Psi_0 = 0\,, \tag{10.57}$$

$$\frac{2i}{\sqrt{6}} \gamma_5 \Big\{ b_5 [D_\lambda(\gamma_\mu \Psi_\mu) - \frac{1}{4}\gamma_\lambda \hat{D}(\gamma_\mu \Psi_\mu)]$$
$$-ib_6 (D_\lambda \Psi_0 - \frac{1}{4}\gamma_\lambda \hat{D}\Psi_0) \Big\} + M\Big[\Psi_\lambda - \frac{1}{4}\gamma_\lambda (\gamma_\mu \Psi_\mu) \Big] = 0\,. \tag{10.58}$$

Equation (10.58) may be re-written as follows

$$\Psi_\lambda - \frac{1}{4}\gamma_\lambda (\gamma_\mu \Psi_\mu)$$
$$= -\frac{2i}{M\sqrt{6}} \gamma_5 \Big\{ b_5 [D_\lambda(\gamma_\mu \Psi_\mu) - \frac{1}{4}\gamma_\lambda \hat{D}(\gamma_\mu \Psi_\mu)] - ib_6 (D_\lambda \Psi_0 - \frac{1}{4}\gamma_\lambda \hat{D}\Psi_0) \Big\}\,,$$

whence it follows

$$D_\lambda \Psi_\lambda - \frac{1}{4}\hat{D}(\gamma_\mu \Psi_\mu)$$
$$= -\frac{2i}{M\sqrt{6}} \gamma_5 D_\lambda \Big\{ b_5 [D_\lambda(\gamma_\mu \Psi_\mu) - \frac{1}{4}\gamma_\lambda \hat{D}(\gamma_\mu \Psi_\mu)] - ib_6 (D_\lambda \Psi_0 - \frac{1}{4}\gamma_\lambda \hat{D}\Psi_0) \Big\}. \tag{10.59}$$

Taking into account the last relation in eq. (10.56), we obtain

$$i\gamma_5 \Big\{ b_1 \hat{D}\underline{(\gamma_\mu \Psi_\mu)} + \frac{4b_2}{\sqrt{6}} \frac{2i}{M\sqrt{6}} \gamma_5 D_\lambda \Big[b_5 [D_\lambda \underline{(\gamma_\mu \Psi_\mu)} - \frac{1}{4}\gamma_\lambda \hat{D}\underline{(\gamma_\mu \Psi_\mu)}]$$
$$-ib_6 (D_\lambda \Psi_0 - \frac{1}{4}\gamma_\lambda \hat{D}\Psi_0) \Big] \Big\} + M(\gamma_\mu \Psi_\mu) = 0\,,$$

or differently

$$\Big\{ M + ib_1 \gamma_5 \hat{D} - \frac{4}{3M} b_2 b_5 (D^2 - \frac{1}{4}\hat{D}\hat{D}) \Big\} (\gamma_\mu \Psi_\mu)$$
$$+i\frac{4}{3M} b_2 b_6 (D^2 - \frac{1}{4}\hat{D}\hat{D})\Psi_0 = 0\,, \tag{10.60}$$

where $D^2 = D_\lambda D_\lambda$.

Taking into account eq. (10.59) in eq. (10.57), we produce

$$-i\frac{4}{3M} b_4 b_5 (D^2 - \frac{1}{4}\hat{D}\hat{D})(\gamma_\mu \Psi_\mu)$$
$$+\Big\{ M + ib_3 \gamma_5 \hat{D} - \frac{4}{3M} b_4 b_6 (D^2 - \frac{1}{4}\hat{D}\hat{D}) \Big\} \Psi_0 = 0\,. \tag{10.61}$$

Now we act on eq. (10.60) by operator

$$M + ib_3 \gamma_5 \hat{D} - \frac{4}{3M} b_4 b_6 (D^2 - \frac{1}{4}\hat{D}\hat{D})\,,$$

and on eq. (10.61) – by operator

$$-i\frac{4}{3M}b_2b_6(D^2-\frac{1}{4}\hat{D}\hat{D}),$$

then sum the results. This yields

$$\begin{aligned}&\Big\{M+i\gamma_5\hat{D}-\frac{4}{3M}(b_2b_5+b_4b_6)(D^2-\frac{1}{4}\hat{D}\hat{D})\\&+\frac{1}{M}b_1b_3\hat{D}\hat{D}-i\frac{4}{3M^2}b_2b_3b_5\gamma_5(\hat{D}D^2-D^2\hat{D})\Big\}(\gamma_\mu\Psi_\mu)\\&-\frac{4}{3M^2}b_2b_3b_6\gamma_5(\hat{D}D^2-D^2\hat{D})\Psi_0=0\,.\end{aligned}\tag{10.62}$$

Now, we act on eq. (10.61) by operator

$$M+ib_1\gamma_5\hat{D}-\frac{4}{3M}b_2b_5(D^2-\frac{1}{4}\hat{D}\hat{D}),$$

and on eq. (10.60) – by operator

$$i\frac{4}{3M}b_4b_5(D^2-\frac{1}{4}\hat{D}\hat{D}),$$

then sum the results. This yields

$$\begin{aligned}&\Big\{M+i\gamma_5\hat{D}-\frac{4}{3M}(b_4b_6+b_2b_5)(D^2-\frac{1}{4}\hat{D}\hat{D})\\&+\frac{1}{M}b_1b_3\hat{D}\hat{D}-i\frac{4}{3M^2}b_1b_4b_6\gamma_5(\hat{D}D^2-D^2\hat{D})\Big\}\Psi_0\\&+\frac{4}{3M^2}b_1b_4b_5\gamma_5(\hat{D}D^2-D^2\hat{D})(\gamma_\mu\Psi_\mu)=0\,.\end{aligned}\tag{10.63}$$

It is readily derived the following identity

$$\hat{D}\hat{D}=D^2-ieF_{\mu\nu}\frac{1}{4}(\gamma_\mu\gamma_\nu-\gamma_\nu\gamma_\mu)\,.\tag{10.64}$$

Thus, we have equations

$$\begin{aligned}&\Big\{M+ib_1\gamma_5\hat{D}-\frac{4}{3M}b_2b_5(D^2-\frac{1}{4}\hat{D}\hat{D})\Big\}(\gamma_\mu\Psi_\mu)\\&+\frac{4i}{3M}b_2b_6(D^2-\frac{1}{4}\hat{D}\hat{D})\Psi_0=0\,,\end{aligned}\tag{10.65}$$

$$-\frac{4i}{3M}b_4b_5(D^2-\frac{1}{4}\hat{D}\hat{D})(\gamma_\mu\Psi_\mu)+\Big\{M+ib_3\gamma_5\hat{D}-\frac{4}{3M}b_4b_6(D^2-\frac{1}{4}\hat{D}\hat{D})\Big\}\Psi_0=0\,,$$

$$\begin{aligned}&\Big\{M+i\gamma_5\hat{D}-\frac{4}{3M}(b_2b_5+b_4b_6)(D^2-\frac{1}{4}\hat{D}\hat{D})\\&+\frac{1}{M}b_1b_3\hat{D}\hat{D}-i\frac{4}{3M^2}b_2b_3b_5\gamma_5(\hat{D}D^2-D^2\hat{D})\Big\}(\gamma_\mu\Psi_\mu)\\&-\frac{4}{3M^2}b_2b_3b_6\gamma_5(\hat{D}D^2-D^2\hat{D})\Psi_0=0\,,\end{aligned}\tag{10.66}$$

$$\Big\{M + i\gamma_5 \hat{D} - \frac{4}{3M}(b_4 b_6 + b_2 b_5)(D^2 - \frac{1}{4}\hat{D}\hat{D}) + \frac{1}{M} b_1 b_3 \hat{D}\hat{D} - i\frac{4}{3M^2} b_1 b_4 b_6 \gamma_5 (\hat{D}D^2 - D^2\hat{D})\Big\}\Psi_0 + \frac{4}{3M^2} b_1 b_4 b_5 \gamma_5 (\hat{D}D^2 - D^2\hat{D})(\gamma_\mu \Psi_\mu) = 0 \,. \tag{10.67}$$

Substituting eq. (10.68) in eq. (10.66), we get

$$\Big\{M + i\gamma_5 \hat{D} - \frac{4}{3M}(b_2 b_5 + b_4 b_6)[D^2 - \frac{1}{4}D^2 + ieF_{[\mu\nu]}\frac{1}{16}(\gamma_\mu\gamma_\nu - \gamma_\nu\gamma_\mu)] + \frac{1}{M} b_1 b_3 [D^2 - ieF_{[\mu\nu]}\frac{1}{4}(\gamma_\mu\gamma_\nu - \gamma_\nu\gamma_\mu)] - \frac{4i}{3M^2} b_2 b_3 b_5 \gamma_5 (\hat{D}D^2 - D^2\hat{D})\Big\}(\gamma_\mu \Psi_\mu) - \frac{4}{3M^2} b_2 b_3 b_6 \gamma_5 (\hat{D}D^2 - D^2\hat{D})\Psi_0 = 0 \,,$$

whence due to identity $D^2(-b_2 b_5 - b_4 b_6 + b_1 b_3) = 0$ (see eq. (10.16)) it follows

$$\Big\{M + i\gamma_5 \hat{D} - \frac{1}{3M} b_1 b_3 ieF_{[\mu\nu]}(\gamma_\mu\gamma_\nu - \gamma_\nu\gamma_\mu) - \frac{4i}{3M^2} b_2 b_3 b_5 \gamma_5 (\hat{D}D^2 - D^2\hat{D})\Big\}(\gamma_\mu \Psi_\mu) - \frac{4}{3M^2} b_2 b_3 b_6 \gamma_5 (\hat{D}D^2 - D^2\hat{D})\Psi_0 = 0 \,. \tag{10.68}$$

Substituting eq. (10.68) in eq. (10.67), we get

$$\Big\{M + i\gamma_5 \hat{D} - \frac{4}{3M}(b_4 b_6 + b_2 b_5)[D^2 - \frac{1}{4}D^2 + ieF_{[\mu\nu]}\frac{1}{16}(\gamma_\mu\gamma_\nu - \gamma_\nu\gamma_\mu)] + \frac{1}{M} b_1 b_3 [D^2 - ieF_{[\mu\nu]}\frac{1}{4}(\gamma_\mu\gamma_\nu - \gamma_\nu\gamma_\mu)] - \frac{4i}{3M^2} b_1 b_4 b_6 \gamma_5 (\hat{D}D^2 - D^2\hat{D})\Big\}\Psi_0 + \frac{4}{3M^2} b_1 b_4 b_5 \gamma_5 (\hat{D}D^2 - D^2\hat{D})(\gamma_\mu \Psi_\mu) = 0 \,.$$

Again, three terms proportional to D^2 cancel each other, thus we obtain

$$\Big\{M + i\gamma_5 \hat{D} - \frac{1}{3M} b_1 b_3 ieF_{[\mu\nu]}(\gamma_\mu\gamma_\nu - \gamma_\nu\gamma_\mu) - \frac{4i}{3M^2} b_1 b_4 b_6 \gamma_5 (\hat{D}D^2 - D^2\hat{D})\Big\}\Psi_0 + \frac{4}{3M^2} b_1 b_4 b_5 \gamma_5 (\hat{D}D^2 - D^2\hat{D})(\gamma_\mu \Psi_\mu) = 0 \,. \tag{10.69}$$

Let us derive equation for the new bispinor function

$$\Phi = b_5(\gamma_\mu \Psi_\mu) - ib_6 \Psi_0 .$$

To this end, we multiply eq. (10.68) by b_5, eq. (10.69) by $(-ib_6)$, and sum the results. In this way, taking in mind identity $b_2 b_3 b_5 = -b_1 b_4 b_6$ (see (10.16)) we find

$$\Big\{i\gamma^5 \hat{D} - \frac{4 b_1 b_3}{M} ieF_{\mu\nu}\frac{\gamma_\mu\gamma_\nu - \gamma_\nu\gamma_\mu}{4} + M\Big\}\Phi(x) = 0 \,. \tag{10.70}$$

It remains to get expressions for bispinor components $\Psi = 0$ and $(\gamma_\mu \Psi_\mu)$ through the main bispinor Φ. Acting in the same way as in the free case, we can derive the following formulas:

$$\gamma_\mu \Psi_\mu = \frac{b_1^2}{b_5(b_1^2 - b_3^2)}\Big\{1 + \frac{4}{3}\Big(\frac{b_1 b_3}{M}\Big)^2 ieF_{\mu\nu}\frac{\gamma_\mu\gamma_\nu - \gamma_\nu\gamma_\mu}{4}\Big\}\Phi \,, \tag{10.71}$$

$$\Psi_0 = -i\frac{b_3^2}{b_6(b_1^2 - b_3^2)}\Big\{1 + \frac{4}{3}\Big(\frac{b_1 b_3}{M}\Big)^2 ieF_{\mu\nu}\frac{\gamma_\mu\gamma_\nu - \gamma_\nu\gamma_\mu}{4}\Big\}\Phi \,. \tag{10.72}$$

10.8 Extension of the model to General relativity

In order to follow the extension of the model from flat Minkowski space to any Riemannian space-time, we should turn back to the system (10.56)–(10.58) and make several simple modifications to it.

1. Taking in mind that in Minkowski space the *ict*-metric was used, however, in Riemannian space we use the metric $g_{\alpha\beta}(x)$, related to signature $(+,-,-,-)$, we must make the following change:

$$M \longrightarrow iM\,. \tag{10.73}$$

2. Now Dirac matrices in spinor basis are determined by the formulas

$$\gamma^0 = \begin{vmatrix} 0 & I \\ I & 0 \end{vmatrix}, \quad \gamma^i = \begin{vmatrix} 0 & -\sigma^i \\ \sigma^i & 0 \end{vmatrix}. \tag{10.74}$$

3. Derivatives are modified according to the rules [55]

$$\begin{gathered} D_\alpha(x) = \partial_\alpha + ieA_\alpha(x) \implies \\ D_\alpha(x) = \nabla_\alpha + \Gamma_\alpha(x) + ieA_\alpha(x)\,, \quad \hat{D} = \gamma^\alpha(x)D_\alpha(x)\,, \end{gathered} \tag{10.75}$$

where $\Gamma_\alpha(x)$ is bispinor connection by Tetrode–Weyl–Fock–Ivanenko [55, 57], and $\gamma^\alpha(x) = \gamma^a e^\alpha_{(a)}(x)$.

4. Note important commutation rules [55]:

$$\begin{gathered} \hat{D}(x) = \gamma^\rho(x)D_\beta = D_\beta\gamma^\rho(x)\,, \qquad D_\sigma(x)g_{\alpha\beta}(x) = g_{\alpha\beta}(x)D_\sigma(x)\,, \\ \hat{D}\hat{D} = D_\alpha D_\beta\Big(\frac{\gamma^\alpha\gamma^\beta + \gamma^\beta\gamma^\alpha}{2} + \frac{\gamma^\alpha\gamma^\beta - \gamma^\beta\gamma^\alpha}{2}\Big) = \Box - \Sigma(x)\,, \\ D^2 = D^\alpha D_\alpha, \quad \Sigma(x) = (-ieF_{\alpha\beta}\sigma^{\alpha\beta}(x) + \frac{R}{4})\,, \end{gathered} \tag{10.76}$$

where $R(x)$ is the Ricci scalar.

5. Note the notations [55]

$$\begin{gathered} \gamma^5(x) = \frac{i}{4!}\epsilon_{\alpha\beta\rho\sigma}(x)\gamma^\alpha(x)\gamma^\beta(x)\gamma^\rho(x)\gamma^\sigma(x)\,, \\ \epsilon^{\alpha\beta\rho\sigma}(x) = \epsilon^{abcd}\, e^\alpha_{(a)}(x)\, e^\beta_{(b)}(x)\, e^\rho_{(c)}(x)\, e^\sigma_{(d)}(x), \quad \epsilon_{0123} = -1\,. \end{gathered} \tag{10.77}$$

Levi-Civita object $\epsilon^{\alpha\beta\rho\sigma}(x)$ changes under tetrad transformations in accordance with the rule [55]

$$\epsilon'^{\alpha\beta\rho\sigma}(x) = -\det[L_{ai}(x)]\, \epsilon^{\alpha\beta\rho\sigma}(x)\,. \tag{10.78}$$

In particular, at the tetrad P-reflection, it transforms as a tetrad pseudoscalar [55]

$$\epsilon^{(p)\alpha\beta\rho\sigma}(x) = (-1)\, \epsilon^{\alpha\beta\rho\sigma}(x)\,. \tag{10.79}$$

We readily derive identity

$$\begin{aligned} \gamma^5(x) &= \frac{i}{4!}\epsilon^{abcd}\, e_{\alpha(a)}e_{\beta(b)}e_{\rho(c)}e_{\sigma(d)}e^\alpha_{(m)}\gamma^m e^\beta_{(n)}\gamma^n e^\rho_{(k)}\gamma^k e^\sigma_{(l)}\gamma^l) \\ &= \frac{i}{4!}\epsilon_{mnkl}\gamma^m\gamma^n\gamma^k\gamma^l = \gamma^5\,. \end{aligned} \tag{10.80}$$

The above analysis for the modified system (remember that $\gamma^5(x) \equiv \gamma^5$)

$$i\gamma^5(x)\Big\{b_1\hat{D}\gamma^\mu(x)\Psi_\mu - \frac{4b_2}{\sqrt{6}}[D^\mu\Psi_\mu - \frac{1}{4}\hat{D}\gamma^\mu(x)\Psi_\mu]\Big\} + iM\gamma^\mu(x)\Psi_\mu = 0, \qquad (10.81)$$

$$i\gamma^5(x)\Big\{b_3\hat{D}\Psi_0 - i\frac{4b_4}{\sqrt{6}}[D^\mu\Psi_\mu - \frac{1}{4}\hat{D}\gamma^\mu(x)\Psi_\mu\,]\Big\} + iM\Psi_0 = 0, \qquad (10.82)$$

$$\frac{2i}{\sqrt{6}}\gamma^5(x)\Big\{b_5[D_\lambda\gamma^\mu(x)\Psi_\mu - \frac{1}{4}\gamma_\lambda(x)\hat{D}\gamma^\mu(x)\Psi_\mu]$$
$$-ib_6[D_\lambda\Psi_0 - \frac{1}{4}\gamma_\lambda(x)\hat{D}\Psi_0]\Big\} + iM\Big\{\Psi_\lambda - \frac{1}{4}\gamma_\lambda(x)\gamma^\mu(x)\Psi_\mu\Big\} = 0 \qquad (10.83)$$

remains in fact the same. We can write down final result without repeating the calculation

$$\Phi = b_5(\gamma_\mu\Psi_\mu) - ib_6\Psi_0\,, \quad \Big\{i\gamma^\alpha(x)\,(\nabla_\alpha + \Gamma_\alpha(x) + ieA_\alpha(x))$$
$$-\frac{4b_1b_3}{iM}\gamma^5\Big[-ieF_{\mu\nu}(x)\sigma^{\mu\nu}(x) + \frac{R(x)}{4}\Big] + i\gamma^5 M\Big\}\Phi(x) = 0\,. \qquad (10.84)$$

Expressions for concomitant components through the main bispinor $\Phi(x)$ are given by the formulas

$$\gamma^\mu(x)\Psi_\mu(x) = \frac{b_1^2}{b_5(b_1^2 - b_3^2)}\Big\{1 - \frac{4}{3}\Big(\frac{b_1b_3}{iM}\Big)^2(-ieF_{\mu\nu}\sigma^{\mu\nu} + \frac{R(x)}{4})\Big\}\Phi\,, \qquad (10.85)$$

$$\Psi_0(x) = -i\frac{b_3^2}{b_6(b_1^2 - b_3^2)}\Big\{1 - \frac{4}{3}\Big(\frac{b_1b_3}{iM}\Big)^2(-ieF_{\mu\nu}\sigma^{\mu\nu} + \frac{R(x)}{4})\Big\}\Phi\,. \qquad (10.86)$$

In absence of external electromagnetic field, eq. (10.84) becomes simpler

$$\Big\{i\gamma^\alpha(x)[\nabla_\alpha + \Gamma_\alpha(x)] - \frac{b_1b_3}{iM}\gamma^5 R(x) + i\gamma^5 M\Big\}\Phi(x) = 0\,. \qquad (10.87)$$

10.9 P-noninvariant particle in the Coulomb field

Let us consider an elementary example, free particle in spherical tetrad of Minkowski space

$$dS^2 = dt^2 - dr^2 - r^2(d\theta^2 + \sin^2\theta d\phi^2)\,,$$

for this case from eq. (10.87), we get the following equation (let $\Phi(x) = r^{-1}\Psi(x)$)

$$\Big(i\gamma^0\partial_t + i\gamma^3\partial_r + \frac{1}{r}\Sigma_{\theta\phi} + i\gamma^5 M\Big)\Psi(x) = 0\,,$$
$$\Sigma_{\theta,\phi} = i\gamma^1\partial_\theta + \gamma^2\frac{i\partial_\phi + i\sigma^{12}}{\sin\theta}\,. \qquad (10.88)$$

We chose substitution for wave function with quantum number ϵ, j, m (we use Wigner D-functions, $D^j_{-m,\sigma}(\phi,\theta,0) \equiv D_\sigma$):

$$\Psi_{\epsilon jm}(x) = \frac{e^{-i\epsilon t}}{r}\left|\begin{array}{c} f_1(r)D_{-1/2} \\ f_2(r)D_{+1/2} \\ f_3(r)D_{-1/2} \\ f_4(r)D_{+1/2} \end{array}\right|. \qquad (10.89)$$

Action of the angular operator is as follows

$$\Sigma_{\theta,\phi}\Psi_{\epsilon jm}(x) = i\nu\frac{e^{-i\epsilon t}}{r}\left|\begin{array}{c} -f_4(r)D_{-1/2} \\ +f_3(r)D_{+1/2} \\ +f_2(r)\,D_{-1/2} \\ -f_1(r)D_{+1/2} \end{array}\right| . \tag{10.90}$$

Futher, we find radial equations (let $\nu = j+1/2, \nu = 1,2,...$ and $\sigma = -1$):

$$\begin{gathered} \epsilon f_3 - i\frac{d}{dr}f_3 - i\frac{\nu}{r}f_4 + iMf_1 = 0, \quad \epsilon f_4 + i\frac{d}{dr}f_4 + i\frac{\nu}{r}f_3 + iMf_2 = 0\,, \\ \epsilon f_1 + i\frac{d}{dr}f_1 + i\frac{\nu}{r}f_2 + i\sigma Mf_3 = 0\,, \quad \epsilon f_2 - i\frac{d}{dr}f_2 - i\frac{\nu}{r}f_1 + i\sigma Mf_4 = 0\,, \end{gathered} \tag{10.91}$$

in comparison with the ordinary Dirac equation here the signs at M in eqs. 3 and 4 are different.

In the case of conventional Dirac equation, we can diagonalise additionally the space reflection operator. In Cartesian tetrad, it is defined by the formula

$$\hat{\Pi}_C = i\gamma^0 \otimes \hat{P}, \quad \hat{\Pi}_{C.} = \left|\begin{array}{cccc} 0 & 0 & i & 0 \\ 0 & 0 & 0 & i \\ i & 0 & 0 & 0 \\ 0 & i & 0 & 0 \end{array}\right| \otimes \hat{P}\,, \qquad \hat{P}\,(\theta,\phi) = (\pi-\theta,\ \phi+\pi),$$

after transforming to spherical basis it takes the form

$$\hat{\Pi}_{sph} = \left|\begin{array}{cccc} 0 & 0 & 0 & -1 \\ 0 & 0 & -1 & 0 \\ 0 & -1 & 0 & 0 \\ -1 & 0 & 0 & 0 \end{array}\right| \otimes \hat{P}\,. \tag{10.92}$$

From eigenvalues equation $\hat{\Pi}_{sph}\Psi_{jm} = \Pi\Psi_{jm}$ (allow for the identity $\hat{P}D^j_{-m,\sigma} = (-1)^j D^j_{-m,-\sigma}$), we find two eigenvalues $\Pi = \delta(-1)^j$ and corresponding restrictions on radial functions:

$$\Pi = \delta\,(-1)^{j+1}\,, \quad \delta = \pm 1\,, \quad f_4 = \delta f_1\,, \quad f_3 = \delta f_2\,. \tag{10.93}$$

However, these restrictions are not consistent with the above system (10.91) of radial equations, because we get (recall that $\sigma = -1$)

$$\begin{gathered} \epsilon f_2 - i\frac{d}{dr}f_2 - i\frac{\nu}{r}f_1 + i\delta Mf_1 = 0\,, \quad \epsilon f_1 + i\frac{d}{dr}f_1 + i\frac{\nu}{r}f_2 + i\delta Mf_2 = 0\,, \\ \epsilon f_1 + i\frac{d}{dr}f_1 + i\frac{\nu}{r}f_2 + i\sigma\delta Mf_2 = 0\,, \quad \epsilon f_2 - i\frac{d}{dr}f_2 - i\frac{\nu}{r}f_1 + i\sigma\delta Mf_1 = 0\,. \end{gathered} \tag{10.94}$$

Thus, the P-noninvariant wave equation provides us with substantially new mathematical tasks when we try to construct solutions for this equation, even in the simplest case of a free particle.

We may add, for instance, the Coulomb potential, so obtaining equations (for simplicity, let the parameter of anomalous magnetic moment vanish)

$$\begin{gathered} (\epsilon + \frac{\alpha}{r})f_3 - i\frac{d}{dr}f_3 - i\frac{\nu}{r}f_4 + iMf_1 = 0\,, \\ (\epsilon + \frac{\alpha}{r})f_4 + i\frac{d}{dr}f_4 + i\frac{\nu}{r}f_3 + iMf_2 = 0\,, \\ (\epsilon + \frac{\alpha}{r})f_1 + i\frac{d}{dr}f_1 + i\frac{\nu}{r}f_2 - iMf_3 = 0\,, \\ (\epsilon + \frac{\alpha}{r})f_2 - i\frac{d}{dr}f_2 - i\frac{\nu}{r}f_1 - iMf_4 = 0\,. \end{gathered} \tag{10.95}$$

Let us try to diagonalise another discrete operator (adding the multiplier γ^5)

$$\hat{\Delta}_{sph} = \left| \begin{array}{cccc} 1 & 0 & 0 & 0 \\ 0 & 1 & 0 & 0 \\ 0 & 0 & -1 & 0 \\ 0 & 0 & 0 & -1 \end{array} \right| \left| \begin{array}{cccc} 0 & 0 & 0 & -1 \\ 0 & 0 & -1 & 0 \\ 0 & -1 & 0 & 0 \\ -1 & 0 & 0 & 0 \end{array} \right| \otimes \hat{P}$$

$$= \left| \begin{array}{cccc} 0 & 0 & 0 & -1 \\ 0 & 0 & -1 & 0 \\ 0 & +1 & 0 & 0 \\ +1 & 0 & 0 & 0 \end{array} \right| \otimes \hat{P} \,. \tag{10.96}$$

The eigenvalue equation $\hat{\Delta}_{sph}\Psi_{jm} = \Delta\ \Psi_{jm}$ gives

$$(-1)^{j+1} f_4 = \Delta f_1\,, \quad (-1)^{j} f_1 = \Delta f_4\,, \quad (-1)^{j+1} f_3 = \Delta f_2\,, \quad (-1)^{j} f_2 = \Delta f_3\,,$$

whence follow two eigenvalues and respective restrictions on radial functions:

$$\delta = \pm i\,, \quad \Delta = \delta(-1)^{j}\,, \quad f_4 = -\delta f_1\,, \quad f_3 = -\delta f_2\,. \tag{10.97}$$

Let us impose these constraints in the radial system (10.91):

$$-\epsilon\delta f_2 + i\frac{d}{dr}\delta f_2 + i\frac{\nu}{r}\delta f_1 + iMf_1 = 0\,, \quad -\epsilon\delta f_1 - i\frac{d}{dr}\delta f_1 - i\frac{\nu}{r}\delta f_2 + iMf_2 = 0\,,$$

$$\epsilon f_1 + i\frac{d}{dr}f_1 + i\frac{\nu}{r}f_2 - i\sigma M\delta f_2 = 0\,, \quad \epsilon f_2 - i\frac{d}{dr}f_2 - i\frac{\nu}{r}f_1 - i\sigma M\delta f_1 = 0\,.$$

We can see that these equations are consistent. Depending on the value of δ, we get

$$\delta = +i, \qquad \begin{array}{l} \epsilon f_2 - i\dfrac{d}{dr}f_2 - i\dfrac{\nu}{r}f_1 - Mf_1 = 0\,, \\ \epsilon f_1 + i\dfrac{d}{dr}f_1 + i\dfrac{\nu}{r}f_2 - Mf_2 = 0\,; \end{array} \tag{10.98}$$

$$\delta = -i, \qquad \begin{array}{l} \epsilon f_2 - i\dfrac{d}{dr}f_2 - i\dfrac{\nu}{r}f_1 + Mf_1 = 0\,, \\ \epsilon f_1 + i\dfrac{d}{dr}f_1 + i\dfrac{\nu}{r}f_2 + Mf_2 = 0\,. \end{array} \tag{10.99}$$

Their more convenient representations are (recall that $\nu = j + 1/2$)

$$\delta = +i, \qquad \begin{array}{l} (i\epsilon + \dfrac{d}{dr})f_2 + (\dfrac{\nu}{r} - iM)f_1 = 0\,, \\ (i\epsilon - \dfrac{d}{dr})f_1 - (\dfrac{\nu}{r} + iM)f_2 = 0\,; \end{array} \tag{10.100}$$

$$\delta = +i, \qquad \begin{array}{l} (i\epsilon + \dfrac{d}{dr})f_2 + (\dfrac{\nu}{r} + iM)f_1 = 0\,, \\ (i\epsilon - \dfrac{d}{dr})f_1 - (\dfrac{\nu}{r} - iM)f_2 = 0\,. \end{array} \tag{10.101}$$

They lead to 2nd-order equations:

$\delta = +i$,

$$\begin{aligned} &\frac{d^2 f_1}{dr^2} + \Big(\frac{-iM}{\nu + iMr} + \frac{1}{r}\Big)\frac{df_1}{dr} + \Big(-\frac{i\epsilon}{r} - \frac{\nu^2}{r^2} - M^2 - \frac{\epsilon M}{\nu + iMr} + \epsilon^2\Big) f_1 = 0 \,, \\ &\frac{d^2 f_2}{dr^2} + \Big(+\frac{iM}{\nu - iMr} + \frac{1}{r}\Big)\frac{df_2}{dr} + \Big(\frac{i\epsilon}{r} - \frac{\nu^2}{r^2} - M^2 - \frac{\epsilon M}{\nu - iMr} + \epsilon^2\Big) f_2 = 0 \,; \end{aligned} \tag{10.102}$$

$\delta = -i$,

$$\begin{aligned} &\frac{d^2 f_1}{dr^2} + \Big(\frac{iM}{\nu - iMr} + \frac{1}{r}\Big)\frac{df_1}{dr} + \Big(-\frac{i\epsilon}{r} - \frac{\nu^2}{r^2} - M^2 + \frac{\epsilon M}{\nu - iMr} + \epsilon^2\Big) f_1 = 0 \,, \\ &\frac{d^2 f_2}{dr^2} + \Big(\frac{-iM}{\nu + iMr} + \frac{1}{r}\Big)\frac{df_2}{dr} + \Big(\frac{i\epsilon}{r} - \frac{\nu^2}{r^2} - M^2 + \frac{\epsilon M}{\nu + iMr} + \epsilon^2\Big) f_2 = 0 \,. \end{aligned} \tag{10.103}$$

Note the symmetry $M \to -M$ between two pairs of equations. In eqs. (10.102) and (10.103), we have the confluent Heun functions.

In presence of the external Coulomb field, instead of eqs. (10.100) and (10.101), we have the following equations

$\delta = +i$,

$$\begin{aligned} &[i(\epsilon + \frac{\alpha}{r}) + \frac{d}{dr}] f_2 + (\frac{\nu}{r} - iM) f_1 = 0 \,, \\ &[i(\epsilon + \frac{\alpha}{r}) - \frac{d}{dr}] f_1 - (\frac{\nu}{r} + iM) f_2 = 0 \,; \end{aligned} \tag{10.104}$$

$\delta = +i$,

$$\begin{aligned} &[i(\epsilon + \frac{\alpha}{r}) + \frac{d}{dr}] f_2 + (\frac{\nu}{r} + iM) f_1 = 0 \,, \\ &[i(\epsilon + \frac{\alpha}{r}) - \frac{d}{dr}] f_1 - (\frac{\nu}{r} - iM) f_2 = 0 \,. \end{aligned} \tag{10.105}$$

Further, we derive the 2nd-order equations

$\delta = +i$,

$$\frac{d^2 f_1}{dr^2} + \Big[\frac{1}{r} + \frac{-iM}{iMr + \nu}\Big]\frac{df_1}{dr}$$

$$\Big[\frac{2\,\epsilon\,\alpha\,\nu - \alpha\,M - i\epsilon\,\nu}{\nu\, r} + \frac{\alpha^2 - \nu^2}{r^2} + \epsilon^2 - M^2 - \frac{M\,(-iM\alpha + \epsilon\,\nu)}{\nu\,(iMr + \nu)}\Big] f_1 = 0 \,,$$

$$\frac{d^2 f_2}{dr^2} + \Big[\frac{1}{r} + \frac{iM}{-iMr + \nu}\Big]\frac{df_2}{dr}$$

$$+\Big[\frac{2\,\epsilon\,\alpha\,\nu - \alpha\,M + i\epsilon\,\nu}{\nu\, r} + \frac{\alpha^2 - \nu^2}{r^2} + \epsilon^2 - M^2 - \frac{M\,(iM\alpha + \epsilon\,\nu)}{\nu\,(-iMr + \nu)}\Big] f_2 = 0 \,;$$

$\delta - -i$,

$$\frac{d^2 f_1}{dr^2} + \Big[\frac{1}{r} + \frac{iM}{-iMr + \nu}\Big]\frac{df_1}{dr}$$

$$+\Big[\frac{2\,\epsilon\,\alpha\,\nu + \alpha\,M - i\epsilon\,\nu}{\nu\, r} + \frac{\alpha^2 - \nu^2}{r^2} + \epsilon^2 - M^2 + \frac{M\,(iM\alpha + \epsilon\,\nu)}{\nu\,(-iMr + \nu)}\Big] f_1 = 0 \,,$$

$$\frac{d^2 f_2}{dr^2} + \Big[\frac{1}{r} + \frac{-iM}{iMr + \nu}\Big]\frac{df_2}{dr}$$

$$+\Big[\frac{2\epsilon\alpha\nu+\alpha M+i\epsilon\nu}{\nu r}+\frac{\alpha^2-\nu^2}{r^2}+\epsilon^2-M^2+\frac{M\,(-iM\alpha+\epsilon\,\nu)}{\nu\,(iMr+\nu)}\Big]f_2=0\,.$$

Here again, we have differential equations for confluent Heun functions. Note the symmetry $M\to -M$ between two pairs of equations. Also, we can see that equations for f_1 and f_2 are complex conjugate to each other. Therefore, we can study only one equation. Let it be the equation for f_1:

$$\frac{d^2f_1}{dr^2}+\Big[\frac{1}{r}+\frac{-iM}{iMr+\nu}\Big]\frac{df_1}{dr}$$

$$\Big[\frac{2\,\epsilon\,\alpha\,\nu-\alpha\,M-i\epsilon\,\nu}{\nu\,r}+\frac{\alpha^2-\nu^2}{r^2}+\epsilon^2-M^2-\frac{M\,(-iM\alpha+\epsilon\,\nu)}{\nu\,(iMr+\nu)}\Big]f_1=0\,.$$

It is convenient to use the variable $x=Mr$, then we get

$$\frac{d^2f_1}{dx^2}+\Big(\frac{1}{x}-\frac{1}{x-i\nu}\Big)\frac{df_1}{dx}$$

$$+\Big(\frac{2E\alpha-\gamma-iE}{x}-\frac{\Gamma^2}{x^2}+E^2-1+\frac{\gamma+iE}{x-i\nu}\Big)f_1=0\,, \tag{10.106}$$

where $E=\epsilon/M$, $\Gamma^2=\nu^2-\alpha^2$, and $\gamma=\alpha/\nu$. We find behaviour of solutions near singular points:

$$x\to i\nu,\quad f_1=(x-i\nu)^\rho,\quad \rho=0,2;\qquad x\to 0,\quad f_1=x^A,\quad A=\pm\Gamma^2\,.$$

In the variable $y=x^{-1}$, eq. (10.106) reads

$$y^4\frac{d^2}{dy^2}+2y^3\frac{d}{dy}-\Big(y-\frac{y}{1-i\nu y}\Big)y^2\frac{d}{dt}$$

$$+\Big((2E\alpha-\gamma-iE)y-\Gamma^2y^2+E^2-1+\frac{(\gamma+iE)y}{1-i\nu y}\Big)f_1=0\,.$$

In vicinity of $y=0$ it becomes simpler

$$\Big(\frac{d^2}{dy^2}+\frac{2}{y}\frac{d}{dy}+\frac{E^2-1}{y^4}\Big)f_1=0\,,$$

therefore, we conclude that the singular point $y=0$ is irregular of the rank 2.

In eq. (10.106) we change the variable, $y=-ix/\nu$:

$$\frac{d^2f_1}{dy^2}+\Big(\frac{1}{y}-\frac{1}{y-1}\Big)\frac{df_1}{dy}$$

$$+\Big[\frac{\nu\,(E+2\,iE\alpha-i\gamma)}{y}-\frac{\Gamma^2}{y^2}-\nu^2\,(-1+E^2)-\frac{\nu\,(-i\gamma+E)}{y-1}\Big]f_1=0\,. \tag{10.107}$$

We search solutions in the form $f_1=y^A(y-1)^Be^{Cy}f(y)$, getting for $f(y)$ the following equation

$$\frac{d^2f}{dy^2}+\Big(2\,C+\frac{2\,A+1}{y}+\frac{2\,B-1}{y-1}\Big)\frac{df}{dy}$$

$$+\Big[\frac{-2\,AB-B+A+\nu\,E-i\nu\,\gamma+2\,i\nu\,E\alpha+2\,AC+C}{y}+\frac{A^2-\Gamma^2}{y^2}$$

$$+C^2-\nu^2E^2+\nu^2+\frac{B-\nu\,E+2\,BC-A+2\,AB+i\nu\,\gamma-C}{y-1}+\frac{B\,(B-2)}{(y-1)^2}\Big]f=0\,.$$

Evidently, we should impose restrictions

$$A = \pm \Gamma\,, \quad B = 0\,,2\,, \quad C = \pm \nu \sqrt{-1 + E^2}\,;$$

to bound states may correspond the values

$$B = 0, \quad A = +\Gamma, \quad C = -i\nu\sqrt{1 - E^2}, \quad f_1 \sim x^{+\Gamma} e^{-\sqrt{1-E^2}x} f(y)\,. \tag{10.108}$$

Equation for $f(y)$ becomes simpler

$$\frac{d^2 f}{dy^2} + \Big(2\,C + \frac{2\,A + 1}{y} + \frac{2\,B - 1}{y - 1}\Big)\frac{df}{dy}$$

$$+\Big[\frac{-2\,AB - B + A + \nu\,E - i\nu\,\gamma + 2\,i\nu\,E\alpha + 2\,AC + C}{y}$$

$$+\frac{B - \nu\,E + 2\,BC - A + 2\,AB + i\nu\,\gamma - C}{y - 1}\Big] f = 0\,;$$

it may be identified with canonical form of the confluent Heun equation

$$\frac{d^2 H}{dz^2} + \Big(- t + \frac{c}{z} + \frac{d}{z-1}\Big)\frac{dH}{dz} + \Big(- \frac{\lambda}{z} + \frac{\lambda - ta}{z - 1}\Big)\; H = 0\,. \tag{10.109}$$

Its parameters are given by relations

$$t = -2C = 2i\nu\sqrt{1 - E^2}\,, \quad c = 2A + 1 = 2\Gamma + 1\,, \quad d = 2B - 1 = -1\,, \tag{10.110}$$

$$\lambda = 2\,AB + B - A - \nu\,E + i\nu\,\gamma - 2\,i\nu\,E\alpha - 2\,AC - C$$

$$= -\Gamma - \nu\,E + i\nu\,\gamma - 2\,i\nu\,E\alpha + 2\Gamma i\nu\sqrt{1 - E^2} + i\nu\sqrt{1 - E^2}\,. \tag{10.111}$$

From relations

$$\lambda - ta = B - \nu\,E + 2\,BC - A + 2\,AB + i\nu\,\gamma - C \quad \longrightarrow$$

$$2i\nu\sqrt{1 - E^2}a = -2\,i\nu\,E\alpha + 2\,\Gamma i\nu\sqrt{1 - E^2}$$

we find expression for parameter a:

$$a = -\frac{E\alpha}{\sqrt{1 - E^2}} + \Gamma, \quad \Gamma = \sqrt{(j + 1/2)^2 - \alpha^2}\,. \tag{10.112}$$

Transcendental Heun functions are determined by additional constraint $a = -n$, which gives the quantization rule for energies

$$\frac{E\alpha}{\sqrt{1 - E^2}} = n + \sqrt{(j + 1/2)^2 - \alpha^2} \equiv N \quad \Longrightarrow \quad E = \frac{1}{\sqrt{1 + \frac{\alpha^2}{N^2}}}\,. \tag{10.113}$$

We can see that the energy spectrum for the P-noninvariant spin 1/2 particle in the external Coulomb field coincides with that for ordinary particle, though the explicit form of the wave function is different.

10.10 Conclusions

Within the theory of relativistic wave equations with extended sets of Lorentz group representations, a new P-noninvariant 20-component wave equation for a spin 1/2 particle is proposed. The presence of an external electromagnetic field and a Riemannian space-time background are taken into account. Due to the internal structure of the particle, additional interaction terms appear, it relates to the anomalous magnetic moment of the particle. Exact solutions of the equation in the presence of the external Coulomb field have been constructed, radial wave functions are expressed in terms of confluent Heun functions.

Bibliography

[1] V.L. Ginzburg and Ya.A. Smorodinsky. On wave equations for particles with variable spin. *Zhurnal Eksperimental'noi i Teoreticheskoi Fiziki*, 13: 274, 1943 (in Russian).

[2] V.L. Ginzburg. To the theory of exited states of elementary particles. *Zhurnal Eksperimental'noi i Teoreticheskoi Fiziki*, 13: 33–58, 1943 (in Russian).

[3] A.S. Davydov. Wave equations of a particle having spin 3/2 in absence of field. *Zhurnal Eksperimental'noi i Teoreticheskoi Fiziki*, 13(9-10): 313–319, 1943 (in Russian).

[4] H.J. Bhabha and Harish-Chandra. On the theory of point particles. *Proceedings of the Royal Society London A.*, 183: 134–141, 1944.

[5] H.J. Bhabha. Relativistic wave equations for the proton. *Proceedings of the Indian Academy of Sciences A.*, 21: 241–264, 1945.

[6] H.J. Bhabha. Relativistic wave equations for elementary particles. *Reviews of Modern Physics* 17(2-3): 200–215, 1945.

[7] H.J. Bhabha. The theory of the elementary particles. *Reports on Progress in Physics*, 10: 253–271, 1946.

[8] Harish-Chandra. On the equations of motion of point particles. *Proceedings of the Royal Society London A.*, 185: 269–287, 1946.

[9] Harish-Chandra. Relativistic equations for elementary particles. *Proceedings of the Royal Society London A.*, 192: 195–218, 1948.

[10] I.M. Gel'fand and A.M. Yaglom. General relativistic invariant equations snd infinitely dimensional representation of the Lorentz group. *Zhurnal Eksperimental'noi i Teoreticheskoi Fiziki*, 18(8): 703–733, 1948 (in Russian).

[11] I.M. Gel'fand and A.M. Yaglom. Pauli theorem for general relativistic invariant wave equations. *Zhurnal Eksperimental'noi i Teoreticheskoi Fiziki*, 18(12): 1096–1104, 1948 (in Russian).

[12] I.M. Gel'fand and A.M. Yaglom. Charge conjugation for general relativistic invariant wave equations. *Zhurnal Eksperimental'noi i Teoreticheskoi Fiziki*, 18(12): 1105–1111, 1948 (in Russian).

[13] H.J. Bhabha. Theory of elementary particles-fields. In *Lectures Delivered at 2nd Summer Seminar Canadian Math. Congress: Held at University of British Columbia*, Aug-1949. pages 1–103.

[14] F.I. Fedorov. To the problem of solving relativistic wave equations. *Doklady AN USSR*, 65(6): 813–814, 1949 (in Russian).

[15] E.E. Fradkin. To the theory of particles with higher spins. *Zhurnal Eksperimental'noi i Teoreticheskoi Fiziki*, 20(1): 27–38, 1950 (in Russian).

[16] F.I. Fedorov. On minimal polynomials of matrices of relativistic wave equations. *Doklady AN USSR*, 79(5): 787–790, 1951 (in Russian).

[17] F.I. Fedorov. To the theory of a spin 2 particle. *Uchionye Zapiski Beloruaasian State University. Ser. Phys.-Mat.*, 12: 156–173, 1951 (in Russian).

[18] H.J. Bhabha. An equation for a particle with two mass states and positive charge density. *Philosophical Magazine Ser VII.*, 43: 33–47, 1952.

[19] F.I. Fedorov. Generalized relativistic wave equations. *Doklady AN USSR*, 82(1): 37–40, 1952 (in Russian).

[20] M. Petras. A contribution of the theory of the Pauli–Fierz's equations a particle with spin 3/2. *Czechoslovak Journal of Physics*, 5(2): 169–170, 1955.

[21] M. Petras. A note to Bhabha's equation for a particle with maximum spin 3/2. *Czechoslovak Journal of Physics*, 5(3): 418–419, 1955.

[22] V.Ya. Fainberg, To the interaction theory of the particles of the higher spins with electromagnetic and meson fields. *Trudy FIAN USSR*, 6: 269–332, 1955 (in Russian).

[23] V.L. Ginzburg. On relativistic wave equations with a mass spectrum. *Acta Phys. Pol.*, 15: 163–175, 1956.

[24] H. Shimazu. A relativistic wave equation for a particle with two mass states of spin 1 and 0. *Progress of Theoretical Physics*, 16(4): 285–298, 1956.

[25] T. Regge. On properties of the particle with spin 2. *Nuovo Cimento*, 5(2): 325–326, 1957.

[26] P.G. Bergmann and A.I. Janis. Subsidiary conditions in covariant theories. *Journals of Physical Review*, 111(4): 1191–1200, 1958.

[27] H.A. Buchdahl. On the compatibility of relativistic wave equations for particles of higher spin in the presence of a gravitational field. *Nuovo Cimemto*, 10: 96–103, 1958.

[28] L.A. Shelepin. Covariant theory of relativictic wave equations. *Nuclear Physics*, 33(4): 580–593, 1962.

[29] A.Z. Capri. First order wave equations for multi-mass fermions. *Nuovo Cimemto B.*, 64(1): 151–158, 1969.

[30] A. Aurilia and H. Umezawa. Theory of high spin fields. *Journals of Physical Review*, 182(5): 1682–1694, 1969.

[31] F.I. Fedorov and V.A. Pletyukhov. Wave equations with repeated representations of the Lorentz group. *Proceedings of the Academy of Sciences of BSSR. Mathematical physics Series*, 6: 81–88, 1969 (in Russian).

[32] V.A. Pletyukhov and F.I. Fedorov. The wave equation with repeated representations for spin 0 particle. *Proceedings of the Academy of Sciences of BSSR. Mathematical physics Series*, 2: 79–85, 1970 (in Russian).

[33] F.I. Fedorov and V.A. Pletyukhov. Wave equations with repeated representations of the Lorentz group. Half-integer spin. *Proceedings of the Academy of Sciences of BSSR. Mathematical physics Series*, 3: 78–83, 1970 (in Russian).

[34] V.A. Pletyukhov, F.I. Fedorov. Wave equation with repeated reprentations for a spin 1 particle. *Proceedings of the Academy of Sciences of BSSR. Mathematical physics Series*, 3: 84–92, 1970 (in Russian).

[35] A. Shamaly and A.Z. Capri. First-order wave equations for integral spin. *Nuovo Cimemto, B.*, 2(2): 235–253, 1971.

[36] V. Amar and U. Dozzio. Finite dimensional Gel'fand–Yaglom equations for arbitrary integral spin. *Nuovo Cimemto B*, 9: 53–63, 1972.

[37] A.Z. Capri. Electromagnetic properties of a new spin-1/2 field. *Progress of Theoretical Physics*, 48(4): 1364–1374, 1972.

[38] A. Shamaly and A.Z. Capri. Unified theories for massive spin 1 fields. *Canadian Journal of Physics*, 51(14): 1467–1470, 1973.

[39] M.A.K. Khalil. Properties of a 20-component spin 1/2 relativistic wave equation. *Journals of Physical Review, D.*, 15(6): 1532–1539, 1977.

[40] A.S. Wightman. Invariant wave equations: general theory and applications to the external field problem. *Lecture Notes in Physics*, 73: 1–101, 1978.

[41] M.A.K. Khalil. An equivalence of relativistic field equations. *Nuovo Cimento. A.*, 45(3): 389–404, 1978.

[42] L. Garding. Mathematics of invariant wave equations. *Lecture Notes in Physics*, 73: 102–164, 1978.

[43] J.P. Gazeau. L'equation de Dirac avec masse et spin arbitrares: une construction simple et naturelle. *Journal of Physics G: Nuclear and Particle Physics*, 6(12): 1459–1475, 1980.

[44] W. Cox. Higher-rank representations for zero-spin filds theories. *Journal of Physics, A*, 15: 627–635, 1982.

[45] W. Cox. First-order formulation of massive spin-2 field theories. *Journal of Physics A*, 15: 253–268, 1982.

[46] P.M. Mathews, B. Vijayalakshmi and M.Sivakuma. On the admissible Lorentz group representations in unique-mass, unique-spin relativistic wave equations. *Journal of Physics A*, 15(11): 1579–1582, 1982.

[47] P.M. Mathews and B. Vijayalakshmi. On inequivalent classes unique-mass-spin relativistic wave equations involving repeated irreducible representations with arbitrary multiplicities. *Journal of Mathematical Physics*, 25(4): 1080–1087, 1984.

[48] W. Cox. On the Lagrangian and Hamiltonian constraint algorithms for the Rarita-Schwinger field coupled to an external electromagnetic field. *Journal of Physics A*, 22(10): 1599–1608, 1989.

[49] S. Deser and A. Waldron. Inconsistencies of massive charged gravitating higher spins. *Nuclear Physics B*, 631: 369–387, 2002.

[50] E.M. Ovsiyuk, V.V. Kisel, Y.A. Voynova, O.V. Veko and V.M. Red'kov. Spin 1/2 particle with anomalous magnetic moment in a uniform magnetic field, exact solutions. *Nonlinear Phenomena in Complex Systems*, 19(2): 153–165, 2016.

[51] V. Kisel, Ya. Voynova, E. Ovsiyuk, V. Balan and V. Red'kov. Spin 1 Particle with Anomalous Magnetic Moment in the External Uniform Magnetic Field. *Nonlinear Phenomena in Complex Systems*, 20(1): 21–39, 2017.

[52] V.V. Kisel, E.M. Ovsiyuk, Ya.A. Voynova and V.M. Red'kov. Quantum mechanics of spin 1 particle with quadrupole moment in external uniform magetic field. *Problems of Physics, Mathematics, and Thechnics*, 32(3): 18–27, 2017.

[53] F.I. Fedorov. *The Lorentz Group.* Science, Moscow, 1979 (in Russian).

[54] I.M. Gel'fand, R.A. Minlos and Z.Ya. Shapiro. *Representations of the Rotation and Lorentz Groups and Their Applications.* Dover Publications, New York, 2018.

[55] V.M. Red'kov. *Fields in Riemannian Space and the Lorentz Group.* Belarusian Science, Minsk, 2009.

[56] V.A. Pletjukhov, V.M. Red'kov and V.I. Strazhev. *Relativistic wave equations and intrinsic degrees of freedom.* Belarusian Science, Minsk, 2015.

[57] V.V. Kisel, E.M. Ovsiyuk, V. Balan, O.V. Veko and V.M. Red'kov. *Elementary Particles with Internal Structure in External Field. Vol. I. General Formalism.* Nova Science Publishers, Inc. USA, 2018; *Vol. II. Physical Problems.* Nova Science Publishers, Inc. USA, 2018.

[58] V.V. Kisel, V.A. Pletyukhov, V.V. Gilewsky, E.M. Ovsiyuk, O.V. Veko and V.M. Red'kov. Spin 1/2 particle with two mass states, interaction with external fields. *Nonlinear Phenomena in Complex Systems*, 20(4): 404–423, 2017.

[59] E.M. Ovsiyuk, O.V. Veko, Ya.A. Voynova, V.V. Kisel, V. Balan and V.M. Red'kov. Spin 1/2 particle with two masses in magnetic field. *Applied Sciences*, 20: 148–166, 2018.

11

Fermion with two mass parameters in the Coulomb field

Generalised wave equation for a spin 1/2 particle with two mass parameters is studied in the presence of an external Coulomb field. After separating the variables, the problem reduces the system to eight differential equations of the 1st order. Taking into account for diagonalization of the space reflection operator, we derive two independent systems of four equations, referring to states of opposite parity. When considering these equations at the large distance from the centre, they take the form of two subsystems for two ordinary Dirac particles in an external Coulomb field, with masses of M_1 and M_2, respectively. To simplify the problem, we perform a transition to the nonrelativistic description of the system. In this way, we derive two systems of linked 2nd-order equations, referring to states with different parities. They lead to 4th-order differential equations for separate functions. Their solutions of the Frobenius type have been constructed, they involve power series with 10-term recurrent relations. Two solutions are appropriate to describe bound states. As a quantization rule, we apply the known transcendency condition; in this way, we derive two analytical formulas for energy spectra. They are similar to nonrelativistic spectra for ordinary spin 1/2 particles, but they are governed by masses M_1 and M_2. Results of constructing solutions and obtaining the energy spectra are extendable to relativistic theory as well.

11.1 Introduction

In [1–3], it was proposed the relativistic equation for a spin 1/2 particle with two mass parameters. In the presence of an external electromagnetic field, there arises the system of equations for two bispinor components. In [3–5], exact solutions of those equations in the presence of the uniform magnetic field were constructed.

In this chapter, we will consider such a particle in an external Coulomb field. The system for two bispinors $\Psi_1(x), \Psi_2(x)$ has the structure

$$\begin{aligned} &\Big\{\gamma^\alpha[i(\partial_\alpha+\Gamma_\alpha)-eA_\alpha]-M_1+b\Lambda_1\Sigma(x)\Big\}\Psi_1(x)-a\Lambda_1\Sigma(x)\Psi_2(x)=0\,,\\ &\Big\{\gamma^\alpha[i(\partial_\alpha+\Gamma_\alpha)-eA_\alpha]-M_2-a\Lambda_2\Sigma(x)\Big\}\Psi_2(x)+b\Lambda_2\Sigma(x)\Psi_1(x)=0\,, \end{aligned} \tag{11.1}$$

where

$$\begin{aligned} \gamma^\alpha(x)=e^\alpha_{(b)}\gamma^b,\quad \Sigma(x)=-ieF_{\alpha\beta}\sigma^{\alpha\beta}(x)\,,\\ \sigma^{\alpha\beta}(x)=\frac{\gamma^\alpha(x)\gamma^\beta(x)-\gamma^\beta(x)\gamma^\alpha(x)}{4}\,. \end{aligned} \tag{11.2}$$

DOI: 10.1201/9781003472377-11

We will apply the following parameters

$$\gamma \in [0, \pi/2], \quad M_1 = \frac{M}{(1+\cos\gamma)/2}, \quad M_2 = \frac{M}{(1-\cos\gamma)/2},$$
$$a = \frac{1}{2}\frac{1}{M}\,(\,4 - 3\sqrt{1+(1/3)\sin^2\gamma} - \cos\gamma\,),$$
$$b = \frac{1}{2}\frac{1}{M}\,(\,4 - 3\sqrt{1+(1/3)\sin^2\gamma} + \cos\gamma\,),$$
$$\Lambda_1 = (1 + \sqrt{1+(1/3)\sin^2\gamma}\,)\,\frac{\cos\gamma - \sqrt{1+(1/3)\sin^2\gamma}}{\cos\gamma(1+\cos\gamma)},$$
$$\Lambda_2 = (1 + \sqrt{1+(1/3)\sin^2\gamma}\,)\,\frac{\cos\gamma + \sqrt{1+(1/3)\sin^2\gamma}}{\cos\gamma(1-\cos\gamma)}. \tag{11.3}$$

Parameter M with dimension of inverse length is arbitrary.

11.2 Separating the variables

We consider this equation in the Coulomb field, using spherical coordinates and tetrad

$$ds^2 = dt^2 - dr^2 - r^2 d\theta^2 - r^2\sin^2\theta d\phi^2, \quad x^\alpha = (t, r, \theta, \phi),$$
$$e^\alpha_{(0)} = (1,0,0,0), \quad e^\alpha_{(3)} = (0,1,0,0),$$
$$e^\alpha_{(1)} = (0,0,\frac{1}{r},0), \quad e^\alpha_{(2)} = (0,0,0,\frac{1}{r\sin\theta}). \tag{11.4}$$

We will apply the notations

$$A_t = -\frac{e}{r}, \quad F_{tr} = -\partial_r A_0 = \frac{e}{r^2}, \quad \Sigma(x) = i\frac{e^2}{r^2}\gamma^0\gamma^3,$$
$$a\Lambda_1 e^2 = \alpha_1, \quad a\Lambda_2 e^2 = \alpha_2, \quad b\Lambda_1 e^2 = \beta_1, \quad b\Lambda_2 e^2 = \beta_2.$$

The system (11.1) takes on the form

$$\Big[\gamma^0(i\partial_t - \frac{\alpha}{r}) + i\gamma^3\partial_r + \frac{1}{r}\Sigma_{\theta\phi} - M_1 + i\frac{\beta_1}{r^2}\gamma^0\gamma^3\Big]\Psi_1 - i\frac{\alpha_1}{r^2}\gamma^0\gamma^3\Psi_2 = 0,$$
$$\Big[\gamma^0(i\partial_t - \frac{\alpha}{r}) + i\gamma^3\partial_r + \frac{1}{r}\Sigma_{\theta\phi} - M_2 - i\frac{\alpha_2}{r^2}\gamma^0\gamma^3\Big]\Psi_2 + i\frac{\beta_2}{r^2}\gamma^0\gamma^3\Psi_1 = 0, \tag{11.5}$$

where

$$\Sigma_{\theta,\phi} = i\gamma^1\partial_\theta + \gamma^2\frac{i\partial_\phi + i\sigma^{12}\cos\theta}{\sin\theta}.$$

Further, we will apply the following expressions for four parameters

$$\alpha_1 = -e^2\frac{1}{3}\,\frac{(1-\cos\gamma)\left(-\cos\gamma\sqrt{12-3\cos^2\gamma} + \cos^2\gamma + 2\right)}{M\cos\gamma\,(1+\cos\gamma)},$$
$$\alpha_2 = e^2\,\frac{2}{3}\,\frac{\sin^2\gamma}{M\,\cos\gamma}, \qquad \beta_1 = -e^2\,\frac{2}{3}\,\frac{\sin^2\gamma}{M\,\cos\gamma} < 0,$$
$$\beta_2 = -\frac{1}{3}\,\frac{e^2\,(1+\cos\gamma)\left(\cos\gamma\sqrt{12-3\cos^2\gamma} + \cos^2\gamma + 2\right)}{M\cos\gamma\,(\cos\gamma - 1)} > 0, \tag{11.6}$$

note two identities

$$\alpha_2 = -\beta_1 \, , \qquad \alpha_1 \beta_2 = -\beta_1{}^2 \, . \tag{11.7}$$

The substitutions for two bispinors with quantum numbers ϵ, j, m have the form

$$\Psi_1(x) = \frac{e^{-i\epsilon t}}{r} \left| \begin{array}{c} f_1(r) D_{-1/2} \\ f_2(r) D_{+1/2} \\ f_3(r) D_{-1/2} \\ f_4(r) D_{+1/2} \end{array} \right| , \qquad \Psi_2(x) = \frac{e^{-i\epsilon t}}{r} \left| \begin{array}{c} g_1(r) D_{-1/2} \\ g_2(r) D_{+1/2} \\ g_3(r) D_{-1/2} \\ g_4(r) D_{+1/2} \end{array} \right| . \tag{11.8}$$

Using the Dirac matrices in spinor basis, we derive eight radial equations

$$\begin{aligned} &\left(\epsilon + \frac{\alpha}{r}\right) f_3 - i\frac{d}{dr} f_3 - i\frac{\nu}{r} f_4 - M_1 f_1 + \frac{i\beta_1}{r^2} f_1 - \frac{i\alpha_1}{r^2} g_1 = 0 \, , \\ &\left(\epsilon + \frac{\alpha}{r}\right) f_4 + i\frac{d}{dr} f_4 + i\frac{\nu}{r} f_3 - M_1 f_2 - \frac{i\beta_1}{r^2} f_2 + \frac{i\alpha_1}{r^2} g_2 = 0 \, , \\ &\left(\epsilon + \frac{\alpha}{r}\right) f_1 + i\frac{d}{dr} f_1 + i\frac{\nu}{r} f_2 - M_1 \delta f_2 - \frac{i\beta_1}{r^2} f_3 + \frac{i\alpha_1}{r^2} g_3 = 0 \, , \\ &\left(\epsilon + \frac{\alpha}{r}\right) f_2 - i\frac{d}{dr} f_2 - i\frac{\nu}{r} f_1 - M_1 f_4 + \frac{i\beta_1}{r^2} f_4 - \frac{i\alpha_1}{r^2} g_4 = 0 \, ; \end{aligned} \tag{11.9}$$

$$\begin{aligned} &\left(\epsilon + \frac{\alpha}{r}\right) g_3 - i\frac{d}{dr} g_3 - i\frac{\nu}{r} g_4 - M_2 g_1 - \frac{i\alpha_2}{r^2} g_1 + \frac{i\beta_2}{r^2} f_1 = 0 \, , \\ &\left(\epsilon + \frac{\alpha}{r}\right) g_4 + i\frac{d}{dr} g_4 + i\frac{\nu}{r} g_3 - M_2 g_2 + \frac{i\alpha_2}{r^2} g_2 - \frac{i\beta_2}{r^2} f_2 = 0 \, , \\ &\left(\epsilon + \frac{\alpha}{r}\right) g_1 + i\frac{d}{dr} g_1 + i\frac{\nu}{r} g_2 - M_2 g_3 + \frac{i\alpha_2}{r^2} g_3 - \frac{i\beta_2}{r^2} \delta f_2 = 0 \, , \\ &\left(\epsilon + \frac{\alpha}{r}\right) g_2 - i\frac{d}{dr} g_2 - i\frac{\nu}{r} g_1 - M_2 g_4 - \frac{i\alpha_2}{r^2} g_4 + \frac{i\beta_2}{r^2} f_4 = 0 \, , \end{aligned} \tag{11.10}$$

where $\nu = j + 1/2$; $j = 1/2, 3/2, \ldots$. The system (11.9)–(11.10) allows for imposing the linear constraints (they follow from diagonalization of the space reflection operator)

$$f_3 = \delta f_2 \, , \quad f_4 = \delta f_1 \, , \quad \delta = \pm 1, \quad g_3 = \delta g_2 \, , \quad g_4 = \delta g_1 \, , \quad \delta = \pm 1 \, ;$$

in this way we obtain

$$\begin{aligned} &\left(\epsilon + \frac{\alpha}{r}\right) \delta f_2 - i\frac{d}{dr} \delta f_2 - i\frac{\nu}{r} \delta f_1 - M_1 f_1 + \frac{i\beta_1}{r^2} f_1 - \frac{i\alpha_1}{r^2} g_1 = 0 \, , \\ &\left(\epsilon + \frac{\alpha}{r}\right) \delta f_1 + i\frac{d}{dr} \delta f_1 + i\frac{\nu}{r} \delta f_2 - M_1 f_2 - \frac{i\beta_1}{r^2} f_2 + \frac{i\alpha_1}{r^2} g_2 = 0 \, , \\ &\left(\epsilon + \frac{\alpha}{r}\right) f_1 + i\frac{d}{dr} f_1 + i\frac{\nu}{r} f_2 - M_1 \delta f_2 - \frac{i\beta_1}{r^2} \delta f_2 + \frac{i\alpha_1}{r^2} \delta g_2 = 0 \, , \\ &\left(\epsilon + \frac{\alpha}{r}\right) f_2 - i\frac{d}{dr} f_2 - i\frac{\nu}{r} f_1 - M_1 \delta f_1 + \frac{i\beta_1}{r^2} \delta f_1 - \frac{i\alpha_1}{r^2} \delta g_1 = 0 \, ; \end{aligned}$$

$$\begin{aligned} &\left(\epsilon + \frac{\alpha}{r}\right) \delta g_2 - i\frac{d}{dr} \delta g_2 - i\frac{\nu}{r} \delta g_1 - M_2 g_1 - \frac{i\alpha_2}{r^2} g_1 + \frac{i\beta_2}{r^2} f_1 = 0 \, , \\ &\left(\epsilon + \frac{\alpha}{r}\right) \delta g_1 + i\frac{d}{dr} \delta g_1 + i\frac{\nu}{r} \delta g_2 - M_2 g_2 + \frac{i\alpha_2}{r^2} g_2 - \frac{i\beta_2}{r^2} f_2 = 0 \, , \\ &\left(\epsilon + \frac{\alpha}{r}\right) g_1 + i\frac{d}{dr} g_1 + i\frac{\nu}{r} g_2 - M_2 \delta g_2 + \frac{i\alpha_2}{r^2} \delta g_2 - \frac{i\beta_2}{r^2} \delta f_2 = 0 \, , \\ &\left(\epsilon + \frac{\alpha}{r}\right) g_2 - i\frac{d}{dr} g_2 - i\frac{\nu}{r} g_1 - M_2 \delta g_1 - \frac{i\alpha_2}{r^2} \delta g_1 + \frac{i\beta_2}{r^2} \delta f_1 = 0 \, . \end{aligned}$$

Preserving only independent equations, we arrive at two subsystems of four equations (depending on $\delta = \pm 1$)

$$\begin{aligned}
&\left(\epsilon + \frac{\alpha}{r}\right) f_1 + i\frac{d}{dr} f_1 + i\frac{\nu}{r} f_2 - M_1 \delta f_2 - \frac{i\beta_1}{r^2}\delta f_2 + \frac{i\alpha_1}{r^2}\delta g_2 = 0\,, \\
&\left(\epsilon + \frac{\alpha}{r}\right) f_2 - i\frac{d}{dr} f_2 - i\frac{\nu}{r} f_1 - M_1 \delta f_1 + \frac{i\beta_1}{r^2}\delta f_1 - \frac{i\alpha_1}{r^2}\delta g_1 = 0\,. \\
&\left(\epsilon + \frac{\alpha}{r}\right) g_1 + i\frac{d}{dr} g_1 + i\frac{\nu}{r} g_2 - M_2 \delta g_2 + \frac{i\alpha_2}{r^2}\delta g_2 - \frac{i\beta_2}{r^2}\delta f_2 = 0\,, \\
&\left(\epsilon + \frac{\alpha}{r}\right) g_2 - i\frac{d}{dr} g_2 - i\frac{\nu}{r} g_1 - M_2 \delta g_1 - \frac{i\alpha_2}{r^2}\delta g_1 + \frac{i\beta_2}{r^2}\delta f_1 = 0\,.
\end{aligned} \tag{11.11}$$

In order to eliminate the presence of imaginary unit, we use the new variables

$$f = (f_2 + f_1), \quad F = i(f_2 - f_1)\,; \quad g = (g_2 + g_1), \quad G = i(g_2 - g_1)\,. \tag{11.12}$$

This results in

$$\begin{aligned}
&\left(\frac{d}{dr} - \frac{\nu}{r} + \delta\frac{\beta_1}{r^2}\right) F - \left(\epsilon + \frac{\alpha}{r} - \delta M_1\right) f - \delta\frac{\alpha_1}{r^2} G = 0\,, \\
&\left(\frac{d}{dr} + \frac{\nu}{r} - \delta\frac{\beta_1}{r^2}\right) f + \left(\epsilon + \frac{\alpha}{r} + \delta M_1\right) F + \delta\frac{\alpha_1}{r^2} g = 0\,, \\
&\left(\frac{d}{dr} - \frac{\nu}{r} - \delta\frac{\alpha_2}{r^2}\right) G - \left(\epsilon + \frac{\alpha}{r} - \delta M_2\right) g + \delta\frac{\beta_2}{r^2} F = 0\,, \\
&\left(\frac{d}{dr} + \frac{\nu}{r} + \delta\frac{\alpha_2}{r^2}\right) g + \left(\epsilon + \frac{\alpha}{r} + \delta M_2\right) G - \delta\frac{\beta_2}{r^2} f = 0\,.
\end{aligned} \tag{11.13}$$

It is convenient to consider the states with different parities separately:

$$\delta = +1, \quad \begin{aligned}
&\left(\frac{d}{dr} - \frac{\nu}{r} + \frac{\beta_1}{r^2}\right) F - \left(\epsilon + \frac{\alpha}{r} - M_1\right) f - \frac{\alpha_1}{r^2} G = 0\,, \\
&\left(\frac{d}{dr} + \frac{\nu}{r} - \frac{\beta_1}{r^2}\right) f + \left(\epsilon + \frac{\alpha}{r} + M_1\right) F + \frac{\alpha_1}{r^2} g = 0\,, \\
&\left(\frac{d}{dr} - \frac{\nu}{r} - \frac{\alpha_2}{r^2}\right) G - \left(\epsilon + \frac{\alpha}{r} - M_2\right) g + \frac{\beta_2}{r^2} F = 0\,, \\
&\left(\frac{d}{dr} + \frac{\nu}{r} + \frac{\alpha_2}{r^2}\right) g + \left(\epsilon + \frac{\alpha}{r} + M_2\right) G - \frac{\beta_2}{r^2} f = 0\,;
\end{aligned} \tag{11.14}$$

$$\delta = -1, \quad \begin{aligned}
&\left(\frac{d}{dr} - \frac{\nu}{r} - \frac{\beta_1}{r^2}\right) F - \left(\epsilon + \frac{\alpha}{r} + M_1\right) f + \frac{\alpha_1}{r^2} G = 0\,, \\
&\left(\frac{d}{dr} + \frac{\nu}{r} + \frac{\beta_1}{r^2}\right) f + \left(\epsilon + \frac{\alpha}{r} - M_1\right) F - \frac{\alpha_1}{r^2} g = 0\,, \\
&\left(\frac{d}{dr} - \frac{\nu}{r} + \frac{\alpha_2}{r^2}\right) G - \left(\epsilon + \frac{\alpha}{r} + M_2\right) g - \frac{\beta_2}{r^2} F = 0\,, \\
&\left(\frac{d}{dr} + \frac{\nu}{r} - \frac{\alpha_2}{r^2}\right) g + \left(\epsilon + \frac{\alpha}{r} - M_2\right) G + \frac{\beta_2}{r^2} f = 0\,.
\end{aligned} \tag{11.15}$$

We can note the symmetry between two systems:

$$M_1, M_2, \alpha_1, \alpha_2, \beta_1, \beta_2 \quad \Longrightarrow \quad -M_1, -M_2, -\alpha_1, -\alpha_2, -\beta_1, -\beta_2\,. \tag{11.16}$$

It may be noticed that at the large r the systems split into independent subsystems:

$\delta = +1$,

$$\begin{aligned}&\Big(\frac{d}{dr} - \frac{\nu}{r}\Big)F - \Big(\epsilon + \frac{\alpha}{r} - M_1\Big)f = 0\,, \quad \Big(\frac{d}{dr} + \frac{\nu}{r}\Big)f + \Big(\epsilon + \frac{\alpha}{r} + M_1\Big)F = 0\,,\\ &\Big(\frac{d}{dr} - \frac{\nu}{r}\Big)G - \Big(\epsilon + \frac{\alpha}{r} - M_2\Big)g = 0\,, \quad \Big(\frac{d}{dr} + \frac{\nu}{r}\Big)g + \Big(\epsilon + \frac{\alpha}{r} + M_2\Big)G = 0\,;\end{aligned} \tag{11.17}$$

$\delta = -1$,

$$\begin{aligned}&\Big(\frac{d}{dr} - \frac{\nu}{r}\Big)F - \Big(\epsilon + \frac{\alpha}{r} + M_1\Big)f = 0\,, \quad \Big(\frac{d}{dr} + \frac{\nu}{r}\Big)f + \Big(\epsilon + \frac{\alpha}{r} - M_1\Big)F = 0\,,\\ &\Big(\frac{d}{dr} - \frac{\nu}{r}\Big)G - \Big(\epsilon + \frac{\alpha}{r} + M_2\Big)g = 0\,, \quad \Big(\frac{d}{dr} + \frac{\nu}{r}\Big)g + \Big(\epsilon + \frac{\alpha}{r} - M_2\Big)G = 0\,.\end{aligned} \tag{11.18}$$

This means that far from the origin $r = 0$, we have equations for two ordinary Dirac particles with different masses M_1 and M_2. In other words, at large distances we observe two unlinked particles with different masses.

11.3 Derivation of the 4th-order equations

With the help two equation in eq. (11.14), one can eliminate the functions $G(r), g(r)$; then we obtain the system of the 2nd-order equations for functions $f(r), F(r)$. Below we write down only their general structure (let $M_1 - M_2 = M$):

$$\begin{aligned}&\Big(\frac{d^2}{dr^2} + (\frac{a_1}{r} + \frac{a_2}{r^2})\frac{d}{dr} + b + \frac{b_1}{r} + ... + \frac{b_4}{r^4}\Big)f\\ &\qquad + \Big(M\frac{d}{dr} + \frac{C_1}{r} + \frac{C_2}{r^2} + \frac{C_3}{r^3}\Big)F = 0\,,\\ &\Big(\frac{d^2}{dr^2} + (\frac{A_1}{r} + \frac{A_2}{r^2})\frac{d}{dr} + B + \frac{B_1}{r} + ... + \frac{B_4}{r^4}\Big)F\\ &\qquad + \Big(M\frac{d}{dr}f + \frac{c_1}{r} + \frac{c_2}{r^2} + \frac{c_3}{r^3}\Big)f = 0\,.\end{aligned} \tag{11.19}$$

Let us introduce special multiplier at f and F, so that

$$\begin{aligned}&F = \Phi\bar{F}, \quad \Phi = x^{\frac{-C_1}{M}} e^{\frac{C_2}{Mx}} e^{\frac{C_3}{2Mx^2}}\,,\\ &\Big(M\frac{d}{dr} + \frac{C_1}{r} + \frac{C_2}{r^2} + \frac{C_3}{r^3}\Big)\Phi\bar{F} = \Phi M\frac{d}{dr}\bar{F},\\ &f = \varphi\bar{f}, \quad \varphi = x^{\frac{-c_1}{M}} e^{\frac{c_2}{Mx}} e^{\frac{c_3}{2Mx^2}}\,,\\ &\Big(M\frac{d}{dr} + \frac{c_1}{r} + \frac{c_2}{r^2} + \frac{c_3}{r^3}\Big)\varphi\bar{F} = \varphi M\frac{d}{dr}\bar{F}\,.\end{aligned} \tag{11.20}$$

Correspondingly the system (11.19)–(11.19) takes the form

$$\frac{1}{\Phi M}\Big(\frac{d^2}{dr^2} + (\frac{a_1}{r} + \frac{a_2}{r^2})\frac{d}{dr} + b + \frac{b_1}{r} + ... \frac{b_4}{r^4}\Big)\varphi\bar{f} + \frac{d}{dr}\bar{F} = 0\,, \tag{11.21}$$

$$\frac{1}{\varphi M}\Big(\frac{d^2}{dr^2} + (\frac{A_1}{r} + \frac{A_2}{r^2})\frac{d}{dr} + B + \frac{B_1}{r} + ... \frac{B_4}{r^4}\Big)\varphi\bar{F} + \frac{d}{dr}\bar{f} = 0\,. \tag{11.22}$$

Let us apply more symmetric notation, $\bar{f}(r) = f_1(r)$, $\bar{F}(r) = -f_2(r)$; then the system (11.21)–(11.22) reads

$$\begin{aligned} \Big(K_2(x)\frac{d^2}{dx^2} + K_1(x)\frac{d}{dx} + K_0(x)\Big) f_1 &= \frac{df_2}{dx} , \\ \Big(L_2(x)\frac{d^2}{dx^2} + L_1(x)\frac{d}{dx} + L_0(x)\Big) f_2 &= \frac{df_1}{dx} . \end{aligned} \tag{11.23}$$

Let us detail the method for obtaining equations of the 4th-order. First, we will eliminate the function f_2:

$$f_2(x) = \int \Big(K_2(x)\frac{d^2}{dx^2} + K_1(x)\frac{d}{dx} + K_0(x)\Big) f_1 ,$$

$$\Big(L_2\frac{d}{dx} + L_1\Big)\Big(K_2\frac{d^2}{dx^2} + K_1\frac{d}{dx} + K_0)\ f_1 + L_0 \int dx(K_2\frac{d^2}{dx^2} + K_1\frac{d}{dx} + K_0\Big) f_1 = 0.$$

Then we divide the second equation by $L_0(x)$ and differentiate the result. In this way, we derive a 4th-order equation for $f_1(x)$:

$$\begin{aligned} \Big\{ \frac{d}{dx}\Big(\frac{L_2}{L_0}\frac{d}{dx} + \frac{L_1}{L_0}\Big)\Big(K_2\frac{d^2}{dx^2} + K_1\frac{d}{dx} + K_0\Big) \\ + \Big(K_2\frac{d^2}{dx^2} + K_1\frac{d}{dx} + K_0\Big)\Big\} f_1(x) = 0. \end{aligned} \tag{11.24}$$

Similarly, we obtain the 4th-order equation for function f_2:

$$\begin{aligned} \Big\{ \frac{d}{dx}\Big(\frac{K_2}{K_0}\frac{d}{dx} + \frac{K_1}{K_0}\Big)\Big(L_2\frac{d^2}{dx^2} + L_1\frac{d}{dx} + L_0\Big) \\ + \Big(L_2\frac{d^2}{dx^2} + L_1\frac{d}{dx} + L_0\Big)\Big\} f_2(x) = 0. \end{aligned} \tag{11.25}$$

11.4 Nonrelativistic approximation

In order to simplify the problem, in the system of radial equations, let us perform the nonrelativistic approximation. We start with

$$\delta = +1, \qquad \begin{aligned} -\Big(\frac{d}{dr} - \frac{\nu}{r} + \frac{\beta_1}{r^2}\Big)F + \Big(\epsilon + \frac{\alpha}{r} - M_1\Big)f + \frac{\alpha_1}{r^2}G = 0 , \\ -\Big(\frac{d}{dr} + \frac{\nu}{r} - \frac{\beta_1}{r^2}\Big)f - \Big(\epsilon + \frac{\alpha}{r} + M_1\Big)F - \frac{\alpha_1}{r^2}g = 0 , \\ -\Big(\frac{d}{dr} - \frac{\nu}{r} - \frac{\alpha_2}{r^2}\Big)G + \Big(\epsilon + \frac{\alpha}{r} - M_2\Big)g - \frac{\beta_2}{r^2}F = 0 , \\ -\Big(\frac{d}{dr} + \frac{\nu}{r} + \frac{\alpha_2}{r^2}\Big)g - \Big(\epsilon + \frac{\alpha}{r} + M_2\Big)G + \frac{\beta_2}{r^2}f = 0 ; \end{aligned} \tag{11.26}$$

$$\delta = -1, \qquad \begin{aligned} -\Big(\frac{d}{dr} - \frac{\nu}{r} - \frac{\beta_1}{r^2}\Big)F + \Big(\epsilon + \frac{\alpha}{r} + M_1\Big)f - \frac{\alpha_1}{r^2}G = 0 , \\ -\Big(\frac{d}{dr} + \frac{\nu}{r} + \frac{\beta_1}{r^2}\Big)f - \Big(\epsilon + \frac{\alpha}{r} - M_1\Big)F + \frac{\alpha_1}{r^2}g = 0 , \\ -\Big(\frac{d}{dr} - \frac{\nu}{r} + \frac{\alpha_2}{r^2}\Big)G + \Big(\epsilon + \frac{\alpha}{r} + M_2\Big)g + \frac{\beta_2}{r^2}F = 0 , \\ -\Big(\frac{d}{dr} + \frac{\nu}{r} - \frac{\alpha_2}{r^2}\Big)g - \Big(\epsilon + \frac{\alpha}{r} - M_2\Big)G - \frac{\beta_2}{r^2}f = 0 . \end{aligned} \tag{11.27}$$

By means of substitutions

$$f, F \sim e^{-i\epsilon t} = e^{-i(M_1+E_1)t}, \quad g, G \sim e^{-i\epsilon t} = e^{-i(M_2+E_2)t}, \quad M_1 + E_1 = M_2 + E_2$$

we introduce two nonrelativistic energies. As a result from eqs. (11.26)–(11.27), we obtain

$$\delta = +1, \qquad \begin{aligned} -\Big(\frac{d}{dr} - \frac{\nu}{r} + \frac{\beta_1}{r^2}\Big)F + \Big(E_1 + \frac{\alpha}{r}\Big)f + \frac{\alpha_1}{r^2}G = 0\,, \\ -\Big(\frac{d}{dr} + \frac{\nu}{r} - \frac{\beta_1}{r^2}\Big)f - \Big(2M_1 + E_1 + \frac{\alpha}{r}\Big)F - \frac{\alpha_1}{r^2}g = 0\,, \\ -\Big(\frac{d}{dr} - \frac{\nu}{r} - \frac{\alpha_2}{r^2}\Big)G + \Big(E_2 + \frac{\alpha}{r}\Big)g - \frac{\beta_2}{r^2}F = 0\,, \\ -\Big(\frac{d}{dr} + \frac{\nu}{r} + \frac{\alpha_2}{r^2}\Big)g - \Big(2M_2 + E_2 + \frac{\alpha}{r}\Big)G + \frac{\beta_2}{r^2}f = 0\,; \end{aligned} \qquad (11.28)$$

$$\delta = -1, \qquad \begin{aligned} -\Big(\frac{d}{dr} - \frac{\nu}{r} - \frac{\beta_1}{r^2}\Big)F + \Big(2M_1 + E_1 + \frac{\alpha}{r}\Big)f - \frac{\alpha_1}{r^2}G = 0\,, \\ -\Big(\frac{d}{dr} + \frac{\nu}{r} + \frac{\beta_1}{r^2}\Big)f - \Big(E_1 + \frac{\alpha}{r}\Big)F + \frac{\alpha_1}{r^2}g = 0\,, \\ -\Big(\frac{d}{dr} - \frac{\nu}{r} + \frac{\alpha_2}{r^2}\Big)G + \Big(2M_2 + E_2 + \frac{\alpha}{r}\Big)g + \frac{\beta_2}{r^2}F = 0\,, \\ -\Big(\frac{d}{dr} + \frac{\nu}{r} - \frac{\alpha_2}{r^2}\Big)g - \Big(E_2 + \frac{\alpha}{r}\Big)G - \frac{\beta_2}{r^2}f = 0\,. \end{aligned} \qquad (11.29)$$

First consider the case $\delta = +1$. Neglecting the nonrelativistic energy in comparison with the rest energy M_1 and M_2

$$2M_1 + E_1 + \frac{\alpha}{r} \approx 2M_1 \quad 2M_2 + E_2 + \frac{\alpha}{r} \approx 2M_2,$$

we get

$$\delta = +1, \qquad \begin{aligned} -\Big(\frac{d}{dr} - \frac{\nu}{r} + \frac{\beta_1}{r^2}\Big)F + \Big(E_1 + \frac{\alpha}{r}\Big)f + \frac{\alpha_1}{r^2}G = 0\,, \\ -\Big(\frac{d}{dr} + \frac{\nu}{r} - \frac{\beta_1}{r^2}\Big)f - 2M_1F - \frac{\alpha_1}{r^2}g = 0\,, \\ -\Big(\frac{d}{dr} - \frac{\nu}{r} - \frac{\alpha_2}{r^2}\Big)G + \Big(E_2 + \frac{\alpha}{r}\Big)g - \frac{\beta_2}{r^2}F = 0\,, \\ -\Big(\frac{d}{dr} + \frac{\nu}{r} + \frac{\alpha_2}{r^2}\Big)g - 2M_2G + \frac{\beta_2}{r^2}f = 0\,. \end{aligned} \qquad (11.30)$$

With the help of second and fourth equations, we eliminate the functions F and G, so obtaining nonrelativistic equations for f and g:

$$\begin{aligned} &\frac{d^2 f}{dr^2} + \Big[\frac{2\,M_1\,\alpha}{r} - \frac{\nu(\nu+1)}{r^2} + \frac{2\beta_1(\nu+1)}{r^3} - \frac{{\beta_1}^2}{r^4} + \frac{M_1\,\alpha_1\,\beta_2}{M_2\,r^4} + 2\,M_1\,E_1\Big]f \\ &+\Big(\frac{\alpha_1}{r^2} - \frac{\alpha_1\,M_1}{r^2 M_2}\Big)\frac{dg}{dr} + \Big(-\frac{\nu\,\alpha_1}{r^3} - \frac{2\,\alpha_1}{r^3} - \frac{M_1\,\alpha_1\,\nu}{M_2\,r^3} + \frac{\alpha_1\,\beta_1}{r^4} - \frac{\alpha_1\,\alpha_2\,M_1}{M_2\,r^4}\Big)g = 0\,, \end{aligned} \qquad (11.31)$$

$$\begin{aligned} &\frac{d^2 g}{dr^2} + \Big[\frac{2\,M_2\,\alpha}{r} - \frac{\nu(\nu+1)}{r^2} - \frac{2\alpha_2(\nu+1)}{r^3} - \frac{{\alpha_2}^2}{r^4} + \frac{M_2\,\alpha_1\,\beta_2}{M_1\,r^4} + 2\,M_2\,E_2\Big]g \\ &+\Big(-\frac{\beta_2}{r^2} + \frac{\beta_2\,M_2}{r^2 M_1}\Big)\frac{df}{dr} + \Big(\frac{\nu\,\beta_2}{r^3} + \frac{2\,\beta_2}{r^3} + \frac{M_2\,\beta_2\,\nu}{M_1\,r^3} + \frac{\alpha_2\,\beta_2}{r^4} - \frac{\beta_1\beta_2 M_2}{M_1\,r^4}\Big)f = 0\,. \end{aligned} \qquad (11.32)$$

These equations are symmetric with respect to the changes eq. (see (11.7))

$$f \Longrightarrow g\,, \quad \alpha_1 \Longrightarrow -\beta_2\,, \quad \alpha_2 \Longrightarrow -\beta_1\,, \quad M_1 \Longrightarrow M_2\,; \tag{11.33}$$

therefore, it suffices to examine only a 4th-order equation for the function f.

Far from the origin $r = 0$, eq. (11.30) simplify

$$\begin{aligned} &-\Big(\frac{d}{dr} - \frac{\nu}{r}\Big)F + \Big(E_1 + \frac{\alpha}{r}\Big)f = 0\,, \quad -\Big(\frac{d}{dr} + \frac{\nu}{r}\Big)f - 2M_1 F = 0\,, \\ &-\Big(\frac{d}{dr} - \frac{\nu}{r}\Big)G + \Big(E_2 + \frac{\alpha}{r}\Big)g = 0\,, \quad -\Big(\frac{d}{dr} + \frac{\nu}{r}\Big)g - 2M_2 G = 0\,; \end{aligned} \tag{11.34}$$

whence follow separate equations for f and g:

$$\delta = +1, \qquad \begin{aligned} &\Big(\frac{d^2}{dr^2} + 2M_1(E_1 + \frac{\alpha}{r}) - \frac{\nu(\nu+1)}{r^2}\Big)f = 0, \\ &\Big(\frac{d^2}{dr^2} + 2M_2(E_2 + \frac{\alpha}{r}) - \frac{\nu(\nu+1)}{r^2}\Big)g = 0. \end{aligned} \tag{11.35}$$

This means that far from the origin, the system looks as two independent nonrelativistic particles with masses of M_1 and M_2, respectively. Their solutions are confluent hypergeometric functions:

$$\delta = +1, \;\; x = 2\sqrt{-2M_1E_1}\, r,$$

$$\begin{aligned} &f(x) = x^{\nu+1}e^{-x/2}F(-n', 2\nu+2, x), \;\; E_1 = -\frac{M_1\alpha^2}{2(n'+\nu+1)^2}, n' = 0, 1, 2, ..., \\ &g(x) = x^{\nu+1}e^{-x/2}F(-n', 2\nu+2, x), \;\; E_2 = -\frac{M_2\alpha^2}{2(n'+\nu+1)^2}, n' = 0, 1, 2, ... \,. \end{aligned} \tag{11.36}$$

The case of opposite parity is considered similarly:

$$\delta = -1, \qquad \begin{aligned} &-\Big(\frac{d}{dr} - \frac{\nu}{r} - \frac{\beta_1}{r^2}\Big)F + 2M_1 f - \frac{\alpha_1}{r^2}G = 0\,, \\ &-\Big(\frac{d}{dr} + \frac{\nu}{r} + \frac{\beta_1}{r^2}\Big)f - \Big(E_1 + \frac{\alpha}{r}\Big)F + \frac{\alpha_1}{r^2}g = 0\,, \\ &-\Big(\frac{d}{dr} - \frac{\nu}{r} + \frac{\alpha_2}{r^2}\Big)G + 2M_2 g + \frac{\beta_2}{r^2}F = 0\,, \\ &-\Big(\frac{d}{dr} + \frac{\nu}{r} - \frac{\alpha_2}{r^2}\Big)g - \Big(E_2 + \frac{\alpha}{r}\Big)G - \frac{\beta_2}{r^2}f = 0\,. \end{aligned} \tag{11.37}$$

Eliminating the functions f and g, we obtain

$$\begin{aligned} &\frac{d^2F}{dr^2} + \Big[\frac{2M_1\alpha}{r} + \frac{\nu(-\nu+1)}{r^2} + \frac{2\beta_1(-\nu+1)}{r^3} - \frac{\beta_1{}^2}{r^4} + \frac{M_1\alpha_1\beta_2}{M_2 r^4} + 2M_1E_1\Big]F \\ &+\Big(\frac{\alpha_1}{r^2} - \frac{\alpha_1 M_1}{r^2 M_2}\Big)\frac{dG}{dr} + \Big(\frac{\nu\alpha_1}{r^3} - \frac{2\alpha_1}{r^3} + \frac{M_1\alpha_1\nu}{M_2 r^3} + \frac{\alpha_1\beta_1}{r^4} - \frac{\alpha_1\alpha_2 M_1}{M_2 r^4}\Big)G = 0\,, \end{aligned} \tag{11.38}$$

$$\begin{aligned} &\frac{d^2G}{dr^2} + \Big[\frac{2M_2\alpha}{r} - \frac{\nu(\nu-1)}{r^2} + \frac{2\alpha_2(\nu-1)}{r^3} - \frac{\alpha_2{}^2}{r^4} + \frac{M_2\alpha_1\beta_2}{M_1 r^4} + 2M_2E_2\Big]G \\ &+\Big(-\frac{\beta_2}{r^2} + \frac{\beta_2 M_2}{r^2 M_1}\Big)\frac{dF}{dr} + \Big(-\frac{\nu\beta_2}{r^3} + \frac{2\beta_2}{r^3} - \frac{M_2\beta_2\nu}{M_1 r^3} + \frac{\alpha_2\beta_2}{r^4} - \frac{\beta_1\beta_2 M_2}{M_1 r^4}\Big)F = 0\,. \end{aligned} \tag{11.39}$$

It should be noted that eqs. (11.38)–(11.39) differ from eqs. (11.31)–(11.32) by formal changes

$$f \quad \Longrightarrow \quad F, \quad g \quad \Longrightarrow \quad G, \quad \nu \quad \Longrightarrow -\nu\,. \tag{11.40}$$

Far from the origin, we get more simple equation

$$\delta=-1, \qquad \begin{array}{l} \Big(\dfrac{d^2}{dr^2}+2M_1(E_1+\dfrac{\alpha}{r})-\dfrac{(\nu-1)\nu}{r^2}\Big)F=0, \\ \Big(\dfrac{d^2}{dr^2}+2M_2(E_2+\dfrac{\alpha}{r})-\dfrac{(\nu-1)\nu}{r^2}\Big)G=0. \end{array} \tag{11.41}$$

Their solutions for bound states are

$\delta=-1,\ x=2\sqrt{-2M_1E_1}\,r,$

$$\begin{array}{ll} F(x)=x^{\nu}e^{-x/2}\,{}_1F_1(-n',2\nu,x), & E_1=-\dfrac{M_1\alpha^2}{2(n'+\nu)^2},\ n'=0,1,2,\dots\,, \\ G(x)=x^{\nu}e^{-x/2}\,{}_1F_1(-n',2\nu,x), & E_2=-\dfrac{M_2\alpha^2}{2(n'+\nu)^2},\ n'=0,1,2,\dots\,. \end{array} \tag{11.42}$$

We can see that the transition from states with parity $\delta=+1$ to states with parity $\delta=-1$ is performed by means of the formal change $\nu+1\Longrightarrow\nu$.

11.5 Solutions of the 4th-order equations

The structure of the 4th-order equation is

$$\begin{aligned} &\frac{d^4g}{dr^4}+\Big(\frac{m_1}{r}+\frac{m_2\,r^3+m_3\,r^2+m_4\,r+m_5}{P}\Big)\frac{d^3g}{dr^3} \\ &+\Big(n_0+\frac{n_1}{r}+\frac{n_2}{r^2}+\frac{n_3}{r^3}+\frac{n_4}{r^4}+\frac{n_5\,r^3+n_6\,r^2+n_7\,r+n_8}{P}\Big)\frac{d^2g}{dr^2} \\ &+\Big(\frac{p_1}{r}+\frac{p_2}{r^2}+\frac{p_3}{r^3}+\frac{p_4}{r^4}+\frac{p_5\,r^3+p_6\,r^2+p_7\,r+p_8}{P}\Big)\frac{dg}{dr} \\ &+\Big(q_0+\frac{q_1}{r}+\frac{q_2}{r^2}+\frac{q_3}{r^3}+\frac{q_4}{r^4}+\frac{q_5}{r^5}+\frac{q_6}{r^6}+\frac{q_7}{r^7}+\frac{q_8}{r^8}+\frac{q_9\,r^3+q_{10}\,r^2+q_{11}\,r+q_{12}}{P}\Big)g=0\,, \end{aligned}$$

where

$$\begin{aligned} P=\;&2\,E_1\,M_2\,(M_1-M_2)^2\,r^4+2\,M_2\,\alpha\,(M_1-M_2)^2\,r^3+2\,M_2\,(\nu+1)\,[(2\,\nu+1)\,M_2+M_1]\,r^2 \\ &+2\,M_2\,(\nu+1)\,[(-3\,\beta_1+\alpha_2)\,M_2+M_1\,(\beta_1+\alpha_2)]\,r \\ &+[\beta_2\,\alpha_1+2\,\beta_1\,(-\alpha_2+\beta_1)]\,{M_2}^2-M_2\,M_1\,\left(-{\alpha_2}^2+{\beta_1}^2+2\,\beta_2\,\alpha_1\right)+{M_1}^2\alpha_1\,\beta_2\,. \end{aligned}$$

First, we make substitution $g=e^{Br}\bar{g}(r)$, which leads to

$$\begin{aligned} &\frac{d^4G}{dr^4}+\Big[4\,B+\frac{m_1}{r}+\frac{m_5+m_2\,r^3+m_3\,r^2+m_4\,r}{P}\Big]\frac{d^3\bar{g}}{dr^3} \\ &+\Big[6\,B^2+n_0+\frac{n_1+3\,m_1\,B}{r}+\frac{n_2}{r^2}+\frac{n_3}{r^3}+\frac{n_4}{r^4} \end{aligned}$$

$$+\frac{(3\,m_2\,B+n_5)\,r^3+(n_6+3\,m_3\,B)\,r^2+(3\,m_4\,B+n_7)\,r+n_8+3\,m_5\,B}{P}\Big]\frac{d^2\bar g}{dr^2}$$

$$+\Big[4\,B^3+2\,n_0\,B+\frac{3\,m_1\,B^2+p_1+2\,n_1\,B}{r}+\frac{p_3+2\,n_3\,B}{r^3}+\frac{p_2+2\,n_2\,B}{r^2}+\frac{2\,n_4\,B+p_4}{r^4}$$

$$\frac{1}{P}\Big(\big(p_5+2\,n_5\,B+3\,m_2\,B^2\big)\,r^3+\big(3\,m_3\,B^2+2\,n_6\,B+p_6\big)\,r^2$$

$$+\big(p_7+2\,n_7\,B+3\,m_4\,B^2\big)\,r+2\,n_8\,B+p_8+3\,m_5\,B^2\Big)\Big]\frac{d\bar g}{dr}$$

$$+\Big[B^4+n_0\,B^2+q_0+\frac{q_1+m_1\,B^3+n_1\,B^2+p_1\,B}{r}+\frac{q_2+n_2\,B^2+p_2\,B}{r^2}+\frac{q_3+n_3\,B^2+p_3\,B}{r^3}$$

$$+\frac{q_4+n_4\,B^2+p_4\,B}{r^4}+\frac{q_5}{r^5}+\frac{q_6}{r^6}+\frac{q_7}{r^7}+\frac{q_8}{r^8}+\frac{1}{P}\Big\{\big(p_5\,B+m_2\,B^3+n_5\,B^2+q_9\big)\,r^3$$

$$+\big(m_3\,B^3+q_{10}+n_6\,B^2+p_6\,B\big)\,r^2+\big(q_{11}+m_4\,B^3+n_7\,B^2+p_7\,B\big)\,r$$

$$+q_{12}+m_5\,B^3+n_8\,B^2+p_8\,B\Big\}\Big]\bar g=0\,.$$

Imposing restriction on parameter B:

$$B^4+n_0\,B^2+q_0=0\,,\quad q_0=4E_1E_2M_1M_2\,,\quad n_0=2E_1M_1+2M_2E_2\,,\tag{11.43}$$

we get four possibilities

$$B=\pm\sqrt{-2M_1E_1}\,,\quad B=\pm\sqrt{-2M_2E_2}\,.\tag{11.44}$$

Now we make substitution

$$\bar g=r^A e^{C/r}G(r)\,,\tag{11.45}$$

the resulting equation for $G(r)$ is cumbersome, by this reason it is not written out explicitly.

Let coefficients at r^{-8} and r^{-7} be vanished

$$\frac{q_8+C^4+n_4\,C^2}{r^8}=0,\tag{11.46}$$

$$\frac{q_7+(-m_1-4\,A+12)\,C^3+n_3\,C^2+2\,n_4\,(1-A)\,C}{r^7}=0\,,\tag{11.47}$$

where

$$m_1=8\,,\quad n_3=2\,(-\alpha_2+\beta_1)\,(\nu+1)\,,\quad n_4=2\,\alpha_1\,\beta_2-{\beta_1}^2-{\alpha_2}^2\,,$$

$$q_7=2\,(-\alpha_2+\beta_1)\,(-\alpha_1\,\beta_2+\alpha_2\,\beta_1)\,(1+\nu)\,,\quad q_8=(-\alpha_1\,\beta_2+\alpha_2\,\beta_1)^2\,.$$

For parameter C, we get the algebraic equation of the 4th-order

$$C^4+\big(2\,\alpha_1\,\beta_2-{\beta_1}^2-{\alpha_2}^2\big)\,C^2+(-\alpha_1\beta_2+\alpha_2\beta_1)^2=0,$$

whence it follows (allowing for eq. (11.7))

$$C^4-4\,C^2{\beta_1}^2=0\qquad\Rightarrow\qquad C=\pm2\beta_1\,,0,\;0\,.\tag{11.48}$$

We consider two first possibilities (see eqs. (11.46) and (11.47))

$$I.\qquad C=+2\beta_1<0\,,\qquad A=\nu+2\,;\tag{11.49}$$

$$II.\qquad C=-2\beta_1>0\,,\qquad A=-\nu\,.\tag{11.50}$$

When $C = 0$, eq. (11.47) holds identically and for defining parameter A, we require that that coefficient at r^{-6} be vanished

$$\frac{(2\beta_2\alpha_1 - \beta_1{}^2 - \alpha_2{}^2)A^2 + (\beta_1{}^2 + \alpha_2{}^2 - 2\beta_2\alpha_1)A + 4\beta_1{}^2\nu(\nu+1)}{r^6} = 0\,,$$

whence it follows

$$III, \quad C = 0, \quad A = \nu + 1\,. \tag{11.51}$$

$$IV, \quad C = 0, \quad A = -\nu\,. \tag{11.52}$$

Only the variants I and III are suitable for describing the bound states.

For states with opposite parity, we have the following four variants (see eq. (11.40))

$$\begin{aligned} &I', \quad C = +2\beta_1 < 0\,, \quad A = -\nu + 2\,;\\ &II', \quad C = -2\beta_1 > 0\,, \quad A = +\nu\,;\\ &III', \quad C = 0, \quad A = -\nu + 1\,;\\ &IV', \quad C = 0, \quad A = +\nu\,. \end{aligned} \tag{11.53}$$

Only the variants I' and IV' are suitable for describing the bound states.

For function G (see eqs. (11.45) and (11.49)), we obtain the equation

$$\begin{aligned} &(P_{10}\,r^{10} + P_7\,r^7 + P_6\,r^6 + P_8\,r^8 + P_9\,r^9)\,\frac{d^4G}{dr^4}\\ &+ (Q_{10}\,r^{10} + Q_9\,r^9 + Q_8\,r^8 + Q_7\,r^7 + Q_6\,r^6 + Q_5\,r^5 + Q_4\,r^4)\,\frac{d^3G}{dr^3}\\ &+ (M_{10}\,r^{10} + M_9\,r^9 + M_8\,r^8 + M_7\,r^7 + M_6\,r^6 + M_5\,r^5 + M_4\,r^4 + M_3\,r^3 + M_2\,r^2)\,\frac{d^2G}{dr^2}\\ &+ (N_{10}\,r^{10} + N_9\,r^9 + N_8\,r^8 + N_7\,r^7 + N_6\,r^6 + N_5\,r^5 + N_4\,r^4 + N_3\,r^3 + N_2\,r^2 + N_1\,r + N_0)\,\frac{dG}{dr}\\ &+ (L_9\,r^9 + L_8\,r^8 + L_7\,r^7 + L_6\,r^6 + L_5\,r^5 + L_4\,r^4 + L_3\,r^3 + L_2\,r^2 + L_1\,r + L_0)\,G = 0\,. \end{aligned}$$

Solutions are searched as power series

$$F = \sum_{l=0}^{\infty} d_l r^l\,, \quad \frac{dF}{dr} = \sum_{l=1}^{\infty} l d_l r^{l-1}\,, \quad \frac{d^2F}{dr^2} = \sum_{l=2}^{\infty} l(l-1) d_l r^{l-2}\,,$$

$$\frac{d^3F}{dr^3} = \sum_{l=3}^{\infty} l(l-1)(l-2) d_l r^{l-3}\,, \quad \frac{d^4F}{dr^4} = \sum_{l=4}^{\infty} l(l-1)(l-2)(l-3) d_l r^{l-4}\,.$$

Further, we obtain

$$\begin{aligned} &P_{10}\sum_{l=4}^{\infty} l(l-1)(l-2)(l-3)d_l r^{l+6} + P_9\sum_{l=4}^{\infty} l(l-1)(l-2)(l-3)d_l r^{l+5}\\ &+P_8\sum_{l=4}^{\infty} l(l-1)(l-2)(l-3)d_l r^{l+4}\\ &+P_7\sum_{l=4}^{\infty} l(l-1)(l-2)(l-3)d_l r^{l+3} + P_6\sum_{l=4}^{\infty} l(l-1)(l-2)(l-3)d_l r^{l+2}\\ &+Q_{10}\sum_{l=3}^{\infty} l(l-1)(l-2)d_l r^{l+7} + Q_9\sum_{l=3}^{\infty} l(l-1)(l-2)d_l r^{l+6} + Q_8\sum_{l=3}^{\infty} l(l-1)(l-2)d_l r^{l+5} \end{aligned}$$

$$+Q_7 \sum_{l=3}^{\infty} l(l-1)(l-2)d_l r^{l+4} + Q_6 \sum_{l=3}^{\infty} l(l-1)(l-2)d_l r^{l+3}$$

$$+Q_5 \sum_{l=3}^{\infty} l(l-1)(l-2)d_l r^{l+2} + Q_4 \sum_{l=3}^{\infty} l(l-1)(l-2)d_l r^{l+1}$$

$$+M_{10} \sum_{l=2}^{\infty} l(l-1)d_l r^{l+8} + M_9 \sum_{l=2}^{\infty} l(l-1)d_l r^{l+7} + M_8 \sum_{l=2}^{\infty} l(l-1)d_l r^{l+6}$$

$$+M_7 \sum_{l=2}^{\infty} l(l-1)d_l r^{l+5} + M_6 \sum_{l=2}^{\infty} l(l-1)d_l r^{l+4} + M_5 \sum_{l=2}^{\infty} l(l-1)d_l r^{l+3}$$

$$+M_4 \sum_{l=2}^{\infty} l(l-1)d_l r^{l+2} + M_3 \sum_{l=2}^{\infty} l(l-1)d_l r^{l+1} + M_2 \sum_{l=2}^{\infty} l(l-1)d_l r^{l}$$

$$+N_{10} \sum_{l=1}^{\infty} l d_l r^{l+9} + N_9 \sum_{l=1}^{\infty} l d_l r^{l+8} + N_8 \sum_{l=1}^{\infty} l d_l r^{l+7} + N_7 \sum_{l=1}^{\infty} l d_l r^{l+6}$$

$$+N_6 \sum_{l=1}^{\infty} l d_l r^{l+5} + N_5 \sum_{l=1}^{\infty} l d_l r^{l+4} + N_4 \sum_{l=1}^{\infty} l d_l r^{l+3} + N_3 \sum_{l=1}^{\infty} l d_l r^{l+2}$$

$$+N_2 \sum_{l=1}^{\infty} l d_l r^{l+1} + N_1 \sum_{l=1}^{\infty} l d_l r^{l} + N_0 \sum_{l=1}^{\infty} l d_l r^{l-1}$$

$$+L_9 \sum_{l=0}^{\infty} d_l r^{l+9} + L_8 \sum_{l=0}^{\infty} d_l r^{l+8} + L_7 \sum_{l=0}^{\infty} d_l r^{l+7} + L_6 \sum_{l=0}^{\infty} d_l r^{l+6} + L_5 \sum_{l=0}^{\infty} d_l r^{l+5}$$

$$+L_4 \sum_{l=0}^{\infty} d_l r^{l+4} + L_3 \sum_{l=0}^{\infty} d_l r^{l+3} + L_2 \sum_{l=0}^{\infty} d_l r^{l+2} + L_1 \sum_{l=0}^{\infty} d_l r^{l+1} + L_0 \sum_{l=0}^{\infty} d_l r^{l} = 0\,.$$

After changing summation indices, we obtain

$$P_{10} \sum_{k=10}^{\infty} (k-6)(k-7)(k-8)(k-9)d_{k-6} r^k + P_9 \sum_{k=9}^{\infty} (k-5)(k-6)(k-7)(k-8)d_{k-5} r^k$$

$$+P_8 \sum_{k=8}^{\infty} (k-4)(k-5)(k-6)(k-7)d_{k-4} r^k + P_7 \sum_{k=7}^{\infty} (k-3)(k-4)(k-5)(k-6)d_{k-3} r^k$$

$$+P_6 \sum_{k=6}^{\infty} (k-2)(k-3)(k-4)(k-5)d_{k-2} r^k + Q_{10} \sum_{k=10}^{\infty} (k-7)(k-8)(k-9)d_{k-7} r^k$$

$$+Q_9 \sum_{k=9}^{\infty} (k-6)(k-7)(k-8)d_{k-6} r^k + Q_8 \sum_{k=8}^{\infty} (k-5)(k-6)(k-7)d_{k-5} r^k$$

$$+Q_7 \sum_{k=7}^{\infty} (k-4)(k-5)(k-6)d_{k-4} r^k + Q_6 \sum_{k=6}^{\infty} (k-3)(k-4)(k-5)d_{k-3} r^k$$

$$+Q_5 \sum_{k=5}^{\infty} (k-2)(k-3)(k-4)d_{k-2} r^k + Q_4 \sum_{k=4}^{\infty} (k-1)(k-2)(k-3)d_{k-1} r^k$$

$$+M_{10} \sum_{k=10}^{\infty} (k-8)(k-9)d_{k-8} r^k + M_9 \sum_{k=9}^{\infty} (k-7)(k-8)d_{k-7} r^k + M_8 \sum_{k=8}^{\infty} (k-6)(k-7)d_{k-6} r^k$$

$$+M_7 \sum_{k=7}^{\infty} (k-5)(k-6)d_{k-5} r^k + M_6 \sum_{k=6}^{\infty} (k-4)(k-5)d_{k-4} r^k + M_5 \sum_{k=5}^{\infty} (k-3)(k-4)d_{k-3} r^k$$

$$+M_4 \sum_{k=4}^{\infty} (k-2)(k-3)d_{k-2} r^k + M_3 \sum_{k=3}^{\infty} (k-1)(k-2)d_{k-1} r^k + M_2 \sum_{k=2}^{\infty} k(k-1)d_k r^k$$

$$+N_{10}\sum_{k=10}^{\infty}(k-9)d_{k-9}r^k+N_9\sum_{k=9}^{\infty}(k-8)d_{k-8}r^k+N_8\sum_{k=8}^{\infty}(k-7)d_{k-7}r^k+N_7\sum_{k=7}^{\infty}(k-6)d_{k-6}r^k$$

$$+N_6\sum_{k=6}^{\infty}(k-5)d_{k-5}r^k+N_5\sum_{k=5}^{\infty}(k-4)d_{k-4}r^k+N_4\sum_{k=4}^{\infty}(k-3)d_{k-3}r^k+N_3\sum_{k=3}^{\infty}(k-2)d_{k-2}r^k$$

$$+N_2\sum_{k=2}^{\infty}(k-1)d_{k-1}r^k+N_1\sum_{k=1}^{\infty}kd_kr^k+N_0\sum_{k=0}^{\infty}(k+1)d_{k+1}r^k$$

$$+L_9\sum_{k=9}^{\infty}d_{k-9}r^k+L_8\sum_{k=8}^{\infty}d_{k-8}r^k+L_7\sum_{k=7}^{\infty}d_{k-7}r^k+L_6\sum_{k=6}^{\infty}d_{k-6}r^k+L_5\sum_{k=5}^{\infty}d_{k-5}r^k$$

$$+L_4\sum_{k=4}^{\infty}d_{k-4}r^k+L_3\sum_{k=3}^{\infty}d_{k-3}r^k+L_2\sum_{k=2}^{\infty}d_{k-2}r^k+L_1\sum_{k=1}^{\infty}d_{k-1}r^k+L_0\sum_{k=0}^{\infty}d_kr^k=0\,.$$

Finally, we get 11-term recurrent relations

$$k=0\,,\qquad N_0\,d_1+L_0\,d_0=0\,,$$

$$k=1\,,\qquad 2\,N_0d_2+(N_1+L_0)\,d_1+L_1\,d_0=0\,,$$

$$\ldots\ldots\ldots\ldots\ldots\ldots\ldots\ldots$$

$$Q_{k-9}d_{k-9}+Q_{k-8}d_{k-8}+\ldots+Q_kd_k+Q_{k+1}d_{k=1}=0\,. \tag{11.54}$$

The known transcendency condition for Frobenius solution has the form

$$Q_{k-9}=0\quad\Longrightarrow\quad N_{10}\,(k-9)+L_9\ =0,\quad k-9=n=1,2,3,\ldots\,; \tag{11.55}$$

explicitly it reads

$$8\Big\{\Big[\alpha\,(E_1+E_2)\,M_1+\Big(\frac{1}{2}\alpha B+E_2\,(-7+A+k)\Big)B\Big]\,M_2$$
$$+\Big[\Big(\frac{1}{2}\alpha B+E_1\,(A+k-9)\Big)M_1+B^2\,(k+A-8)\Big]B\Big\}E_1M_2\,(M_1-M_2)^2=0.$$

Let

$$A=\nu+2,\quad B=-\sqrt{-2M_1E_1}, \tag{11.56}$$

then the above equation takes the form

$$E_1\,(M_1-M_2)\,M_2\,\Big(\sqrt{2}\sqrt{-E_1M_1}(k+\nu-5)-\alpha M_1\Big)\,(E_1M_1-E_2M_2)=0.$$

This equation has three roots

$$E_1=0,\quad E_1=-\frac{\alpha^2M_1}{2(k-5+\nu)^2},\quad E_2=\frac{M_1}{M_2}E_1.$$

As physical, we consider only the second root

$$E_1=-\frac{\alpha^2M_1}{2(k-5+\nu)^2},\quad k-9=n=1,2,3,\ldots\,. \tag{11.57}$$

Now, let

$$A=\nu+2,\quad B=-\sqrt{-2M_2E_2}, \tag{11.58}$$

then eq. (11.56) takes the form

$$E_1\left(M_1-M_2\right)M_2\left(E_1M_1-E_2M_2\right)\left(\alpha M_2-\sqrt{2}\sqrt{-E_2M_2}(k+\nu-7)\right)=0.$$

As physical, we consider only the root

$$E_2=-\frac{1}{2}\frac{\alpha^2 M_2}{(k-7+\nu)^2}\qquad k-9=n=1,2,3,\dots\,. \tag{11.59}$$

Now, we consider the case *III*: $C=0, A=\nu+1$. Here we have the following equation

$$\left(P_9\,r^9+P_8\,r^8+P_7\,r^7+P_6\,r^6+P_5\,r^5\right)\frac{d^4G}{dr^4}$$

$$+\left(Q_9\,r^9+Q_8\,r^8+Q_7\,r^7+Q_6\,r^6+Q_5\,r^5+Q_4\,r^4\right)\frac{d^3G}{dr^3}$$

$$+\left(M_9\,r^9+M_8\,r^8+M_7\,r^7+M_6\,r^6+M_5\,r^5+M_4\,r^4+M_3\,r^3+M_2\,r^2+M_1\,r\right)\frac{d^2G}{dr^2}$$

$$+\left(N_9\,r^9+N_8\,r^8+N_7\,r^7+N_6\,r^6+N_5\,r^5+N_4\,r^4+N_3\,r^3+N_2\,r^2+N_1\,r+N_0\right)\frac{dG}{dr}$$

$$+\left(L_8\,r^8+L_7\,r^7+L_6\,r^6+L_5\,r^5+L_4\,r^4+L_3\,r^3+L_2\,r^2+L_1\,r+L_0\right)G=0\,.$$

Solutions are searched as power series

$$F=\sum_{l=0}^{\infty}d_l r^l\,,\quad \frac{dF}{dr}=\sum_{l=1}^{\infty}l d_l r^{l-1}\,,\quad \frac{d^2F}{dr^2}=\sum_{l=2}^{\infty}l(l-1)d_l r^{l-2}\,,$$

$$\frac{d^3F}{dr^3}=\sum_{l=3}^{\infty}l(l-1)(l-2)d_l r^{l-3}\,,\quad \frac{d^4F}{dr^4}=\sum_{l=4}^{\infty}l(l-1)(l-2)(l-3)d_l r^{l-4}\,.$$

Further, we get

$$P_9\sum_{l=4}^{\infty}l(l-1)(l-2)(l-3)d_l r^{l+5}+P_8\sum_{l=4}^{\infty}l(l-1)(l-2)(l-3)d_l r^{l+4}+P_7\sum_{l=4}^{\infty}l(l-1)(l-2)(l-3)d_l r^{l+3}$$

$$+P_6\sum_{l=4}^{\infty}l(l-1)(l-2)(l-3)d_l r^{l+2}+P_5\sum_{l=4}^{\infty}l(l-1)(l-2)(l-3)d_l r^{l+1}+Q_9\sum_{l=3}^{\infty}l(l-1)(l-2)d_l r^{l+6}$$

$$+Q_8\sum_{l=3}^{\infty}l(l-1)(l-2)d_l r^{l+5}+Q_7\sum_{l=3}^{\infty}l(l-1)(l-2)d_l r^{l+4}+Q_6\sum_{l=3}^{\infty}l(l-1)(l-2)d_l r^{l+3}$$

$$+Q_5\sum_{l=3}^{\infty}l(l-1)(l-2)d_l r^{l+2}+Q_4\sum_{l=3}^{\infty}l(l-1)(l-2)d_l r^{l+1}+M_9\sum_{l=2}^{\infty}l(l-1)d_l r^{l+7}$$

$$+M_8\sum_{l=2}^{\infty}l(l-1)d_l r^{l+6}+M_7\sum_{l=2}^{\infty}l(l-1)d_l r^{l+5}+M_6\sum_{l=2}^{\infty}l(l-1)d_l r^{l+4}+M_5\sum_{l=2}^{\infty}l(l-1)d_l r^{l+3}$$

$$+M_4\sum_{l=2}^{\infty}l(l-1)d_l r^{l+2}+M_3\sum_{l=2}^{\infty}l(l-1)d_l r^{l+1}+M_2\sum_{l=2}^{\infty}l(l-1)d_l r^{l}+M_1\sum_{l=2}^{\infty}l(l-1)d_l r^{l-1}$$

$$+N_9\sum_{l=1}^{\infty}l d_l r^{l+8}+N_8\sum_{l=1}^{\infty}l d_l r^{l+7}+N_7\sum_{l=1}^{\infty}l d_l r^{l+6}+N_6\sum_{l=1}^{\infty}l d_l r^{l+5}+N_5\sum_{l=1}^{\infty}l d_l r^{l+4}$$

$$+N_4\sum_{l=1}^{\infty}l d_l r^{l+3}+N_3\sum_{l=1}^{\infty}l d_l r^{l+2}+N_2\sum_{l=1}^{\infty}l d_l r^{l+1}+N_1\sum_{l=1}^{\infty}l d_l r^{l}+N_0\sum_{l=1}^{\infty}l d_l r^{l-1}$$

$$+L_8\sum_{l=0}^{\infty} d_l r^{l+8} + L_7\sum_{l=0}^{\infty} d_l r^{l+7} + L_6\sum_{l=0}^{\infty} d_l r^{l+6} + L_5\sum_{l=0}^{\infty} d_l r^{l+5} + L_4\sum_{l=0}^{\infty} d_l r^{l+4}$$

$$+L_3\sum_{l=0}^{\infty} d_l r^{l+3} + L_2\sum_{l=0}^{\infty} d_l r^{l+2} + L_1\sum_{l=0}^{\infty} d_l r^{l+1} + L_0\sum_{l=0}^{\infty} d_l r^{l} = 0\,.$$

Changing the summation indices we get

$$P_9\sum_{k=9}^{\infty}(k-5)(k-6)(k-7)(k-8)d_{k-5}r^k + P_8\sum_{k=8}^{\infty}(k-4)(k-5)(k-6)(k-7)d_{k-4}r^k$$

$$+P_7\sum_{k=7}^{\infty}(k-3)(k-4)(k-5)(k-6)d_{k-3}r^k + P_6\sum_{k=6}^{\infty}(k-2)(k-3)(k-4)(k-5)d_{k-2}r^k$$

$$+P_5\sum_{k=5}^{\infty}(k-1)(k-2)(k-3)(k-4)d_{k-1}r^k + Q_9\sum_{k=9}^{\infty}(k-6)(k-7)(k-8)d_{k-6}r^k$$

$$+Q_8\sum_{k=8}^{\infty}(k-5)(k-6)(k-7)d_{k-5}r^k + Q_7\sum_{k=7}^{\infty}(k-4)(k-5)(k-6)d_{k-4}r^k$$

$$+Q_6\sum_{k=6}^{\infty}(k-3)(k-4)(k-5)d_{k-3}r^k$$

$$+Q_5\sum_{k=5}^{\infty}(k-2)(k-3)(k-4)d_{k-2}r^k + Q_4\sum_{k=4}^{\infty}(k-1)(k-2)(k-3)d_{k-1}r^k + M_9\sum_{k=9}^{\infty}(k-7)(k-8)d_{k-7}r^k$$

$$+M_8\sum_{k=8}^{\infty}(k-6)(k-7)d_{k-6}r^k + M_7\sum_{k=7}^{\infty}(k-5)(k-6)d_{k-5}r^k + M_6\sum_{k=6}^{\infty}(k-4)(k-5)d_{k-4}r^k$$

$$+M_5\sum_{k=5}^{\infty}(k-3)(k-4)d_{k-3}r^k + M_4\sum_{k=4}^{\infty}(k-2)(k-3)d_{k-2}r^k + M_3\sum_{k=3}^{\infty}(k-1)(k-2)d_{k-1}r^k$$

$$+M_2\sum_{k=2}^{\infty}k(k-1)d_k r^k + M_1\sum_{k=1}^{\infty}(k+1)k d_{k+1}r^k + N_9\sum_{k=9}^{\infty}(k-8)d_{k-8}r^k + N_8\sum_{k=8}^{\infty}(k-7)d_{k-7}r^k$$

$$+N_7\sum_{k=7}^{\infty}(k-6)d_{k-6}r^k + N_6\sum_{k=6}^{\infty}(k-5)d_{k-5}r^k + N_5\sum_{k=5}^{\infty}(k-4)d_{k-4}r^k$$

$$+N_4\sum_{k=4}^{\infty}(k-3)d_{k-3}r^k + N_3\sum_{k=3}^{\infty}(k-2)d_{k-2}r^k + N_2\sum_{k=2}^{\infty}(k-1)d_{k-1}r^k$$

$$+N_1\sum_{k=1}^{\infty}k d_k r^k + N_0\sum_{k=0}^{\infty}(k+1)d_{k+1}r^k$$

$$+L_8\sum_{k=8}^{\infty}d_{k-8}r^k + L_7\sum_{k=7}^{\infty}d_{k-7}r^k + L_6\sum_{k=6}^{\infty}d_{k-6}r^k + L_5\sum_{k=5}^{\infty}d_{k-5}r^k + L_4\sum_{k=4}^{\infty}d_{k-4}r^k$$

$$+L_3\sum_{k=3}^{\infty}d_{k-3}r^k + L_2\sum_{k=2}^{\infty}d_{k-2}r^k + L_1\sum_{k=1}^{\infty}d_{k-1}r^k + L_0\sum_{k=0}^{\infty}d_k r^k = 0\,.$$

So we arrive at 10-term recurrent relation

$$Q_{k-8}d_{k-8} + Q_{k-7}d_{k-7} + \ldots + Q_k d_k + Q_{k+1}d_{k+1} = 0. \tag{11.60}$$

Transcendency condition for Frobenius solution has the form

$$Q_{k-8} = 0 \quad \Longrightarrow \quad N_9\,(k-8) + L_8 = 0\,, \quad k-8 = n = 1, 2, 3, \ldots\,;$$

explicitly it reads

$$8\Big\{\Big[\alpha\,(E_1+E_2)\,M_1+\Big(\frac{1}{2}\alpha B+E_2\,(-6+A+k)\Big)B\Big]\,M_2$$
$$+\Big[\Big(\frac{1}{2}\alpha B+E_1\,(A+k-8)\Big)M_1+B^2\,(k+A-7)\Big]B\Big\}E_1M_2\,(M_1-M_2)^2=0.$$

Let

$$A=\nu+1,\quad B=-\sqrt{-2M_1E_1}, \tag{11.61}$$

the above equation take the form

$$E_1\,(M_1-M_2)\,M_2\,(E_1M_1-E_2M_2)\left(\sqrt{2}\sqrt{-E_1M_1}(k+\nu-5)-\alpha M_1\right)=0;$$

its solutions are

$$E_1=0,\quad E_1=-\frac{\alpha^2M_1}{2(k+\nu-5)^2},\quad E_1=\frac{E_2M_2}{M_1};$$

as physical, we take the root

$$E_1=-\frac{1}{2}\frac{\alpha^2M_1}{(k-5+\nu)^2}\qquad k-8=n=1,2,3,\ldots\,. \tag{11.62}$$

Now let

$$A=\nu+1,\quad B=-\sqrt{-2M_2E_2}, \tag{11.63}$$

the eq. (11.61) reads

$$E_1\,(M_1-M_2)\,M_2\,(E_1M_1-E_2M_2)\left(\alpha M_2-\sqrt{2}\sqrt{-E_2M_2}(k+\nu-7)\right)=0.$$

Its physical solutions is

$$E_2=-\frac{1}{2}\frac{M_2\,\alpha^2}{(k-7+\nu)^2}\qquad k-8=n=1,2,3,\ldots\,. \tag{11.64}$$

Thus, we have found the following spectra

$$\begin{aligned}
&1.\ A=\nu+2, B=-\sqrt{-2M_1E_1}, E_1=-\frac{1}{2}\frac{\alpha^2M_1}{(n+4+\nu)^2}, k-9=n=1,2,\ldots;\\
&2.\ A=\nu+2, B=-\sqrt{-2M_2E_2}, E_2=-\frac{1}{2}\frac{\alpha^2M_2}{(n+2+\nu)^2}, k-9=n=1,2\ldots;\\
&3.\ A=\nu+1, B=-\sqrt{-2M_1E_1}, E_1=-\frac{1}{2}\frac{\alpha^2M_1}{(n+3+\nu)^2}, k-8=n=1,2,\ldots;\\
&4.\ A=\nu+1, B=-\sqrt{-2M_2E_2}, E_2=-\frac{1}{2}\frac{M_2\,\alpha^2}{(n+1+\nu)^2}, k-8=n=1,2,\ldots\,.
\end{aligned} \tag{11.65}$$

For states with opposite parity, we have the following spectra (changing the notations $\nu+2\Longrightarrow\nu, \nu+1\Longrightarrow\nu$)

$$\begin{aligned}
&A=\nu, B=-\sqrt{-2M_1E_1},\quad E_1=-\frac{1}{2}\frac{\alpha^2M_1}{[k-7+\nu]^2};\\
&A=\nu, B=-\sqrt{-2M_2E_2},\quad E_2=-\frac{1}{2}\frac{\alpha^2M_2}{[k-9+\nu]^2};\\
&A=\nu, B=-\sqrt{-2M_1E_1},\quad E_1=-\frac{1}{2}\frac{\alpha^2M_1}{[k-6+\nu]^2};\\
&A=\nu, B=-\sqrt{-2M_2E_2},\quad E_2=-\frac{1}{2}\frac{M_2\,\alpha^2}{[k-8+\nu]^2}.
\end{aligned} \tag{11.66}$$

Let us examine numerically two spectra

$$E_1 = -\frac{\alpha^2 M_1}{2}\frac{1}{(n+\nu)^2}, \qquad E_2 = -\frac{\alpha^2 M_2}{2}\frac{1}{(n+\nu)^2}. \tag{11.67}$$

they are governed by the factor (let $\nu = 1$, $n = \overline{0, 10}$, see Fig. 11.1)

$$-\frac{1}{(n+\nu)^2} \quad \Longrightarrow \quad -1, -\frac{1}{4}, -\frac{1}{9}, -\frac{1}{16}, -\frac{1}{25}, -\frac{1}{36}, -\frac{1}{49}, -\frac{1}{64}, -\frac{1}{81}, -\frac{1}{100}, -\frac{1}{121}.$$

Relation between two spectra is described by the formula (see Fig. 11.2)

$$\frac{E_1}{E_2} = \frac{M_1}{M_2} = \frac{1-\cos\gamma}{1+\cos\gamma}, \qquad \gamma \in [0, \pi/2]. \tag{11.68}$$

When $\nu = 3$, two spectra are governed by the factor

$$-\frac{1}{(n+\nu)^2} \quad \Longrightarrow \quad -\frac{1}{9}, -\frac{1}{16}, -\frac{1}{25}, -\frac{1}{36}, -\frac{1}{49}, -\frac{1}{64}, -\frac{1}{81}, -\frac{1}{100}, -\frac{1}{121}, -\frac{1}{144}, -\frac{1}{169}.$$

The difference of two spectra $\Delta = E_1^{(\nu=3)} - E_1^{(\nu=1)}$ is illustrated in Fig. 11.3.

Let us detail the structure of nonrelativistic wave functions. Taking in mind the approximate equalities

$$\delta = +1, \qquad \begin{array}{ll} f_1 = f + iF \approx f, & f_2 = f - iF \approx f, \\ g_1 = g + iG \approx g, & g_2 = g - iG \approx g; \end{array} \tag{11.69}$$

$$\delta = -1, \qquad \begin{array}{ll} f_1 = f + iF \approx iF, & f_2 = f - iF \approx -iF, \\ g_1 = g + iG \approx iG, & g_2 = g - iG \approx -iG, \end{array} \tag{11.70}$$

we get

$$\delta = +1, \quad \Psi_{\delta=+1} = \left| \begin{array}{c} f(r)D_{-1/2} \\ f(r)D_{+1/2} \end{array} \right|, \qquad \left| \begin{array}{c} g(r)D_{-1/2} \\ g(r)D_{+1/2} \end{array} \right|; \tag{11.71}$$

$$\delta = -1, \quad \Psi_{\delta=-1} = i \left| \begin{array}{c} F(r)D_{-1/2} \\ -F(r)D_{+1/2} \end{array} \right|, \qquad \left| \begin{array}{c} iG(r)D_{-1/2} \\ -G(r)D_{+1/2} \end{array} \right|. \tag{11.72}$$

11.6 Solutions in relativistic case

In relativistic case, the structure of the resulting 4th-order equation for $F(r)$ is as follows

$$\frac{d^4F}{dr^4} + \left(\frac{m_1}{r} + \frac{m_2}{r^2} + \frac{m_3 r^5 + m_4 r^4 + m_5 r^3 + m_6 r^2 + m_7 r + m_8}{P}\right)\frac{d^3F}{dr^3}$$

$$+\left(n_0 + \frac{n_1}{r} + \frac{n_2}{r^2} + \frac{n_3}{r^3} + \frac{n_4}{r^4} + \frac{n_5 r^5 + n_6 r^4 + n_7 r^3 + n_8 r^2 + n_9 r + n_{10}}{P}\right)\frac{d^2F}{dr^2}$$

$$+\left(\frac{p_1}{r} + \frac{p_2}{r^2} + \frac{p_3}{r^3} + \frac{p_4}{r^4} + \frac{p_5}{r^5} + \frac{p_6 r^5 + p_7 r^4 + p_8 r^3 + p_9 r^2 + p_{10} r + p_{11}}{P}\right)\frac{dF}{dr}$$

$$+\Big(q_0 + \frac{q_1}{r} + \frac{q_2}{r^2} + \frac{q_3}{r^3} + \frac{q_4}{r^4} + \frac{q_5}{r^5} + \frac{q_6}{r^6}$$
$$+\frac{q_7\, r^5 + q_8\, r^4 + q_9\, r^3 + q_{10}\, r^2 + q_{11}\, r + q_{12}}{P}\Big)F = 0\,,$$

where

$$P = (M_1 - M_2)^2\,(\epsilon + M_2)\,(\epsilon - M_1)\,r^6 - (M_1 - M_2)^2\,\alpha\,(-M_2 - 2\,\epsilon + M_1)\,r^5$$
$$+\Big[\left(4\,\nu^2 - 6\,\nu + 2 + \alpha^2\right)M_1{}^2 + \left(\left(-6 - 8\,\nu^2 + 14\,\nu\right)\epsilon - 2\,M_2\,\left(-1 + \nu + \alpha^2\right)\right)M_1$$
$$+4\,\left(-1 + \nu\right)^2\epsilon^2 + 2\,M_2\,\left(-1 + \nu\right)\epsilon + M_2{}^2\alpha^2\Big]\,r^4$$
$$-8\,[\beta_1\,M_1 - \epsilon\,\beta_1 + (-\frac{1}{2} + \nu)\alpha]\,(-1 + \nu)\,(-\epsilon + M_1)\,r^3$$
$$+[(-\alpha - 2\,\epsilon\,\beta_1 + 2\,\nu\,\alpha + \beta_1\,M_1 + M_2\,\beta_1)^2$$
$$+2\,(-\alpha - 2\,\epsilon\,\beta_1 + 2\,\nu\,\alpha + \beta_1\,M_1 + M_2\,\beta_1)(M_1 - M_2)\beta_1$$
$$+2\,\alpha\,\beta_1\,(M_1 - M_2)(2 + 2\,\nu) - 6\,(M_1 - M_2)\alpha\,\beta_1$$
$$-4\,(-2\,\epsilon + 2\,\nu\,\epsilon - \nu\,M_1 + 2\,M_1 - M_2\,\nu)\alpha\,\beta_1\,]r^2$$
$$-8\,\beta_1\,[(-\frac{1}{2} + \nu)\alpha + \beta_1\,(-\epsilon + M_1]\alpha\,r + 4\,\alpha^2\beta_1{}^2\,. \tag{11.73}$$

First, we make substitution $F = e^{K\,r}\,\bar{F}(r)$, which yields

$$\frac{d^4\bar{F}}{dr^4} + \Big[4\,K + \frac{m_1}{r} + \frac{m_2}{r^2} + \frac{m_3\,r^5 + m_4\,r^4 + m_5\,r^3 + m_6\,r^2 + m_7\,r + m_8}{P}\Big]\frac{d^3\bar{F}}{dr^3}$$
$$+\Big\{n_0 + 6\,K^2 + \frac{3\,m_1\,K + n_1}{r} + \frac{n_2 + 3\,m_2\,K}{r^2} + \frac{n_3}{r^3} + \frac{n_4}{r^4}$$
$$+\frac{1}{P}\Big[\,(n_5 + 3\,m_3\,K)\,r^5 + (3\,m_4\,K + n_6)\,r^4 + (n_7 + 3\,m_5\,K)\,r^3$$
$$+\,(n_8 + 3\,m_6\,K)\,r^2 + (3\,m_7\,K + n_9)\,r + 3\,m_8\,K + n_{10}\Big]\Big\}\frac{d^2\bar{F}}{dr^2}$$
$$+\Big\{2\,n_0\,K + 4\,K^3 + \frac{p_1 + 3\,m_1\,K^2 + 2\,n_1\,K}{r} + \frac{p_2 + 3\,m_2\,K^2 + 2\,n_2\,K}{r^2} + \frac{p_3 + 2\,n_3\,K}{r^3}$$
$$+\frac{p_4 + 2\,n_4\,K}{r^4} + \frac{p_5}{r^5} + \frac{1}{P}\,[(3\,m_3\,K^2 + 2\,n_5\,K + p_6)\,r^5 + \left(3\,m_4\,K^2 + p_7 + 2\,n_6\,K\right)r^4$$
$$+\,\left(p_8 + 3\,m_5\,K^2 + 2\,n_7\,K\right)r^3 + \left(2\,n_8\,K + p_9 + 3\,m_6\,K^2\right)r^2$$
$$+\,\left(2\,n_9\,K + 3\,m_7\,K^2 + p_{10}\right)r + 2\,n_{10}\,K + p_{11} + 3\,m_8\,K^2]\,\Big\}\frac{d\bar{F}}{dr}$$
$$+\Big\{n_0\,K^2 + K^4 + q_0 + \frac{q_1 + p_1\,K + m_1\,K^3 + n_1\,K^2}{r} + \frac{q_2 + p_2\,K + m_2\,K^3 + n_2\,K^2}{r^2}$$
$$+\frac{q_3 + p_3\,K + n_3\,K^2}{r^3} + \frac{q_4 + p_4\,K + n_4\,K^2}{r^4} + \frac{q_5 + p_5\,K}{r^5} + \frac{q_6}{r^6}$$
$$+\frac{1}{P}\,[\left(q_7 + p_6\,K + n_5\,K^2 + m_3\,K^3\right)r^5 + \left(q_8 + p_7\,K + m_4\,K^3 + n_6\,K^2\right)r^4$$
$$+\,\left(p_8\,K + q_9 + n_7\,K^2 + m_5\,K^3\right)r^3 + \left(n_8\,K^2 + p_9\,K + m_6\,K^3 + q_{10}\right)r^2$$

$$+ \left(n_9\, K^2 + q_{11} + p_{10}\, K + m_7\, K^3\right) r + m_8\, K^3 + p_{11}\, K + n_{10}\, K^2 + q_{12}\big] \Big\}\bar{F} = 0\,.$$

Impose restriction on parameter K:

$$K^4 + n_0\, K^2 + q_0 = 0\,, \quad n_0 = -{M_1}^2 - {M_2}^2 + 2\,\epsilon^2\,, \quad q_0 = \left(\epsilon^2 - M_1^2\right)\left(\epsilon^2 - M_2^2\right)\,,$$

$$\left(K^2 - {M_1}^2 + \epsilon^2\right)\left(K^2 - {M_2}^2 + \epsilon^2\right) = 0\,,$$

in this way we find four possible solutions (below we follow only the negative values)

$$K_1 = -\sqrt{{M_1}^2 - \epsilon^2} < 0\,, \quad K_2 = -\sqrt{{M_2}^2 - \epsilon^2} < 0\,. \tag{11.74}$$

Further, we make the second substitution

$$\bar{F} = r^H e^{L/r} \bar{F}\,, \tag{11.75}$$

we will not write out the resulting equation because of its bulkiness. Let the coefficient at r^{-8} be vanished

$$\frac{L^2\left(L^2 - m_2\, L + n_4\right)}{r^8} = 0\,, \qquad m_2 = 0\,, \quad n_4 = -4\beta_1^2\,.$$

There exist four variants

$$I. \qquad L = 2\beta_1 < 0\,, \quad H = \nu\ > 0; \tag{11.76}$$

$$II. \qquad L = -2\beta_1 > 0\,, \quad H = 1 - \nu \le 0\,; \tag{11.77}$$

$$III. \qquad L = 0\,, \quad H = \sqrt{\nu^2 - \alpha^2} > 0\,; \tag{11.78}$$

$$IV. \qquad L = 0\,, \quad H = -\sqrt{\nu^2 - \alpha^2} < 0\,. \tag{11.79}$$

Only variants I and III are suitable for describing bound states. Allowing for this restriction we arrive at the following equation for $\bar{F}$:

$$\left(P_{12}\, r^{12} + P_{11}\, r^{11} + P_{10}\, r^{10} + P_9\, r^9 + P_8\, r^8 + P_7\, r^7 + P_6\, r^6\right) \frac{d^4 \bar{F}}{dr^4}$$

$$+ \left(Q_{12}\, r^{12} + Q_{11}\, r^{11} + Q_{10}\, r^{10} + Q_9\, r^9 + Q_8\, r^8 + Q_7\, r^7 + Q_6\, r^6 + Q_5\, r^5 + Q_4\, r^4\right) \frac{d^3 \bar{F}}{dr^3}$$

$$+(M_{12}r^{12} + M_{11}r^{11} + M_{10}r^{10} + M_9 r^9 + M_8 r^8 + M_7 r^7$$

$$+M_6 r^6 + M_5 r^5 + M_4 r^4 + M_3 r^3 + M_2 r^2)\frac{d^2 \bar{F}}{dr^2}$$

$$+(N_{12}\, r^{12} + N_{11}\, r^{11} + N_{10}\, r^{10} + N_9\, r^9 + N_8\, r^8 + N_7\, r^7$$

$$+N_6\, r^6 + N_5\, r^5 + N_4\, r^4 + N_3\, r^3 + N_2\, r^2 + N_1\, r + N_0)\frac{d\bar{F}}{dr}$$

$$+(L_{11}\, r^{11} + L_{10}\, r^{10} + L_9\, r^9 + L_8\, r^8 + L_7\, r^7 + L_6\, r^6 + L_5\, r^5$$

$$+L_4\, r^4 + L_3\, r^3 + L_2\, r^2 + L_1\, r + L_0)\bar{F} = 0\,.$$

Solutions are searched as power series

$$\bar{F} = \sum_{l=0}^{\infty} d_l r^l\,, \quad \frac{d\bar{F}}{dr} = \sum_{l=1}^{\infty} l d_l r^{l-1}\,, \quad \frac{d^2\bar{F}}{dr^2} = \sum_{l=2}^{\infty} l(l-1) d_l r^{l-2}\,,$$

$$\frac{d^3\bar{F}}{dr^3} = \sum_{l=3}^{\infty} l(l-1)(l-2)d_l r^{l-3}, \quad \frac{d^4\bar{F}}{dr^4} = \sum_{l=4}^{\infty} l(l-1)(l-2)(l-3)d_l r^{l-4},$$

further, we get

$$P_{12}\sum_{l=4}^{\infty} l(l-1)(l-2)(l-3)d_l r^{l+8} + P_{11}\sum_{l=4}^{\infty} l(l-1)(l-2)(l-3)d_l r^{l+7}$$
$$+P_{10}\sum_{l=4}^{\infty} l(l-1)(l-2)(l-3)d_l r^{l+6} + P_9\sum_{l=4}^{\infty} l(l-1)(l-2)(l-3)d_l r^{l+5}$$
$$+P_8\sum_{l=4}^{\infty} l(l-1)(l-2)(l-3)d_l r^{l+4} + P_7\sum_{l=4}^{\infty} l(l-1)(l-2)(l-3)d_l r^{l+3}$$
$$+P_6\sum_{l=4}^{\infty} l(l-1)(l-2)(l-3)d_l r^{l+2} + Q_{12}\sum_{l=3}^{\infty} l(l-1)(l-2)d_l r^{l+9} + Q_{11}\sum_{l=3}^{\infty} l(l-1)(l-2)d_l r^{l+8}$$
$$+Q_{10}\sum_{l=3}^{\infty} l(l-1)(l-2)d_l r^{l+7} + Q_9\sum_{l=3}^{\infty} l(l-1)(l-2)d_l r^{l+6} + Q_8\sum_{l=3}^{\infty} l(l-1)(l-2)d_l r^{l+5}$$
$$+Q_7\sum_{l=3}^{\infty} l(l-1)(l-2)d_l r^{l+4} + Q_6\sum_{l=3}^{\infty} l(l-1)(l-2)d_l r^{l+3} + Q_5\sum_{l=3}^{\infty} l(l-1)(l-2)d_l r^{l+2}$$
$$+Q_4\sum_{l=3}^{\infty} l(l-1)(l-2)d_l r^{l+1} + M_{12}\sum_{l=2}^{\infty} l(l-1)d_l r^{l+10} + M_{11}\sum_{l=2}^{\infty} l(l-1)d_l r^{l+9}$$
$$+M_{10}\sum_{l=2}^{\infty} l(l-1)d_l r^{l+8} + M_9\sum_{l=2}^{\infty} l(l-1)d_l r^{l+7} + M_8\sum_{l=2}^{\infty} l(l-1)d_l r^{l+6} + M_7\sum_{l=2}^{\infty} l(l-1)d_l r^{l+5}$$
$$+M_6\sum_{l=2}^{\infty} l(l-1)d_l r^{l+4} + M_5\sum_{l=2}^{\infty} l(l-1)d_l r^{l+3} + M_4\sum_{l=2}^{\infty} l(l-1)d_l r^{l+2} + M_3\sum_{l=2}^{\infty} l(l-1)d_l r^{l+1}$$
$$+M_2\sum_{l=2}^{\infty} l(l-1)d_l r^{l} + N_{12}\sum_{l=1}^{\infty} l d_l r^{l+11} + N_{11}\sum_{l=1}^{\infty} l d_l r^{l+10} + N_{10}\sum_{l=1}^{\infty} l d_l r^{l+9} + N_9\sum_{l=1}^{\infty} l d_l r^{l+8}$$
$$+N_8\sum_{l=1}^{\infty} l d_l r^{l+7} + N_7\sum_{l=1}^{\infty} l d_l r^{l+6} + N_6\sum_{l=1}^{\infty} l d_l r^{l+5} + N_5\sum_{l=1}^{\infty} l d_l r^{l+4} + N_4\sum_{l=1}^{\infty} l d_l r^{l+3}$$
$$+N_3\sum_{l=1}^{\infty} l d_l r^{l+2} + N_2\sum_{l=1}^{\infty} l d_l r^{l+1} + N_1\sum_{l=1}^{\infty} l d_l r^{l} + N_0\sum_{l=1}^{\infty} l d_l r^{l-1} + L_{11}\sum_{l=0}^{\infty} d_l r^{l+11} + L_{10}\sum_{l=0}^{\infty} d_l r^{l+10}$$
$$+L_9\sum_{l=0}^{\infty} d_l r^{l+9} + L_8\sum_{l=0}^{\infty} d_l r^{l+8} + L_7\sum_{l=0}^{\infty} d_l r^{l+7} + L_6\sum_{l=0}^{\infty} d_l r^{l+6} + L_5\sum_{l=0}^{\infty} d_l r^{l+5}$$
$$+L_4\sum_{l=0}^{\infty} d_l r^{l+4} + L_3\sum_{l=0}^{\infty} d_l r^{l+3} + L_2\sum_{l=0}^{\infty} d_l r^{l+2} + L_1\sum_{l=0}^{\infty} d_l r^{l+1} + L_0\sum_{l=0}^{\infty} d_l r^{l} = 0.$$

We change summation indices

$$P_{12}\sum_{k=12}^{\infty}(k-8)(k-9)(k-10)(k-11)d_{k-8}r^k + P_{11}\sum_{k=11}^{\infty}(k-7)(k-8)(k-9)(k-10)d_{k-7}r^k$$
$$+P_{10}\sum_{k=10}^{\infty}(k-6)(k-7)(k-8)(k-9)d_{k-6}r^k + P_9\sum_{k=9}^{\infty}(k-5)(k-6)(k-7)(k-8)d_{k-5}r^k$$
$$+P_8\sum_{k=8}^{\infty}(k-4)(k-5)(k-6)(k-7)d_{k-4}r^k + P_7\sum_{k=7}^{\infty}(k-3)(k-4)(k-5)(k-6)d_{k-3}r^k$$

$$+P_6 \sum_{k=6}^{\infty}(k-2)(k-3)(k-4)(k-5)d_{k-2}r^k + Q_{12} \sum_{k=12}^{\infty}(k-9)(k-10)(k-11)d_{k-9}r^k$$

$$+Q_{11} \sum_{k=11}^{\infty}(k-8)(k-9)(k-10)d_{k-8}r^k + Q_{10} \sum_{k=10}^{\infty}(k-7)(k-8)(k-9)d_{k-7}r^k$$

$$+Q_9 \sum_{k=9}^{\infty}(k-6)(k-7)(k-8)d_{k-6}r^k + Q_8 \sum_{k=8}^{\infty}(k-5)(k-6)(k-7)d_{k-5}r^k$$

$$+Q_7 \sum_{k=7}^{\infty}(k-4)(k-5)(k-6)d_{k-4}r^k + Q_6 \sum_{k=6}^{\infty}(k-3)(k-4)(k-5)d_{k-3}r^k + Q_5 \sum_{k=5}^{\infty}(k-2)(k-3)(k-4)d_{k-2}r^k$$

$$+Q_4 \sum_{k=4}^{\infty}(k-1)(k-2)(k-3)d_{k-1}r^k + M_{12} \sum_{k=12}^{\infty}(k-10)(k-11)d_{k-10}r^k + M_{11} \sum_{k=11}^{\infty}(k-9)(k-10)d_{k-9}r^k$$

$$+M_{10} \sum_{k=10}^{\infty}(k-8)(k-9)d_{k-8}r^k + M_9 \sum_{k=9}^{\infty}(k-7)(k-8)d_{k-7}r^k + M_8 \sum_{k=8}^{\infty}(k-6)(k-7)d_{k-6}r^k$$

$$+M_7 \sum_{k=7}^{\infty}(k-5)(k-6)d_{k-5}r^k + M_6 \sum_{k=6}^{\infty}(k-4)(k-5)d_{k-4}r^k + M_5 \sum_{k=5}^{\infty}(k-3)(k-4)d_{k-3}r^k$$

$$+M_4 \sum_{k=4}^{\infty}(k-2)(k-3)d_{k-2}r^k + M_3 \sum_{k=3}^{\infty}(k-1)(k-2)d_{k-1}r^k + M_2 \sum_{k=2}^{\infty}k(k-1)d_k r^k$$

$$+N_{12} \sum_{k=12}^{\infty}(k-11)d_{k-11}r^k + N_{11} \sum_{k=11}^{\infty}(k-10)d_{k-10}r^k + N_{10} \sum_{k=10}^{\infty}(k-9)d_{k-9}r^k + N_9 \sum_{k=9}^{\infty}(k-8)d_{k-8}r^k$$

$$+N_8 \sum_{k=8}^{\infty}(k-7)d_{k-7}r^k + N_7 \sum_{k=7}^{\infty}(k-6)d_{k-6}r^k + N_6 \sum_{k=6}^{\infty}(k-5)d_{k-5}r^k + N_5 \sum_{k=5}^{\infty}(k-4)d_{k-4}r^k$$

$$+N_4 \sum_{k=4}^{\infty}(k-3)d_{k-3}r^k + N_3 \sum_{k=3}^{\infty}(k-2)d_{k-2}r^k + N_2 \sum_{k=2}^{\infty}(k-1)d_{k-1}r^k + N_1 \sum_{k=1}^{\infty}kd_k r^k$$

$$+N_0 \sum_{k=0}^{\infty}(k+1)d_{k+1}r^k + L_{11} \sum_{k=11}^{\infty} d_{k-11}r^k + L_{10} \sum_{k=10}^{\infty} d_{k-10}r^k + L_9 \sum_{k=9}^{\infty} d_{k-9}r^k$$

$$+L_8 \sum_{k=8}^{\infty} d_{k-8}r^k + L_7 \sum_{k=7}^{\infty} d_{k-7}r^k + L_6 \sum_{k=6}^{\infty} d_{k-6}r^k + L_5 \sum_{k=5}^{\infty} d_{k-5}r^k + L_4 \sum_{k=4}^{\infty} d_{k-4}r^k$$

$$+L_3 \sum_{k=3}^{\infty} d_{k-3}r^k + L_2 \sum_{k=2}^{\infty} d_{k-2}r^k + L_1 \sum_{k=1}^{\infty} d_{k-1}r^k + L_0 \sum_{k=0}^{\infty} d_k r^k = 0\,.$$

We arrive at the 13-term recurrent formula

$$Q_{k-11}d_{k-11} + Q_{k-10}d_{k-10} + ... + Q_k d_k + Q_{k+1}d_{k+1} = 0. \tag{11.80}$$

Transcendency condition for Frobenius solutions has the form

$$Q_{k-11} = L_{11} + N_{12}\,(k-11) = 0\,, \qquad k - 11 = n \geq 0\,,$$

in explicit form it reads

$$-4\,(M_1-\epsilon)\,(M_2+\epsilon)\,(M_1-M_2)^2 \Big\{ (k-10+H)\,K^3 + \alpha\,\epsilon\,K^2$$
$$+\Big[\epsilon^2\,(H-10+k) + (\frac{9}{2} - \frac{1}{2}\,k - \frac{1}{2}\,H)M_1{}^2 - \frac{1}{2}\,(H-11+k)\,M_2{}^2\Big]K$$
$$-\frac{1}{2}\,\alpha\,\epsilon\,\left(M_1{}^2 + M_2{}^2 - 2\,\epsilon^2\right) \Big\} = 0\,. \tag{11.81}$$

Let

$$K = -\sqrt{M_1^2 - \epsilon^2}\,, \quad H = \nu, \tag{11.82}$$

then eq. (11.81) has the form

$$-2\,(M_1 - \epsilon)\,(M_2 + \epsilon)\,(M_1 - M_2)^2 \Big\{ (-2\,k + 20 - 2\,\nu)\,\big({M_1}^2 - \epsilon^2\big)^{3/2}$$

$$+ \Big[(-2\,k + 20 - 2\,\nu)\,\epsilon^2 + (k - 9 + \nu)\,{M_1}^2$$

$$+ (k - 11 + \nu)\,{M_2}^2 \Big] \sqrt{{M_1}^2 - \epsilon^2} + \alpha\,\epsilon\,\big(M_1^2 - M_2^2\big) \Big\} = 0\,,$$

whence we find three roots

$$\epsilon = +M_1\,, \; -M_2\,, \quad \epsilon = \pm \frac{M_1}{\sqrt{1 + \alpha^2/(k - 11 + \nu)^2}}\,.$$

Now let

$$K = -\sqrt{M_2^2 - \epsilon^2}, \quad H = \nu, \tag{11.83}$$

in this case, from eq. (11.81) we find the roots

$$\epsilon = M_1\,, \; -M_2\,, \quad \epsilon = \pm \frac{M_2}{\sqrt{1 + \alpha^2/(k - 9 + \nu)^2}}\,. \tag{11.84}$$

Let us consider the variant (11.78). Here we have the following equation for $\bar{F}$:

$$\big(P_{11}\,r^{11} + P_{10}\,r^{10} + P_9\,r^9 + P_8\,r^8 + P_7\,r^7 + P_6\,r^6 + P_5\,r^5\big)\,\frac{d^4\bar{F}}{dr^4}$$

$$+ \big(Q_{11}\,r^{11} + Q_{10}\,r^{10} + Q_9\,r^9 + Q_8\,r^8 + Q_7\,r^7 + Q_6\,r^6 + Q_5\,r^5 + Q_4\,r^4 + Q_3\,r^3\big)\,\frac{d^3\bar{F}}{dr^3}$$

$$+ (M_{11}r^{11} + M_{10}r^{10} + M_9r^9 + M_8r^8 + M_7r^7 + M_6r^6$$

$$+ M_5r^5 + M_4r^4 + M_3r^3 + M_2r^2 + M_1r)\frac{d^2\bar{F}}{dr^2}$$

$$+ (N_{11}r^{11} + N_{10}r^{10} + N_9r^9 + N_8r^8 + N_7r^7 + N_6r^6$$

$$+ N_5r^5 + N_4r^4 + N_3r^3 + N_2r^2 + N_1r + N_0)\frac{d\bar{F}}{dr}$$

$$+ (L_{10}\,r^{10} + L_9\,r^9 + L_8\,r^8 + L_7\,r^7 + L_6\,r^6$$

$$+ L_5\,r^5 + L_4\,r^4 + L_3\,r^3 + L_2\,r^2 + L_1\,r + L_0)\bar{F} = 0\,.$$

Solutions are searched as power series

$$\bar{F} = \sum_{l=0}^{\infty} d_l r^l\,, \quad \frac{d\bar{F}}{dr} = \sum_{l=1}^{\infty} l d_l r^{l-1}\,, \quad \frac{d^2\bar{F}}{dr^2} = \sum_{l=2}^{\infty} l(l-1) d_l r^{l-2}\,,$$

$$\frac{d^3\bar{F}}{dr^3} = \sum_{l=3}^{\infty} l(l-1)(l-2) d_l r^{l-3}\,, \quad \frac{d^4\bar{F}}{dr^4} = \sum_{l=4}^{\infty} l(l-1)(l-2)(l-3) d_l r^{l-4}\,.$$

Further we get

$$
\begin{aligned}
&P_{11}\sum_{k=11}^{\infty}(k-7)(k-8)(k-9)(k-10)d_{k-7}r^k+P_{10}\sum_{k=10}^{\infty}(k-6)(k-7)(k-8)(k-9)d_{k-6}r^k\\
&+P_9\sum_{k=9}^{\infty}(k-5)(k-6)(k-7)(k-8)d_{k-5}r^k+P_8\sum_{k=8}^{\infty}(k-4)(k-5)(k-6)(k-7)d_{k-4}r^k\\
&+P_7\sum_{k=7}^{\infty}(k-3)(k-4)(k-5)(k-6)d_{k-3}r^k+P_6\sum_{k=6}^{\infty}(k-2)(k-3)(k-4)(k-5)d_{k-2}r^k\\
&+P_5\sum_{k=5}^{\infty}(k-1)(k-2)(k-3)(k-4)d_{k-1}r^k+Q_{11}\sum_{k=11}^{\infty}(k-8)(k-9)(k-10)d_{k-8}r^k\\
&+Q_{10}\sum_{k=10}^{\infty}(k-7)(k-8)(k-9)d_{k-7}r^k+Q_9\sum_{k=9}^{\infty}(k-6)(k-7)(k-8)d_{k-6}r^k\\
&+Q_8\sum_{k=8}^{\infty}(k-5)(k-6)(k-7)d_{k-5}r^k+Q_7\sum_{k=7}^{\infty}(k-4)(k-5)(k-6)d_{k-4}r^k\\
&+Q_6\sum_{k=6}^{\infty}(k-3)(k-4)(k-5)d_{k-3}r^k+Q_5\sum_{k=5}^{\infty}(k-2)(k-3)(k-4)d_{k-2}r^k\\
&+Q_4\sum_{k=4}^{\infty}(k-1)(k-2)(k-3)d_{k-1}r^kQ_3\sum_{k=3}^{\infty}k(k-1)(k-2)d_kr^k\\
&+M_{11}\sum_{k=11}^{\infty}(k-9)(k-10)d_{k-9}r^k+M_{10}\sum_{k=10}^{\infty}(k-8)(k-9)d_{k-8}r^k\\
&+M_9\sum_{k=9}^{\infty}(k-7)(k-8)d_{k-7}r^k+M_8\sum_{k=8}^{\infty}(k-6)(k-7)d_{k-6}r^k\\
&+M_7\sum_{k=7}^{\infty}(k-5)(k-6)d_{k-5}r^k+M_6\sum_{k=6}^{\infty}(k-4)(k-5)d_{k-4}r^k\\
&+M_5\sum_{k=5}^{\infty}(k-3)(k-4)d_{k-3}r^k+M_4\sum_{k=4}^{\infty}(k-2)(k-3)d_{k-2}r^k\\
&+M_3\sum_{k=3}^{\infty}(k-1)(k-2)d_{k-1}r^k+M_2\sum_{k=2}^{\infty}k(k-1)d_kr^k+M_1\sum_{k=1}^{\infty}(k+1)kd_{k+1}r^k\\
&+N_{11}\sum_{k=11}^{\infty}(k-10)d_{k-10}r^k+N_{10}\sum_{k=10}^{\infty}(k-9)d_{k-9}r^k+N_9\sum_{k=9}^{\infty}(k-8)d_{k-8}r^k\\
&+N_8\sum_{k=8}^{\infty}(k-7)d_{k-7}r^k+N_7\sum_{k=7}^{\infty}(k-6)d_{k-6}r^k+N_6\sum_{k=6}^{\infty}(k-5)d_{k-5}r^k\\
&+N_5\sum_{k=5}^{\infty}(k-4)d_{k-4}r^k+N_4\sum_{k=4}^{\infty}(k-3)d_{k-3}r^k+N_3\sum_{k=3}^{\infty}(k-2)d_{k-2}r^k
\end{aligned}
$$

$$+N_2 \sum_{k=2}^{\infty}(k-1)d_{k-1}r^k + N_1 \sum_{k=1}^{\infty} k d_k r^k + N_0 \sum_{k=0}^{\infty}(k+1)d_{k+1}r^k$$

$$+L_{10} \sum_{k=10}^{\infty} d_{k-10}r^k + L_9 \sum_{k=9}^{\infty} d_{k-9}r^k + L_8 \sum_{k=8}^{\infty} d_{k-8}r^k$$

$$+L_7 \sum_{k=7}^{\infty} d_{k-7}r^k + L_6 \sum_{k=6}^{\infty} d_{k-6}r^k + L_5 \sum_{k=5}^{\infty} d_{k-5}r^k$$

$$+L_4 \sum_{k=4}^{\infty} d_{k-4}r^k + L_3 \sum_{k=3}^{\infty} d_{k-3}r^k + L_2 \sum_{k=2}^{\infty} d_{k-2}r^k + L_1 \sum_{k=1}^{\infty} d_{k-1}r^k + L_0 \sum_{k=0}^{\infty} d_k r^k = 0\,.$$

We arrive at the 12-term recurrent formula

$$Q_{k-10}d_{k-10} + Q_{k-9}d_{k-9} + ... + Q_k d_k + Q_{k+1}d_{k+1} = 0. \tag{11.85}$$

Transcendency condition for Frobenius solutions has the form

$$Q_{k-10} = L_{10} + N_{11}\,(k-10) = 0\,, \qquad k - 10 = n \geq 0$$

in explicit form it reads

$$-4\,(M_1-\epsilon)\,(M_2+\epsilon)\,(M_1-M_2)^2 \Big\{ (k-9+H)\,K^3 + \alpha\,\epsilon\,K^2$$

$$+\Big[(k-9+H)\,\epsilon^2 + (4 - \frac{1}{2}\,k - \frac{1}{2}\,H) {M_1}^2 - \frac{1}{2}\,(k-10+H)\,{M_2}^2 \Big] K$$

$$-\frac{1}{2}\,\alpha\,\epsilon\,\left({M_1}^2 + {M_2}^2 - 2\,\epsilon^2\right) \Big\} = 0\,. \tag{11.86}$$

Let

$$K = -\sqrt{M_1^2 - \epsilon^2}\,, \quad H = \sqrt{\nu^2 - \alpha^2},$$

then eq. (11.86) takes the form

$$-4\,(M_1-\epsilon)(M_2+\epsilon)(M_1-M_2)^2 \Big\{ -(k-9+\sqrt{\nu^2-\alpha^2})({M_1}^2-\epsilon^2)^{3/2}$$

$$+({M_1}^2-\epsilon^2)\alpha\,\epsilon - [(k-9+\sqrt{\nu^2-\alpha^2})\epsilon^2 + (4 - \frac{1}{2}\,k - \frac{1}{2}\,\sqrt{\nu^2-\alpha^2}){M_1}^2$$

$$-\frac{1}{2}\,{M_2}^2(k-10+\sqrt{\nu^2-\alpha^2})]\sqrt{{M_1}^2-\epsilon^2} - \frac{1}{2}\,\alpha\,\epsilon\,({M_1}^2+{M_2}^2-2\,\epsilon^2\Big\} = 0\,,$$

whence we find the root

$$\epsilon = M_1\,,\ -M_2\,, \quad \epsilon = \pm\frac{M_1}{\sqrt{1+\alpha^2/\left(k-10+\sqrt{\nu^2-\alpha^2}\right)^2}}\,. \tag{11.87}$$

Let

$$K = -\sqrt{M_2^2 - \epsilon^2}, \quad H = \sqrt{\nu^2 - \alpha^2}, \tag{11.88}$$

then from eq. (11.86) we obtain the roots

$$\epsilon = M_1\,,\ -M_2\,, \quad \epsilon = \pm\frac{M_2}{\sqrt{1+\alpha^2/\left(k-8+\sqrt{\nu^2-\alpha^2}\right)^2}}\,. \tag{11.89}$$

11.7 Conclusions

Generalised wave equation for a spin 1/2 particle with two mass parameters is studied in the presence of an external Coulomb field. After separating the variables, the problem reduces the system to eight differential equations of the 1st order. Taking into account diagonalization of the space reflection operator, we derive two independent systems of four equations, referring to states of opposite parity. When considering these equations at a large distance from the centre, they take the form of two subsystems for two ordinary Dirac particles in external Coulomb field, with masses of M_1 and M_2, respectively. To simplify the problem, we perform a transition to the nonrelativistic description of the system. In this way, we derive two systems of linked 2nd-order equations, referring to states with different parities. They lead to 4th-order differential equations for separate functions. Their solutions of the Frobenius type have been constructed; they involve power series with 10-term recurrent relations. Two solutions are appropriate to describe bound states. As a quantization rule, we apply the known transcendency condition; in this way, we derive two analytical formulas for energy spectra. They are similar to nonrelativistic spectra for ordinary spin 1/2 particles, but they are governed by masses M_1 and M_2. Results of constructing solutions and obtaining the energy spectra are extendable to relativistic theory as well.

11.8 Figures

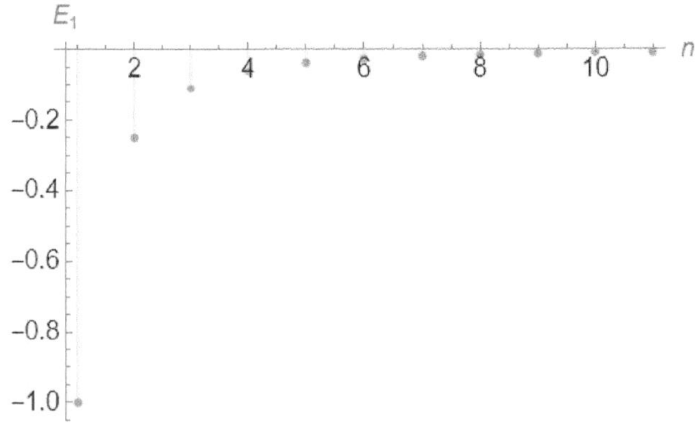

FIGURE 11.1
Energy levels $E_1(n)$ up to a factor.

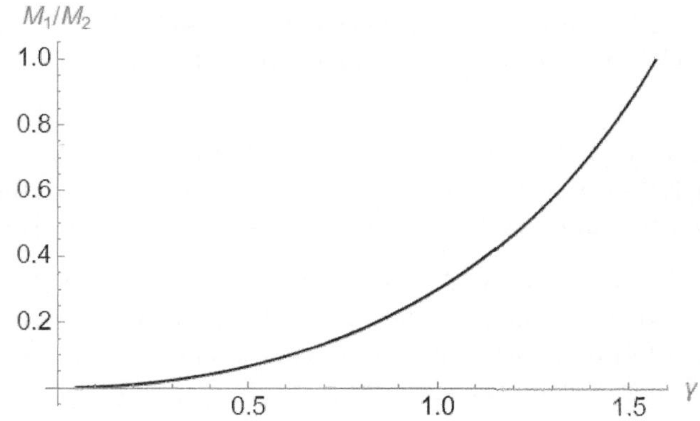

FIGURE 11.2
Relative coefficient $E_1/E_2 = M_1/M_2$.

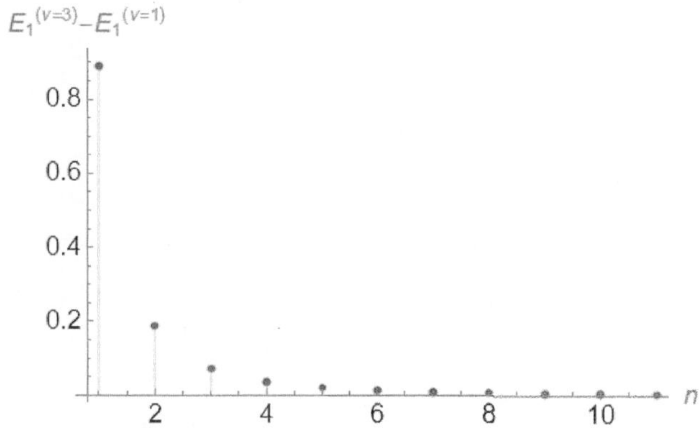

FIGURE 11.3
The difference of two spectra $E_1^{(\nu=3)} - E_1^{(\nu=1)}$.

Bibliography

[1] V.V. Kisel, V.A. Pletyukhov, V.V. Gilewsky, E.M. Ovsiyuk, O.V. Veko and V.M. Red'kov. Spin 1/2 particle with two mass states. interaction with external fields. *Nonlinear Dynamics and Applications*, 23: 210–230, 2017.

[2] V.V. Kisel, V.A. Pletyukhov, V.V. Gilewsky, E.M. Ovsiyuk, O.V. Veko and V.M. Red'kov. Spin 1/2 particle with two mass states, interaction with external fields. *Nonlinear Phenomena in Complex Systems*, 20(4): 404–423, 2017.

[3] V.V. Kisel, E.M. Ovsiyuk, O.V. Veko and V.M. Red'kov. Fermion with intrinsic mass spectrum in external fields. *Proceedings of the Komi Science Cenre, Ural Branch of Russiabn Academy of Sciences. Physical and Mathematical Sciences*, 33: 81–88, 2018.

[4] E.M. Ovsiyuk, O.V. Veko, Ya.A. Voynova, V.V. Kisel, V. Balan and V.M. Red'kov. Spin 1/2 particle with two masses in magnetic field. *Applied Sciences*, 20: 148–166, 2018.

[5] E.M. Ovsiyuk, O.V. Veko, Ya.A. Voynova, V.M. Red'kov, V.V. Kisel and N.V. Samsonenko. Spin 1/2 particle with two masses in external magnetic field. *Journal of Mechanics of Continua and Mathematical Sciences, Special Issue*, 1: 651–660, 2019.

[6] E.M. Ovsiyuk, Ya.A. Voynova, V.V. Kisel, V.A. Pletyukhov, V.V. Gilewsky and V.M. Red'kov. On modeling neutrinos oscillations by geometry methods in the frames of the theory for a fermion with three mass parameters. *Journal of Physics: Conference Series*, 1416: Paper 012040 [9 pages], 2019.

12

On modelling neutrinos oscillations by geometry methods

In this chapter, starting from the general Gel'fand-Yaglom approach, a new wave equation for spin 1/2 fermion, which is characterised by three mass parameters, is derived. On the basis of the 20-component wave function, three auxiliary bispinors are introduced, in the absence of an external field, these bispinors obey to three separate Dirac-like equations with different masses M_1, M_2, M_3. It is shown that in the presence of external fields, electromagnetic field, or gravitational non-Euclidean background with non-vanishing Ricci scalar curvature, the main equation is not split into separate equations, instead a quite definite mixing of three Dirac-like equations arises. It is shown that a generalised equation for a Majorana particle with three mass parameters exists as well; such a generalised Majorana equation is not split into three separate equations in a curved background if the Ricci scalar of the space-time model does not vanish.

12.1 Fermion with three mass parameters

In the context of existence of the similar neutrinos of different masses, we examine a possibility within the theory of relativistic wave equations to describe a spin 1/2 particle with three mass parameters. In general, existence of more general wave equations than commonly used ones is well known within the so-called Gel'fand–Yaglom formalism – see references [1–53] and also books [54–58].

Recently, a model for a spin 1/2 particle with two mass parameters was developed [60, 61]. The main properties of that model are as follows: The main wave equation for a 16-component field is presented in spin-tensor form and is based on the use of the usual Dirac matrices. For two auxiliary bispinors – they determine the initial 16-component wave function – in the absence of external fields, two separate Dirac-like equations are derived, they differ in masses of M_1 and M_2. However, in the presence of an external electromagnetic field or gravitational one with a non-vanishing Ricci scalar, the wave equation does not split into separated equations, instead a quite definite mixing of two Dirac-like equations arises. It is shown that a generalised equation for a Majorana particle with two mass parameters exists as well, such a generalised Majorana equation is not trivial if Ricci scalar does not vanish.

In fact, in the present paper, we extend the analysis from [60] to a fermion with three mass parameters. A model for such a fermion is based on an extended set of irreducible representations of the Lorentz group (we adhere to notation from [58])

$$T = (0, \frac{1}{2}) \oplus (\frac{1}{2}, 0) \oplus (0, \frac{1}{2})' \oplus (\frac{1}{2}, 0)' \oplus (1, \frac{1}{2}) \oplus (\frac{1}{2}, 1) . \qquad (12.1)$$

DOI: 10.1201/9781003472377-12

After performing rather laborious analysis (it is omitted for technical reason), we derive the system of equations in spin-tensor form

$$
\begin{gathered}
c_1\hat{\partial}(\gamma_\mu\Psi_\mu) + \frac{c_3}{\sqrt{6}}\Big(\hat{\partial}(\gamma_\mu\Psi_\mu) - 4(\partial_\mu\Psi_\mu)\Big) + M(\gamma_\mu\Psi_\mu) = 0\,,\\
c_2\hat{\partial}\Psi_0 - i\frac{4c_4}{\sqrt{6}}\Big(-\frac{1}{4}\hat{\partial}(\gamma_\mu\Psi_\mu) + (\partial_\mu\Psi_\mu)\Big) + M\Psi_0 = 0\,,\\
-\frac{2fc_3^*}{\sqrt{6}}\Big(\partial_\lambda(\gamma_\mu\Psi_\mu) - \frac{1}{4}\gamma_\lambda\hat{\partial}(\gamma_\mu\Psi_\mu)\Big) + i\frac{2gc_4^*}{\sqrt{6}}\Big(\partial_\lambda\Psi_0 - \frac{1}{4}\gamma_\lambda\hat{\partial}\Psi_0\Big)\\
+M\Big(\Psi_\lambda - \frac{1}{4}\gamma_\lambda(\gamma_\mu\Psi_\mu)\Big) = 0\,,
\end{gathered} \tag{12.2a}
$$

where 20-component wave function consists of bispinor Ψ_0 and vector-bispinor Ψ_μ; the system contains a number of numerical parameters: c_1, c_2 are real, c_3, c_4 are complex, and $f, g \in \{\pm 1\}$. The physical sense of these parameters will be clear below. Dirac matrices are specified in the Weyl spinor basis (we use *ict*-metric in Minkowski space):

$$
\gamma_\mu = \frac{1}{i}\begin{vmatrix} 0 & \sigma^\mu \\ \sigma^\mu & 0 \end{vmatrix}, \sigma^4 = \begin{vmatrix} i & 0 \\ 0 & i \end{vmatrix}, \sigma^1 = \begin{vmatrix} 0 & 1 \\ 1 & 0 \end{vmatrix}, \sigma^2 = \begin{vmatrix} 0 & -i \\ i & 0 \end{vmatrix}, \sigma^3 = \begin{vmatrix} 1 & 0 \\ 0 & -1 \end{vmatrix}.
$$

12.2 Reformulation of the initial equations

The system (12.2a) may be presented differently, which is substantial point for further studying. Let us act on the first equation in eq. (12.2a) by operator $\frac{1}{4}\gamma_\lambda$, it yields

$$
\frac{1}{4}c_1\gamma_\lambda\hat{\partial}(\gamma_\mu\psi_\mu) + \frac{c_3}{\sqrt{6}}\Big(\frac{1}{4}\gamma_\lambda\hat{\partial}(\gamma_\mu\psi_\mu) - \gamma_\lambda(\partial_\mu\psi_\mu)\Big) + \frac{M}{4}\gamma_\lambda(\gamma_\mu\psi_\mu) = 0\,.
$$

Summing this equation with the third one in eq. (12.2a), we obtain

$$
\begin{gathered}
-\frac{2fc_3^*}{\sqrt{6}}\Big(\partial_\lambda(\gamma_\mu\psi_\mu) - \frac{1}{4}\gamma_\lambda\hat{\partial}(\gamma_\mu\psi_\mu)\Big) + i\frac{2gc_4^*}{\sqrt{6}}\Big(\partial_\lambda\psi_0 - \frac{1}{4}\gamma_\lambda\hat{\partial}\psi_0\Big)\\
+\frac{1}{4}c_1\gamma_\lambda\hat{\partial}(\gamma_\mu\psi_\mu) - \frac{c_3}{\sqrt{6}}\Big(\gamma_\lambda(\partial_\mu\psi_\mu) - \frac{1}{4}\gamma_\lambda\hat{\partial}(\gamma_\mu\psi_\mu)\Big) + M\psi_\lambda = 0\,.
\end{gathered}
$$

In fact, this equation, together with the second equation in eq. (12.2a), makes the system that is equivalent to the initial system (12.2a):

$$
\begin{gathered}
c_2\hat{\partial}\psi_0 - i\frac{4c_4}{\sqrt{6}}\Big(-\frac{1}{4}\hat{\partial}(\gamma_\mu\psi_\mu) + (\partial_\mu\psi_\mu)\Big) + M\psi_0 = 0\,,\\
-\frac{2fc_3^*}{\sqrt{6}}(\partial_\lambda\Big(\gamma_\mu\psi_\mu) - \frac{1}{4}\gamma_\lambda\hat{\partial}(\gamma_\mu\psi_\mu)\Big) + i\frac{2gc_4^*}{\sqrt{6}}\Big(\partial_\lambda\psi_0 - \frac{1}{4}\gamma_\lambda\hat{\partial}\psi_0\Big)\\
+\frac{1}{4}c_1\gamma_\lambda\hat{\partial}(\gamma_\mu\psi_\mu) - \frac{c_3}{\sqrt{6}}\Big(\gamma_\lambda(\partial_\mu\psi_\mu) - \frac{1}{4}\gamma_\lambda\hat{\partial}(\gamma_\mu\psi_\mu)\Big) + M\psi_\lambda = 0\,.
\end{gathered} \tag{12.2}
$$

To prove this, we multiply the second equation in eq. (12.2a) by the matrix γ_λ and take into account identity $\gamma_\lambda\gamma_\lambda = 4$. Then we derive

$$
c_1\hat{\partial}(\gamma_\mu\psi_\mu) + \frac{c_3}{\sqrt{6}}\Big(\hat{\partial}(\gamma_\mu\psi_\mu) - 4(\partial_\mu\psi_\mu)\Big) + M(\gamma_\mu\psi_\mu) = 0\,,
$$

which coincides with the first equation in eq. (12.2a).

Because the vector-bispinor Ψ_μ is determined through bispinors, Ψ_0, $(\gamma_\mu\Psi_\mu)$, $(\partial_\mu\Psi_\mu)$ (see the second equation in eq. (12.2a)), let us re-formulate the main system so that it contains only these three bispinors: $(\gamma_\mu\Psi_\mu)$, Ψ_0, and $(\partial_\mu\Psi_\mu)$. To this end, let us act on the second equation in eq. (12.2a) by operator ∂_λ, so we get

$$-f\frac{\sqrt{6}}{4}c_3^*\partial_\lambda\partial_\lambda(\gamma_\mu\Psi_\mu)+ig\frac{\sqrt{6}}{4}c_4^*\partial_\lambda\partial_\lambda\Psi_0+\frac{1}{4}c_1\partial_\lambda\partial_\lambda(\gamma_\mu\Psi_\mu)$$
$$-\frac{c_3}{\sqrt{6}}\hat{\partial}(\partial_\mu\Psi_\mu)+\frac{c_3}{4\sqrt{6}}\partial_\lambda\partial_\lambda(\gamma_\mu\Psi_\mu)+M(\partial_\mu\Psi_\mu)=0\,. \tag{12.3}$$

Now, let us act on the first equation in eq. (12.2a) by operator $\hat{\partial}$, this yields

$$c_1\partial_\lambda\partial_\lambda(\gamma_\mu\Psi_\mu)+\frac{c_3}{\sqrt{6}}\partial_\lambda\partial_\lambda(\gamma_\mu\Psi_\mu)=\frac{4c_3}{\sqrt{6}}\hat{\partial}(\partial_\mu\Psi_\mu)-M\hat{\partial}(\gamma_\mu\Psi_\mu)\,,$$

with this in mind eq. (12.3) can be re-written as

$$-f\frac{\sqrt{6}}{4}c_3^*\partial_\lambda\partial_\lambda(\gamma_\mu\Psi_\mu)+ig\frac{\sqrt{6}}{4}c_4^*\partial_\lambda\partial_\lambda\Psi_0+M(\partial_\mu\Psi_\mu)-\frac{M}{4}\hat{\partial}(\gamma_\mu\Psi_\mu)=0\,. \tag{12.4}$$

Now, acting on eq. (12.3) by operator ∂_λ we obtain

$$\partial_\lambda\partial_\lambda(\gamma_\mu\Psi_\mu)=\frac{1}{c_1+c_3/\sqrt{6}}\Big[\frac{4c_3}{\sqrt{6}}\hat{\partial}(\partial_\mu\Psi_\mu)-M\hat{\partial}(\gamma_\mu\Psi_\mu)\Big]. \tag{12.5}$$

With eq. (12.5) in mind, we reduce eq. (12.3) to the form

$$-\frac{f|c_3|^2}{c_1+c_3/\sqrt{6}}\hat{\partial}(\partial_\mu\Psi_\mu)+M\frac{\sqrt{6}}{4}\frac{fc_3^*}{c_1+c_3/\sqrt{6}}$$
$$+ig\frac{\sqrt{6}}{4}c_4^*\partial_\lambda\partial_\lambda\Psi_0+M(\partial_\mu\Psi_\mu)-\frac{M}{4}\hat{\partial}(\gamma_\mu\Psi_\mu)=0\,. \tag{12.6}$$

Also, acting on eq. (12.2a) by $\hat{\partial}$ we derive

$$c_2\partial_\lambda\partial_\lambda\Psi_0+i\frac{c_4}{\sqrt{6}}\partial_\lambda\partial_\lambda(\gamma_\mu\Psi_\mu)=\hat{\partial}\Big[i\frac{4}{\sqrt{6}}c_4(\partial_\mu\Psi_\mu)-M\Psi_0\Big]\,,$$

which (taking into account (12.5)) can be re-written as

$$c_2\partial_\lambda\partial_\lambda\Psi_0+i\frac{c_4}{\sqrt{6}}\frac{1}{c_1+c_3/\sqrt{6}}\Big[\frac{4c_3}{\sqrt{6}}\hat{\partial}(\partial_\mu\Psi_\mu)-M\hat{\partial}(\gamma_\mu\Psi_\mu)\Big]=\hat{\partial}\Big[i\frac{4}{\sqrt{6}}c_4(\partial_\mu\Psi_\mu)-M\Psi_0\Big]\,,$$

whence for the term $\partial_\lambda\partial_\lambda\Psi_0$ we derive

$$\partial_\lambda\partial_\lambda\Psi_0=\frac{i}{c_2}\Big[-\frac{2}{3}\frac{c_3c_4}{c_1+\frac{c_3}{\sqrt{6}}}+\frac{4c_4}{\sqrt{6}}\Big]\hat{\partial}(\partial_\mu\Psi_\mu)+i\frac{M}{c_2\sqrt{6}}\cdot\frac{c_4}{c_1+\frac{c_3}{\sqrt{6}}}\hat{\partial}(\gamma_\mu\Psi_\mu)-\frac{M}{c_2}\hat{\partial}\Psi_0\,.$$

Therefore, eq. (12.6) reads

$$-\frac{fc_2|c_3|^2+gc_1|c_4|^2}{c_2(c_1+\frac{c_3}{\sqrt{6}})}\hat{\partial}(\partial_\mu\Psi_\mu)-igM\frac{\sqrt{6}}{4}\cdot\frac{c_4^*}{c_2}\hat{\partial}\Psi_0+M(\partial_\mu\Psi_\mu)$$

$$+\frac{M}{4}\frac{1}{c_2(c_1+\frac{c_3}{\sqrt{6}})}\Big[\sqrt{6}fc_2c_3^*-g|c_4|^2-c_2(c_1+\frac{c_3}{\sqrt{6}})\Big]\hat{\partial}(\gamma_\mu\Psi_\mu)=0\,. \tag{12.7}$$

Equation (12.7) is one of equations we need – it does not contain the vector-bispinor Ψ_λ. To derive another equation of that type, it suffices to express the term $(\partial_\mu \Psi_\mu)$ from eq. (12.7) and substitute the result in the first equation of the system (12.2a). In this way we get

$$\frac{1}{c_2(c_1+\frac{c_3}{\sqrt{6}})}\Big[c_1c_2(c_1+\frac{c_3}{\sqrt{6}})+fc_2|c_3|^2-\frac{g}{\sqrt{6}}c_3|c_4|^2\Big]\hat{\partial}(\gamma_\mu\Psi_\mu)$$
$$-\frac{4c_3}{M\sqrt{6}}\frac{fc_2|c_3|^2+gc_1|c_4|^2}{c_2(c_1+\frac{c_3}{\sqrt{6}})}\hat{\partial}(\partial_\mu\Psi_\mu)-ig\frac{c_3c_4^*}{c_2}\hat{\partial}\Psi_0+M(\gamma_\mu\Psi_\mu)=0\,. \tag{12.8}$$

Similarly, expressing the term $(\partial_\mu \Psi_\mu)$ from eq. (12.3) and substituting it into the second equation in eq. (12.2a), we derive

$$\frac{c_2^2+g|c_4|^2}{c_2}\hat{\partial}\Psi_0-i\frac{4c_4}{M\sqrt{6}}\frac{fc_2|c_3|^2+gc_1|c_4|^2}{c_2(c_1+\frac{c_3}{\sqrt{6}})}\hat{\partial}(\partial_\mu\Psi_\mu)$$
$$+\frac{i}{\sqrt{6}}\frac{c_4}{c_2(c_1+\frac{c_3}{\sqrt{6}})}\Big[\sqrt{6}fc_2c_3^*-g|c_4|^2\Big]\hat{\partial}(\gamma_\mu\Psi_\mu)+M\Psi_0=0\,. \tag{12.9}$$

Thus, the needed form of the main system with respect to bispinors $(\gamma_\mu\Psi_\mu)$, Ψ_0, $(\partial_\mu\Psi_\mu)$ is as follows:

$$\frac{1}{c_2(c_1+c_3/\sqrt{6})}\Big\{c_1c_2(c_1+c_3/\sqrt{6})+fc_2|c_3|^2-\frac{g}{\sqrt{6}}c_3|c_4|^2\Big\}\hat{\partial}(\gamma_\mu\Psi_\mu)$$
$$-ig\frac{c_3c_4^*}{c_2}\hat{\partial}\Psi_0-\frac{4c_3}{M\sqrt{6}}\cdot\frac{fc_2|c_3|^2+gc_1|c_4|^2}{c_2(c_1+c_3/\sqrt{6})}\hat{\partial}(\partial_\mu\Psi_\mu)+M(\gamma_\mu\Psi_\mu)=0\,, \tag{12.10}$$

$$\frac{i}{\sqrt{6}}\frac{c_4}{c_2(c_1+c_3/\sqrt{6})}(\sqrt{6}fc_2c_3^*-g|c_4|^2)\hat{\partial}(\gamma_\mu\Psi_\mu)+\frac{c_2^2+g|c_4|^2}{c_2}\hat{\partial}\Psi_0$$
$$-i\frac{4c_4}{M\sqrt{6}}\frac{fc_2|c_3|^2+gc_1|c_4|^2}{c_2(c_1+c_3/\sqrt{6})}\hat{\partial}(\partial_\mu\Psi_\mu)+M\Psi_0=0\,, \tag{12.11}$$

$$+\frac{M}{4}\frac{1}{c_2(c_1+c_3/\sqrt{6})}\Big\{\sqrt{6}fc_2c_3^*-g|c_4|^2-c_2(c_1+c_3/\sqrt{6})\Big\}\hat{\partial}(\gamma_\mu\Psi_\mu)$$
$$-igM\frac{\sqrt{6}}{4}\frac{c_4^*}{c_2}\hat{\partial}\Psi_0-\frac{fc_2|c_3|^2+gc_1|c_4|^2}{c_2(c_1+c_3/\sqrt{6})}\hat{\partial}(\partial_\mu\Psi_\mu)+M(\partial_\mu\Psi_\mu)=0\,. \tag{12.12}$$

12.3 Characteristic equation, possible values of masses

Let us find the characteristic equation for the matrix of the 1st-order system (12.10)–(12.12) with respect to bispinors $\gamma_\mu\Psi_\mu$, Ψ_0, and $\partial_\mu\Psi_\mu$; it has the form $\det F=0$, where the matrix F is

$$\begin{vmatrix} \frac{1}{c_2(c_1+\frac{c_3}{\sqrt{6}})}\{c_1c_2(c_1+\frac{c_3}{\sqrt{6}})+fc_2|c_3|^2-\frac{g}{\sqrt{6}}c_3|c_4|^2\}-\lambda & -ig\frac{c_3c_4^*}{c_2} & -\frac{4c_3}{M\sqrt{6}}\cdot\frac{fc_2|c_3|^2+gc_1|c_4|^2}{c_2(c_1+\frac{c_3}{\sqrt{6}})} \\ \frac{i}{\sqrt{6}}\cdot\frac{c_4}{c_2(c_1+\frac{c_3}{\sqrt{6}})}\{\sqrt{6}fc_2c_3^*-g|c_4|^2\} & \frac{c_2^2+g|c_4|^2}{c_2}-\lambda & -i\frac{4c_4}{M\sqrt{6}}\cdot\frac{fc_2|c_3|^2+gc_1|c_4|^2}{c_2(c_1+\frac{c_3}{\sqrt{6}})} \\ \frac{M}{4}\cdot\frac{1}{c_2(c_1+\frac{c_3}{\sqrt{6}})}\{\sqrt{6}fc_2c_3^*-g|c_4|^2-c_2(c_1+\frac{c_3}{\sqrt{6}})\} & -igM\frac{\sqrt{6}}{4}\cdot\frac{c_4^*}{c_2} & -\frac{fc_2|c_3|^2+gc_1|c_4|^2}{c_2(c_1+\frac{c_3}{\sqrt{6}})}-\lambda \end{vmatrix}.$$

In this way we get

$$\lambda^3 - \lambda^2(c_1 + c_2) + \lambda(c_1 c_2 - f|c_3|^2 - g|c_4|^2) + (f c_2|c_3|^2 + g c_1|c_4|^2) = 0 . \qquad (12.13)$$

The matrix of the system (12.10)–(12.12) can be transformed to a diagonal form

$$K = \begin{vmatrix} A_1 & B_1 & R_1 \\ A_2 & B_2 & R_2 \\ A_3 & B_3 & R_3 \end{vmatrix} \Longrightarrow \begin{vmatrix} \lambda_1 & 0 & 0 \\ 0 & \lambda_2 & 0 \\ 0 & 0 & \lambda_3 \end{vmatrix} , \qquad (12.14)$$

and the system will take the form of three separated equations of the Dirac type but with different masses:

$$M_1 = \frac{M}{\lambda_1} , \quad M_2 = \frac{M}{\lambda_2} , \quad M_3 = \frac{M}{\lambda_3} . \qquad (12.15)$$

That transformation is done in accordance with the following procedure

$$(K - M)\Psi = 0 , \quad \Psi' = S\Psi , \quad SKS^{-1} = \hat{K}' = \begin{vmatrix} \lambda_1 & 0 & 0 \\ 0 & \lambda_2 & 0 \\ 0 & 0 & \lambda_3 \end{vmatrix} . \qquad (12.16)$$

The matrix S obeys the following relations

$$S = \begin{vmatrix} a_1 & a_2 & a_3 \\ b_1 & b_2 & b_3 \\ r_1 & r_2 & r_3 \end{vmatrix} , \quad \begin{vmatrix} a_1 & a_2 & a_3 \\ b_1 & b_2 & b_3 \\ r_1 & r_2 & r_3 \end{vmatrix} \begin{vmatrix} A_1 & B_1 & R_1 \\ A_2 & B_2 & R_2 \\ A_3 & B_3 & R_3 \end{vmatrix} = \begin{vmatrix} \lambda_1 & 0 & 0 \\ 0 & \lambda_2 & 0 \\ 0 & 0 & \lambda_3 \end{vmatrix} \begin{vmatrix} a_1 & a_2 & a_3 \\ b_1 & b_2 & b_3 \\ r_1 & r_2 & r_3 \end{vmatrix} ,$$

whence three linear subsystems follow

$$\begin{aligned} a_1 A_1 + a_2 A_2 + a_3 A_3 &= \lambda_1 a_1 , \\ a_1 B_1 + a_2 B_2 + a_3 B_3 &= \lambda_1 a_2 , \\ a_1 R_1 + a_2 R_2 + a_3 R_3 &= \lambda_1 a_3 ; \end{aligned} \qquad (12.17)$$

$$\begin{aligned} b_1 A_1 + b_2 A_2 + b_3 A_3 &= \lambda_2 b_1 , \\ b_1 B_1 + b_2 B_2 + b_3 B_3 &= \lambda_2 b_2 , \\ b_1 R_1 + b_2 R_2 + b_3 R_3 &= \lambda_2 b_3 ; \end{aligned} \qquad (12.18)$$

$$\begin{aligned} r_1 A_1 + r_2 A_2 + r_3 A_3 &= \lambda_3 r_1 , \\ r_1 B_1 + r_2 B_2 + r_3 B_3 &= \lambda_3 r_2 , \\ r_1 R_1 + r_2 R_2 + r_3 R_3 &= \lambda_3 r_3 . \end{aligned} \qquad (12.19)$$

These systems are solved easily. For instance, we get

$$\begin{aligned} a_2 &= a_1 \frac{-igc_3 c_4^*(c_1 + \frac{c_3}{\sqrt{6}})\lambda_1}{\lambda_1 c_2(\lambda_1 - c_2) + (\lambda_1 - c_2)[f c_2|c_3|^2 + g c_1|c_4|^2] - \lambda_1(c_1 + \frac{c_3}{\sqrt{6}})g|c_4|^2} , \\ a_3 &= -a_1 \frac{4}{\sqrt{6}} \frac{1}{M} \frac{c_3(\lambda_1 - c_2)[f c_2|c_3|^2 + g c_1|c_4|^2}{\lambda_1 c_2(\lambda_1 - c_2) + (\lambda_1 - c_2)[f c_2|c_3|^2 + g c_1|c_4|^2] - \lambda_1(c_1 + \frac{c_3}{\sqrt{6}})g|c_4|^2} . \end{aligned} \qquad (12.20)$$

At these a_2, a_3, the first equation in (12.17) reduces to the above characteristic equation

$$c_2^2(c_1 + \frac{c_3}{\sqrt{6}})^2\{\lambda_1^3 - \lambda_1^2(c_1 + c_2) + \lambda_1(c_1c_2 - f|c_3|^2 - g|c_4|^2) + (fc_2|c_3|^2 + gc_1|c_4|^2\} = 0\,,$$

where $\lambda = \lambda_1$.

Similarly we get solutions of the system (12.18):

$$\begin{aligned} b_2 &= b_1 \frac{-igc_3c_4^*(c_1 + \frac{c_3}{\sqrt{6}})\lambda_2}{\lambda_2 c_2(\lambda_2 - c_2) + (\lambda_2 - c_2)[fc_2|c_3|^2 + gc_1|c_4|^2] - \lambda_2(c_1 + \frac{c_3}{\sqrt{6}})g|c_4|^2}\,, \\ b_3 &= -b_1 \frac{4}{\sqrt{6}}\frac{1}{M}\frac{c_3(\lambda_2 - c_2)[fc_2|c_3|^2 + gc_1|c_4|^2}{\lambda_2 c_2(\lambda_2 - c_2) + (\lambda_2 - c_2)[fc_2|c_3|^2 + gc_1|c_4|^2] - \lambda_2(c_1 + \frac{c_3}{\sqrt{6}})g|c_4|^2}\,, \end{aligned} \tag{12.21}$$

and the first equation in eq. (12.18) reduces to characteristic equation with $\lambda = \lambda_2$.

Similar results are for the system (12.19):

$$\begin{aligned} r_2 &= r_1 \frac{-igc_3c_4^*(c_1 + \frac{c_3}{\sqrt{6}})\lambda_3}{\lambda_3 c_2(\lambda_3 - c_2) + (\lambda_3 - c_2)[fc_2|c_3|^2 + gc_1|c_4|^2] - \lambda_3(c_1 + \frac{c_3}{\sqrt{6}})g|c_4|^2}\,, \\ r_3 &= -r_1 \frac{4}{\sqrt{6}}\frac{1}{M}\frac{c_3(\lambda_3 - c_2)[fc_2|c_3|^2 + gc_1|c_4|^2}{\lambda_3 c_2(\lambda_3 - c_2) + (\lambda_3 - c_2)[fc_2|c_3|^2 + gc_1|c_4|^2] - \lambda_3(c_1 + \frac{c_3}{\sqrt{6}})g|c_4|^2}\,. \end{aligned} \tag{12.22}$$

12.4 On solutions of the characteristic equation

Characteristic equation may be presented in the short form $\lambda^3 + \lambda^2 a + \lambda b + c = 0$, where (recall that $f, g \in \{-1, +1\}$)

$$a = -(c_1 + c_2), \quad b = c_1c_2 - f|c_3|^2 - g|c_4|^2, \quad c = fc_2|c_3|^2 + gc_1|c_4|^2\,. \tag{12.23}$$

After standard change of the variable, we get to a more simple form

$$y^3 + py + q = 0\,, \quad y = \lambda - \frac{c_1 + c_3}{3} = \lambda + \frac{a}{3}\,,$$

where

$$p = -\frac{a^3}{3} + b = -\frac{1}{3}(c_1 + c_3)^3 + c_1c_2 - f|c_3|^2 - g|c_4|^2\,,$$

$$q = 2(\frac{a}{3})^3 - \frac{ab}{3} + c = -\frac{2}{27}c_1 + c_3{}^3 + \frac{c_1 + c_2}{3}\{c_1c_2 - f|c_3|^2 - g|c_4|^2\} + \{f|c_3|^2 + g|c_4|^2\}\,.$$

By physical reason, we assume real-valuedness of all three roots. It is possible if the following inequalities are valid

$$p < 0\,, \qquad Q < 0\,, \quad \text{where} \quad Q = (\frac{p}{3})^3 + (\frac{q}{2})^2\,. \tag{12.24}$$

Solutions can be presented in trigonometric form:

$$\begin{aligned} \lambda_1 &= y_1 + \frac{c_1 + c_2}{3} = 2\sqrt{-\frac{p}{3}}\cos\frac{\alpha}{3} + \frac{c_1 + c_2}{3}\,, \\ \lambda_2 &= y_2 + \frac{c_1 + c_2}{3} = -2\sqrt{-\frac{p}{3}}\cos(\frac{\alpha}{3} - \frac{\pi}{3}) + \frac{c_1 + c_2}{3}\,, \\ \lambda_3 &= y_3 + \frac{c_1 + c_2}{3} = -2\sqrt{-\frac{p}{3}}\cos(\frac{\alpha}{3} + \frac{\pi}{3}) + \frac{c_1 + c_2}{3}\,. \end{aligned} \tag{12.25}$$

We may readily derive relations (remember on real-valuedness of the roots):

$$\lambda_1 + \lambda_2 + \lambda_3 = c_1 + c_2 \,, \quad \lambda_1^2 + \lambda_2^2 + \lambda_3^2 = c_1^2 + c_2^2 + 2f|c_3|^2 + 2g|c_4|^2 > 0 \,,$$

$$\lambda_1\lambda_2 + \lambda_1\lambda_3 + \lambda_2\lambda_2 = c_1c_2 - f|c_3|^2 - g|c_4|^2 \,, \quad \lambda_1\lambda_2\lambda_3 = -fc_2|c_3|^2 - gc_1|c_4|^2 \,,$$

$$p = \frac{1}{6}\{(\lambda_1 + \lambda_2 + \lambda_3)^2 - 3(\lambda_1^2 + \lambda_2^2 + \lambda_3^2)\} < 0 \,,$$

$$q = \frac{1}{6}\{\frac{5}{9}(\lambda_1 + \lambda_2 + \lambda_3)^2 - (\lambda_1^2 + \lambda_2^2 + \lambda_3^2)\}(\lambda_1 + \lambda_2 + \lambda_3) - \lambda_1\lambda_2\lambda_3 \,,$$

$$\cos\frac{\alpha}{3} = \frac{2\lambda_1 - \lambda_2 - \lambda_3}{\sqrt{-2[(\lambda_1 + \lambda_2 + \lambda_3)^2 - 3(\lambda_1^2 + \lambda_2^2 + \lambda_3^2)]}} \,,$$

$$\cos(\frac{\alpha}{3} - \frac{\pi}{3}) = \frac{2\lambda_2 - \lambda_1 - \lambda_3}{\sqrt{-2[(\lambda_1 + \lambda_2 + \lambda_3)^2 - 3(\lambda_1^2 + \lambda_2^2 + \lambda_3^2)]}} \,,$$

$$\cos(\frac{\alpha}{3} + \frac{\pi}{3}) = \frac{2\lambda_3 - \lambda_1 - \lambda_2}{\sqrt{-2[(\lambda_1 + \lambda_2 + \lambda_3)^2 - 3(\lambda_1^2 + \lambda_2^2 + \lambda_3^2)]}} \,.$$

In the first place, we are interested in positive roots. Evidently, three positive roots may arise only if

$$c_1 > 0 \,, \quad c_2 > 0 \,, \quad f = -1 \,, \quad g = -1 \,. \tag{12.26}$$

Let us introduce notation $|c_4|^2 = a^2$, $|c_3|^2 = b^2$, and consider two constraints from above

$$\lambda_1 + \lambda_2 = c_1 + c_2 - \lambda_3 \,, \quad \lambda_1\lambda_2 = \frac{c_1a^2 + c_2b^2}{\lambda_3} \,,$$

whence expressions for λ_1, λ_2 follow

$$\lambda_{1,2} = \frac{c_1 + c_2 - \lambda_3}{2} \pm \sqrt{(\frac{c_1 + c_2 - \lambda_3}{2})^2 - \frac{c_1a^2 + c_2b^2}{\lambda_3}} \,, \tag{12.27}$$

under square root we should have positive term. Note that due to inequality

$$\lambda_1 + \lambda_2 = c_1 + c_2 - \lambda_3 > 0 \,,$$

we have obligatory inequalities $0 < \lambda_3 < c_1 + c_2$. Therefore, the following parametrization is possible:

$$\lambda_3 = (c_1 + c_2)\cos\alpha \,, \qquad \alpha \in (0, \frac{\pi}{2}) \,, \quad \Gamma = \frac{c_1a^2 + c_2b^2}{(c_1 + c_2)^3} \,. \tag{12.28}$$

Correspondingly, for the roots we get expressions

$$\lambda_{1,2} = \lambda_3 \; \frac{1}{\cos\alpha}\Big(\sin^2(\alpha/2) \pm \sqrt{\sin^4(\alpha/2) - \Gamma/\cos\alpha} \,\Big) . \tag{12.29}$$

The term under the square root should be positive: $\cos\alpha \; \sin^4(\alpha/2) > \Gamma$. Further, we obtain expressions for three possible mass parameters:

$$M_3 = \frac{M}{\lambda_3} = \frac{M}{(c_1 + c_2)}\frac{1}{\cos\alpha} = \mu\,\frac{1}{\cos\alpha} \,,$$

$$M_1 = \frac{M}{\lambda_2} = \frac{\mu}{\sin^2(\alpha/2) - \sqrt{\sin^4(\alpha/2) - \Gamma/\cos\alpha}} \,, \tag{12.30}$$

$$M_2 = \frac{M}{\lambda_2} = \frac{\mu}{\sin^2(\alpha/2) + \sqrt{\sin^4(\alpha/2) - \Gamma/\cos\alpha}} \,.$$

There exists a special case:

$$a \to 0, b \to 0 \Longrightarrow \Gamma \to 0, \quad M_3 = \frac{\mu}{\cos\alpha}, \quad M_1 \to \infty, M_2 = \frac{\mu}{1-\cos^2\alpha}, \tag{12.31}$$

if additionally $\alpha = 0$ or $\alpha = \pi/2$ then yet another mass will be infinite.

Let us study inequality $\cos\alpha \ \sin^4(\alpha/2) > \Gamma$, here we have the problem

$$f(\alpha) = \cos\alpha \ \frac{(1-\cos^2\alpha)^2}{4} > \Gamma \,, \quad f(\alpha = 0) = 0 \,, \quad f(\alpha = \frac{\pi}{2} = 0) \,.$$

In the variable $x = \cos\alpha$, we have

$$f(x) = \frac{1}{4}x(1-x^2)^2 > \Gamma, \quad x \in (0,1) \,,$$

in the interval $x \in (0,1)$ there exists a point of a local maximum α_0:

$$\frac{df}{dx} = \frac{(1-x^2)(1-5x^2)}{4}, \qquad \cos\alpha_0 = \frac{1}{\sqrt{5}}, \quad f(\alpha_0) = \frac{4}{25\sqrt{5}} \,.$$

Thus, we have the following constraint for $f(\alpha)$:

$$\Gamma < \Gamma_0 = \frac{4}{25\sqrt{5}}, \qquad f(\alpha) > \Gamma \,,$$

and there exists some finite interval containing the point α_0. The value $\Gamma_0 = 4/25\sqrt{5}$ is peculiar because the interval for α degenerates into a single point: $\alpha = \alpha_0$, and we have situation when two masses are fixed and equal to each other:

$$\cos\alpha_0 = \frac{1}{\sqrt{5}}, M_3 = \frac{\mu}{\cos\alpha_0} = \mu\sqrt{5}, M_{1,2} = \frac{\mu}{(1-\cos\alpha_0)/2} = \mu\frac{2\sqrt{5}}{\sqrt{5}-1}. \tag{12.32}$$

12.5 Diagonalization in the case of a free particle

Let us introduce shortening notations

$$A_1 = \frac{1}{c_2(c_1 + c_3/\sqrt{6})}\left[c_2(c_1^2 + f|c_3|^2) + \frac{c_3}{\sqrt{6}}(c_1c_2 - g|c_4|^2\right],$$

$$A_2 = \frac{i}{\sqrt{6}}\frac{c_4}{c_2(c_1 + c_3/\sqrt{6})}[f\sqrt{6}c_2c_3^* - g|c_4|^2],$$

$$A_3 = \frac{M}{4}\frac{1}{c_2(c_1 + c_3/\sqrt{6})}\left[f\sqrt{6}c_2c_3^* - g|c_4|^2 - c_2(c_1 + c_3/\sqrt{6})\right],$$

$$B_1 = -ig\frac{c_3c_4^*}{c_2}, \qquad B_2 = \frac{c_2^2 + g|c_4|^2}{c_2}, \qquad B_3 = -igM\frac{\sqrt{6}}{4}\frac{c_4^*}{c_2},$$

$$R_1 = -\frac{4}{M}\frac{c_3}{\sqrt{6}}\frac{fc_2|c_3|^2 + gc_1|c_4|^2}{c_2(c_1 + c_3/\sqrt{6})}, \quad R_2 = -\frac{4i}{M}\frac{c_4}{\sqrt{6}}\frac{fc_2|c_3|^2 + gc_1|c_4|^2}{c_2(c_1 + c_3/\sqrt{6})},$$

$$R_3 = -\frac{fc_2|c_3|^2 + gc_1|c_4|^2}{c_2(c_1 + c_3/\sqrt{6})}.$$

With the use of the above notations, three equations (12.10)–(12.12) read

$$A_1\hat{\partial}\ (\gamma_\mu\Psi_\mu) + B_1\hat{\partial}\ \Psi_0 + R_1\hat{\partial}\ (\partial_\mu\Psi_\mu) + M(\gamma_\mu\Psi_\mu) = 0\ ,$$
$$A_2\hat{\partial}\ (\gamma_\mu\Psi_\mu) + B_2\hat{\partial}\ \Psi_0 + R_2\hat{\partial}\ (\partial_\mu\Psi_\mu) + M\Psi_0 = 0\ , \tag{12.33}$$
$$A_3\hat{\partial}\ (\gamma_\mu\Psi_\mu) + B_3\hat{\partial}\ \Psi_0 + R_3\hat{\partial}\ (\partial_\mu\Psi_\mu) + M(\partial_\mu\Psi_\mu) = 0\ .$$

These equations may be transformed into equations for separate bispinors. To this end, first we sum three equations with coefficients a_1, a_2, a_3, this yields

$$(a_1A_1 + a_2A_2 + a_2A_3)\hat{\partial}\ (\gamma_\mu\Psi_\mu) + (a_1B_1 + a_2B_2 + a_3B_3)\hat{\partial}\ \Psi_0$$
$$+(a_1R_1 + a_2R_2 + a_3R_3)\hat{\partial}\ (\partial_\mu\Psi_\mu) + M\{a_1(\gamma_\mu\Psi_\mu) + a_2\Psi_0 + a_3(\partial_\mu\Psi_\mu)\} = 0\,,$$

taking in mind identities

$$a_1A_1 + a_2A_2 + a_3A_3 = \lambda_1 a_1\ ,$$
$$a_1B_1 + a_2B_2 + a_3B_3 = \lambda_1 a_2\ , \tag{12.34}$$
$$a_1R_1 + a_2R_2 + a_3R_3 = \lambda_1 a_3\ ,$$

the last equation is re-written as

$$\lambda_1 a_1\hat{\partial}(\gamma_\mu\Psi_\mu) + \lambda_1 a_2\hat{\partial}\Psi_0 + \lambda_1 a_3\hat{\partial}(\partial_\mu\Psi_\mu) + M\{a_1(\gamma_\mu\Psi_\mu) + a_2\Psi_0 + a_3(\partial_\mu\Psi_\mu)\} = 0\,.$$

Thus, we produce equation

$$\Phi_1 = a_1(\gamma_\mu\Psi_\mu) + a_2\Psi_0 + a_3(\partial_\mu\Psi_\mu), (\hat{\partial} + \frac{M}{\lambda_1})\Phi_1 = 0, M_1 = M/\lambda_1\, m. \tag{12.35}$$

Similarly, the root λ_2 enters the system

$$b_1A_1 + b_2A_2 + b_3A_3 = \lambda_2 b_1\,, b_1B_1 + b_2B_2 + b_3B_3 = \lambda_2 b_2\,, b_1R_1 + b_2R_2 + b_3R_3 = \lambda_2 b_3\,,$$

its solution is known. Turning to eq. (12.33), we sum them with coefficients b_1, b_2, b_3. In this way, we derive an equation for new bispinor

$$\Phi_2 = b_1(\gamma_\mu\Psi_\mu) + b_2\Psi_0 + b_3(\partial_\mu\Psi_\mu), (\hat{\partial} + \frac{M}{\lambda_2})\Phi_2 = 0, M_2 = M/\lambda_2. \tag{12.36}$$

Finally, turning to the system which contains the root λ_3:

$$r_1A_1 + r_2A_2 + r_3A_3 = \lambda_3 r_1, r_1B_1 + r_2B_2 + r_3B_3 = \lambda_3 r_2, r_1R_1 + r_2R_2 + r_3R_3 = \lambda_3 r_3,$$

and acting as above, we obtain an equation for the third bispinor

$$\Phi_3 = r_1(\gamma_\mu\Psi_\mu) + r_2\Psi_0 + r_3(\partial_\mu\Psi_\mu), (\hat{\partial} + \frac{M}{\lambda_3})\Phi_3 = 0, M_3 = M/\lambda_3. \tag{12.37}$$

12.6 Presence of electromagnetic fields

The presence of electromagnetic fields is taken into account by modifying the derivative: $D_\mu = \partial_\mu - ieA_\mu(x)$. We proceed with eq. (12.2a), but now in presence of external field (let $\hat{D} = \gamma_\mu D_\mu$) we should work only with second and third equations:

$$c_2\hat{D}\Psi_0 - \frac{4ic_4}{\sqrt{6}}\Big[(D_\mu\Psi_\mu) - \frac{1}{4}\hat{D}(\gamma_\mu\Psi_\mu)\Big] + M\Psi_0 = 0\,, \tag{12.38}$$

$$-\frac{2fc_3^*}{\sqrt{6}}\Big[D_\lambda(\gamma_\mu\Psi_\mu)-\frac{1}{4}\gamma_\lambda\hat{D}(\gamma_\mu\Psi_\mu)\Big]+\frac{2igc_4^*}{\sqrt{6}}\Big[D_\lambda\Psi_0-\frac{1}{4}\gamma_\lambda\hat{D}\Psi_0\Big]+\frac{c_1}{4}\gamma_\lambda\hat{D}(\gamma_\mu\Psi_\mu)$$
$$-\frac{c_3}{\sqrt{6}}\Big[\gamma_\lambda(D_\mu\Psi_\mu)-\frac{1}{4}\gamma_\lambda\hat{D}(\gamma_\mu\Psi_\mu)\Big]+M\Psi_\lambda=0\,. \qquad (12.39)$$

We have to perform some transformation over this system. First, let us act on eq. (12.39) by the matrix γ_λ, this results in

$$(c_1+\frac{c_3}{\sqrt{6}})\hat{D}(\gamma_\mu\Psi_\mu)-\frac{4c_3}{\sqrt{6}}(D_\mu\Psi_\mu)+M(\gamma_\mu\Psi_\mu)=0\,. \qquad (12.40)$$

Now, let us us act on eq. (12.39) by operator D_λ:

$$-\frac{2fc_3^*}{\sqrt{6}}\Big[D^2(\gamma_\mu\Psi_\mu)-\frac{1}{4}\hat{D}\hat{D}(\gamma_\mu\Psi_\mu)\Big]+\frac{2igc_4^*}{\sqrt{6}}\Big[D^2\Psi_0-\frac{1}{4}\hat{D}\hat{D}\Psi_0\Big]+\frac{c_1}{4}\hat{D}\hat{D}(\gamma_\mu\Psi_\mu)$$
$$-\frac{c_3}{\sqrt{6}}\Big[\hat{D}(D_\mu\Psi_\mu)-\frac{1}{4}\hat{D}\hat{D}(\gamma_\mu\Psi_\mu)\Big]+M(D_\lambda\Psi_\lambda)=0\,, \qquad (12.41)$$

where $D^2=D_\lambda D_\lambda$. It is readily proved the following relations

$$\hat{D}\hat{D}=D_\lambda D_\rho\,[\,\frac{1}{2}(\gamma_\lambda\gamma_\rho-\gamma_\rho\gamma_\lambda)+\frac{1}{2}(\gamma_\lambda\gamma_\rho+\gamma_\rho\gamma_\lambda)\,]$$
$$=[\,D^2+(D_\lambda D_\rho-D_\rho D_\lambda)\sigma_{\lambda\rho}\,]\,. \qquad (12.42)$$

With this identity (12.41), eq. (12.41) reads

$$-\frac{2fc_3^*}{\sqrt{6}}\Big[\frac{3}{4}D^2(\gamma_\mu\Psi_\mu)-\frac{1}{4}(D_\lambda D_\rho-D_\rho D_\lambda)\sigma_{\lambda\rho}(\gamma_\mu\Psi_\mu)\Big]+M(D_\lambda\Psi_\lambda)$$
$$+\frac{2igc_4^*}{\sqrt{6}}\Big[\frac{3}{4}D^2\Psi_0-\frac{1}{4}(D_\lambda D_\rho-D_\rho D_\lambda)\sigma_{\lambda\rho}\Psi_0\Big]$$
$$+\frac{c_1}{4}[(D_\lambda D_\rho-D_\rho D_\lambda)\sigma_{\lambda\rho}+D^2](\gamma_\mu\Psi_\mu)$$
$$-\frac{c_3}{\sqrt{6}}\Big[\hat{D}(D_\mu\Psi_\mu)-\frac{1}{4}[(D_\lambda D_\rho-D_\rho D_\lambda)\sigma_{\lambda\rho}+D^2](\gamma_\mu\Psi_\mu)\Big]=0\,. \qquad (12.43)$$

Note the commutator $D_\lambda D_\rho-D_\rho D_\lambda=(-ieF_{\lambda\rho})$. Equation (12.43) can be re-written as

$$-\frac{2fc_3^*}{\sqrt{6}}\Big[\frac{3}{4}D^2\,(\gamma_\mu\Psi_\mu)-\frac{1}{4}(-ieF_{\lambda\rho})\,\sigma_{\lambda\rho}\,(\gamma_\mu\Psi_\mu)\Big]$$
$$+\frac{2igc_4^*}{\sqrt{6}}\Big[\frac{3}{4}D^2\,\Psi_0-\frac{1}{4}(-ieF_{\lambda\rho})\,\sigma_{\lambda\rho}\,\Psi_0\Big]-\frac{c_3}{\sqrt{6}}\hat{D}\,(D_\mu\Psi_\mu)$$
$$+\frac{1}{4}(c_1+\frac{c_3}{\sqrt{6}})\,[\,D^2+(-ieF_{\lambda\rho})\,\sigma_{\lambda\rho}\,]\,(\gamma_\mu\Psi_\mu)+M\,(D_\lambda\Psi_\lambda)=0\,. \qquad (12.44)$$

Acting on eq. (12.40) by operator $\hat{D}$, we get

$$(c_1+\frac{c_3}{\sqrt{6}})[\,D^2+(-ieF_{\lambda\rho})\sigma_{\lambda\rho}\,]\,(\gamma_\mu\Psi_\mu)=\frac{4c_3}{\sqrt{6}}\hat{D}(D_\mu\Psi_\mu)-M\hat{D}(\gamma_\mu\Psi_\mu)\,. \qquad (12.45)$$

With this relation in mind, eq. (12.44) reduces to the form (we introduce a new numeration of equations):

equation III

$$-\frac{2fc_3^*}{\sqrt{6}}\Big[\frac{3}{4}D^2-\frac{1}{4}(-ieF_{\lambda\rho})\,\sigma_{\lambda\rho}\,\Big](\gamma_\mu\Psi_\mu)$$
$$+\frac{2igc_4^*}{\sqrt{6}}\Big[\frac{3}{4}D^2-\frac{1}{4}(-ieF_{\lambda\rho})\,\sigma_{\lambda\rho}\Big]\Psi_0-\frac{M}{4}\hat{D}(\gamma_\mu\Psi_\mu)+M\,(D_\lambda\Psi_\lambda)=0\,. \qquad (12.46)$$

Let us write down two other equations:

equation I

$$c_2 \hat{D}\Psi_0 - \frac{4ic_4}{\sqrt{6}} \Big[(D_\mu \Psi_\mu) - \frac{1}{4} \hat{D}(\gamma_\mu \Psi_\mu) \Big] + M\Psi_0 = 0 \,, \tag{12.47}$$

equation II

$$(c_1 + \frac{c_3}{\sqrt{6}}) \hat{D}(\gamma_\mu \Psi_\mu) - \frac{4c_3}{\sqrt{6}} (D_\mu \Psi_\mu) + M(\gamma_\mu \Psi_\mu) = 0 \,. \tag{12.48}$$

We see that only equation III contains the 2nd-order operator D^2. The task is to get the system of 1st-order for three bispinors

$$(\gamma_\mu \Psi_\mu) \,, \qquad \Psi_0 \,, \qquad (D_\mu \Psi_\mu) \,.$$

Therefore, we have to exclude operator D^2 from eq. (12.46). First, let us re-write eq. (12.46), distinguishing separate terms

$$- \frac{\sqrt{6} f c_3^*}{4} D^2 (\gamma_\mu \Psi_\mu) + \frac{f c_3^*}{2\sqrt{6}} (-ieF_{\lambda\rho}) \, \sigma_{\lambda\rho} \, (\gamma_\mu \Psi_\mu)$$
$$+ \frac{i\sqrt{6} g c_4^*}{4} D^2 \Psi_0 - \frac{igc_4^*}{2\sqrt{6}} (-ieF_{\lambda\rho}) \, \sigma_{\lambda\rho} \Psi_0 - \frac{1}{4} M \hat{D}(\gamma_\mu \Psi_\mu) + M(D_\lambda \Psi_\lambda) = 0 \,.$$

With the help of eq. (12.45) in the form

$$\begin{aligned} D^2(\gamma_\mu \Psi_\mu) \;\; &= \;\; (+ieF_{\lambda\rho}) \sigma_{\lambda\rho} (\gamma_\mu \Psi_\mu) \\ &+ \frac{4}{\sqrt{6}} \frac{c_3}{(c_1 + c_3/\sqrt{6})} \hat{D}(D_\mu \Psi_\mu) \frac{M}{(c_1 + c_3/\sqrt{6})} \hat{D}(\gamma_\mu \Psi_\mu) \,, \end{aligned} \tag{12.49}$$

we transform in eq. (12.46) the first term

$$- \frac{\sqrt{6} f c_3^*}{4} \frac{4}{\sqrt{6}} \frac{c_3}{(c_1 + c_3/\sqrt{6})} \hat{D}(D_\mu \Psi_\mu) + \frac{M}{4} \frac{\sqrt{6} f c_3^*}{(c_1 + c_3/\sqrt{6})} \hat{D}(\gamma_\mu \Psi_\mu)$$
$$+ \frac{i\sqrt{6} g c_4^*}{4} D^2 \Psi_0 - \frac{igc_4^*}{2\sqrt{6}} (-ieF_{\lambda\rho}) \, \sigma_{\lambda\rho} \Psi_0$$
$$+ f c_3^* \, (\frac{\sqrt{6}}{4} + \frac{1}{2\sqrt{6}}) \, (-ieF_{\lambda\rho}) \, \sigma_{\lambda\rho} \, (\gamma_\mu \Psi_\mu) - \frac{1}{4} M \hat{D}(\gamma_\mu \Psi_\mu) + M(D_\lambda \Psi_\lambda) = 0 \,.$$

In this way, we transform eq. (12.46) to the form

$$\begin{aligned} &- \frac{f|c_3|^2}{(c_1 + c_3/\sqrt{6})} \hat{D}(D_\mu \Psi_\mu) + \frac{M}{4} \, [\, \frac{\sqrt{6} f c_3^*}{(c_1 + c_3/\sqrt{6})} - 1 \,] \, \hat{D}(\gamma_\mu \Psi_\mu) \\ &+ \frac{i\sqrt{6} g c_4^*}{4} D^2 \Psi_0 - \frac{igc_4^*}{2\sqrt{6}} (-ieF_{\lambda\rho}) \sigma_{\lambda\rho} \Psi_0 \\ &+ \frac{2}{\sqrt{6}} f c_3^* (-ieF_{\lambda\rho}) \sigma_{\lambda\rho} (\gamma_\mu \Psi_\mu) + M(D_\lambda \Psi_\lambda) = 0 \,. \end{aligned} \tag{12.50}$$

In order to transform the term $D^2 \Psi_0$, we turn to eq. (12.47) and act on it by operator $\hat{D}$:

$$c_2 \hat{D}\hat{D}\Psi_0 - \frac{4ic_4}{\sqrt{6}} \Big[\hat{D}(D_\mu \Psi_\mu) - \frac{1}{4} \hat{D}\hat{D}(\gamma_\mu \Psi_\mu) \Big] + M\hat{D}\Psi_0 = 0 \,;$$

whence with the use of identity $\hat{D}\hat{D} = D^2 + (-ieF_{\lambda\rho})\sigma_{\lambda\rho}$ we derive

$$D^2\Psi_0 = -(-ieF_{\lambda\rho})\sigma_{\lambda\rho}\Psi_0 + \frac{4ic_4}{\sqrt{6}c_2}\hat{D}(D_\mu\Psi_\mu)$$
$$-\frac{ic_4}{\sqrt{6}c_2}D^2(\gamma_\mu\Psi_\mu) - \frac{ic_4}{\sqrt{6}c_2}(-ieF_{\lambda\rho})\sigma_{\lambda\rho}(\gamma_\mu\Psi_\mu) - \frac{M}{c_2}\hat{D}\Psi_0 .$$

Because the term $D^2(\gamma_\mu\Psi_\mu)$ may be expressed with the use of eq. (12.49) as follows

$$D^2(\gamma_\mu\Psi_\mu) = (+ieF_{\lambda\rho})\sigma_{\lambda\rho}(\gamma_\mu\Psi_\mu)$$
$$+\frac{4}{\sqrt{6}}\frac{c_3}{(c_1 + c_3/\sqrt{6})}\hat{D}(D_\mu\Psi_\mu) - \frac{M}{(c_1 + c_3/\sqrt{6})}\hat{D}(\gamma_\mu\Psi_\mu) ,$$

then the previous relation takes the form

$$D^2\Psi_0 = -\frac{M}{c_2}\hat{D}\Psi_0 - (-ieF_{\lambda\rho})\sigma_{\lambda\rho}\Psi_0$$
$$+\frac{4i}{\sqrt{6}}\frac{c_1c_4}{c_2(c_1 + c_3/\sqrt{6})}\hat{D}(D_\mu\Psi_\mu) + iM\frac{1}{\sqrt{6}}\frac{c_4}{c_2(c_1 + c_3/\sqrt{6})}\hat{D}(\gamma_\mu\Psi_\mu) . \tag{12.51}$$

With the help of this, we can exclude the term $D^2\Psi_0$ in eq. (12.50), so deriving

$$\Big[-\frac{f|c_3|^2}{(c_1 + c_3/\sqrt{6})} + \frac{4i}{\sqrt{6}}\frac{c_1c_4}{c_2(c_1 + c_3/\sqrt{6})}\Big]\hat{D}(D_\mu\Psi_\mu)$$
$$+\Big[\frac{M}{4}\frac{\sqrt{6}fc_3^* - (c_1 + c_3/\sqrt{6})}{(c_1 + c_3/\sqrt{6})} - \frac{M}{4}\frac{gc_4^*c_4}{c_2(c_1 + c_3/\sqrt{6})}\Big]\hat{D}(\gamma_\mu\Psi_\mu)$$
$$-\frac{i\sqrt{6}gc_4^*}{4}\frac{M}{c_2}\hat{D}\Psi_0 - \Big(\frac{i\sqrt{6}gc_4^*}{4} + \frac{igc_4^*}{2\sqrt{6}}\Big)(-ieF_{\lambda\rho})\sigma_{\lambda\rho}\Psi_0$$
$$+\frac{2}{\sqrt{6}}fc_3^*\,(-ieF_{\lambda\rho})\sigma_{\lambda\rho}\,(\gamma_\mu\Psi_\mu) + M\,(D_\mu\Psi_\mu) = 0 .$$

Thus, equation III reads

$$+\frac{M}{4}\frac{\sqrt{6}fc_2c_3^* - c_2(c_1 + c_3/\sqrt{6}) - g|c_4|^2}{c_2(c_1 + c_3/\sqrt{6})}\,\hat{D}(\gamma_\mu\Psi_\mu) - i\frac{\sqrt{6}}{4}M\,\frac{gc_4^*}{c_2}\,\hat{D}\Psi_0$$
$$-\frac{fc_2|c_3|^2 + gc_1|c_4|^2}{c_2(c_1 + c_3/\sqrt{6})}\,\hat{D}(D_\mu\Psi_\mu) - i\frac{2}{\sqrt{6}}\,gc_4^*\,(-ieF_{\lambda\rho})\sigma_{\lambda\rho}\,\Psi_0$$
$$+\frac{2}{\sqrt{6}}fc_3^*\,(-ieF_{\lambda\rho})\sigma_{\lambda\rho}\,(\gamma_\mu\Psi_\mu) + M\,(D_\mu\Psi_\mu) = 0 , \tag{12.52}$$

it does not contain the 2nd-order operator D^2.

12.7 Quasidiagonal form of the system

We start with three equations

$$c_2 \hat{D}\Psi_0 - \frac{4ic_4}{\sqrt{6}} \Big[(D_\mu \Psi_\mu) - \frac{1}{4}\hat{D}(\gamma_\mu \Psi_\mu) \Big] + M\Psi_0 = 0\,, \tag{12.53}$$

$$(c_1 + \frac{c_3}{\sqrt{6}})\hat{D}(\gamma_\mu \Psi_\mu) - \frac{4c_3}{\sqrt{6}}(D_\mu \Psi_\mu) + M(\gamma_\mu \Psi_\mu) = 0\,, \tag{12.54}$$

$$\begin{aligned} &\frac{M}{4}\frac{\sqrt{6}fc_2c_3^* - c_2(c_1 + c_3/\sqrt{6}) - g|c_4|^2}{c_2(c_1 + c_3/\sqrt{6})}\hat{D}(\gamma_\mu \Psi_\mu) - i\frac{\sqrt{6}}{4}M\,\frac{gc_4^*}{c_2}\,\hat{D}\Psi_0 \\ &-\frac{fc_2|c_3|^2 + gc_1|c_4|^2}{c_2(c_1 + c_3/\sqrt{6})}\hat{D}(D_\mu \Psi_\mu) - i\frac{2}{\sqrt{6}}\,gc_4^*\,(-ieF_{\lambda\rho})\sigma_{\lambda\rho}\,\Psi_0 \\ &+\frac{2}{\sqrt{6}}fc_3^*\,(-ieF_{\lambda\rho})\sigma_{\lambda\rho}\,(\gamma_\mu \Psi_\mu) + M\,(D_\mu \Psi_\mu) = 0\,. \end{aligned} \tag{12.55}$$

The next task is to transform the system into a form that is similar to that existing for free particles, but modified in the presence of an external field. To this end, first let us consider eq. (12.53):

$$c_2\hat{D}\Psi_0 + \frac{ic_4}{\sqrt{6}}\hat{D}(\gamma_\mu \Psi_\mu) - \frac{4ic_4}{\sqrt{6}}(D_\mu \Psi_\mu) + M\Psi_0 = 0\,,$$

The term $(D_\mu \Psi_\mu)$ is excluded with eq. (12.55), so we obtain

$$\begin{aligned} &\frac{c_2^2 + g|c_4|^2}{c_2}\hat{D}\Psi_0 + \frac{ic_4}{\sqrt{6}}\frac{\sqrt{6}fc_2c_3^* - g|c_4|^2}{c_2(c_1 + c_3/\sqrt{6})}\hat{D}(\gamma_\mu \Psi_\mu) \\ &-\frac{4ic_4}{\sqrt{6}}\frac{1}{M}\frac{fc_2|c_3|^2 + gc_1|c_4|^2}{c_2(c_1 + c_3/\sqrt{6})}\hat{D}(D_\mu \Psi_\mu) + \frac{4}{3}\frac{1}{M}g|c_4|^2(-ieF_{\lambda\rho})\sigma_{\lambda\rho}\Psi_0 \\ &+\frac{4i}{3}\frac{1}{M}fc_3^*c_4(-ieF_{\lambda\rho})\sigma_{\lambda\rho}\,(\gamma_\mu \Psi_\mu) + M\Psi_0 = 0\,. \end{aligned} \tag{12.56}$$

Similarly, from equation

$$(c_1 + \frac{c_3}{\sqrt{6}})\hat{D}(\gamma_\mu \Psi_\mu) - \frac{4c_3}{\sqrt{6}}(D_\mu \Psi_\mu) + M(\gamma_\mu \Psi_\mu) = 0\,,$$

we can exclude the term $(D_\mu \Psi_\mu)$ as well

$$\begin{aligned} &\frac{1}{c_2(c_1 + c_3/\sqrt{6})}\Big[c_2(c_1^2 + f|c_3|^2) + \frac{c_3}{\sqrt{6}}(c_1c_2 - g_4|c^2)\Big]\hat{D}(\gamma_\mu \Psi_\mu) \\ &-\frac{4}{\sqrt{6}}\frac{c_3}{M}\frac{fc_2|c_3|^2 + gc_1|c_4|^2}{c_2(c_1 + c_3/\sqrt{6})}\hat{D}(D_\mu \Psi_\mu) - i\frac{gc_3c_4^*}{c_2}\,\hat{D}\Psi_0 - \frac{4ic_3}{3}\frac{1}{M}gc_4^*(-ieF_{\lambda\rho})\sigma_{\lambda\rho}\,\Psi_0 \\ &+\frac{4c_3}{3}\frac{1}{M}fc_3^*(-ieF_{\lambda\rho})\sigma_{\lambda\rho}\,(\gamma_\mu \Psi_\mu) + M(\gamma_\mu \Psi_\mu) = 0\,. \end{aligned} \tag{12.57}$$

Let us introduce notations:

$$\gamma_\mu \Psi_\mu = \bar{\Phi}_1, \quad \bar{\Phi}_2 = \Psi_0, \quad \bar{\Phi}_3 = D_\mu \Psi_\mu\,, \quad (-ieF_{\lambda\rho})\sigma_{\lambda\rho} = \Sigma\,.$$

Thus, we have the following three equations

$$\frac{1}{c_2(c_1+c_3/\sqrt{6})}\left[c_2(c_1^2+f|c_3|^2)+\frac{c_3}{\sqrt{6}}(c_1c_2-g_4|c^2)\right]\hat{D}\bar{\Phi}_1$$

$$-i\frac{gc_3c_4^*}{c_2}\hat{D}\bar{\Phi}_2-\frac{4}{\sqrt{6}}\frac{c_3}{M}\frac{fc_2|c_3|^2+gc_1|c_4|^2}{c_2(c_1+c_3/\sqrt{6})}\hat{D}\bar{\Phi}_3$$

$$+\frac{4c_3}{3}\frac{1}{M}fc_3^*\Sigma\,\bar{\Phi}_1-\frac{4ic_3}{3}\frac{1}{M}gc_4^*\Sigma\,\bar{\Phi}_2+M\bar{\Phi}_1=0\,,$$

$$\frac{ic_4}{\sqrt{6}}\frac{\sqrt{6}fc_2c_3^*-g|c_4|^2}{c_2(c_1+c_3/\sqrt{6})}\hat{D}\bar{\Phi}_1+\frac{c_2^2+g|c_4|^2}{c_2}\hat{D}\bar{\Phi}_2-\frac{4ic_4}{\sqrt{6}}\frac{1}{M}\frac{fc_2|c_3|^2+gc_1|c_4|^2}{c_2(c_1+c_3/\sqrt{6})}\hat{D}\bar{\Phi}_3$$

$$+\frac{4i}{3}\frac{1}{M}fc_3^*c_4\Sigma\,\bar{\Phi}_1+\frac{4}{3}\frac{1}{M}g|c_4|^2\Sigma\bar{\Phi}_2+M\Psi_0=0,$$

$$+\frac{M}{4}\frac{\sqrt{6}fc_2c_3^*-c_2(c_1+c_3/\sqrt{6})-g|c_4|^2}{c_2(c_1+c_3/\sqrt{6})}\hat{D}\bar{\Phi}_1$$

$$+i\frac{\sqrt{6}}{4}M\frac{gc_4^*}{c_2}\hat{D}\bar{\Phi}_2-\frac{fc_2|c_3|^2+gc_1|c_4|^2}{c_2(c_1+c_3/\sqrt{6})}\hat{D}\bar{\Phi}_3$$

$$+\frac{2}{\sqrt{6}}fc_3^*\,\Sigma\,\bar{\Phi}_1-i\frac{2}{\sqrt{6}}\,gc_4^*\,\Sigma\bar{\Phi}_2+M\,\bar{\Phi}_3=0\,.$$

Having used the previously introduce notations $A_i, B_i, R_i,\ i=1,2,3$, we may re-write the system in a shorter form:

$$A_1\hat{D}\,\bar{\Phi}_1+B_1\hat{D}\,\bar{\Phi}_2+R_1\hat{D}\,\bar{\Phi}_3+M\,\bar{\Phi}_1+\frac{4}{3}\frac{f|c_3|^2}{M}\Sigma\,\bar{\Phi}_1-\frac{4i}{3}\frac{gc_3c_4^*}{M}\Sigma\,\bar{\Phi}_2=0\,,$$

$$A_2\hat{D}\,\bar{\Phi}_1+B_2\hat{D}\,\bar{\Phi}_2+R_2\hat{D}\,\bar{\Phi}_3+M\,\bar{\Phi}_2+\frac{4i}{3}\frac{fc_3^*c_4}{M}\Sigma\,\bar{\Phi}_1+\frac{4}{3}\frac{g|c_4|^2}{M}\Sigma\,\bar{\Phi}_2=0\,,$$

$$A_3\hat{D}\,\bar{\Phi}_1+B_3\hat{D}\,\bar{\Phi}_2+R_3\hat{D}\,\bar{\Phi}_3+M\,\bar{\Phi}_3+\frac{2}{\sqrt{6}}fc_3^*\,\Sigma\,\bar{\Phi}_1-i\frac{2}{\sqrt{6}}\,gc_4^*\Sigma\,\bar{\Phi}_2=0\,.$$

Now, we act in the same manner as for the free field case. We multiply equations 1,2,3 by a_1, a_2, a_3 and sum the results:

$$(a_1A_1+a_2A_2+a_3A_3)\hat{D}\bar{\Phi}_1+(a_1B_1+a_2B_2+a_3B_3)\hat{D}\bar{\Phi}_2$$

$$+(a_1R_1+a_2R_2+a_3R_3)\hat{D}\bar{\Phi}_3+M(a_1\bar{\Phi}_1+a_2\bar{\Phi}_2+a_3\bar{\Phi}_3)$$

$$+(a_1\frac{4}{3}\frac{f|c_3|^2}{M}+a_2\frac{4i}{3}\frac{fc_3^*c_4}{M}+a_3\frac{2}{\sqrt{6}}fc_3^*)\Sigma\,\bar{\Phi}_1$$

$$+(-a_1\frac{4i}{3}\frac{gc_3c_4^*}{M}+a_2\frac{4}{3}\frac{g|c_4|^2}{M}-a_3\frac{2i}{\sqrt{6}}\,gc_4^*)\Sigma\,\bar{\Phi}_2=0\,,$$

whence taking in mind three relations

$$a_1A_1+a_2A_2+a_3A_3=\lambda_1a_1\,,\ a_1B_1+a_2B_2+a_3B_3=\lambda_1a_2\,,\ a_1R_1+a_2R_2+a_3R_3=\lambda_1a_3\,,$$

we get

$$\lambda_1\hat{D}(a_1\bar{\Phi}_1+a_2\bar{\Phi}_2+a_3\bar{\Phi}_3)+M(a_1\bar{\Phi}_1+a_2\bar{\Phi}_2a_3\bar{\Phi}_3)$$

$$+(a_1 \frac{4}{3}\frac{f|c_3|^2}{M} + a_2 \frac{4i}{3}\frac{f c_3^* c_4}{M} + a_3 \frac{2}{\sqrt{6}} f c_3^*) \Sigma \bar{\Phi}_1$$

$$+(-a_1 \frac{4i}{3}\frac{g c_3 c_4^*}{M} + a_2 \frac{4}{3}\frac{g|c_4|^2}{M} - a_3 \frac{2i}{\sqrt{6}} g c_4^*) \Sigma \bar{\Phi}_2 = 0 . \qquad (12.58)$$

Similarly, we obtain yet two equations (in fact, changing a_i on b_i and on r_i)

$$\lambda_1 \hat{D}(b_1 \bar{\Phi}_1 + b_2 \bar{\Phi}_2 + b_3 \bar{\Phi}_3) + M(b_1 \bar{\Phi}_1 + b_2 \bar{\Phi}_2 + b_3 \bar{\Phi}_3)$$

$$+(b_1 \frac{4}{3}\frac{f|c_3|^2}{M} + b_2 \frac{4i}{3}\frac{f c_3^* c_4}{M} + b_3 \frac{2}{\sqrt{6}} f c_3^*) \Sigma \bar{\Phi}_1$$

$$+(-b_1 \frac{4i}{3}\frac{g c_3 c_4^*}{M} + b_2 \frac{4}{3}\frac{g|c_4|^2}{M} - b_3 \frac{2i}{\sqrt{6}} g c_4^*) \Sigma \bar{\Phi}_2 = 0 ; \qquad (12.59)$$

$$\lambda_1 \hat{D}(r_1 \bar{\Phi}_1 + r_2 \bar{\Phi}_2 + r_3 \bar{\Phi}_3) + M(r_1 \bar{\Phi}_1 + r_2 \bar{\Phi}_2 + r_3 \bar{\Phi}_3)$$

$$+(r_1 \frac{4}{3}\frac{f|c_3|^2}{M} + r_2 \frac{4i}{3}\frac{f c_3^* c_4}{M} + r_3 \frac{2}{\sqrt{6}} f c_3^*) \Sigma \bar{\Phi}_1$$

$$+(-r_1 \frac{4i}{3}\frac{g c_3 c_4^*}{M} + r_2 \frac{4}{3}\frac{g|c_4|^2}{M} - r_3 \frac{2i}{\sqrt{6}} g c_4^*) \Sigma \bar{\Phi}_2 = 0 . \qquad (12.60)$$

By definition, we introduce three new bispinors

$$\begin{aligned} \Phi_1 &= a_1 \bar{\Phi}_1 + a_2 \bar{\Phi}_2 + a_3 \bar{\Phi}_3 , \\ \Phi_2 &= b_1 \bar{\Phi}_1 + b_2 \bar{\Phi}_2 + b_3 \bar{\Phi}_3 , \\ \Phi_3 &= r_1 \bar{\Phi}_1 + r_2 \bar{\Phi}_2 + r_3 \bar{\Phi}_3 , \end{aligned} \qquad (12.61)$$

then the above system is presented shorter

$$\lambda_1 \hat{D} \Phi_1 + M \Phi_1$$

$$+(a_1 \frac{4}{3}\frac{f|c_3|^2}{M} + a_2 \frac{4i}{3}\frac{f c_3^* c_4}{M} + a_3 \frac{2}{\sqrt{6}} f c_3^*) \Sigma\ \bar{\Phi}_1$$

$$+(-a_1 \frac{4i}{3}\frac{g c_3 c_4^*}{M} + a_2 \frac{4}{3}\frac{g|c_4|^2}{M} - a_3 \frac{2i}{\sqrt{6}}\ g c_4^*) \Sigma\ \bar{\Phi}_2 = 0 , \qquad (12.62)$$

$$\lambda_1 \hat{D} \Phi_2 + M \Phi_2$$

$$+(b_1 \frac{4}{3}\frac{f|c_3|^2}{M} + b_2 \frac{4i}{3}\frac{f c_3^* c_4}{M} + b_3 \frac{2}{\sqrt{6}} f c_3^*) \Sigma\ \bar{\Phi}_1$$

$$+(-b_1 \frac{4i}{3}\frac{g c_3 c_4^*}{M} + b_2 \frac{4}{3}\frac{g|c_4|^2}{M} - b_3 \frac{2i}{\sqrt{6}}\ g c_4^*) \Sigma\ \bar{\Phi}_2 = 0 , \qquad (12.63)$$

$$\lambda_1 \hat{D} \Phi_3 + M \Phi_3$$

$$+(r_1 \frac{4}{3}\frac{f|c_3|^2}{M} + r_2 \frac{4i}{3}\frac{f c_3^* c_4}{M} + r_3 \frac{2}{\sqrt{6}} f c_3^*) \Sigma\ \bar{\Phi}_1$$

$$+(-r_1 \frac{4i}{3}\frac{g c_3 c_4^*}{M} + r_2 \frac{4}{3}\frac{g|c_4|^2}{M} - r_3 \frac{2i}{\sqrt{6}}\ g c_4^*) \Sigma\ \bar{\Phi}_2 = 0 . \qquad (12.64)$$

12.8 Mixing the components in the system

Relationship between two sets of bispinors $\bar{\Phi}_i$ and Φ_i is determined by the formula

$$\begin{vmatrix} \Phi_1 \\ \Phi_2 \\ \Phi_3 \end{vmatrix} = \begin{vmatrix} a_1 & a_2 & a_3 \\ b_1 & b_2 & b_3 \\ r_1 & r_2 & r_3 \end{vmatrix} \begin{vmatrix} \bar{\Phi}_1 \\ \bar{\Phi}_2 \\ \bar{\Phi}_3 \end{vmatrix}, \qquad \Phi = S\bar{\Phi}. \tag{12.65}$$

In fact, we need the inverse matrix:

$$\bar{\Phi} = S^{-1}\Phi, \qquad S^{-1} = \frac{1}{\det S} \begin{vmatrix} S_{11} & S_{21} & S_{31} \\ S_{12} & S_{22} & S_{32} \\ S_{13} & S_{22} & S_{33} \end{vmatrix},$$

$$S^{-1} = \frac{1}{\det S} \begin{vmatrix} (b_2r_3 - b_3r_2) & -(a_2r_3 - a_3r_2) & (a_2b_3 - a_3b_2) \\ -(b_1r_3 - b_3r_1) & (a_1r_3 - a_3r_1) & -(a_1b_3 - a_3b_1) \\ (b_1r_2 - b_2r_1) & -(a_1r_2 - a_2r_1) & (a_1b_2 - a_2b_1) \end{vmatrix},$$

$$\det S = r_1(a_2b_3 - a_3b_2) - r_2(a_1b_3 - a_3b_1) + r_3(a_1b_2 - a_2b_1). \tag{12.66}$$

The elements of the matrix S were given above. Because a_1, b_1, and r_1 are arbitrary, let us fix them so that elements look most simple (we collect them into three sets)

$$a_1 = \lambda_1\, c_2(\lambda_1 - c_2) + (\lambda_1 - c_2)\,[fc_2|c_3|^2 + gc_1|c_4|^2] - \lambda_1\,(c_1 + c_3/\sqrt{6})\,g|c_4|^2\,,$$
$$b_1 = \lambda_2\, c_2(\lambda_2 - c_2) + (\lambda_2 - c_2)\,[fc_2|c_3|^2 + gc_1|c_4|^2] - \lambda_2\,(c_1 + c_3/\sqrt{6})\,g|c_4|^2\,,$$
$$r_1 = \lambda_3\, c_2(\lambda_3 - c_2) + (\lambda_3 - c_2)\,[fc_2|c_3|^2 + gc_1|c_4|^2] - \lambda_3\,(c_1 + c_3/\sqrt{6})\,g|c_4|^2\,;$$

$$a_2 = -igc_3c_4^*\,(c_1 + c_3/\sqrt{6})\lambda_1,$$
$$b_2 = -igc_3c_4^*\,(c_1 + c_3/\sqrt{6})\lambda_2\,,$$
$$r_2 = -igc_3c_4^*\,(c_1 + c_3/\sqrt{6})\lambda_3\,;$$

$$a_3 = -\frac{4}{\sqrt{6}}\frac{1}{M}\,c_3(\lambda_1 - c_2)\,[fc_2|c_3|^2 + gc_1|c_4|^2]\,,$$
$$b_3 = -\frac{4}{\sqrt{6}}\frac{1}{M}\,c_3(\lambda_2 - c_2)\,[fc_2|c_3|^2 + gc_1|c_4|^2]\,,$$
$$r_3 = -\frac{4}{\sqrt{6}}\frac{1}{M}\,c_3(\lambda_3 - c_2)\,[fc_2|c_3|^2 + gc_1|c_4|^2]\,.$$

We note symmetry in these expressions with respect to indexes 1, 2, and 3. With notation

$$K = [fc_2|c_3|^2 + gc_1|c_4|^2]\,, \qquad L = (c_1 + c_3/\sqrt{6})\,, \tag{12.67}$$

the formulas read shorter

$$a_1 = \lambda_1\, c_2(\lambda_1 - c_2) + (\lambda_1 - c_2)\,K - \lambda_1\,L\,g|c_4|^2\,,$$
$$b_1 = \lambda_2\, c_2(\lambda_2 - c_2) + (\lambda_2 - c_2)\,K - \lambda_2\,L\,g|c_4|^2\,,$$
$$r_1 = \lambda_3\, c_2(\lambda_3 - c_2) + (\lambda_3 - c_2)\,K - \lambda_3\,L\,g|c_4|^2;$$

$$a_2 = -igc_3c_4^* \, L \, \lambda_1 \, , \quad b_2 = -igc_3c_4^* \, L \, \lambda_2 \, , \quad r_2 = -igc_3c_4^* \, L \, \lambda_3 \, ;$$

$$a_3 = -\frac{4}{\sqrt{6}} \frac{1}{M} c_3(\lambda_1 - c_2)K \, , \; b_3 = -\frac{4}{\sqrt{6}} \frac{1}{M} c_3(\lambda_2 - c_2)K \, , \; r_3 = -\frac{4}{\sqrt{6}} \frac{1}{M} \, c_3(\lambda_3 - c_2)K \, .$$

Let us collect together results related to the matrix S^{-1}.

$$\det S = -\frac{2\,i\,\sqrt{6}}{3} \, \frac{c_2{}^2 c_3{}^2 c_4^* \, (\lambda_1 - \lambda_3) \, (\lambda_1 - \lambda_2) \, (\lambda_2 - \lambda_3) \, KLg}{M} \, , \tag{12.68}$$

and the complete inverse matrix S^{-1} is

$$S^{-1}$$

$$= \left| \begin{array}{ccc} \frac{1}{(\lambda_1-\lambda_3)(\lambda_1-\lambda_2)c_2} & -\frac{1}{(\lambda_2-\lambda_3)(\lambda_1-\lambda_2)c_2} & \frac{1}{(\lambda_2-\lambda_3)(\lambda_1-\lambda_3)c_2} \\ \frac{iLg(|c_4|)^2+i(-\lambda_2+c_2)(-\lambda_3+c_2)}{(\lambda_1-\lambda_3)(\lambda_1-\lambda_2)Lc_4^* \, gc_3 \, c_2} & \frac{-iLg(|c_4|)^2-i(-\lambda_1+c_2)(-\lambda_3+c_2)}{(\lambda_2-\lambda_3)(\lambda_1-\lambda_2)Lc_4^* \, gc_3 \, c_2} & \frac{iLg(|c_4|)^2+i(-\lambda_1+c_2)(-\lambda_2+c_2)}{(\lambda_2-\lambda_3)(\lambda_1-\lambda_3)Lc_4^* \, gc_3 \, c_2} \\ \frac{1}{4} \, \frac{M\sqrt{6}(\lambda_2 \, \lambda_3+K)}{c_3 \, c_2 \, (\lambda_1-\lambda_3)(\lambda_1-\lambda_2)K} & -\frac{1}{4} \, \frac{M\sqrt{6}(\lambda_1 \, \lambda_3+K)}{c_3 \, c_2 \, (\lambda_2-\lambda_3)(\lambda_1-\lambda_2)K} & \frac{1}{4} \, \frac{M\sqrt{6}(\lambda_1 \, \lambda_2+K)}{c_3 \, c_2 \, (\lambda_2-\lambda_3)(\lambda_1-\lambda_3)K} \end{array} \right| . \tag{12.69}$$

Let turn to eqs. (12.62)–(12.64) and calculate the term

$$\Big\{a_1 \frac{4}{3} \frac{f|c_3|^2}{M} + a_2 \frac{4i}{3} \frac{fc_3^*c_4}{M} + a_3 \frac{2}{\sqrt{6}} fc_3^* \Big\} \Sigma \, \bar{\Phi}_1 = \lambda_1(\lambda_1 - c_2) \frac{4}{3M} fc_2|c_3|^2 \, \Sigma \, \bar{\Phi}_1 .$$

Let us calculate the term

$$\Big\{ - a_1 \frac{4i}{3} \frac{gc_3c_4^*}{M} + a_2 \frac{4}{3} \frac{g|c_4|^2}{M} - a_3 \frac{2i}{\sqrt{6}} \, gc_4^* \Big\} \Sigma \, \bar{\Phi}_2 = -\lambda_1(\lambda_1 - c_2) \frac{4i}{3} \frac{1}{M} \, gc_2c_3c_4^* \, \Sigma \, \bar{\Phi}_2 .$$

Therefore, eq. (12.62) takes the form

$$\lambda_1 \hat{D} \Phi_1 + M\Phi_1 + \frac{4c_2c_3}{3M} \lambda_1(\lambda_1 - c_2) \, \Sigma \, (fc_3^* \, \bar{\Phi}_1 - igc_4^* \bar{\Phi}_2) = 0 \, . \tag{12.70}$$

Similarly calculate the term

$$\Big\{b_1 \frac{4}{3} \frac{f|c_3|^2}{M} + b_2 \frac{4i}{3} \frac{fc_3^*c_4}{M} + b_3 \frac{2}{\sqrt{6}} fc_3^* \Big\} \Sigma \, \bar{\Phi}_1 = \lambda_2(\lambda_2 - c_2) \frac{4}{3M} fc_2|c_3|^2 \, \Sigma \, \bar{\Phi}_1 \, ,$$

calculate the term

$$\Big\{ - b_1 \frac{4i}{3} \frac{gc_3c_4^*}{M} + b_2 \frac{4}{3} \frac{g|c_4|^2}{M} - b_3 \frac{2i}{\sqrt{6}} \, gc_4^* \Big\} \Sigma \, \bar{\Phi}_2 = -\lambda_2(\lambda_2 - c_2) \frac{4i}{3M} gc_2c_3c_4^* \Sigma \, \bar{\Phi}_2 \, .$$

Therefore, eq. (12.63) takes the form

$$\lambda_2 \hat{D} \Phi_2 + M\Phi_2 + \frac{4c_2c_3}{3M} \lambda_2(\lambda_2 - c_2) \, \Sigma \, (fc_3^* \, \bar{\Phi}_1 - igc_4^* \bar{\Phi}_2) = 0 \, . \tag{12.71}$$

The third eq. (12.64) is presented as

$$\lambda_3 \hat{D} \Phi_3 + M\Phi_3 + \frac{4c_2c_3}{3M} \lambda_3(\lambda_3 - c_2) \, \Sigma \, (fc_3^* \, \bar{\Phi}_1 - igc_4^* \bar{\Phi}_2) = 0 \, . \tag{12.72}$$

Let us collect three equations together:

$$\begin{aligned}
&\lambda_1 \hat{D}\Phi_1 + M\Phi_1 + \frac{4c_2c_3}{3M}\lambda_1(\lambda_1 - c_2)\,\Sigma\left[fc_3^*\,\bar{\Phi}_1 - igc_4^*\bar{\Phi}_2\right] = 0\,,\\
&\lambda_2 \hat{D}\Phi_2 + M\Phi_2 + \frac{4c_2c_3}{3M}\lambda_2(\lambda_2 - c_2)\,\Sigma\left[fc_3^*\,\bar{\Phi}_1 - igc_4^*\bar{\Phi}_2\right] = 0\,,\\
&\lambda_3 \hat{D}\Phi_3 + M\Phi_3 + \frac{4c_2c_3}{3M}\lambda_3(\lambda_3 - c_2)\,\Sigma\left[fc_3^*\,\bar{\Phi}_1 - igc_4^*\bar{\Phi}_2\right] = 0\,.
\end{aligned} \tag{12.73}$$

Taking in mind the explicit form of inverse matrix S^{-1}, the functions $\bar{\Phi}_1$ and $\bar{\Phi}_2$ are expressed in terms of Φ_1, Φ_2, and Φ_3 by the formulas

$$\bar{\Phi}_1 = \frac{1}{(\lambda_1 - \lambda_3)\,(\lambda_1 - \lambda_2)\,c_2}\Phi_1 - \frac{1}{(\lambda_2 - \lambda_3)\,(\lambda_1 - \lambda_2)\,c_2}\Phi_2 + \frac{1}{(\lambda_2 - \lambda_3)\,(\lambda_1 - \lambda_3)\,c_2}\Phi_3\,,$$

and

$$\bar{\Phi}_2 = \frac{iLg\,(|c_4|)^2 + i\,(-\lambda_2 + c_2)\,(-\lambda_3 + c_2)}{(\lambda_1 - \lambda_3)\,(\lambda_1 - \lambda_2)\,Lc_4^*\,gc_3\,c_2}\Phi_1$$

$$+\frac{-iLg\,(|c_4|)^2 - i\,(-\lambda_1 + c_2)\,(-\lambda_3 + c_2)}{(\lambda_2 - \lambda_3)\,(\lambda_1 - \lambda_2)\,Lc_4^*\,gc_3\,c_2}\Phi_2 + \frac{iLg\,(|c_4|)^2 + i\,(-\lambda_1 + c_2)\,(-\lambda_2 + c_2)}{(\lambda_2 - \lambda_3)\,(\lambda_1 - \lambda_3)\,Lc_4^*\,gc_3\,c_2}\Phi_3\,.$$

Further, calculate the term

$$fc_3^*\,\bar{\Phi}_1(x) - igc_4^*\bar{\Phi}_2(x) = L_1\Phi_1(x) + L_2\Phi_2(x) + L_3\Phi_3(x) = \Phi(x)\,, \tag{12.74}$$

where expressions for L_i turn to be rather symmetrical:

$$\begin{aligned}
L_1 &= \frac{gL|c_4|^2 + fL|c_3|^2 + {c_2}^2 - c_2(\lambda_2 + \lambda_3) + \lambda_2\,\lambda_3}{Lc_2\,c_3(\lambda_1 - \lambda_2)(\lambda_1 - \lambda_3)}\,,\\
L_2 &= \frac{gL|c_4|^2 + fL|c_3|^2 + {c_2}^2 - c_2(\lambda_3 + \lambda_1) + \lambda_3\,\lambda_1}{Lc_2\,c_3(\lambda_2 - \lambda_3)(\lambda_2 - \lambda_1)}\,,\\
L_3 &= \frac{gL|c_4|^2 + fL|c_3|^2 + {c_2}^2 - c_2(\lambda_1 + \lambda_2) + \lambda_1\,\lambda_2}{Lc_2\,c_3(\lambda_3 - \lambda_1)(\lambda_3 - \lambda_2)}\,.
\end{aligned} \tag{12.75}$$

Recall the notation $L = (c_1 + c_3/\sqrt{6})$; as shown above, positive value for three masses appear only when $f = -1$ and $g = -1$. With notations

$$Y_i = \frac{4c_2c_3}{3M}(\lambda_i - c_2), \quad i = 1, 2, 3\,,$$

eq. (12.73) may be presented as

$$\begin{aligned}
&\hat{D}\Phi_1(x) + M_1\Phi_1(x) + Y_1\,\Sigma(x)\Phi(x) = 0\,,\\
&\hat{D}\Phi_2(x) + M_2\Phi_2(x) + Y_2\,\Sigma(x)\Phi(x) = 0\,,\\
&\hat{D}\Phi_3(x) + M_3\Phi_3(x) + Y_3\,\Sigma(x)\Phi(x) = 0\,,
\end{aligned} \tag{12.76}$$

where $\Sigma(x) = -ieF_{ab}(x)\sigma_{ab}$.

12.9 Extension to general relativity

In order to follow the extension of the model from flat Minkowski space to any Riemannian space-time, we should turn back to the system (12.53)–(12.55) and make several simple modifications to it.

1. Taking in mind that in Minkowski space the *ict*-metric was used; however, in Riemannian space we use the metric $g_{\alpha\beta}(x)$, related to signature $(+,-,-,-)$, we must make the following change:

$$M \Longrightarrow iM\,. \tag{12.77}$$

2. Now Dirac matrices in spinor basis are determined by the formulas

$$\gamma^0 = \begin{vmatrix} 0 & I \\ I & 0 \end{vmatrix}, \quad \gamma^i = \begin{vmatrix} 0 & -\sigma^i \\ \sigma^i & 0 \end{vmatrix}. \tag{12.78}$$

3. Derivatives are modified according to the rules [57]

$$\begin{gathered} D_\alpha(x) = \partial_\alpha + ieA_\alpha(x) \Longrightarrow \\ D_\alpha(x) = \nabla_\alpha + \Gamma_\alpha(x) + ieA_\alpha(x)\,, \quad \hat{D} = \gamma^\alpha(x) D_\alpha(x)\,, \end{gathered} \tag{12.79}$$

where $\Gamma_\alpha(x)$ is bispinor connection by Tetrode-Weyl-Fock-Ivanenko [57], and $\gamma^\alpha(x) = \gamma^a e^\alpha_{(a)}(x)$.

4. Note important commutation rules [57]:

$$\begin{gathered} \gamma^\rho(x) D_\beta = D_\beta \gamma^\rho(x)\,, \qquad D_\sigma(x) g_{\alpha\beta}(x) = g_{\alpha\beta}(x) D_\sigma(x)\,, \\ \hat{D}\hat{D} = D_\alpha D_\beta\Big(\frac{\gamma^\alpha\gamma^\beta + \gamma^\beta\gamma^\alpha}{2} + \frac{\gamma^\alpha\gamma^\beta - \gamma^\beta\gamma^\alpha}{2}\Big) = \Box - \Sigma(x)\,, \\ D^2 = D^\alpha D_\alpha, \quad \Sigma(x) = (-ieF_{\alpha\beta}\sigma^{\alpha\beta}(x) + \frac{R}{4})\,, \end{gathered} \tag{12.80}$$

where $R(x)$ is the Ricci scalar.

After that all above analysis but now for the modified system (12.53)–(12.55):

$$\begin{gathered} c_2 \hat{D}\Psi_0 - \frac{4ic_4}{\sqrt{6}}\Big[(D^\mu \Psi_\mu) - \frac{1}{4}\hat{D}(\gamma^\mu(x)\Psi_\mu)\Big] \\ + iM\Psi_0 = 0\,, \end{gathered} \tag{12.81}$$

$$(c_1 + \frac{c_3}{\sqrt{6}})\hat{D}(\gamma^\mu(x)\Psi_\mu) - \frac{4c_3}{\sqrt{6}}(D^\mu\Psi_\mu) + iM(\gamma^\mu(x)\Psi_\mu) = 0\,, \tag{12.82}$$

$$\begin{gathered} + \frac{iM}{4}\frac{\sqrt{6}fc_2c_3^* - c_2(c_1 + c_3/\sqrt{6}) - g|c_4|^2}{c_2(c_1 + c_3/\sqrt{6})}\,\hat{D}(\gamma^\mu(x)\Psi_\mu) - i\frac{\sqrt{6}}{4}iM\,\frac{gc_4^*}{c_2}\,\hat{D}\Psi_0 \\ \frac{fc_2|c_3|^2 + gc_1|c_4|^2}{c_2(c_1 + c_3/\sqrt{6})}\,\hat{D}(D^\mu\Psi_\mu) - i\frac{2}{\sqrt{6}}\,gc_1^*\,(-ieF_{\lambda\rho}\sigma_{\lambda\rho}(x) + \frac{R}{4})\,\Psi_0 \\ + \frac{2}{\sqrt{6}}fc_3^*\,(-ieF_{\lambda\rho}\sigma_{\lambda\rho} + \frac{R}{4})\,(\gamma^\mu(x)\Psi_\mu) + iM\,(D^\mu\Psi_\mu) = 0 \end{gathered} \tag{12.83}$$

is repeated with no substantial changes, and we arrive at the generally covariant system

$$\begin{gathered} i\hat{D}(x)\Phi_1(x) - M_1\Phi_1(x) + iY_1\,\Sigma(x)\Phi(x) = 0\,, \\ i\hat{D}(x)\Phi_2(x) - M_2\Phi_2(x) + iY_2\,\Sigma(x)\Phi(x) = 0\,, \\ i\hat{D}(x)\Phi_3(x) - M_3\Phi_3(x) + iY_3\,\Sigma(x)\Phi(x) = 0\,, \end{gathered} \tag{12.84}$$

where

$$\begin{gathered} \hat{D}(x) = \gamma^\alpha(x) D_\alpha\,, \quad D_\alpha(x) = \nabla_\alpha + \Gamma_\alpha(x) + ieA_\alpha(x)\,, \\ Y_i = \frac{4c_2c_3}{3iM}(\lambda_i - c_2), \quad \Phi = L_1\Phi_1(x) + L_2\Phi_2(x) + L_3\Phi_3(x)\,, \\ \Sigma(x) = -ieF_{\lambda\rho}(x)\sigma_{\lambda\rho}(x) + \frac{R(x)}{4}\,. \end{gathered} \tag{12.85}$$

Note that because bispinors are scalars with respect to general relativity, the covariant derivative ∇_α acts as an ordinary one ∂_α.

One important remark should be made. Evidently, that the system (12.84)–(12.85) allows for restriction to Majorana case (that is when $e = 0$), then instead of eqs. (12.84)–(12.85) we have

$$\begin{aligned} &i\gamma^\alpha(x)(\nabla_\alpha + \Gamma_\alpha)\Phi_1(x) - M_1\Phi_1(x) + Y_1' \frac{R(x)}{4}\Phi(x) = 0\,, \\ &i\gamma^\alpha(x)(\nabla_\alpha + \Gamma_\alpha)\Phi_2(x) - M_2\Phi_2(x) + Y_2' \frac{R(x)}{4}\Phi(x) = 0\,, \\ &i\gamma^\alpha(x)(\nabla_\alpha + \Gamma_\alpha)\Phi_3(x) - M_3\Phi_3(x) + Y_3' \frac{R(x)}{4}\Phi(x) = 0\,, \end{aligned} \tag{12.86}$$

where

$$Y_i' = \frac{4c_2c_3}{3M}(\lambda_i - c_2), \quad \Phi = L_1\Phi_1(x) + L_2\Phi_2(x) + L_3\Phi_3(x)\,, \tag{12.87}$$

In any Majorana basis, the properties hold $[i\gamma^\alpha(x)]^* = i\gamma^\alpha(x)$ $[\Gamma_\alpha(x)]^* = \Gamma_\alpha(x)$, therefore bispinors $\Phi_1(x), \Phi_2(x), \Phi_3(x)$, and $\Phi(x)$ may be real or imaginary.

12.10 Model example

For simplicity let us follow 1-dimensional case $(t, x, y = 0, z = 0)$. So we start with the system (the curved background is taken into account by the constant Ricci scalar R)

$$\begin{aligned} &(i\gamma^0\partial_t + i\gamma^1\partial_1 - M_1)\Phi_1 + d_1(L_1\Phi_1 + L_2\Phi_2 + L_3\Phi_3) = 0\,, \\ &(i\gamma^0\partial_t + i\gamma^1\partial_1 - M_2)\Phi_2 + d_2(L_1\Phi_1 + L_2\Phi_2 + L_3\Phi_3) = 0\,, \\ &(i\gamma^0\partial_t + i\gamma^1\partial_1 - M_3)\Phi_3 + d_3(L_1\Phi_1 + L_2\Phi_2 + L_3\Phi_3) = 0\,, \end{aligned} \tag{12.88}$$

where $d_i = Y_i\,\frac{R}{4}$, $i = 1, 2, 3$. The system (12.88) is transformed to the matrix form

$$(i\gamma^0\partial_t + i\gamma^1\partial_1 - M)\left|\begin{array}{c} \Phi_1 \\ \Phi_2 \\ \Phi_3 \end{array}\right|$$

$$= -\left|\begin{array}{ccc} M - M_1 + d_1L_1 & d_1L_2 & d_1L_3 \\ d_2L_1 & M - M_2 + d_2L_2 & d_2L_3 \\ d_3L_1 & d_3L_2 & M - M_3 + d_3L_3 \end{array}\right| \left|\begin{array}{c} \Phi_1 \\ \Phi_2 \\ \Phi_3 \end{array}\right|,$$

or in brief

$$\Delta\,\Phi = T\,\Phi\,, \quad \Delta = -(i\gamma^0\partial_t + i\gamma^1\partial_1 - M)\,, \tag{12.89}$$

where

$$T = \left|\begin{array}{ccc} M - M_1 + d_1L_1 & d_1L_2 & d_1L_3 \\ d_2L_1 & M - M_2 + d_2L_2 & d_2L_3 \\ d_3L_1 & d_3L_2 & M - M_3 + d_3L_3 \end{array}\right|. \tag{12.90}$$

The 3-column Φ in eq. (12.89) is subject to linear transformation to diagonalise the mixing matrix T:

$$\bar{\Phi} = S\Phi\,, \quad STS^{-1} = T_0 = \begin{vmatrix} \mu_1 & 0 & 0 \\ 0 & \mu_2 & 0 \\ 0 & 0 & \mu_3 \end{vmatrix}\,. \tag{12.91}$$

After that we will have three separate Dirac-like equations with different masses $\bar{M}_1$:

$$\begin{aligned} &[\, i\gamma^0\partial_t + i\gamma^1\partial_1 - \bar{M}_1\,]\,\bar{\Phi}_1 = 0\,, \quad \bar{M}_1 = M + \mu_1\,; \\ &[i\gamma^0\partial_t + i\gamma^1\partial_1 - \bar{M}_2\,]\,\bar{\Phi}_2 = 0\,, \quad \bar{M}_2 = M + \mu_2\,; \\ &[i\gamma^0\partial_t + i\gamma^1\partial_1 - \bar{M}_3\,]\,\bar{\Phi}_3 = 0\,, \quad \bar{M}_3 = M + \mu_3\,, \end{aligned} \tag{12.92}$$

by physical reason we assume the real-valuedness of μ_i, and positiveness of $M + \mu_i$. To find the transformation S, we should solve the equation $ST = T_0S$, where

$$T = \begin{vmatrix} s_{11} & s_{12} & s_{13} \\ s_{21} & s_{22} & s_{23} \\ s_{31} & s_{32} & s_{33} \end{vmatrix}\,.$$

It leads to three linear subsystems

$$\begin{aligned} &(M - M_1 + d_1L_1)\; s_{11} + d_2L_1\; s_{12} + d_3L_1\; s_{13} = \mu_1 s_{11}\,, \\ &d_1L_2\; s_{11} + (M - M_2 + d_2L_2)\; s_{12} + d_3L_2\; s_{13} = \mu_1 s_{12}\,, \\ &d_1L_3\; s_{11} + d_2L_3\; s_{12} + (M - M_3 + d_3L_3)\; s_{13} = \mu_1 s_{13}\,; \end{aligned}$$

$$\begin{aligned} &(M - M_1 + d_1L_1)\; s_{21} + d_2L_1\; s_{22} + d_3L_1\; s_{23} = \mu_2 s_{21}\,, \\ &d_1L_2\; s_{21} + (M - M_2 + d_2L_2)\; s_{22} + d_3L_2\; s_{23} = \mu_2 s_{22}\,, \\ &d_1L_3\; s_{21} + d_2L_3\; s_{22} + (M - M_3 + d_3L_3)\; s_{23} = \mu_2 s_{23}\,; \end{aligned}$$

$$\begin{aligned} &(M - M_1 + d_1L_1)\; s_{31} + d_2L_1\; s_{32} + d_3L_1\; s_{33} = \mu_3 s_{31}\,, \\ &d_1L_2\; s_{31} + (M - M_2 + d_2L_2)\;\; s_{32} + d_3L_2\; s_{33} = \mu_3 s_{32}\,, \\ &d_1L_3\; s_{31} + d_2L_3\; s_{32} + (M - M_3 + d_3L_3)\; s_{33} = \mu_3 s_{33}\,. \end{aligned}$$

Each row of the matrix S may be fixed up to an arbitrary multiplier. For diagonal elements in T_0, we get a cubic algebraic equation

$$\begin{vmatrix} M - M_1 + d_1L_1 - \mu & d_2L_1 & d_3L_1 \\ d_1L_2 & M - M_2 + d_2L_2 - \mu & d_3L_2 \\ d_1L_3 & d_2L_3 & M - M_3 + d_3L_3 - \mu \end{vmatrix} = 0$$

or explicitly

$$\begin{aligned} &-\mu^3 + (3\,M - M_1 - M_2 - M_3 + d_1\,L_1 + d_2\,L_2 + d_3\,L_3)\,\mu^2 \\ &+[-L_1d_1\,(2M - M_2 - M_3) - d_2L_2\,(2M - M_1 - M_3) - d_3\,L_3\,(2M - M_2 - M_1) \\ &-3M^2 + 2\,(M_1 + M_2 + M_3)\,M - (M_1M_2 + M_1M_3 + M_2M_3)]\mu \\ &+L_1d_1(M - M_2)(M - M_3) + L_2d_2(M - M_3)(M - M_1) \\ &+L_3d_3(M - M_1)(M - M_2) + (M - M_1)(M - M_2)(M - M_3) = 0\,. \end{aligned}$$

To get a more simple form of cubic equation, we are to make several steps. The first step is the substitution $M_i = \frac{M}{\lambda_i}$, where M is arbitrary. We may simplify the task without loss of generality by setting $c_1 = c_2 = 1$, then the cubic equation for λ_i becomes simpler

$$\lambda^3 - 2\lambda^2 + (1+k)\lambda - k = 0\,, \quad k = a^2 + b^2,$$

and its roots are

$$\lambda_3 = 1, \quad \lambda_{1,2} = \frac{1}{2} \mp \frac{1}{2}\sqrt{1-4k}, \quad k \in (0, \frac{1}{4})\,. \tag{12.93}$$

Correspondingly, the masses M_i equal to

$$M_3 = M\,, \quad M_{1,2} = \frac{2M}{1 \mp \sqrt{1-4k}}\,, \quad k \in (0, \frac{1}{4})\,. \tag{12.94}$$

Besides, we may reduce explicit expressions for coefficients L_i in the formulas

$$\Phi(x) = L_1\Phi_1(x) + L_2\Phi_2(x) + L_3\Phi_3(x),$$

$$L_1 = \frac{1}{b}\frac{1}{(\lambda_1 - 1)(\lambda_1 - \lambda_2)}\Big\{ -2k + \frac{1}{2L}(-1 + 2\lambda_2\Big\},$$

$$L_2 = \frac{1}{b}\frac{1}{(\lambda_2 - 1)(\lambda_2 - \lambda_1)}\Big\{ -2k + \frac{1}{2L}(-1 + 2\lambda_1\Big\},$$

$$L_3 = \frac{1}{b}\,\frac{1}{(1-\lambda_1)(1-\lambda_2)}\Big\{ -2k + \frac{1}{2L}(1 - 2\lambda_1 - 2\lambda_2 + 4\lambda_1\,\lambda_2)\Big\},$$

where

$$L = 1 + \frac{b}{\sqrt{6}}, \quad (a^2 + b^2) = k < \frac{1}{4}, \quad 0 < 2b < 1\,.$$

In turn, the coefficients d_i are written as

$$d_1 = \frac{Rb}{6M}\,(\lambda_1 - \frac{1}{2}),\; d_2 = \frac{Rb}{6M}\,(\lambda_2 - \frac{1}{2}),\; d_3 = \frac{Rb}{6M}\,(\lambda_3 - \frac{1}{2})\,. \tag{12.95}$$

It is possible to introduce dimensionless parameters (scalar curvature R and M^2 have the same dimension of meter^{-2});

$$R = 6r\,M^2 \;\Longrightarrow\; d_i = M\;rb(\lambda_i - \frac{1}{2}) = M\;D_i.$$

Finally, the roots may be done dimensionless

$$\mu_i = M\;\Delta_i, \qquad \Delta_i \;\text{ is dimensionless.} \tag{12.96}$$

In this way, we arrive at the following cubic equation for Δ_i:

$$\Delta^3 + \frac{2k-1}{k}\Delta^2 + \Big[1 + \Big(1 - \frac{\sqrt{6}}{2(\sqrt{6}+b)}\,\frac{1-4k}{2k}\Big)r\Big]\Delta$$
$$+\Big(1 + \frac{\sqrt{6}}{2(\sqrt{6}+b)}\,\frac{1-4k}{2k}\Big)r = 0\,, \tag{12.97}$$

where $0 < k < \frac{1}{4}, 0 < b < 2$. Taking in mind the relationship $R = 6rM^2$, we expect dimensionless parameter r is small because the effects of geometry in the model under consideration should be small. Besides, there exist two physically different possibilities:

$r > 0$ at positive curvature, and $r < 0$ at negative curvature. We have followed several cases of weak and strong gravitation of different curvature sings:

$$\begin{aligned} &r = +10^{-30}\,, r = +10^{-5}, r = +10^{-3}, r = +10^{-2}, r = +1\,; \\ &r = -10^{-30}\,, r = -10^{-5}, r = -10^{-3}, r = -10^{-2}, r = -1\,. \end{aligned} \tag{12.98}$$

The cases $r = \pm 10^{-2}, \pm 1$ correspond to a very strong curvature of space. A numerical study showed that the dependence of the roots Δ_i upon parameter $b \in (0, 2)$ is very inappreciable, by this reason below we take the value $b = 0$.

$\underline{b = 0,\ r = -10^{-5},}$

$$\begin{array}{llll}
k = 0.23, & \Delta_1 = 0.0000109, & \Delta_2 = 0.559 & \Delta_3 = 1.789\,; \\
k = 0.22, & \Delta_1 = 0.0000114, & \Delta_2 = 0.485 & \Delta_3 = 2.060\,; \\
k = 0.21, & \Delta_1 = 0.0000119, & \Delta_2 = 0.429 & \Delta_3 = 2.333\,; \\
k = 0.20, & \Delta_1 = 0.0000125, & \Delta_2 = 0.382 & \Delta_3 = 2.618\,; \\
k = 0.18, & \Delta_1 = 0.0000139, & \Delta_2 = 0.308 & \Delta_3 = 3.248\,; \\
k = 0.16, & \Delta_1 = 0.0000156, & \Delta_2 = 0.250 & \Delta_3 = 4.000\,; \\
k = 0.14, & \Delta_1 = 0.0000179, & \Delta_2 = 0.202 & \Delta_3 = 4.940\,; \\
k = 0.12, & \Delta_1 = 0.0000208, & \Delta_2 = 0.162 & \Delta_3 = 6.171\,; \\
k = 0.10, & \Delta_1 = 0.0000250, & \Delta_2 = 0.127 & \Delta_3 = 7.873\,; \\
k = 0.08, & \Delta_1 = 0.0000313\,, & \Delta_2 = 0.096, & \Delta_3 = 10.404\,; \\
k = 0.06, & \Delta_1 = 0.0000417, & \Delta_2 = 0.068, & \Delta_3 = 14.598\,; \\
k = 0.04, & \Delta_1 = 0.0000626, & \Delta_2 = 0.043 & \Delta_3 = 22.956\,; \\
k = 0.02, & \Delta_1 = 0.0001257, & \Delta_2 = 0.021, & \Delta_3 = 47.979\,.
\end{array}$$

$\underline{b = 0,\ r = +10^{-5},}$

$$\begin{array}{llll}
k = 0.23, & \Delta_1 = 0.559, & \Delta_2 = 1.788, & \Delta_3 = -0.00001\,; \\
k = 0.22, & \Delta_1 = 0.485, & \Delta_2 = 2.060, & \Delta_3 = -0.00001\,; \\
k = 0.21, & \Delta_1 = 0.429, & \Delta_2 = 2.333, & \Delta_3 = -0.00001\,; \\
k = 0.20, & \Delta_1 = 0.382, & \Delta_2 = 2.618, & \Delta_3 = -0.00001\,; \\
k = 0.18, & \Delta_1 = 0.308, & \Delta_2 = 2.248, & \Delta_3 = -0.00001\,; \\
k = 0.16, & \Delta_1 = 0.250, & \Delta_2 = 3.999, & \Delta_3 = -0.00002\,; \\
k = 0.12, & \Delta_1 = 0.162, & \Delta_2 = 6.171, & \Delta_3 = -0.00002\,; \\
k = 0.10, & \Delta_1 = 0.127, & \Delta_2 = 7.873, & \Delta_3 = -0.00002\,; \\
k = 0.08, & \Delta_1 = 0.096, & \Delta_2 = 10.404, & \Delta_3 = -0.00003\,; \\
k = 0.06, & \Delta_1 = 0.069, & \Delta_2 = 14.598, & \Delta_3 = -0.00004\,; \\
k = 0.04, & \Delta_1 = 0.044, & \Delta_2 = 22.956, & \Delta_3 = -0.00006\,; \\
k = 0.02, & \Delta_1 = 0.021, & \Delta_2 = 47.979, & \Delta_3 = -0.00012\,.
\end{array}$$

At the change of curvature sign, one root Δ_3 becomes negative, though very small. Let the curvature increase:

$\underline{b = 0,\ r = -10^{-3},}$

$$\begin{array}{llll}
k = 0.23, & \Delta_1 = 0.001, & \Delta_2 = 0.557, & \Delta_3 = 1.790\,; \\
k = 0.22, & \Delta_1 = 0.001, & \Delta_2 = 0.483, & \Delta_3 = 2.061\,; \\
k = 0.21, & \Delta_1 = 0.001, & \Delta_2 = 0.427, & \Delta_3 = 2.334\,; \\
k = 0.20, & \Delta_1 = 0.001, & \Delta_2 = 0.380, & \Delta_3 = 2.619\,; \\
k = 0.18, & \Delta_1 = 0.001, & \Delta_2 = 0.306, & \Delta_3 = 3.248\,; \\
k = 0.16, & \Delta_1 = 0.002, & \Delta_2 = 0.248, & \Delta_3 = 4.000\,; \\
k = 0.14, & \Delta_1 = 0.002, & \Delta_2 = 0.200, & \Delta_3 = 4.941\,; \\
k = 0.12, & \Delta_1 = 0.002, & \Delta_2 = 0.160, & \Delta_3 = 6.171\,; \\
k = 0.10, & \Delta_1 = 0.003, & \Delta_2 = 0.124, & \Delta_3 = 7.873\,; \\
k = 0.08, & \Delta_1 = 0.003, & \Delta_2 = 0.093, & \Delta_3 = 10.404\,; \\
k = 0.06, & \Delta_1 = 0.004, & \Delta_2 = 0.064, & \Delta_3 = 14.598\,; \\
k = 0.04, & \Delta_1 = 0.008, & \Delta_2 = 0.036, & \Delta_3 = 22.956\,; \\
k = 0.02, & \Delta_1 = 47.98, & \Delta_2 = 0.0105 + 0.01i, & \Delta_3 = 0.0105 - 0.01i\,;
\end{array}$$

There appear physically senseless complex roots.

$\underline{b = 0,\ r = +10^{-3},}$

$$
\begin{array}{llll}
k = 0.23, & \Delta_1 = 0.561, & \Delta_2 = 1.788, & \Delta_3 = -0.001\,; \\
k = 0.22, & \Delta_1 = 0.487, & \Delta_2 = 2.059, & \Delta_3 = -0.001\,; \\
k = 0.21, & \Delta_1 = 0.430, & \Delta_2 = 2.333, & \Delta_3 = -0.001\,; \\
k = 0.20, & \Delta_1 = 0.384, & \Delta_2 = 2.617, & \Delta_3 = -0.001\,; \\
k = 0.18, & \Delta_1 = 0.310, & \Delta_2 = 3.247, & \Delta_3 = -0.001\,; \\
k = 0.16, & \Delta_1 = 0.252, & \Delta_2 = 3.999, & \Delta_3 = -0.002\,; \\
k = 0.14, & \Delta_1 = 0.204, & \Delta_2 = 4.940, & \Delta_3 = -0.002\,; \\
k = 0.12, & \Delta_1 = 0.164, & \Delta_2 = 6.171, & \Delta_3 = -0.002\,; \\
k = 0.10, & \Delta_1 = 0.129, & \Delta_2 = 7.873, & \Delta_3 = -0.002\,; \\
k = 0.08, & \Delta_1 = 0.099, & \Delta_2 = 10.404, & \Delta_3 = -0.003\,; \\
k = 0.06, & \Delta_1 = 0.072, & \Delta_2 = 14.598, & \Delta_3 = -0.003\,; \\
k = 0.04, & \Delta_1 = 0.049, & \Delta_2 = 22.957, & \Delta_3 = -0.005\,; \\
k = 0.02, & \Delta_1 = 0.029, & \Delta_2 = 47.979, & \Delta_3 = -0.008\,;
\end{array}
$$

The root Δ_3 becomes negative, though small.

$\underline{b = 0,\ r = -10^{-2},}$

$$
\begin{array}{llll}
k = 0.23, & \Delta_1 = 0.011, & \Delta_2 = 0.536, & \Delta_3 = 1.801\,; \\
k = 0.22, & \Delta_1 = 0.012, & \Delta_2 = 0.465, & \Delta_3 = 2.069\,; \\
k = 0.20, & \Delta_1 = 0.013, & \Delta_2 = 0.363, & \Delta_3 = 2.624\,; \\
k = 0.18, & \Delta_1 = 0.015, & \Delta_2 = 0.290, & \Delta_3 = 3.251\,; \\
k = 0.16, & \Delta_1 = 0.017, & \Delta_2 = 0.231, & \Delta_3 = 4.002\,; \\
k = 0.14, & \Delta_1 = 0.020, & \Delta_2 = 0.181, & \Delta_3 = 4.942\,; \\
k = 0.12, & \Delta_1 = 0.025, & \Delta_2 = 0.137, & \Delta_3 = 6.172\,; \\
k = 0.10, & \Delta_1 = 0.034, & \Delta_2 = 0.093, & \Delta_3 = 7.873\,; \\
k = 0.08, & \Delta_1 = 10.403, & \Delta_2 = 0.485 + 0.256i, & \Delta_3 = 0.485 - 0.256i\,; \\
k = 0.06, & \Delta_1 = 14.597, & \Delta_2 = 0.349 + 0.405i, & \Delta_3 = 0.349 - 0.405i\,; \\
k = 0.04, & \Delta_1 = 22.955, & \Delta_2 = 0.226 + 0.470i, & \Delta_3 = 0.226 - 0.470i\,; \\
k = 0.02, & \Delta_1 = 47.977, & \Delta_2 = 0.115 + 0.497i, & \Delta_3 = 0.115 - 0.497i\,; \\
k = 0.01, & \Delta_1 = 97.987, & \Delta_2 = 0.626 + 0.501i, & \Delta_3 = 0.626 - 0.501i\,.
\end{array}
$$

We can see complex roots.

$\underline{b = 0,\ r = +10^{-2},}$

$$
\begin{array}{llll}
k = 0.24, & \Delta_1 \simeq 0.697, & \Delta_2 = 1.480, & \Delta_3 = -0.010\,; \\
k = 0.23, & \Delta_1 \simeq 0.582, & \Delta_2 = 1.776, & \Delta_3 = -0.011\,; \\
k = 0.22, & \Delta_1 \simeq 0.505, & \Delta_2 = 2.051, & \Delta_3 = -0.011\,; \\
k = 0.21, & \Delta_1 \simeq 0.447, & \Delta_2 = 2.326, & \Delta_3 = -0.011\,; \\
k = 0.20, & \Delta_1 \simeq 0.399, & \Delta_2 = 2.613, & \Delta_3 = -0.012\,; \\
k = 0.18, & \Delta_1 \simeq 0.325, & \Delta_2 = 3.244, & \Delta_3 = -0.013\,; \\
k = 0.16, & \Delta_1 \simeq 0.267, & \Delta_2 = 3.998, & \Delta_3 = -0.015\,; \\
k = 0.14, & \Delta_1 \simeq 0.220, & \Delta_2 = 4.939, & \Delta_3 = -0.016\,; \\
k = 0.12, & \Delta_1 \simeq 0.181, & \Delta_2 = 6.171, & \Delta_3 = -0.019\,; \\
k = 0.10, & \Delta_1 \simeq 0.148, & \Delta_2 = 7.873, & \Delta_3 = -0.021\,; \\
k = 0.08, & \Delta_1 \simeq 0.120, & \Delta_2 = 10.405, & \Delta_3 = -0.025\,; \\
k = 0.06, & \Delta_1 \simeq 0.097, & \Delta_2 = 14.599, & \Delta_3 = -0.030\,; \\
k = 0.04, & \Delta_1 \simeq 0.077, & \Delta_2 = 22.958, & \Delta_3 = -0.035\,; \\
k = 0.02, & \Delta_1 \simeq 0.061, & \Delta_2 = 47.981, & \Delta_3 = -0.043\,; \\
k = 0.01, & \Delta_1 \simeq 0.055, & \Delta_2 = 97.992, & \Delta_3 = -0.047\,.
\end{array}
$$

The root Δ_3 becomes negative.

If the curvature increases further, we see complex roots at both negative and positive curvatures

<u>$b = 0,\ r = +1,$</u>

$$
\begin{array}{llll}
k = 0.24, & \Delta_1 = 1.265 + 1.13i, & \Delta_2 = -0.362, & \Delta_3 = 1.265 - 1.13i\,; \\
k = 0.18, & \Delta_1 = 1.173, & \Delta_2 = 2.804, & \Delta_3 = -0.422\,; \\
k = 0.12, & \Delta_1 = 0.695, & \Delta_2 = 6.128, & \Delta_3 = -0.489\,; \\
k = 0.06, & \Delta_1 = 0.503, & \Delta_2 = 14.727, & \Delta_3 = -0.563\,; \\
k = 0.01, & \Delta_1 = 0.406, & \Delta_2 = 98.221, & \Delta_3 = -0.627\,.
\end{array}
$$

<u>$b = 0,\ r = -1,$</u>

$$
\begin{array}{llll}
k = 0.24, & \Delta_1 = 2.339, & \Delta_2 = -0.86 + 0.66i, & \Delta_3 = -0.86 - 0.66i\,; \\
k = 0.18, & \Delta_1 = 3.556, & \Delta_2 = -0.247 + 0.63i, & \Delta_3 = -0.24 - 0.63i\,; \\
k = 0.12, & \Delta_1 = 6.213, & \Delta_2 = 0.60 + 0.58i, & \Delta_3 = 0.60 - 0.58i\,; \\
k = 0.06, & \Delta_1 = 14.468, & \Delta_2 = 0.996 + 0.53i, & \Delta_3 = 0.996 - 0.53i\,; \\
k = 0.01, & \Delta_1 = 97.757, & \Delta_2 = 0.12 + 0.49i, & \Delta_3 = 0.12 - 0.49i\,.
\end{array}
$$

This means that at these values of curvature, the model becomes non-interpreted.

Now we consider the general structure and particular examples of transformation S, with the help of which the mixing matrix in eq. (12.88) reduces to diagonal form. Because the rows of the relevant matrix are fixed up to arbitrary multipliers, we may construct them within the following form

$$
\bar{\Phi}_i = S_{ij}\Phi_j, \quad S = \begin{vmatrix} s_{11} & s_{12} & 1 \\ s_{21} & s_{22} & 1 \\ s_{31} & s_{32} & 1 \end{vmatrix}, \quad s_{13} = s_{23} = s_{33} = +1\,;
$$

that is,

$$
\bar{\Phi}_1 = s_{11}\Phi_1 + s_{12}\Phi_2 + \Phi_3\,, \ \bar{\Phi}_2 = s_{21}\Phi_1 + s_{22}\Phi_2 + \Phi_3\,, \ \bar{\Phi}_3 = s_{31}\Phi_1 + s_{32}\Phi_2 + \Phi_3\,. \quad (12.99)
$$

For unknown elements we have three subsystems

$$
\begin{aligned}
(M - M_1 + d_1L_1 - \mu_{(1)}) \cdot s_{(1)1} + d_2L_1 \cdot s_{(1)2} &= -d_3L_1\,, \\
d_1L_2 \cdot s_{(1)1} + (M - M_2 + d_2L_2 - \mu_{(1)}) \cdot s_{(1)2} &= -d_3L_2\,; \\
(M - M_1 + d_1L_1 - \mu_{(2)}) \cdot s_{(2)1} + d_2L_1 \cdot s_{(2)2} &= -d_3L_1\,, \\
d_1L_2 \cdot s_{(2)1} + (M - M_2 + d_2L_2 - \mu_{(2)}) \cdot s_{(2)2} &= -d_3L_2\,; \\
(M - M_1 + d_1L_1 - \mu_{(3)}) \cdot s_{(3)1} + d_2L_1 \cdot s_{(3)2} &= -d_3L_1\,, \\
d_1L_2 \cdot s_{(3)1} + (M - M_2 + d_2L_2 - \mu_{(3)}) \cdot s_{(3_2} &= -d_3L_2\,.
\end{aligned}
$$

For each $\mu = \mu_{(i)}, i = 1, 2, 3$ we have three determinants:

$$
O^{(i)} = \begin{vmatrix} (M - M_1 + d_1L_1 - \mu_i) & d_2L_1 \\ d_1L_2 & (M - M_2 + d_2L_2 - \mu_i) \end{vmatrix},
$$

$$
O_1^{(i)} = \begin{vmatrix} -d_3L_1 & d_2L_1 \\ -d_3L_2 & (M - M_2 + d_2L_2 - \mu_i) \end{vmatrix}, O_2^{(i)} = \begin{vmatrix} (M - M_1 + d_1L_1 - \mu_i) & -d_3L_1 \\ d_1L_2 & -d_3L_2 \end{vmatrix},
$$

and therefore three sets of solutions:

$$
s_{(1)1} = \frac{O_1^{(1)}}{O^{(1)}},\ s_{(1)2} = \frac{O_2^{(1)}}{O^{(1)}},\quad s_{(2)1} = \frac{O_1^{(2)}}{O^{(2)}},\ s_{(2)2} = \frac{O_2^{(2)}}{O^{(2)}},\ s_{(3)1} = \frac{O_1^{(3)}}{O^{(3)}},\ s_{(3)2} = \frac{O_2^{(3)}}{O^{(3)}}\,.
$$

Taking in mind notations from the above $d_i = M\ D_i,\ \mu_i = M\Delta_i$, the determinants may be re-written in dimensionless form

$$O^{(i)} = M^2 \begin{vmatrix} (1-\lambda_1^{-1}+D_1L_1-\Delta_i) & D_2L_1 \\ D_1L_2 & (1-\lambda_2^{-1}+D_2L_2-\Delta_i) \end{vmatrix},$$

$$O_1^{(i)} = M^2 \begin{vmatrix} -D_3L_1 & D_2L_1 \\ -D_3L_2 & (1-\lambda_2^{-1}+D_2L_2-\Delta_i) \end{vmatrix},$$

$$O_2^{(i)} = M^2 \begin{vmatrix} (1-\lambda_1^{-1}+D_1L_1-\Delta_i) & -D_3L_1 \\ D_1L_2 & -D_3L_2 \end{vmatrix}.$$

Recall expressions for involved quantities (in dimensionless form)

$$L_1 = \frac{2}{(1+\sqrt{1-4k})\sqrt{1-4k}}[\,-2k+\frac{1}{2L}\sqrt{1-4k}\,],$$

$$L_2 = \frac{2}{(1-\sqrt{1-4k})\sqrt{1-4k}}[\,+2k+\frac{1}{2L}\sqrt{1-4k}\,],$$

$$L_3 = \frac{1}{k}\,[\,-2k+\frac{1}{2L}(4k-1)\,], \quad L = 1+\frac{b}{\sqrt{6}}, \quad \text{let } b=0,$$

$$D_1 = -\frac{r}{2}\,\sqrt{1-4k}, \quad D_2 = +\frac{r}{2}\,\sqrt{1-4k}, \quad D_3 = \frac{r}{2}.$$

Asymptotical behaviour of $L_i(k)$ is given below:

$$k \to \frac{1}{4}, \quad L_1 \to -\infty, \quad L_2 \to +\infty, \quad L_3 \to -2,$$
$$k \to 0, \quad L_1 \to +0.5, \quad L_2 \to +\infty, \quad L_3 \to -\infty. \tag{12.100}$$

Let us write down several particular matrices from $\Phi = S^{-1}\bar{\Phi}$:

$$k=0.20, r=+10^{-5}, \qquad S^{-1} = \begin{vmatrix} -8693394.736 & 4986681.114 & 3706713.621 \\ -76752.901 & 18059.506 & 58693.395 \\ -2.612 & 2.644 & 0.968 \end{vmatrix},$$

$$k=0.20, r=-10^{-5}, \qquad S^{-1} = \begin{vmatrix} -3864009.458 & 9059949.095 & -5195939.637 \\ -60689.893 & -60689.898 & -20708.455 \\ 0.999 & -2.684 & 2.685 \end{vmatrix},$$

$$k=0.10, r=+10^{-5}, \qquad S^{-1} = \begin{vmatrix} 36454126.28 & -1.850 & -1.797 \\ -12200.458 & 420.705 & 11779.752 \\ -2.080 & 2.071 & 1.008 \end{vmatrix},$$

$$k=0.10, r=-10^{-5}, \qquad S^{-1} = \begin{vmatrix} 17976761.23 & -36475931.28 & 18499170.04 \\ -11796.328 & 12217.625 & -421.297 \\ 1.010 & -2.081 & 2.072 \end{vmatrix},$$

$$k=0.01, r=+10^{-5}, \qquad S^{-1} = \begin{vmatrix} 6331779.188 & -6032189.243 & -299589.945 \\ -0.304 & -2.673 & 2.976 \\ -0.052 & 1.049 & 0.002 \end{vmatrix},$$

$$k=0.01, r=-10^{-5}, \qquad S^{-1} = \begin{vmatrix} 40282943.48 & -34879043.03 & -5403900.450 \\ -1.913 & 10.448 & -8.536 \\ 0.330 & 0.714 & -0.044 \end{vmatrix}.$$

12.11 Plane wave solutions of Majorana type

Let us consider briefly solutions of three separate equations

$$\begin{aligned} &[\, i\gamma^0\partial_t + i\gamma^1\partial_1 - \bar{M}_1 \,]\, \bar{\Phi}_1 = 0\,, \quad \bar{M}_1 = M(1+\Delta_1)\,; \\ &[i\gamma^0\partial_t + i\gamma^1\partial_1 - \bar{M}_2 \,]\, \bar{\Phi}_2 = 0\,, \quad \bar{M}_2 = M(1+\Delta_2)\,; \\ &[i\gamma^0\partial_t + i\gamma^1\partial_1 - \bar{M}_3 \,]\, \bar{\Phi}_3 = 0\,, \quad \bar{M}_3 = M(1+\Delta_3)\,. \end{aligned} \tag{12.101}$$

It is convenient to employ a fixed Majorana basis for Dirac matrices

$$\gamma^0_M = \left| \begin{array}{cccc} 0 & -i & 0 & 0 \\ i & 0 & 0 & 0 \\ 0 & 0 & 0 & i \\ 0 & 0 & -i & 0 \end{array} \right|, \quad \gamma^1_M = \left| \begin{array}{cccc} -i & 0 & 0 & 0 \\ 0 & i & 0 & 0 \\ 0 & 0 & -i & 0 \\ 0 & 0 & 0 & i \end{array} \right|,$$

$$\gamma^2_M = \left| \begin{array}{cccc} 0 & 0 & 0 & i \\ 0 & 0 & -i & 0 \\ 0 & -i & 0 & 0 \\ i & 0 & 0 & 0 \end{array} \right|, \quad \gamma^3_M = \left| \begin{array}{cccc} 0 & i & 0 & 0 \\ i & 0 & 0 & 0 \\ 0 & 0 & 0 & i \\ 0 & 0 & i & 0 \end{array} \right|.$$

We search solutions in the form

$$\bar{\Phi}_1 = e^{-iEt+ikx} \left| \begin{array}{c} A_1 \\ A_2 \\ A_3 \\ A_4 \end{array} \right|, \qquad \bar{\Phi}_2 = e^{-iEt+ikx} \left| \begin{array}{c} B_1 \\ B_2 \\ B_3 \\ B_4 \end{array} \right|, \qquad \bar{\Phi}_3 = e^{-iEt+ikx} \left| \begin{array}{c} C_1 \\ C_2 \\ C_3 \\ C_4 \end{array} \right|,$$

complex numerics are decomposed into real and imaginary parts:

$$A_i = a_i + ia'_i, \qquad B_i = b_i + ib'_i, \qquad C_i = c_i + ic'_i\,. \tag{12.102}$$

Because eqs. (12.101) have the same structure, it suffices to follow one of them

$$[\, i\gamma^0\partial_t + i\gamma^1\partial_1 - \bar{M}_1]\, \bar{\Phi}_1 = 0\,. \tag{12.103}$$

From eq. (12.103) we get an algebraic system

$$\left| \begin{array}{cccc} k - i\bar{M}_1 & -E & 0 & 0 \\ E & -k - i\bar{M}_1 & 0 & 0 \\ 0 & 0 & k - i\bar{M}_1 & E \\ 0 & 0 & -E & -k - i\bar{M}_1 \end{array} \right| \left| \begin{array}{c} a_1 + ia'_1 \\ a_2 + ia'_2 \\ a_3 + ia'_3 \\ a_4 + ia'_4 \end{array} \right| = 0\,. \tag{12.104}$$

Let us diagonalise the known helicity operator

$$(\gamma^2\gamma^3\partial_1)\, \bar{\phi}_1 = \sigma\, \bar{\phi}_1\,; \tag{12.105}$$

its eigenvectors are

$$\bar{\Phi}_1(\sigma = +k) = (\cos\varphi - i\sin\varphi) \left| \begin{array}{c} (a_1 + ia'_1) \\ (a_2 + ia'_2) \\ +i(a_1 + ia'_1) \\ -i(a_2 + ia'_2) \end{array} \right|,$$

$$\bar{\Phi}_1(\sigma = -k) = (\cos\varphi - i\sin\varphi) \left| \begin{array}{c} (a_1 + ia'_1) \\ (a_2 + ia'_2) \\ -i(a_1 + ia'_1) \\ +i(a_2 + ia'_2) \end{array} \right|.$$

Taking into account these restrictions in the Dirac system (12.104), we obtain Dirac solutions with fixed helicities (let $k = +\sqrt{E^2 - \bar{M}^2}$):

$$\bar{\Phi}_{\sigma=+k} = (\cos\varphi - i\sin\varphi) \left| \begin{array}{c} A_1 \\ \frac{k-i\bar{M}_1}{E} A_1 \\ +iA_1 \\ -i\frac{k-i\bar{M}_1}{E} A_1 \end{array} \right| ,$$

$$\bar{\Phi}_{\sigma=-k} = (\cos\varphi - i\sin\varphi) \left| \begin{array}{c} A_1 \\ \frac{k-i\bar{M}_1}{E} A_1 \\ -iA_1 \\ +i\frac{k-i\bar{M}_1}{E} A_1 \end{array} \right| . \tag{12.106}$$

Each of them is decomposed into Majorana constituents $\bar{\Phi}^{\pm}$:

$$\bar{\Phi}^{+}_{\sigma=+k} = \left| \begin{array}{c} (a_1\cos\varphi + a_1'\sin\varphi) \\ E^{-}[\,(k\cos\varphi - \bar{M}_1\sin\varphi)a_1 + (\bar{M}_1\cos\varphi + k\sin\varphi)a_1'\,] \\ +(-a_1'\cos\varphi + a_1\sin\varphi) \\ E^{-1}[\,-(\bar{M}_1\cos\varphi + k\sin\varphi)a_1 + (k\cos\varphi - \bar{M}_1\sin\varphi)a_1'\,] \end{array} \right| ,$$

$$\bar{\Phi}^{-}_{\sigma=+k} = i \left| \begin{array}{c} (a_1'\cos\varphi - a_1\sin\varphi) \\ E^{-}[\,-(\bar{M}_1\cos\varphi + k\sin\varphi)a_1 + (k\cos\varphi - \bar{M}_1\sin\varphi)a_1'\,] \\ +(a_1\cos\varphi + a_1'\sin\varphi) \\ -E^{-1}[\,(k\cos\varphi - \bar{M}_1\sin\varphi)a_1 + (\bar{M}_1\cos\varphi + k\sin\varphi)a_1'\,] \end{array} \right| .$$

and

$$\bar{\Phi}^{+}_{\sigma=-k} = \left| \begin{array}{c} (a_1\cos\varphi + a_1'\sin\varphi) \\ E^{-}[\,(k\cos\varphi - \bar{M}_1\sin\varphi)a_1 + (\bar{M}_1\cos\varphi + k\sin\varphi)a_1'\,] \\ -(-a_1'\cos\varphi + a_1\sin\varphi) \\ -E^{-1}[\,-(\bar{M}_1\cos\varphi + k\sin\varphi)a_1 + (k\cos\varphi - \bar{M}_1\sin\varphi)a_1'\,] \end{array} \right| ,$$

$$\bar{\Phi}^{-}_{\sigma=-k} = i \left| \begin{array}{c} (a_1'\cos\varphi - a_1\sin\varphi) \\ E^{-}[\,-(\bar{M}_1\cos\varphi + k\sin\varphi)a_1 + (k\cos\varphi - \bar{M}_1\sin\varphi)a_1'\,] \\ -(a_1\cos\varphi + a_1'\sin\varphi) \\ E^{-1}[\,(k\cos\varphi - \bar{M}_1\sin\varphi)a_1 + (\bar{M}_1\cos\varphi + k\sin\varphi)a_1'\,] \end{array} \right| .$$

Solutions for all three fields from eq. (12.92) are similar, they differ only in the mass parameters $\bar{M}_i$.

Let us recall that with the use of the transformation $S = (s_{ij})$, these Majorana solutions can be decomposed in linear combinations in terms of solutions with physical masses M_1, M_2, M_3:

$$\begin{aligned} \bar{\Phi}_1 &= s_{(1)1}\Phi_1^{M_1} + s_{(1)2}\Phi_2^{M_2} + \Phi_3^{M_3} , \\ \bar{\Phi}_1 &= s_{(2)1}\Phi_1^{M_1} + s_{(2)2}\Phi_2^{M_2} + \Phi_3^{M_3} , \\ \bar{\Phi}_1 &= s_{(3)1}\Phi_1^{M_1} + s_{(3)2}\Phi_2^{M_2} + \Phi_3^{M_3} . \end{aligned} \tag{12.107}$$

In turn, the Majorana states with physical; masses can be decomposed in combinations of states with modified masses:

$$\begin{aligned} \Phi_1^{M_1} &= s_{11}^{-1}\bar{\Phi}_1 + s_{12}^{-1}\bar{\Phi}_2 + s_{13}^{-1}\bar{\Phi}_3 , \\ \Phi_2^{M_2} &= s_{21}^{-1}\bar{\Phi}_1 + s_{22}^{-1}\bar{\Phi}_2 + s_{23}^{-1}\bar{\Phi}_3 , \\ \Phi_3^{M_3} &= s_{31}^{-1}\bar{\Phi}_1 + s_{32}^{-1}\bar{\Phi}_2 + s_{33}^{-1}\bar{\Phi}_3 , \end{aligned} \tag{12.108}$$

The used approximations when an external cosmological background is taken into account by a constant Ricci parameter and the Cartesian coordinates are used, are not necessary. Both of these simplifications may be withdrawn. For instance, we might take de Sitter's cosmological background (of the 1-st or 2-nd type) and apply de Sitter's static coordinates.

12.12 Conclusions

In this chapter, starting from the general Gel'fand-Yaglom approach, a new wave equation for spin 1/2 fermion, which is characterised by three mass parameters, is derived. On the basis of the 20-component wave function, three auxiliary bispinors are introduced. In the absence of an external field, these bispinors obey three separate Dirac-like equations with different masses M_1, M_2, M_3. It is shown that in the presence of external fields, electromagnetic fields, or gravitational non-Euclidean background with non-vanishing Ricci scalar curvature. The main equation is not split into separate equations; instead, a quite definite mixing of three Dirac-like equations arises. It is shown that a generalised equation for Majorana particle with three mass parameters exists as well; such a generalised Majorana equation is not split into three separate equations in any curved background with a nonzero Ricci scalar.

Bibliography

[1] V.L. Ginzburg and Ya.A. Smorodinsky. On wave equations for particles with variable spin. *Zhurnal Eksperimental'noi i Teoreticheskoi Fiziki*, 13: 274, 1943.(in Russian).

[2] V.L. Ginzburg. To the theory of exited states of elementary particles. *Zhurnal Eksperimental'noi i Teoreticheskoi Fiziki*, 13: 33–58, 1943 (in Russian).

[3] A.S. Davydov. Wave equations of a particle having spin 3/2 in absence of field. *Zhurnal Eksperimental'noi i Teoreticheskoi Fiziki*, 13(9-10): 313–319, 1943 (in Russian).

[4] H.J. Bhabha and Harish-Chandra. On the theory of point particles. *Proceedings of the Royal Society London A.*, 183: 134–141, 1944.

[5] H.J. Bhabha. Relativistic wave equations for the proton. *Proceedings of the Indian National Science A.*, 21: 241–264, 1945.

[6] H.J. Bhabha. Relativistic wave equations for elementary particles. *Reviews of Modern Physics*, 17(2-3): 200–215, 1945.

[7] H.J. Bhabha. The theory of the elementary particles. *Reports on Progress in Physics*, 10: 253–271, 1946.

[8] Harish-Chandra. On the equations of motion of point particles. *Proceedings of the Royal Society London A.*, 185: 269–287, 1946.

[9] Harish-Chandra, Relativistic equations for elementary particles. *Proceedings of the Royal Society London A.*, 192: 195–218, 1948.

[10] I.M. Gel'fand and A.M. Yaglom. General relativistic invariant equations snd infinitely dimensional representation of the Lorentz group. *Zhurnal Eksperimental'noi i Teoreticheskoi Fiziki*, 18(8): 703–733, 1948 (in Russian).

[11] I.M. Gel'fand and A.M. Yaglom. Pauli theorem for general relativistic invariant wave equations. *Zhurnal Eksperimental'noi i Teoreticheskoi Fiziki*, 18(12): 1096–1104, 1948 (in Russian).

[12] I.M. Gel'fand and A.M. Yaglom. Charge conjugation for general relativistic invariant wave equations. *Zhurnal Eksperimental'noi i Teoreticheskoi Fiziki*, 18(12): 1105–1111, 1948.

[13] H.J. Bhabha. Theory of elementary particles-fields. *Lectures Delivered at 2nd Summer Seminar Canadian Math. Congress: held at University of British Columbia*, Aug-1949. pages 1–103.

[14] F.I. Fedorov. To the problem of solving relativistic wave equations. *Doklady AN USSR*, 65(6): 813–814, 1949 (in Russian).

[15] E.E. Fradkin. To the theory of particles with higher spins. *Zhurnal Eksperimental'noi i Teoreticheskoi Fiziki*, 20(1): 27–38, 1950 (in Russian).

[16] F.I. Fedorov. On minimal polynomials of matrices of relativistic wave equations. *Doklady AN USSR*, 79(5): 787–790, 1951 (in Russian).

[17] F.I. Fedorov. To the theory of a spin 2 particle. *Scientific Notes of Belorussian State University, Ser. Physics and Mathematics* 12: 156–173, 1951 (in Russian).

[18] H.J. Bhabha. An equation for a particle with two mass states and positive charge density. *Philosophical Magazine Series. VII*, 43: 33–47, 1952.

[19] F.I. Fedorov. Generalised relativistic wave equations. *Doklady AN USSR*, 82(1): 37–40, 1952 (in Russian).

[20] M. Petras. A contribution of the theory of the Pauli–Fierz's equations a particle with spin 3/2. *Czechoslovak Journal of Physics*, 5(2): 169–170, 1955.

[21] M. Petras. A note to Bhabha's equation for a particle with maximum spin 3/2. *Czechoslovak Journal of Physics*, 5(3): 418–419, 1955.

[22] V.Ya. Fainberg, To the interactionb theory of the particles of the higher spins with electromagnetic and meson fields. *Trudy FIAN USSR*, 6: 269–332, 1955 (in Russian).

[23] V.L. Ginzburg. On relativistic wave equations with a mass spectrum. *Acta Physica Polonica*, 15: 163–175, 1956.

[24] H. Shimazu. A relativistic wave equation for a particle with two mass states of spin 1 and 0. *Progress of Theoretical Physics*, 16(4): 285–298, 1956.

[25] T. Regge. On properties of the particle with spin 2. *Nuovo Cimento*, 5(2): 325–326, 1957.

[26] P.G. Bergmann, A.I. Janis. Subsidiary conditions in covariant theories. *Journals of Physical Review*, 111(4): 1191–1200, 1958.

[27] H.A. Buchdahl. On the compatibility of relativistic wave equations for particles of higher spin in the presence of a gravitational field. *Nuovo Cimemto*, 10: 96–103, 1958.

[28] L.A. Shelepin. Covariant theory of relativictic wave equations. *Nuclear Physics*, 33(4): 580–593, 1962.

[29] A.Z. Capri. First order wave equations for multi-mass fermions. *Nuovo Cimento B.*, 64(1): 151–158, 1969.

[30] A. Aurilia and H. Umezawa. Theory of high spin fields. *Journals of Physical Review*, 182(5): 1682–1694, 1969.

[31] F.I. Fedorov and V.A. Pletyukhov. Wave equations with repeated representations of the Lorentz group. *Proceedings of the Academy of Sciences of BSSR. Mathematical physics Series*, 6: 81–88, 1969 (in Russian).

[32] V.A. Pletyukhov and F.I. Fedorov. The wave equation with repeated representations for spin 0 particle. *Proceedings of the Academy of Sciences of BSSR. Mathematical physics Series*, 2: 79–85, 1970 (in Russian).

[33] F.I. Fedorov and V.A. Pletyukhov. Wave equations with repeated representations of the Lorentz group. Half-integer spin. *Proceedings of the Academy of Sciences of BSSR. Mathematical physics Series*, 3: 78–83, 1970 (in Russian).

[34] V.A. Pletyukhov, F.I. Fedorov. Wave equation with repeated reprentations for a spin 1 particle. *Proceedings of the Academy of Sciences of BSSR. Mathematical physics Series*, 3: 84–92, 1970 (in Russian).

[35] A. Shamaly and A.Z. Capri. First-order wave equations for integral spin. *Nuovo Cimento B.*, 2(2): 235–253, 1971.

[36] V. Amar and U. Dozzio. Finite dimensional Gel'fand–Yaglom equations for arbitrary integral spin. *Nuovo Cimento B.*, 9: 53–63, 1972.

[37] A.Z. Capri. Electromagnetic properties of a new spin-1/2 field. *Progress of Theoretical Physics*, 48(4): 1364–1374, 1972.

[38] A. Shamaly and A.Z. Capri. Unified theories for massive spin 1 fields. *Canadian Journal of Physics*, 51(14): 1467–1470, 1973.

[39] M.A.K. Khalil. Properties of a 20-component spin 1/2 relativistic wave equation. *Journals of Physical Review D*, 15(6): 1532–1539, 1977.

[40] A.S. Wightman. Invariant wave equations: general theory and applications to the external field problem. *Lecture Notes in Physics*, 73: 1–101, 1978.

[41] M.A.K. Khalil. An equivalence of relativistic field equations. *Nuovo Cimento A.*, 45(3): 389–404, 1978.

[42] L. Garding. Mathematics of invariant wave equations. *Lecture Notes in Physics*, 73: 102–164, 1978.

[43] J.P. Gazeau. L'equation de Dirac avec masse et spin arbitrares: une construction simple et naturelle. *Journal of Physics G: Nuclear and Particle Physics*, 6(12): 1459–1475, 1980.

[44] W. Cox. Higher-rank representations for zero-spin filds theories. *Journal of Physics A*, 15: 627–635, 1982.

[45] W. Cox. First-order formulation of massive spin-2 field theories. *Journal of Physics A*, 15: 253–268, 1982.

[46] P.M. Mathews, B. Vijayalakshmi and M.Sivakuma. On the admissible Lorentz group representations in unique-mass, unique-spin relativistic wave equations. *Journal of Physics A*, 15(11): 1579–1582, 1982.

[47] P.M. Mathews and B. Vijayalakshmi. On inequivalent classes unique-mass-spin relativistic wave equations involving repeated irreducible representations with arbitrary multiplicities. *Journal of Mathematical Physics*, 25(4): 1080–1087, 1984.

[48] W. Cox. On the Lagrangian and Hamiltonian constraint algorithms for the Rarita-Schwinger field coupled to an external electromagnetic field. *Journal of Physics A*, 22(10): 1599–1608, 1989.

[49] S. Deser and A. Waldron. Inconsistencies of massive charged gravitating higher spins. *Nuclear Physics B*, 631: 369–387, 2002.

[50] V. Simulik. Relativistic wave equations of arbitrary spin in quantum mechanics and field theory, example spin $S = 2$. *Journal of Physics: Conference Series*, 804(1): paper 012040, 2017.

[51] E.M. Ovsiyuk, V.V. Kisel, Y.A. Voynova, O.V. Veko and V.M. Red'kov. Spin 1/2 particle with anomalous magnetic moment in a uniform magnetic field, exact solutions. *Nonlinear Phenomena in Complex Systems*, 19(2): 153–165, 2016.

[52] V. Kisel, Ya. Voynova, E. Ovsiyuk, V. Balan and V. Red'kov. Spin 1 Particle with Anomalous Magnetic Moment in the External Uniform Magnetic Field. *Nonlinear Phenomena in Complex Systems*, 20(1): 21–39, 2017.

[53] E.M. Ovsiyuk, Ya.A. Voynova, V.V. Kisel, V. Balan and V.M. Red'kov. Spin 1 Particle with Anomalous Magnetic Moment in the External Uniform Electric Field. In *Quaternions: Theory and Applications. Editor: Sandra Griffin*, pages 47–84. Nova Science Publishers, Inc. USA, 2017.

[54] V.V. Kisel, E.M. Ovsiyuk, Ya.A. Voynova and V.M. Red'kov. Quantum mechanics of spin 1 particle with quadrupole moment in external uniform magetic field. *Problems of Physics, Mathematics, and Thechnics*, 32(3): 18–27, 2017.

[55] F.I. Fedorov. *The Lorentz Group.* Science, Moscow, 1979 (in Russian).

[56] I.M. Gel'fand, R.A. Minlos and Z.Ya. Shapiro. *Representations of the Rotation and Lorentz Groups and Their Applications.* Dover Publications, New York, 2018.

[57] V.M. Red'kov. *Fields in Riemannian Space and the Lorentz Group.* Belarusian Science, Minsk, 2009.

[58] V.A. Pletjukhov, V.M. Red'kov and V.I. Strazhev. *Relativistic wave equations and intrinsic degrees of freedom.* Belarusian Science, Minsk, 2015.

[59] V.V. Kisel, E.M. Ovsiyuk, V. Balan, O.V. Veko and V.M. Red'kov. *Elementary Particles with Internal Structure in External Field. Vol. I. General Formalism.* Nova Science Publishers, Inc. USA, 2018; *Vol. II. Physical Problems.* Nova Science Publishers, Inc. USA, 2018.

[60] V.V. Kisel, V.A. Pletyukhov, V.V. Gilewsky, E.M. Ovsiyuk, O.V. Veko and V.M. Red'kov. Spin 1/2 particle with two mass states, interaction with external fields. *Nonlinear Phenomena in Complex Systems*, 20(4): 404–423, 2017.

[61] E.M. Ovsiyuk, O.V. Veko, Ya.A. Voynova, V.V. Kisel, V. Balan and V.M. Red'kov. Spin 1/2 particle with two masses in magnetic field. *Applied Sciences*, 20: 148–166, 2018.

13

Helicity operator for a spin 2 particle in magnetic field

Explicit form of the helicity operator for symmetric 2nd rank tensor describing the spin 2 particle is specified in cylindrical coordinates. After separating the variables the system of ten differential 1st-order equations is derived. It is split into two independent subsystem of four and six equations. The system of four equations is solved straightforwardly in terms of confluent hypergeometric functions; there are corresponding eigenvalues and eigenfunctions. A subsystem of six equations can be reduced to one ordinary differential equation of the 4th-order. Corresponding 4th-order operator is factorised into permutable 2nd-order operators, so the problem reduces to solving two differential equations of the 2nd-order. Their solutions are constructed in terms of Bessel functions. This analysis is extended to the presence of an external uniform magnetic field, when solutions are constructed in terms of confluent hypergeometric functions.

The chapter is based on [1–31].

13.1 Introduction

It is known that the eigenvalue states of the helicity operator play a substantial role in studying any spin particle in external electromagnetic (or gravitational) fields with cylindric symmetry. In the present work, we specify this problem for a spin 2 particle in Minkowski space-time

$$\Sigma^{cart} = J^{23}\frac{\partial}{\partial x} + J^{31}\frac{\partial}{\partial y} + J^{12}\frac{\partial}{\partial z}, \qquad \Sigma^{cart} H^{cart} = \sigma H^{cart}, \tag{13.1}$$

where $H^{cart}(x,y,z)$ consists of ten components of the symmetric 2nd-rank tensor referring to the spin 2 particle.

13.2 Helicity operator in cylindric basis, separating the variables

We will apply the description of this field in cylindric coordinates $x^\alpha = (t, r, \phi, z)$ and the corresponding tetrad

$$dS^2 = dt^2 - dr^2 - r^2 d\phi^2 - dz^2, \quad e^{\beta}_{(a)}(x) = \left| \begin{array}{cccc} 1 & 0 & 0 & 0 \\ 0 & 1 & 0 & 0 \\ 0 & 0 & 1/r & 0 \\ 0 & 0 & 0 & 1 \end{array} \right| . \tag{13.2}$$

DOI: 10.1201/9781003472377-13

Transition from Cartesian tetrad to a cylindric one is performed by means of the local transformation (belonging to the Lorentz group)

$$L_b{}^a(x) = e_{(b)}^{\ \ '\beta'}(x') \frac{\partial x^{\ \alpha}}{\partial x^{\beta'}} e_\alpha{}^{(a)}(x) = e_{(b)}^{\ \ '\beta'}(x') \frac{\partial x^{\ a}}{\partial x^{\beta'}}, \tag{13.3}$$

or in explicit form

$$L_b{}^a(x) = \begin{vmatrix} 1 & 0 & 0 & 0 \\ 0 & 1 & 0 & 0 \\ 0 & 0 & 1/r & 0 \\ 0 & 0 & 0 & 1 \end{vmatrix} \begin{vmatrix} 1 & 0 & 0 & 0 \\ 0 & \frac{\partial x}{\partial r} & \frac{\partial y}{\partial r} & 0 \\ 0 & \frac{\partial x}{\partial \phi} & \frac{\partial y}{\partial \phi} & 0 \\ 0 & 0 & 0 & 1 \end{vmatrix} = \begin{vmatrix} 1 & 0 & 0 & 0 \\ 0 & \cos\phi & \sin\phi & 0 \\ 0 & -\sin\phi & \cos\phi & 0 \\ 0 & 0 & 0 & 1 \end{vmatrix}.$$

Therefore, the tensor of the second rank H transforms according to the rule

$$H^{cart} = [H^{cart}_{(ab)}], \quad H^{cyl} = (L \otimes L) H^{cart} = L H^{cart} \tilde{L}\,.$$

Correspondingly, the helicity operator transforms to cylindric tetrad as follows (in 10-dimensional representation, the S is a 10×10 matrix)

$$H^{cyl} = (L \otimes L) H^{cart} \quad \Longrightarrow \quad H^{cyl} = S H^{cart},$$

$$\Sigma^{cyl} = S(\phi) \Big[J^{23} \big(\cos\phi \frac{\partial}{\partial r} - \frac{\sin\phi}{r} \frac{\partial}{\partial \phi} \big) + J^{31} \big(\sin\phi \frac{\partial}{\partial r} + \frac{\cos\phi}{r} \frac{\partial}{\partial \phi} \big) + J^{12} \frac{\partial}{\partial z} \Big] S(-\phi)\,;$$

whence we get

$$\Sigma^{cyl} = \Big[S(\phi) J^{23} S_2(-\phi) \cos\phi + S_2(\phi) J^{31} S(-\phi) \sin\phi \Big] \frac{\partial}{\partial r}$$

$$+ \frac{1}{r} \Big[- S(\phi) J^{23} S(-\phi) \sin\phi + S_2(\phi) J^{31} S_2(-\phi) \cos\phi \Big] \frac{\partial}{\partial \phi}$$

$$+ \frac{1}{r} \Big[- \sin\phi S(\phi) J^{23} \frac{\partial S(\ \ \phi)}{\partial \phi} + \cos\phi S(\phi) J^{31} \frac{\partial S(-\phi)}{\partial \phi} \Big] + S(\phi) J^{12} S(-\phi) \frac{\partial}{\partial z}\,.$$

Taking in mind the structure of the field function (it contains the multipliers $e^{im\phi}$ and e^{ikz}), we can change the last relation to the form

$$\Sigma^{cyl} = \Big[S(\phi) J^{23} S(-\phi) \cos\phi + S(\phi) J^{31} S(-\phi) \sin\phi \Big] \frac{d}{dr}$$

$$+ \frac{im}{r} \Big[- S(\phi) J^{23} S(-\phi) \sin\phi + S(\phi) J^{31} S(-\phi) \cos\phi \Big]$$

$$+ \frac{1}{r} \Big[- \sin\phi S(\phi) J^{23} \frac{\partial S(-\phi)}{\partial \phi} + \cos\phi S(\phi) J^{31} \frac{\partial S(-\phi)}{\partial \phi} \Big] + ik S(\phi) J^{12} S(-\phi)\,,$$

the inverse matrix equals $S^{-1} = S(-\phi)$.

Let us find expression for ten components of H in the cylindric tetrad basis. We will apply the following notations

$$H^{cyl} = \text{the column}(f_1, f_2, f_3, c_1, c_2, c_3, d_1, d_2, d_3, f_0),$$

$$S(\phi) = \begin{vmatrix} \cos^2\phi & \sin^2\phi & \cdot & \cdot & \cdot & \sin 2\phi & \cdot & \cdot & \cdot & \cdot \\ \sin^2\phi & \cos^2\phi & \cdot & \cdot & \cdot & -\sin 2\phi & \cdot & \cdot & \cdot & \cdot \\ \cdot & \cdot & 1 & \cdot & \cdot & \cdot & \cdot & \cdot & \cdot & \cdot \\ \cdot & \cdot & \cdot & \cos\phi & -\sin\phi & \cdot & \cdot & \cdot & \cdot & \cdot \\ \cdot & \cdot & \cdot & \sin\phi & \cos\phi & \cdot & \cdot & \cdot & \cdot & \cdot \\ -\cos\phi\sin\phi & \cos\phi\sin\phi & \cdot & \cdot & \cdot & \cos 2\phi & \cdot & \cdot & \cdot & \cdot \\ \cdot & \cdot & \cdot & \cdot & \cdot & \cdot & \cos\phi & \sin\phi & \cdot & \cdot \\ \cdot & \cdot & \cdot & \cdot & \cdot & \cdot & -\sin\phi & \cos\phi & \cdot & \cdot \\ \cdot & \cdot & \cdot & \cdot & \cdot & \cdot & \cdot & \cdot & 1 & \cdot \\ \cdot & \cdot & \cdot & \cdot & \cdot & \cdot & \cdot & \cdot & \cdot & 1 \end{vmatrix}.$$

After simple calculation we derive the following expression for helicity operator in cylindric basis

$$\Sigma^{cyl} = \begin{vmatrix} 0 & 0 & 0 & 0 & 0 & 0 & 0 & 0 & 0 & 0 \\ 0 & 0 & 0 & -2 & 0 & 0 & 0 & 0 & 0 & 0 \\ 0 & 0 & 0 & 2 & 0 & 0 & 0 & 0 & 0 & 0 \\ 0 & 1 & -1 & 0 & 0 & 0 & 0 & 0 & 0 & 0 \\ 0 & 0 & 0 & 0 & 0 & 1 & 0 & 0 & 0 & 0 \\ 0 & 0 & 0 & 0 & -1 & 0 & 0 & 0 & 0 & 0 \\ 0 & 0 & 0 & 0 & 0 & 0 & 0 & 0 & 0 & 0 \\ 0 & 0 & 0 & 0 & 0 & 0 & 0 & 0 & -1 & 0 \\ 0 & 0 & 0 & 0 & 0 & 0 & 0 & 1 & 0 & 0 \\ 0 & 0 & 0 & 0 & 0 & 0 & 0 & 0 & 0 & 0 \end{vmatrix} \frac{d}{dr}$$

$$+ \frac{im}{r} \begin{vmatrix} 0 & 0 & 0 & 0 & 2 & 0 & 0 & 0 & 0 & 0 \\ 0 & 0 & 0 & 0 & 0 & 0 & 0 & 0 & 0 & 0 \\ 0 & 0 & 0 & 0 & -2 & 0 & 0 & 0 & 0 & 0 \\ 0 & 0 & 0 & 0 & 0 & -1 & 0 & 0 & 0 & 0 \\ -1 & 0 & 1 & 0 & 0 & 0 & 0 & 0 & 0 & 0 \\ 0 & 0 & 0 & 1 & 0 & 0 & 0 & 0 & 0 & 0 \\ 0 & 0 & 0 & 0 & 0 & 0 & 0 & 0 & 1 & 0 \\ 0 & 0 & 0 & 0 & 0 & 0 & 0 & 0 & 0 & 0 \\ 0 & 0 & 0 & 0 & 0 & 0 & -1 & 0 & 0 & 0 \\ 0 & 0 & 0 & 0 & 0 & 0 & 0 & 0 & 0 & 0 \end{vmatrix}$$

$$+ \frac{1}{r} \begin{vmatrix} 0 & 0 & 0 & -2 & 0 & 0 & 0 & 0 & 0 & 0 \\ 0 & 0 & 0 & 0 & 0 & 0 & 0 & 0 & 0 & 0 \\ 0 & 0 & 0 & 2 & 0 & 0 & 0 & 0 & 0 & 0 \\ -1 & 1 & 0 & 0 & 0 & 0 & 0 & 0 & 0 & 0 \\ 0 & 0 & 0 & 0 & 0 & 2 & 0 & 0 & 0 & 0 \\ 0 & 0 & 0 & 0 & 1 & 0 & 0 & 0 & 0 & 0 \\ 0 & 0 & 0 & 0 & 0 & 0 & 0 & 0 & 0 & 0 \\ 0 & 0 & 0 & 0 & 0 & 0 & 0 & 0 & 0 & 0 \\ 0 & 0 & 0 & 0 & 0 & 0 & 0 & 1 & 0 & 0 \\ 0 & 0 & 0 & 0 & 0 & 0 & 0 & 0 & 0 & 0 \end{vmatrix} + ik \begin{vmatrix} 0 & 0 & 0 & 0 & 0 & -2 & 0 & 0 & 0 & 0 \\ 0 & 0 & 0 & 0 & 0 & 2 & 0 & 0 & 0 & 0 \\ 0 & 0 & 0 & 0 & 0 & 0 & 0 & 0 & 0 & 0 \\ 0 & 0 & 0 & 0 & 1 & 0 & 0 & 0 & 0 & 0 \\ 0 & 0 & 0 & -1 & 0 & 0 & 0 & 0 & 0 & 0 \\ 1 & -1 & 0 & 0 & 0 & 0 & 0 & 0 & 0 & 0 \\ 0 & 0 & 0 & 0 & 0 & 0 & 0 & -1 & 0 & 0 \\ 0 & 0 & 0 & 0 & 0 & 0 & 1 & 0 & 0 & 0 \\ 0 & 0 & 0 & 0 & 0 & 0 & 0 & 0 & 0 & 0 \\ 0 & 0 & 0 & 0 & 0 & 0 & 0 & 0 & 0 & 0 \end{vmatrix},$$

taking in mind explicit form of generators for tensor we obtain its shorter form

$$\Sigma^{cyl} = J^{23}\frac{d}{dr} + \frac{im}{r}J^{31} + ikJ^{12} + \frac{1}{r} \begin{vmatrix} 0 & 0 & 0 & -2 & 0 & 0 & 0 & 0 & 0 & 0 \\ 0 & 0 & 0 & 0 & 0 & 0 & 0 & 0 & 0 & 0 \\ 0 & 0 & 0 & 2 & 0 & 0 & 0 & 0 & 0 & 0 \\ -1 & 1 & 0 & 0 & 0 & 0 & 0 & 0 & 0 & 0 \\ 0 & 0 & 0 & 0 & 0 & 2 & 0 & 0 & 0 & 0 \\ 0 & 0 & 0 & 0 & 1 & 0 & 0 & 0 & 0 & 0 \\ 0 & 0 & 0 & 0 & 0 & 0 & 0 & 0 & 0 & 0 \\ 0 & 0 & 0 & 0 & 0 & 0 & 0 & 0 & 0 & 0 \\ 0 & 0 & 0 & 0 & 0 & 0 & 0 & 1 & 0 & 0 \\ 0 & 0 & 0 & 0 & 0 & 0 & 0 & 0 & 0 & 0 \end{vmatrix}.$$

It is convenient to use the cyclic representation, in which the generator J^{12} becomes diagonal, which leads to

$$
\bar{\Sigma}^{cyl} = \bar{J}^{23}\frac{d}{dr} + \frac{im}{r}\bar{J}^{31} + ik\bar{J}^{12} + \frac{1}{r}\begin{vmatrix}
0 & 0 & 0 & 0 & 0 & i\sqrt{2} & 0 & 0 & 0 & 0 \\
0 & 0 & 0 & -i\sqrt{2} & 0 & -i\sqrt{2} & 0 & 0 & 0 & 0 \\
0 & 0 & 0 & i\sqrt{2} & 0 & 0 & 0 & 0 & 0 & 0 \\
0 & 0 & -i\sqrt{2} & 0 & 0 & 0 & 0 & 0 & 0 & 0 \\
0 & 0 & 0 & -\frac{i}{\sqrt{2}} & 0 & -\frac{i}{\sqrt{2}} & 0 & 0 & 0 & 0 \\
-i\sqrt{2} & 0 & 0 & 0 & 0 & 0 & 0 & 0 & 0 & 0 \\
0 & 0 & 0 & 0 & 0 & 0 & 0 & 0 & 0 & 0 \\
0 & 0 & 0 & 0 & 0 & 0 & -\frac{i}{\sqrt{2}} & 0 & -\frac{i}{\sqrt{2}} & 0 \\
0 & 0 & 0 & 0 & 0 & 0 & 0 & 0 & 0 & 0 \\
0 & 0 & 0 & 0 & 0 & 0 & 0 & 0 & 0 & 0
\end{vmatrix},
$$

where [31]

$$
\bar{J}^{23} = \frac{1}{\sqrt{2}}\begin{vmatrix}
. & . & . & . & . & -2i & . & . & . & . \\
. & . & . & -2i & . & -2i & . & . & . & . \\
. & . & . & -2i & . & . & . & . & . & . \\
. & -i & -i & . & -i & . & . & . & . & . \\
. & . & . & -i & . & -i & . & . & . & . \\
-i & -i & . & . & -i & . & . & . & . & . \\
. & . & . & . & . & . & . & -i & . & . \\
. & . & . & . & . & . & -i & . & -i & . \\
. & . & . & . & . & . & . & -i & . & . \\
. & . & . & . & . & . & . & . & . & .
\end{vmatrix},
$$

$$
\bar{J}^{31} = \frac{1}{\sqrt{2}}\begin{vmatrix}
. & . & . & . & . & -2 & . & . & . & . \\
. & . & . & -2 & . & 2 & . & . & . & . \\
. & . & . & 2 & . & . & . & . & . & . \\
. & 1 & -1 & . & 1 & . & . & . & . & . \\
. & . & . & -1 & . & 1 & . & . & . & . \\
1 & -1 & . & . & -1 & . & . & . & . & . \\
. & . & . & . & . & . & . & -1 & . & . \\
. & . & . & . & . & . & 1 & . & -1 & . \\
. & . & . & . & . & . & . & 1 & . & . \\
. & . & . & . & . & . & . & . & . & .
\end{vmatrix},
$$

$$
\bar{J}^{12} = \begin{vmatrix}
-2i & . & . & . & . & . & . & . & . & . \\
. & 0 & . & . & . & . & . & . & . & . \\
. & . & +2i & . & . & . & . & . & . & . \\
. & . & . & +i & . & . & . & . & . & . \\
. & . & . & . & 0 & . & . & . & . & . \\
. & . & . & . & . & -i & . & . & . & . \\
. & . & . & . & . & . & -i & . & . & . \\
. & . & . & . & . & . & . & 0 & . & . \\
. & . & . & . & . & . & . & . & +i & . \\
. & . & . & . & . & . & . & . & . & 0
\end{vmatrix},
$$

The eigenvalue equation $\bar{\Sigma}\bar{H} = \sigma\bar{H}$ gives the following system

$$
\begin{aligned}
2c_3' &= -\frac{2(m-1)}{r}c_3 + i\sqrt{2}(\sigma - 2k)f_1,\\
2\left(c_1' + c_3'\right) &= -\frac{2(m+1)}{r}c_1 + \frac{2(m-1)}{r}c_3 + i\sqrt{2}\sigma f_2,\\
2c_1' &= i\sqrt{2}(2k+\sigma)f_3 + \frac{2(m+1)}{r}c_1,\\
c_2' + f_2' + f_3' &= i\sqrt{2}(k+\sigma)c_1 + \frac{m}{r}\left(c_2 + f_2\right) - \frac{m+2}{r}f_3,\\
c_1' + c_3' &= -\frac{m+1}{r}c_1 + i\sqrt{2}\sigma c_2 + \frac{m-1}{r}c_3,\\
c_2' + f_1' + f_2' &= i\sqrt{2}(\sigma - k)c_3 + \frac{m-2}{r}f_1 - \frac{m}{r}\left(c_2 + f_2\right),\\
d_2' &= i\sqrt{2}(\sigma - k)d_1 - \frac{m}{r}d_2,\\
d_1' + d_3' &= \frac{m-1}{r}d_1 + i\sqrt{2}\sigma d_2 - \frac{m+1}{r}d_3,\\
d_2' &= i\sqrt{2}(k+\sigma)d_3 + \frac{m}{r}d_2,\\
\sqrt{2}\sigma f_0 &= 0.
\end{aligned}
\tag{13.4}
$$

Let us apply special notations for eight differential operators:

$$
\frac{1}{\sqrt{2}}\left(\frac{d}{dr} \pm \frac{m}{r}\right) = a_m^{\pm}, \quad \frac{1}{\sqrt{2}}\left(\frac{d}{dr} \pm \frac{m+1}{r}\right) = a_{m+1}^{\pm}, \quad \frac{1}{\sqrt{2}}\left(\frac{d}{dr} \pm \frac{m-1}{r}\right) = a_{m-1}^{\pm},
$$

$$
\frac{1}{\sqrt{2}}\left(\frac{d}{dr} + \frac{m+2}{r}\right) = a_{m+2}^{+}, \qquad \frac{1}{\sqrt{2}}\left(\frac{d}{dr} - \frac{m-2}{r}\right) = a_{m-2}^{-}.
$$

Then the above system may be presented as three independent subsystems:

$$
\begin{aligned}
I \qquad a_m^+ d_2 &= i(\sigma - k)d_1,\\
a_{m-1}^- d_1 + a_{m+1}^+ d_3 &= i\sigma d_2,\\
a_m^- d_2 &= i(\sigma + k)d_3;
\end{aligned}
\tag{13.5}
$$

$$
\begin{aligned}
II \qquad a_{m-1}^+ c_3 &= i\left(\frac{\sigma}{2} - k\right)f_1,\\
a_{m+1}^+ c_1 + a_{m-1}^- c_3 &= i\frac{\sigma}{2}f_2,\\
a_{m+1}^- c_1 &= i\left(\frac{\sigma}{2} + k\right)f_3,\\
a_m^- c_2 + a_m^- f_2 + a_{m+2}^+ f_3 &= i(\sigma + k)c_1,\\
a_{m+1}^+ c_1 + a_{m-1}^- c_3 &= i\sigma c_2,\\
a_m^+ c_2 + a_{m-2}^- f_1 + a_m^+ f_2 &= i(\sigma - k)c_3;
\end{aligned}
\tag{13.6}
$$

$$
III \qquad \sigma f_0 = 0 \quad \Longrightarrow \quad \sigma \neq 0, f_0 = 0. \tag{13.7}
$$

Consider the system I; after eliminating the variables d_1 and d_3, we get the equation for d_2:

$$
\left[\frac{1}{i(\sigma - k)}a_{m-1}^- a_m^+ + \frac{1}{i(\sigma + k)}a_{m+1}^+ a_m^- - i\sigma\right] d_2 = 0. \tag{13.8}
$$

Similarly, in the system II one can eliminate the variables f_1, f_2, f_3:

$$f_1 = \frac{2}{i(\sigma - 2k)} a^+_{m-1} c_3, \quad f_2 = \frac{2}{i\sigma} a^+_{m+1} c_1 + \frac{2}{i\sigma} a^-_{m-1} c_3, \quad f_3 = \frac{2}{i(\sigma + 2k)} a^-_{m+1} c_1,$$

this results in the system for c_1, c_2, c_3:

$$a^-_m c_2 + \frac{2}{i\sigma} a^-_m a^+_{m+1} c_1 + \frac{2}{i\sigma} a^-_m a^-_{m-1} c_3 + \frac{2}{i(\sigma + 2k)} a^+_{m+2} a^-_{m+1} c_1 = i(\sigma + k) c_1,$$

$$a^+_m c_2 + \frac{2}{i(\sigma - 2k)} a^-_{m-2} a^+_{m-1} c_3 + \frac{2}{i\sigma} a^+_m a^+_{m+1} c_1 + \frac{2}{i\sigma} a^+_m a^-_{m-1} c_3 = i(\sigma - k) c_3,$$

$$a^+_{m+1} c_1 + a^-_{m-1} c_3 = i\sigma c_2.$$

With the help of third equation, we can eliminate the variable c_2, so deriving the system for c_1, c_2:

$$\begin{gathered} \Big[3a^-_m a^+_{m+1} + \frac{2\sigma}{(\sigma + 2k)} a^+_{m+2} a^-_{m+1} + \sigma(\sigma + k) \Big] c_1 + 3a^-_m a^-_{m-1} c_3 = 0, \\ \Big[3a^+_m a^-_{m-1} c_3 + \frac{2\sigma}{(\sigma - 2k)} a^-_{m-2} a^+_{m-1} + \sigma(\sigma - k) \Big] c_3 + 3a^+_m a^+_{m+1} c_1 = 0. \end{gathered} \tag{13.9}$$

In order to take into account the presence of external magnetic field, it suffices to make one change, $m \Longrightarrow m + eBr^2/2$, this leads to new eight operators

$$a^{\pm}_m = \frac{1}{\sqrt{2}} \Big(\frac{d}{dr} \pm \frac{m + eBr^2/2}{r} \Big),$$

$$a^{\pm}_{m+1} = \frac{1}{\sqrt{2}} \Big(\frac{d}{dr} \pm \frac{m + eBr^2/2 + 1}{r} \Big), \quad a^{\pm}_{m-1} = \frac{1}{\sqrt{2}} \Big(\frac{d}{dr} \pm \frac{m + eBr^2/2 - 1}{r} \Big),$$

$$a^+_{m+2} = \frac{1}{\sqrt{2}} \Big(\frac{d}{dr} + \frac{m + eBr^2/2 + 2}{r} \Big), \quad a^-_{m-2} = \frac{1}{\sqrt{2}} \Big(\frac{d}{dr} - \frac{m + eBr^2/2 - 2}{r} \Big).$$

13.3 The system I, the free particle

Let us turn to the case I, first consider eq. (13.8) for a free particle. Allowing for the identities

$$a^-_{m-1} a^+_m = \frac{1}{2} \frac{d^2}{dr^2} + \frac{1}{2r} \frac{d}{dr} - \frac{m^2}{2r^2}, \qquad a^+_{m-1} a^-_m = \frac{1}{2} \frac{d^2}{dr^2} - \frac{1}{2r} \frac{d}{dr} + \frac{(2 - m)m}{2r^2},$$

we get

$$\left[\frac{d^2}{dr^2} + \frac{k}{\sigma r} \frac{d}{dr} - \frac{mk + m(m-1)\sigma}{\sigma r^2} - \left(k^2 - \sigma^2\right) \right] d_2 = 0. \tag{13.10}$$

Near the point $r = 0$, this equation becomes simpler

$$\left[\frac{d^2}{dr^2} + \frac{k}{\sigma r} \frac{d}{dr} - \frac{mk + m(m-1)\sigma}{\sigma r^2} \right] d_2 = 0, \quad d_2 = r^A;$$

so that

$$A(A-1) + \frac{k}{\sigma} A - \frac{mk + m(m-1)\sigma}{\sigma} = 0, \quad A = m, -\frac{k}{\sigma} - m + 1. \tag{13.11}$$

Near the point $r \to \infty$ we have

$$\left[\frac{d^2}{dr^2} + \frac{k}{\sigma r}\frac{d}{dr} - (k^2 - \sigma^2)\right] d_2 = 0, \quad d_2 = e^{\pm i\sqrt{\sigma^2 - k^2} r}. \tag{13.12}$$

Solutions should be searched in the form $d_2 = r^A e^{Cr} f$, which yields

$$f'' + \left(2C + \frac{2A}{r} + \frac{k}{\sigma r}\right) f' + \left[\frac{A(A-1)}{\sigma r^2} + \frac{kA}{\sigma}\frac{1}{r^2} - \frac{mk + m(m-1)\sigma}{\sigma}\frac{1}{r^2}\right] f$$
$$+ \left[C^2 - (k^2 - \sigma^2) + \frac{2AC}{r} + \frac{kC}{\sigma}\frac{1}{r}\right] f = 0.$$

Imposing the known restrictions

$$A(A-1) + \frac{k}{\sigma}A - \frac{mk + m(m-1)\sigma}{\sigma} = 0, \quad C^2 - (k^2 - \sigma^2) = 0,$$

we reduce the above equation to the form

$$r\frac{d^2 f}{dr^2} + \left(2A + \frac{k}{\sigma} + 2Cr\right)\frac{df}{dr} + \left(2AC + \frac{k}{\sigma}C\right) f = 0;$$

the possible expressions for A and C are known. Changing the variable, $z = -2Cr$, we reduce the equation to hypergeometric form (for definiteness let $C = +i\sqrt{\sigma^2 - k^2}$):

$$z\frac{d^2 f}{dz^2} + (2A + \frac{k}{\sigma} - z)\frac{df}{dz} - (A + \frac{k}{2\sigma}) f = 0,$$
$$d_2 = z^A e^{-z/2}\Phi(c, a, z), \quad a = A + \frac{k}{2\sigma}, \ c = 2A + \frac{k}{\sigma} = 2a. \tag{13.13}$$

Solutions will be regular in the point $r = 0$, if we take positive values for A:

$$(a) \quad m > 0, \quad d_2 \sim z^A = z^m \longrightarrow 0;$$
$$(b) \quad m < 0, \quad d_2 \sim z^A = z^{1-m-\frac{k}{\sigma}}, \quad 1 - m > k/\sigma. \tag{13.14}$$

These solutions may be presented in terms of Bessel functions. Indeed, starting with the eq. (13.10), let us make the substitution $d_2(r) = \varphi(r)\bar{d}_2(r)$:

$$\frac{d}{dr}d_2 = \varphi'\bar{d}_2 + \varphi\bar{d}_2', \quad \frac{d^2}{dr^2}d_2 = \varphi''\bar{d}_2 + 2\varphi'\bar{d}_2' + \varphi\bar{d}_2'',$$

then we get

$$\bar{d}_2'' + \left(2\frac{\varphi'}{\varphi} + \frac{k}{\sigma r}\right)\bar{d}_2' + \left(\frac{k}{\sigma r}\frac{\varphi'}{\varphi} + \frac{\varphi''}{\varphi} - \frac{mk + m(m-1)\sigma}{\sigma r^2} - (k^2 - \sigma^2)\right)\bar{d}_2 = 0.$$

We fix the function φ as follows

$$2\frac{\varphi'}{\varphi} + \frac{k}{\sigma r} = \frac{1}{r} \quad \Longrightarrow \quad \frac{\varphi'}{\varphi} = \frac{1}{r}(1 - k/\sigma)/2 \quad \Longrightarrow \quad \varphi = r^{(1-k/\sigma)/2}; \tag{13.15}$$

accordingly the above equation takes the form

$$\bar{d}_2'' + \frac{1}{r}\bar{d}_2' + \left(-(k^2 - \sigma^2) - \frac{(m - 1/2 + k/2\sigma)^2}{r^2}\right)\bar{d}_2 = 0.$$

In the variable $y = i\sqrt{k^2 - \sigma^2}\, r$, it has the Bessel structure

$$\left(\frac{d^2}{dy^2} + \frac{1}{y}\frac{d}{dy} + 1 - \frac{\nu^2}{y^2}\right)\bar{d}_2' = 0, \quad \nu = m - \frac{1}{2} + \frac{k}{2\sigma},$$
$$\bar{d}_2'(y) = C_1 J_{+\nu}(y) + C_2 J_{-\nu}(y), \quad d_2(y) = r^{\frac{1-k/\sigma}{2}}\bar{d}_2(z). \tag{13.16}$$

13.4 The case I, particle in magnetic field

In the presence of the magnetic field, the equation for d_2 takes the form (for brevity we simplify the notation, $eB \Longrightarrow B$)

$$\Big[\frac{d^2}{dr^2} + \frac{k}{r\sigma}\frac{d}{dr} + \frac{-2\sigma\left(B(2m-1)+2k^2\right)+2Bk+4\sigma^3}{4\sigma} - \frac{B^2}{4}r^2 + \frac{-km-(m-1)m\sigma}{r^2\sigma}\Big] d_2 = 0. \tag{13.17}$$

Near the point $r = 0$, we get

$$\Big[\frac{d^2}{dr^2} + \frac{k}{r\sigma}\frac{d}{dr} - \frac{km+(m-1)m\sigma}{\sigma r^2\sigma}\Big] d_2 = 0,$$

which coincides with what we have in the case of a free particle. Near the point $r = \infty$ we get

$$\Big(\frac{d^2}{dr^2} + \frac{k}{r\sigma}\frac{d}{dr} - \frac{1}{4}B^2r^2\Big) d_2 = 0, \quad d_2 = e^{Cr^2}, \quad C = \pm\frac{B}{4}.$$

Making the change of variable $r^2 = x$, we transform eq. (13.17) to the form

$$\Big[\frac{d^2}{dx^2} + \frac{(k+\sigma)}{2\sigma x}\frac{d}{dx} - \frac{1}{16}B^2 + \frac{B(k-2m\sigma+\sigma)}{8\sigma x} - \frac{(k^2-\sigma^2)}{4x} + \frac{m(\sigma-k)}{4\sigma x^2} - \frac{m^2}{4x^2}\Big] d_2 = 0. \tag{13.18}$$

Its solutions are searched in the form $d_2 = x^A e^{Cx} f$, and for $f(x)$, we obtain the equation

$$f'' + \Big(\frac{2A}{x} + \frac{(k+\sigma)}{2\sigma x} + 2C\Big) f' + \Big[\frac{4A(A-1)-m^2}{4x^2} + \frac{m(\sigma-k)+2A(k+\sigma)}{4\sigma x^2}$$
$$+ \frac{B(k-2m\sigma+\sigma)+4C(k+\sigma)}{8\sigma x} - \frac{(k^2-\sigma^2)-8AC}{4x} + C^2 - \frac{1}{16}B^2\Big] f = 0.$$

Imposing the constraints

$$4A(A-1)\sigma - \sigma m^2 + m(\sigma-k) + 2A(k+\sigma) = 0, \quad C^2 - \frac{1}{16}B^2 = 0;$$

we get the expected results

$$A = \frac{m}{2}, \frac{1-m}{2} - \frac{k}{2\sigma}; \quad C = -\frac{B}{4}, +\frac{B}{4}; \tag{13.19}$$

below assume $C = -\frac{B}{4}$ $(B > 0)$. The main equation becomes simpler

$$xf'' + \Big(2A + \frac{(k+\sigma)}{2\sigma} + 2Cx\Big) f'$$
$$+ \Big[\frac{B(k-2m\sigma+\sigma)+4C(k+\sigma)}{8\sigma} - \frac{(k^2-\sigma^2)-8AC}{4}\Big] f = 0.$$

In the variable $2Cx = -z$, it reads as an equation of confluent hypergeometric type

$$zf'' + (c-z)f' - af = 0,$$
$$z\frac{d^2f''}{dz^2} + \Big(2A + \frac{k+\sigma}{2\sigma} - z\Big)\frac{df}{dz} - \Big(\frac{m}{2} + \frac{k^2-\sigma^2}{2B} + A\Big) f = 0. \tag{13.20}$$

Imposing the usual constraint to get polynomials

$$a = \frac{m}{2} + \frac{k^2 - \sigma^2}{2B} + A = -n, \quad n = 0, 1, 2, 3, \ldots;$$

we find expressions for σ:

$$\sigma = \pm\sqrt{k^2 + (A + \frac{m}{2} + n)2B}\,. \tag{13.21}$$

Depending on the sign of m, we have two possibilities

$$\begin{aligned} m > 0, \quad & \sigma = \pm\sqrt{k^2 + (m+n)2B}; \\ m < 0, \quad & \sigma = \pm\sqrt{k^2 + (n + \frac{1 - k/\sigma}{2})2B}, \end{aligned} \tag{13.22}$$

the second equation determines σ in inexplicit form. Assuming that solutions are regular at the point $z = 0$, we follow two possibilities depending on the sign of m:

$$m > 0, \quad d_2 \sim z^{m/2} e^{-z/2} \Phi(-n, c, z), \quad c = m + \frac{1}{2} + \frac{k}{2\sigma}\,; \tag{13.23}$$

$$m < 0, \quad d_2 \sim z^{\frac{1-m}{2} - \frac{k}{2\sigma}} e^{-z/2} \Phi(-n, c', z), \quad c = \frac{3}{2} - m, \frac{1-m}{2} > \frac{k}{2\sigma}. \tag{13.24}$$

Equation (13.24) leads to the cubic equation

$$\sigma^3 - \sigma[(k^2 + B(2n+1)] + Bk = 0, \ B > 0, \tag{13.25}$$

with the roots

$$\begin{aligned} \sigma_1 &= \frac{\sqrt[3]{2}\,(\beta + k^2 + 2\beta n)}{\phi} + \frac{\phi}{3\sqrt[3]{2}}, \\ \sigma_2 &= \frac{i\,(\sqrt{3} + i)\,\phi}{6\sqrt[3]{2}} - \frac{i\,(\sqrt{3} - i)\,(\beta + k^2 + 2\beta n)}{2^{2/3}\phi}, \\ \sigma_3 &= \frac{i\,(\sqrt{3} + i)\,(\beta + k^2 + 2\beta n)}{2^{2/3}\phi} - \frac{(1 + i\sqrt{3})\,\phi}{6\sqrt[3]{2}}, \end{aligned} \tag{13.26}$$

where $\phi = \sqrt[3]{\sqrt{729\beta^2 k^2 - 108\,(\beta + k^2 + 2\beta n)^3} - 27\beta k}$. In the dimensionless variables $\sigma/k = x, \ \ B/k^2 = b$, we have

$$\begin{aligned} x^3 - x[1 + b(2n+1)] + b = 0, \quad x_1 &= \frac{\sqrt[3]{2}(2bn + b + 1)}{\psi} + \frac{\psi}{3\sqrt[3]{2}}, \\ x_2 &= \frac{i\,(\sqrt{3} + i)\,\psi}{6\sqrt[3]{2}} - \frac{i\,(\sqrt{3} - i)\,(2bn + b + 1)}{2^{2/3}\psi}, \\ x_3 &= \frac{i\,(\sqrt{3} + i)\,(2bn + b + 1)}{2^{2/3}\psi} - \frac{(1 + i\sqrt{3})\,\psi}{6\sqrt[3]{2}}, \end{aligned} \tag{13.27}$$

where $\psi = \sqrt[3]{\sqrt{729b^2 - 108(2bn + b + 1)^3} - 27b}$.

TABLE 13.1
The values of x_i $(i = 1, 2, 3)$ and $b = 1$

n	$x_1(n)$	$x_2(n)$	$x_3(n)$
1	1.861	-2.115	0.2541
2	2.361	-2.529	0.1674
3	2.764	-2.889	0.1252
4	3.111	-3.211	0.1001
5	3.422	-3.505	0.08338
6	3.705	-3.777	0.07145
7	3.968	-4.031	0.06252
8	4.215	-4.270	0.05557
9	4.447	-4.497	0.05001
10	4.668	-4.713	0.04546
11	4.878	-4.920	0.04167
12	5.080	-5.118	0.03846
13	5.274	-5.309	0.03572
14	5.460	-5.494	0.03333
15	5.641	-5.672	0.03125
16	5.816	-5.846	0.02941
17	5.986	-6.014	0.02778
18	6.151	-6.178	0.02632
19	6.312	-6.337	0.02500
20	6.469	-6.493	0.02381

Let $b = 1, 10$, then the roots behave as shown in two tables below where

$$b = 1, \qquad x_1(n) = ..., \qquad x_2(n) = ..., \quad x_3(n) = ...$$

$$b = 10, \qquad x_1(n) = ..., \qquad x_2(n) = ..., \quad x_3(n) = ...$$

13.5 The case II, free particle

Let us turn to the case II for a free particle, the system (13.9) in explicit form reads

$$\left[\frac{3}{2}\left(\frac{d^2}{dr^2} + \frac{1}{r}\frac{d}{dr} - \frac{(m+1)^2}{r^2}\right) + \frac{\sigma}{(\sigma+2k)}\left(\frac{d^2}{dr^2} + \frac{1}{r}\frac{d}{dr} - \frac{(m+1)^2}{r^2}\right) + \sigma(\sigma+k)\right]c_1$$
$$+\frac{3}{2}\left(\frac{d^2}{dr^2} + \frac{(1-2m)}{r}\frac{d}{dr} + \frac{(m^2-1)}{r^2}\right)c_3 = 0,$$
$$\left[\frac{3}{2}\left(\frac{d^2}{dr^2} + \frac{1}{r}\frac{d}{dr} - \frac{(m-1)^2}{r^2}\right) + \frac{\sigma}{(\sigma-2k)}\left(\frac{d^2}{dr^2} + \frac{1}{r}\frac{d}{dr} - \frac{(m-1)^2}{r^2}\right) + \sigma(\sigma-k)\right]c_3$$
$$+\frac{3}{2}\left(\frac{d^2}{dr^2} + \frac{(1+2m)}{r}\frac{d}{dr} + \frac{(m^2-1)}{r^2}\right)c_1 = 0.$$

TABLE 13.2
The values of x_i $(i = 1, 2, 3)$ and $b = 10$

n	$x_1(n)$	$x_2(n)$	$x_3(n)$
1	3.048	-3.230	0.1827
2	3.935	-4.044	0.1097
3	4.648	-4.726	0.07838
4	5.263	-5.324	0.06099
5	5.812	-5.862	0.04992
6	6.313	-6.355	0.04225
7	6.777	-6.813	0.03662
8	7.210	-7.243	0.03232
9	7.619	-7.648	0.02892
10	8.007	-8.033	0.02617
11	8.377	-8.401	0.02389
12	8.731	-8.753	0.02198
13	9.071	-9.092	0.02036
14	9.399	-9.418	0.01895
15	9.716	-9.734	0.01773
16	10.02	-10.04	0.01666
17	10.32	-10.34	0.01571
18	10.61	-10.62	0.01486
19	10.89	-10.91	0.01410
20	11.17	-11.18	0.01341

These equations can be transformed to other form

$$\Big[(3 + \frac{2\sigma}{\sigma + 2k})\Big(\frac{d^2}{dr^2} + \frac{1}{r}\frac{d}{dr} - \frac{(m+1)^2}{r^2}\Big) + 2\sigma(\sigma + k)\Big] c_1$$
$$+3\Big(\frac{d^2}{dr^2} + \frac{(1-2m)}{r}\frac{d}{dr} + \frac{(m^2-1)}{r^2}\Big) c_3 = 0, \tag{13.28}$$

$$\Big[(3 + \frac{2\sigma}{\sigma - 2k})\Big(\frac{d^2}{dr^2} + \frac{1}{r}\frac{d}{dr} - \frac{(m-1)^2}{r^2}\Big) + 2\sigma(\sigma - k)\Big] c_3$$
$$+3\Big(\frac{d^2}{dr^2} + \frac{(1+2m)}{r}\frac{d}{dr} + \frac{(m^2-1)}{r^2}\Big) c_1 = 0. \tag{13.29}$$

This system may be considered as linear with respect to the variables c_1'', c_3'':

$$\begin{vmatrix} 3 + \frac{2\sigma}{2k+\sigma} & 3 \\ 3 & 3 - \frac{2\sigma}{2k-\sigma} \end{vmatrix} \begin{vmatrix} c_1'' \\ c_3'' \end{vmatrix}$$

$$= - \begin{vmatrix} \Big[(3 + \frac{2\sigma}{\sigma+2k})(\frac{1}{r}\frac{d}{dr} - \frac{(m+1)^2}{r^2}) + 2\sigma(k+\sigma)\Big] c_1 + 3\Big(\frac{(1-2m)}{r}\frac{d}{dr} + \frac{(m^2-1)}{r^2}\Big) c_3 \\ 3\Big(\frac{(1+2m)}{r}\frac{d}{dr} + \frac{(m^2-1)}{r^2}\Big) c_1 + \Big[(3 + \frac{2\sigma}{\sigma-2k})\Big(\frac{1}{r}\frac{d}{dr} - \frac{(-m+1)^2}{r^2}\Big) - 2\sigma(k-\sigma)\Big] c_3 \end{vmatrix},$$

evidently it is symmetric under the change

$$c_3 \Leftrightarrow c_1, \quad m \Leftrightarrow -m, \quad \sigma \Leftrightarrow -\sigma. \tag{13.30}$$

Its solution has the form

$$c_1'' = \frac{1}{8}\frac{(9m-8)\sigma^2 - 36\,k^2 m}{\sigma^2 r}\, c_1'$$

$$
\begin{aligned}
&+\Big[\frac{1}{8}\,\frac{8\,\sigma\,k^2+12\,k^3-9\,\sigma^2 k-5\,\sigma^3}{\sigma}-\frac{1}{8}\,\frac{(m+1)(-17\,\sigma^2 m-8\,\sigma^2+36\,k^2 m)}{\sigma^2 r^2}\Big]c_1\\
&\quad-\frac{3}{8}\,\frac{m(-4\,\sigma\,k+12\,k^2-5\,\sigma^2)}{\sigma^2 r}\,c_3'+\Big[\frac{3}{8}\,\frac{(2\,k^2+\sigma^2-3\,\sigma\,k)(\sigma+2\,k)}{\sigma}\\
&\quad+\frac{3}{8}\,\frac{m(-4\,\sigma\,mk+12\,k^2 m+5\,\sigma^2+4\,\sigma\,k-12\,k^2-5\,\sigma^2 m)}{\sigma^2 r^2}\Big]c_3\,,
\end{aligned}
$$

$$
\begin{aligned}
c_3''&=\frac{3}{8}\,\frac{m(4\,\sigma\,k+12\,k^2-5\,\sigma^2)}{\sigma^2 r}\,c_1'\\
&+\Big[-\frac{3}{8}\,\frac{(\sigma+k)(-\sigma^2+4\,k^2)}{\sigma}+\frac{3}{8}\,\frac{m(6\,k+5\,\sigma)(-\sigma+2\,k)(m+1)}{\sigma^2 r^2}\Big]c_1\\
&\quad+\frac{1}{8}\,\frac{36\,k^2 m-9\,\sigma^2 m-8\,\sigma^2}{\sigma^2 r}\,c_3'\\
&+\Big[-\frac{1}{8}\,\frac{-9\,\sigma^2 k+5\,\sigma^3-8\,\sigma\,k^2+12\,k^3}{\sigma}-\frac{1}{8}\,\frac{(-17\,\sigma^2 m+8\,\sigma^2+36\,k^2 m)(m-1)}{\sigma^2 r^2}\Big]c_3\,;
\end{aligned}
$$

the above symmetry may be noted in these formulas as well. Shortly we can write

$$
\begin{aligned}
\frac{d^2}{dr^2}c_1&=K_1\frac{d}{dr}c_1+(\frac{L_1}{r^2}+M_1)c_1+(\frac{F_1}{r}\frac{d}{dr}+\frac{G_1}{r^2}+H_1)c_3,\\
\frac{d^2}{dr^2}c_3&=K_3\frac{d}{dr}c_3+(\frac{L_3}{r^2}+M_3)c_3+(\frac{F_3}{r}\frac{d}{dr}+\frac{G_3}{r^2}+H_3)c_1.
\end{aligned}
\tag{13.31}
$$

Let us eliminate from this system the variable c_3. To this end, first we present this function as $c_3=\varphi\bar{c_3}$, and require that

$$
(\frac{F_1}{r}\frac{d}{dr}+\frac{G_1}{r^2}+H_1)\varphi\bar{c_3}=\varphi\,\frac{F_1}{r}\frac{d}{dr}\,\bar{c_3}.
$$

Whence it follows

$$
\frac{F_1}{r}\frac{\varphi'}{\varphi}\bar{c_3}+\frac{F_1}{r}\bar{c_3'}+\frac{G_1}{r^2}c_3+H_1\bar{c_3}=\frac{F_1}{r}\frac{d}{dr}\,\bar{c_3},
$$

or differently

$$
\frac{F_1}{r}\bar{c_3'}+\Big(\frac{F_1}{r}\frac{\varphi'}{\varphi}+\frac{G_1}{r^2}+H_1\Big)\bar{c_3}=\frac{F_1}{r}\frac{d}{dr}\,\bar{c_3}.
$$

Further, we derive equation for determining the function φ:

$$
\frac{F_1}{r}\frac{\varphi'}{\varphi}=-\frac{G_1}{r^2}-H_1\quad\Longrightarrow\quad\frac{d}{dr}\ln\varphi=-\frac{G_1}{F_1}\frac{1}{r}-\frac{H_1}{F_1}r,
$$

so we obtain

$$
\ln\varphi=-\frac{G_1}{F_1}\ln r-\frac{H_1}{2F_1}r^2\quad\Longrightarrow\quad\ln\Big(\varphi r^{G_1/F_1}\Big)=-\frac{H_1}{2F_1}r^2,
$$

whence it follows

$$
\varphi r^{G_1/F_1}=e^{-\frac{H_1}{2F_1}r^2}\quad\Longrightarrow\quad\varphi=r^{\;G_1/F_1}e^{-\frac{H_1}{2F_1}r^2}.
\tag{13.32}
$$

Therefore, the initial system reads

$$
\begin{aligned}
\frac{d^2}{dr^2}c_1&=K_1\frac{d}{dr}c_1+(\frac{L_1}{r^2}+M_1)c_1+\varphi\,\frac{F_1}{r}\frac{d}{dr}\,\bar{c_3},\quad c_3=\varphi\bar{c_3},\\
\frac{d^2}{dr^2}\varphi\bar{c_3}&=K_3\frac{d}{dr}\varphi\bar{c_3}+(\frac{L_3}{r^2}+M_3)\varphi\bar{c_3}+(\frac{F_3}{r}\frac{d}{dr}+\frac{G_3}{r^2}+H_3)c_1.
\end{aligned}
\tag{13.33}
$$

From eq. (13.33), we express the derivative $\frac{d}{dr}\bar{c}_3$:

$$\frac{d}{dr}\bar{c}_3 = \frac{1}{F_1}\frac{r}{\varphi}\Big(\frac{d^2}{dr^2}c_1 - K_1\frac{d}{dr}c_1 - (\frac{L_1}{r^2} + M_1)c_1\Big),$$

below we will need the second derivative as well

$$\frac{d^2}{dr^2}\bar{c}_3 = \frac{d}{dr}\Big\{\frac{1}{F_1}\frac{r}{\varphi}\Big(\frac{d^2}{dr^2}c_1 - K_1\frac{d}{dr}c_1 - (\frac{L_1}{r^2} + M_1)c_1\Big)\Big\}.$$

Expression for function c_3 itself is found by integrating

$$\bar{c}_3 = \int \frac{1}{F_1}\frac{r}{\varphi}\Big(\frac{d^2}{dr^2}c_1 - K_1\frac{d}{dr}c_1 - (\frac{L_1}{r^2} + M_1)c_1\Big)dr. \qquad (13.34)$$

Now we turn to the second equation from eq. (13.33)

$$\frac{d^2}{dr^2}\varphi\bar{c}_3 = K_3\frac{d}{dr}\varphi\bar{c}_3 + (\frac{L_3}{r^2} + M_3)\varphi\bar{c}_3 + (\frac{F_3}{r}\frac{d}{dr} + \frac{G_3}{r^2} + H_3)c_1;$$

taking into account two identities

$$\frac{d^2}{dr^2}\varphi\bar{c}_3 = \frac{d}{dr}\Big[\varphi'\bar{c}_3 + \varphi\bar{c}_3'\Big] = \varphi''\bar{c}_3 + 2\varphi'\bar{c}_3' + \varphi\bar{c}_3'', \quad K_3\frac{d}{dr}\varphi\bar{c}_3 = K_3\Big[\varphi'\bar{c}_3 + \varphi\bar{c}_3'\Big],$$

we transform the above equation to other form

$$\varphi''\bar{c}_3 + 2\varphi'\bar{c}_3' + \varphi\bar{c}_3'' = K_3\varphi'\bar{c}_3 + K_3\varphi\bar{c}_3' + (\frac{L_3}{r^2} + M_3)\varphi\bar{c}_3 + \frac{F_3}{r}c_1' + \frac{G_3}{r^2}c_1 + H_3c_1.$$

Whence after re-grouping the terms we obtain

$$\varphi\bar{c}_3'' + (2\varphi' - K_3\varphi)\bar{c}_3' + \Big(\varphi'' - K_3\varphi' - (\frac{L_3}{r^2} + M_3)\varphi\Big)\bar{c}_3 - \frac{F_3}{r}c_1' - \frac{G_3}{r^2}c_1 - H_3c_1 = 0,$$

which is equivalent to

$$\Big[\bar{c}_3'' + (2\frac{\varphi'}{\varphi} - K_3)\bar{c}_3'\Big] + \Big(\frac{\varphi''}{\varphi} - K_3\frac{\varphi'}{\varphi} - (\frac{L_3}{r^2} + M_3)\Big)\bar{c}_3 - \Big[\frac{1}{\varphi}\frac{F_3}{r}c_1' + \frac{1}{\varphi}\frac{G_3}{r^2}c_1 + \frac{1}{\varphi}H_3c_1\Big] = 0.$$

With the help of the notation

$$\Delta(r) = \Big(\frac{\varphi''}{\varphi} - K_3\frac{\varphi'}{\varphi} - (\frac{L_3}{r^2} + M_3)\Big),$$

we re-write the last equation differently

$$\frac{1}{\Delta(r)}\Big[\bar{c}_3'' + (2\frac{\varphi'}{\varphi} - K_3)\bar{c}_3'\Big] + \bar{c}_3 - \frac{1}{\Delta(r)}\Big[\frac{1}{\varphi}\frac{F_3}{r}c_1' + \frac{1}{\varphi}\frac{G_3}{r^2}c_1 + \frac{1}{\varphi}H_3c_1\Big] = 0.$$

After differentiating this equation we obtain

$$\frac{d}{dr}\Big\{\frac{1}{\Delta(r)}\Big[\bar{c}_3'' + (2\frac{\varphi'}{\varphi} - K_3)\bar{c}_3'\Big]\Big\} + \bar{c}_3'$$
$$-\frac{d}{dr}\Big\{\frac{1}{\Delta(r)}\Big[\frac{1}{\varphi}\frac{F_3}{r}c_1' + \frac{1}{\varphi}\frac{G_3}{r^2}c_1 + \frac{1}{\varphi}H_3c_1\Big]\Big\} = 0, \qquad (13.35)$$

this is the 4th-order equation for function $c_1(r)$. Recall that

$$\bar{c}_3' = \frac{1}{F_1}\frac{r}{\varphi}\Big(\frac{d^2}{dr^2}c_1 - K_1\frac{d}{dr}c_1 - (\frac{L_1}{r^2} + M_1)c_1\Big),$$
$$\bar{c}_3'' = \frac{d}{dr}\Big\{\frac{1}{F_1}\frac{r}{\varphi}\Big(\frac{d^2}{dr^2}c_1 - K_1\frac{d}{dr}c_1 - (\frac{L_1}{r^2} + M_1)c_1\Big)\Big\}. \qquad (13.36)$$

In similar manner we can derive a 4th-order equation for function c_3.

Equations of the 4th-order for two (initial) functions $c_1(r)$, $c_3(r)$ in explicit form read

$$\frac{d^4 c_1}{dr^4} + \frac{2}{r}\frac{d^3 c_1}{dr^3} + \Big(\frac{5}{4}\sigma^2 - 2k^2 + \frac{-3-4m-2m^2}{r^2}\Big)\frac{d^2 c_1}{dr^2}$$
$$+\Big(\frac{1}{4}\frac{5\sigma^2 - 8k^2}{r} + \frac{3+4m+2m^2}{r^3}\Big)\frac{dc_1}{dr}$$
$$+\Big[-\frac{5}{4}\sigma^2 k^2 + \frac{1}{4}\sigma^4 + k^4 + \frac{1}{4}\frac{(-5\sigma^2 + 8k^2)(m+1)^2}{r^2}$$
$$+\frac{(m+3)(m-1)(m+1)^2}{r^4}\Big]c_1 = 0,$$

$$\frac{d^4 c_3}{dr^4} + \frac{2}{r}\frac{d^3 c_3}{dr^3} + \Big(\frac{5}{4}\sigma^2 - 2k^2 + \frac{-3+4m-2m^2}{r^2}\Big)\frac{d^2 c_3}{dr^2}$$
$$+\Big(\frac{1}{4}\frac{5\sigma^2 - 8k^2}{r} + \frac{3-4m+2m^2}{r^3}\Big)\frac{dc_3}{dr}$$
$$+\Big[-\frac{5}{4}\sigma^2 k^2 + \frac{1}{4}\sigma^4 + k^4 + \frac{1}{4}\frac{(-5\sigma^2 + 8k^2)(m-1)^2}{r^2}$$
$$+\frac{(m-3)(m+1)(m-1)^2}{r^4}\Big]c_3 = 0.$$

Both equations may be factorised:

$$\Big\{\frac{d^2}{dr^2} + \frac{1}{r}\frac{d}{dr} + \Big[-k^2 + \frac{1}{4}\sigma^2 - \frac{(m+1)^2}{r^2}\Big]\Big\}\Big\{\frac{d^2}{dr^2} + \frac{1}{r}\frac{d}{dr} + \Big[-k^2 + \sigma^2 - \frac{(m+1)^2}{r^2}\Big]\Big\}c_1 = 0,$$
$$\Big\{\frac{d^2}{dr^2} + \frac{1}{r}\frac{d}{dr} + \Big[-k^2 + \frac{1}{4}\sigma^2 - \frac{(m-1)^2}{r^2}\Big]\Big\}\Big\{\frac{d^2}{dr^2} + \frac{1}{r}\frac{d}{dr} + \Big[-k^2 + \sigma^2 - \frac{(m-1)^2}{r^2}\Big]\Big\}c_3 = 0,$$

where two multipliers are permutable.

It suffices to solve two 2nd-order equations for c_1:

$$I, \quad \Big[\frac{d^2}{dr^2} + \frac{1}{r}\frac{d}{dr} - k^2 + \sigma^2 - \frac{(m+1)^2}{r^2}\Big]c_1 = 0;$$
$$II, \quad \Big[\frac{d^2}{dr^2} + \frac{1}{r}\frac{d}{dr} - k^2 + \frac{1}{4}\sigma^2 - \frac{(m+1)^2}{r^2}\Big]c_1 = 0; \qquad (13.37)$$

they reduce to Bessel equations

$$I, \quad x = i\sqrt{k^2 - \sigma^2}, \quad \Big[\frac{d^2}{dx^2} + \frac{1}{x}\frac{d}{dx} + 1 - \frac{(m+1)^2}{x^2}\Big]c_1^I(x) = 0;$$
$$II, \quad y = i\sqrt{k^2 - \sigma^2/4}, \quad \Big[\frac{d^2}{dy^2} + \frac{1}{y}\frac{d}{dy} + 1 - \frac{(m+1)^2}{y^2}\Big]c_1^{II}(y) = 0 \qquad (13.38)$$

Similarly, for function c_3 we get

$$I', \quad x = i\sqrt{k^2 - \sigma^2}, \quad \Big[\frac{d^2}{dx^2} + \frac{1}{x}\frac{d}{dx} + 1 - \frac{(m-1)^2}{x^2}\Big]c_3^I(x) = 0;$$
$$II', \quad y = i\sqrt{k^2 - \sigma^2/4}, \quad \Big[\frac{d^2}{dy^2} + \frac{1}{y}\frac{d}{dy} + 1 - \frac{(m-1)^2}{y^2}\Big]c_3^{II}(y) = 0. \qquad (13.39)$$

13.6 The case II, the presence of magnetic field

Here, we have the following system of equations for the variables c_1, c_3 (for brevity we change the notation $eB \Longrightarrow B$):

$$\Big\{\frac{3}{2}\Big[\frac{d^2}{dr^2}+\frac{1}{r}\frac{d}{dr}-\frac{(m+Br^2/2+1)^2}{r^2}+B\Big]$$

$$+\frac{\sigma}{(\sigma+2k)}\Big[\frac{d^2}{dr^2}+\frac{1}{r}\frac{d}{dr}-\frac{(m+Br^2/2+1)^2}{r^2}-B\Big]+\sigma(\sigma+k)\Big\}c_1$$

$$+\frac{3}{2}\Big\{\frac{d^2}{dr^2}+\frac{1-2(m+Br^2/2)}{r}\frac{d}{dr}+\frac{(m+Br^2/2)^2-1}{r^2}-B\Big\}c_3=0,$$

$$\Big\{\frac{3}{2}\Big[\frac{d^2}{dr^2}+\frac{1}{r}\frac{d}{dr}-\frac{(m+Br^2/2-1)^2}{r^2}-B\Big]$$

$$+\frac{\sigma}{(\sigma-2k)}\Big[\frac{d^2}{dr^2}+\frac{1}{r}\frac{d}{dr}-\frac{(m+Br^2/2-1)^2}{r^2}+B\Big]+\sigma(\sigma-k)\Big\}c_3$$

$$+\frac{3}{2}\Big\{\frac{d^2}{dr^2}+\frac{1+2(m+Br^2/2)}{r}\frac{d}{dr}+\frac{(m+Br^2/2)^2-1}{r^2}+B\Big\}c_1=0.$$

They can be transformed to other form

$$\Big[\Big(3+\frac{2\sigma}{\sigma+2k}\Big)\Big(\frac{d^2}{dr^2}+\frac{1}{r}\frac{d}{dr}-\frac{(m+Br^2/2+1)^2}{r^2}\Big)+\Big(3-\frac{2\sigma}{\sigma+2k}\Big)B+2\sigma(\sigma+k)\Big]c_1$$

$$+3\Big[\frac{d^2}{dr^2}+\frac{1-2(m+Br^2/2)}{r}\frac{d}{dr}+\frac{(m+Br^2/2)^2-1}{r^2}-B\Big]c_3=0,$$

$$\Big[\Big(3+\frac{2\sigma}{\sigma-2k}\Big)\Big(\frac{d^2}{dr^2}+\frac{1}{r}\frac{d}{dr}-\frac{(m+Br^2/2-1)^2}{r^2}\Big)+\Big(-3+\frac{2\sigma}{\sigma-2k}\Big)B+2\sigma(\sigma-k)\Big]c_3$$

$$+3\Big[\frac{d^2}{dr^2}+\frac{1+2(m+Br^2/2)}{r}\frac{d}{dr}+\frac{(m+Br^2/2)^2-1}{r^2}+B\Big]c_1=0.$$

Whence it follows 4th-order equations for c_1 and c_3:

$$c_1''''+\frac{2}{r}c_1'''+\left(-\frac{e^2B^2r^2}{2}+\frac{5\sigma^3-8\sigma k^2-8\sigma mB+8\sigma B+12kB}{4\sigma}-\frac{2m^2+4m+3}{r^2}\right)c_1''$$

$$+\left(-\frac{3}{2}B^2r+\frac{1}{4}\frac{5\sigma^3-8\sigma k^2-8\sigma mB+8\sigma B+12kB}{r\sigma}+\frac{2m^2+4m+3}{r^3}\right)c_1'$$

$$+\Bigg[\frac{1}{16}B^4r^4-\frac{1}{16}\frac{B^2\left(5\sigma^3-8\sigma k^2-8\sigma mB+8\sigma B+12kB\right)r^2}{\sigma}$$

$$-\frac{1}{4}\frac{(m+1)^2\left(5\sigma^3-8\sigma k^2-8\sigma mB+8\sigma B+12kB\right)}{r^2\sigma}+\frac{(m+3)(m-1)(m+1)^2}{r^4}$$

$$-\frac{1}{4\sigma}\left(-\sigma^5+5\sigma^3Bm-5\sigma^3B+5\sigma^3k^2-21\sigma^2kB-6\sigma m^2B^2\right.$$

$$+4\,\sigma B^2m+16B^2\sigma-8\sigma k^2Bm+8\sigma k^2B$$

$$\left.-4\sigma k^4+12kB^2m-12kB^2+12k^3B\right)\Bigg]c_1=0;$$

$$c_3'''' + \frac{2}{r}c_3''' + \left(-\frac{1}{2}B^2r^2 + \frac{1}{4}\frac{5\sigma^3 - 8\sigma k^2 - 8\sigma mB - 8\sigma B + 12kB}{\sigma} - \frac{2\,m^2 - 4\,m + 3}{r^2}\right)c_3''$$
$$+\left(-\frac{3}{2}B^2r + \frac{1}{4}\frac{5\sigma^3 - 8\sigma k^2 - 8\sigma mB - 8\sigma B + 12kB}{r\sigma} + \frac{2m^2 - 4m + 3}{r^3}\right)c_3'$$
$$+\left[\frac{1}{16}B^4r^4 - \frac{1}{16}\frac{B^2\left(5\sigma^3 - 8\sigma k^2 - 8\sigma mB - 8\sigma B + 12kB\right)r^2}{\sigma}\right.$$
$$-\frac{1}{4}\frac{(m-1)^2\left(5\sigma^3 - 8\sigma k^2 - 8\sigma mB - 8\sigma B + 12kB\right)}{r^2\sigma} + \frac{(m-3)(m+1)(m-1)^2}{r^4}$$
$$-\frac{1}{4\sigma}\left(-\sigma^5 + 5\sigma^3Bm + 5\sigma^3B + 5\sigma^3k^2 - 21\sigma^2kB - 6\sigma m^2B^2\right.$$
$$-4\sigma B^2m + 16B^2\sigma - 8\sigma k^2Bm - 8\sigma k^2B - 4\sigma k^4$$
$$\left.\left.+12kB^2 + 12kB^2m + 12k^3B\right)\right]c_3 = 0.$$

Both equations can be factorised. For the variable c_1 we get

$$\left\{\frac{d^2}{dr^2} + \frac{1}{r}\frac{d}{dr} - \frac{1}{4}B^2r^2 - \frac{(m+1)^2}{r^2}\right.$$
$$\left.+\frac{\sigma\left(5\sigma^2 - 8k^2\right) - 8B\sigma(m-1) + 12kB \pm 3\sqrt{\sigma^6 - 24kB\sigma^3 + 16\left(2\sigma^2 + k^2\right)B^2}}{8\sigma}\right\}$$
$$\times\left\{\frac{d^2}{dr^2} + \frac{1}{r}\frac{d}{dr} - \frac{1}{4}B^2r^2 - \frac{(m+1)^2}{r^2}\right.$$
$$\left.+\frac{\sigma\left(5\sigma^2 - 8k^2\right) - 8B\sigma(m-1) + 12kB \mp 3\sqrt{\sigma^6 - 24kB\sigma^3 + 16\left(2\sigma^2 + k^2\right)B^2}}{8\sigma}\right\}c_1 = 0.$$

For the variable c_3, we have slightly different equation (note the change $m - 1 \Longrightarrow m + 1$)

$$\left\{\frac{d^2}{dr^2} + \frac{1}{r}\frac{d}{dr} - \frac{1}{4}B^2r^2 - \frac{(m-1)^2}{r^2}\right.$$
$$\left.+\frac{\sigma\left(5\sigma^2 - 8k^2\right) - 8B\sigma(m+1) + 12kB \pm 3\sqrt{\sigma^6 - 24kB\sigma^3 + 16\left(2\sigma^2 + k^2\right)B^2}}{8\sigma}\right\}$$
$$\times\left\{\frac{d^2}{dr^2} + \frac{1}{r}\frac{d}{dr} - \frac{1}{4}B^2r^2 - \frac{(m-1)^2}{r^2}\right.$$
$$\left.+\frac{\sigma\left(5\sigma^2 - 8k^2\right) - 8B\sigma(m+1) + 12kB \mp 3\sqrt{\sigma^6 - 24kB\sigma^3 + 16\left(2\sigma^2 + k^2\right)B^2}}{8\sigma}\right\}c_3 = 0.$$

It suffices to examine one equation

$$\left\{\frac{d^2}{dr^2} + \frac{1}{r}\frac{d}{dr} - \frac{1}{4}B^2r^2 - \frac{(m+1)^2}{r^2}\right.$$
$$\left.+\frac{\sigma\left(5\sigma^2 - 8k^2\right) - 8B\sigma(m-1) + 12kB \pm 3\sqrt{\sigma^6 - 24kB\sigma^3 + 16\left(2\sigma^2 + k^2\right)B^2}}{8\sigma}\right\}c_1 = 0.$$

In the variable $x = \frac{1}{2}Br^2$, the last equation reads

$$\frac{d^2c_1}{dx^2} + \frac{1}{x}\frac{dc_1}{dx} + \left[-\frac{1}{4} - \frac{1}{4}\frac{(m+1)^2}{x^2}\right.$$

$$+\frac{\sigma\left(5\sigma^{2}-8k^{2}\right)-8B\sigma\left(m-1\right)+12kB\pm3\sqrt{\sigma^{6}-24kB\sigma^{3}+16\left(2\sigma^{2}+k^{2}\right)B^{2}}}{16\sigma Bx}\Bigg]c_{1}=0.$$

With the use of substitution $c_1(x)=e^{Ax}x^{C}F(x)$, we get the following equation for $F(x)$:

$$x\frac{d^{2}F}{dx^{2}}+(2Ax+2C+1)\frac{dF}{dx}+\Big\{(A^{2}-\frac{1}{4})x+\frac{1}{4}\frac{4C^{2}-(m+1)^{2}}{x}$$

$$+\frac{1}{16}\frac{1}{\sigma B}\Big[16A\sigma(1+2C)B+\sigma(5\sigma^{2}-8k^{2})-8B\sigma\left(m-1\right)+12kB$$

$$\pm3\sqrt{\sigma^{6}-24kB\sigma^{3}+16\left(2\sigma^{2}+k^{2}\right)B^{2}}\Big]\Big\}F=0.$$

Imposing restrictions on parameters $A=-\frac{1}{2}, C=\pm|\frac{m+1}{2}|$, we simplify the above equation to the form

$$x\frac{d^{2}F}{dx^{2}}+(2C+1-x)\frac{dF}{dx}$$

$$+\frac{1}{16}\frac{1}{\sigma B}\Big[-8\sigma(1+2C)B+\sigma(5\sigma^{2}-8k^{2})-8B\sigma\left(m-1\right)+12kB$$

$$\pm3\sqrt{\sigma^{6}-24kB\sigma^{3}+16\left(2\sigma^{2}+k^{2}\right)B^{2}}\Big]F=0.$$

It is identified with the confluent hypergeometric equation with parameters

$$\alpha_{\pm}=-\frac{1}{16}\frac{1}{\sigma eB}\Big[-8\sigma(1+2C)B+\sigma(5\sigma^{2}-8k^{2})-8B\sigma(m-1)+12kB$$

$$\pm3\sqrt{\sigma^{6}-24kB\sigma^{3}+16(2\sigma^{2}+k^{2})B^{2}}\Big],\qquad\gamma=2C+1.$$

The known polynomial condition ($\alpha_{\pm}=-n,\ n=0,1,2,\ ...$) gives the quantization rule

$$\frac{1}{16}\frac{1}{\sigma B}\Big[-8\sigma B(1+2C)-8B\sigma(m-1)+\sigma(5\sigma^{2}-8k^{2})+12kB$$

$$\pm3\sqrt{\sigma^{6}-24kB\sigma^{3}+16\left(2\sigma^{2}+k^{2}\right)B^{2}}\Big]=n,\quad n=0,1,2,...$$

or differently (let $C=+|\frac{m+1}{2}|$)

$$\sigma(5\sigma^{2}-8k^{2})+12kB$$

$$\pm3\sqrt{\sigma^{6}-24kB\sigma^{3}+16(2\sigma^{2}+k^{2})B^{2}}=8\sigma B(2n+m+|m+1|). \tag{13.40}$$

Let us change the notation $2n+m+|m+1|=N$, then we have

$$\pm3\sqrt{\sigma^{6}-24kB\sigma^{3}+16(2\sigma^{2}+k^{2})B^{2}}=8\sigma BN-\sigma\left(5\sigma^{2}-8k^{2}\right)-12kB.$$

After squaring the above equation, we obtain

$$\sigma\Big\{\sigma^{5}-5\left(BN+k^{2}\right)\sigma^{3}+21kB\sigma^{2}$$

$$+\Big[4\left(BN+k^{2}\right)^{2}-18B^{2}\Big]\sigma-12kB\left(BN+k^{2}\right)\Big\}=0; \tag{13.41}$$

the root $\sigma=0$ is nonphysical. So we get 5th-order equation

$$BN+k^{2}=\gamma,\quad\sigma^{5}-5\gamma\sigma^{3}+21kB\sigma^{2}+\left[4\gamma^{2}-18B^{2}\right]\sigma-12kB\gamma=0. \tag{13.42}$$

In dimensionless variables

$$x = \frac{\sigma}{k}, \quad b = \frac{B}{k^2}, \quad \Gamma = \frac{\gamma}{\sigma^2} = bN + 1,$$

we have the equation

$$x^5 - 5\Gamma x^3 + 21bx^2 + (4\Gamma^2 - 18b^2)x - 12b\Gamma = 0. \tag{13.43}$$

Thus we have constructed the following solutions

$$c_1(x) = e^{-x/2} x^C G(\alpha_\pm = -n, \gamma, x), \quad C = \pm\frac{|m+1|}{2}, x = \frac{1}{2}Br^2. \tag{13.44}$$

Having known the functions $c_1(x)$, one can find an explicit form of the concomitant function $c_3(x)$ (we should use a formula of type (13.34)). When solving the 2nd-order equation for function c_3, we obtain a similar result with one formal change $m+1 \Longrightarrow m-1$. Therefore, in this case, we have constructed the following solutions

$$\begin{gathered} c_3(x) = e^{-x/2} x^{C'} G(x), \quad C' = \pm\frac{|m-1|}{2}, \quad m' = m-2, \\ 2n + m' + |m'+1| = N', \quad bN' + 1 = \Gamma', \\ X^5 - 5\Gamma' X^3 + 21bX^2 + (4\Gamma'^2 - 18b^2)X - 12b\Gamma' = 0. \end{gathered} \tag{13.45}$$

We can see that solutions (13.45) and (13.44) provide us with one the same spectrum for σ.

13.7 Numerical study of the 5th-order equation

We start with the equation

$$x^5 - 5\Gamma x^3 + 21bx^2 + (4\Gamma^2 - 18b^2)x - 12b\Gamma = 0, \tag{13.46}$$

where

$$\Gamma = \frac{\gamma}{o^2} = bN + 1, \ x = \frac{\sigma}{k}.$$

The five roots of eq. (13.46) at different values of b are given in the tables below

$b = 1/20$

N=	1.	$x=$	-2.17999	-0.962925	0.142914	1.1127	1.8873
N=	3.	$x=$	-2.26543	-1.01463	0.130574	1.14949	2.
N=	5.	$x=$	-2.34817	-1.06388	0.120171	1.18679	2.10508
N=	7.	$x=$	-2.42843	-1.11095	0.111288	1.22404	2.20406
N=	9.	$x=$	-2.50641	-1.15609	0.10362	1.26091	2.29798
N=	11.	$x=$	-2.58228	-1.19952	0.0969351	1.29726	2.38761
N=	13.	$x=$	-2.65619	-1.2414	0.0910573	1.333	2.47354
N=	15.	$x=$	-2.72827	-1.28189	0.0858495	1.36809	2.55622
N=	17.	$x=$	-2.79865	-1.3211	0.0812038	1.40254	2.63601
N=	19.	$x=$	-2.86743	-1.35916	0.0770342	1.43633	2.71322
N=	21.	$x=$	-2.93471	-1.39614	0.0732713	1.4695	2.78809
N=	23.	$x=$	-3.00059	-1.43215	0.0698584	1.50204	2.86084
N=	25.	$x=$	-3.06514	-1.46724	0.0667491	1.534	2.93164
N=	27.	$x=$	-3.12844	-1.50149	0.0639046	1.56538	3.00065
N=	29.	$x=$	-3.19055	-1.53496	0.0612925	1.59622	3.068
N=	31.	$x=$	-3.25154	-1.56769	0.0588855	1.62653	3.13381

$b = 1/10$

N=							
N=	1.	$x =$	−2.33231	−0.941056	0.273369	1.27639	1.72361
N=	3.	$x =$	−2.48561	−1.04282	0.232139	1.29629	2.
N=	5.	$x =$	−2.63167	−1.13638	0.201333	1.34651	2.22021
N=	7.	$x =$	−2.77131	−1.22319	0.17761	1.4046	2.4123
N=	9.	$x =$	−2.90523	−1.30439	0.158835	1.46478	2.586
N=	11.	$x =$	−3.034	−1.38085	0.143627	1.52497	2.74626
N=	13.	$x =$	−3.15814	−1.45327	0.131067	1.5843	2.89603
N=	15.	$x =$	−3.27806	−1.52218	0.120522	1.64242	3.0373
N=	17.	$x =$	−3.39415	−1.58804	0.111546	1.69918	3.17146
N=	19.	$x =$	−3.50674	−1.6512	0.103813	1.75457	3.29955
N=	21.	$x =$	−3.61609	−1.71196	0.0970826	1.8086	3.42237
N=	23.	$x =$	−3.72247	−1.77057	0.091172	1.86131	3.54056
N=	25.	$x =$	−3.82609	−1.82724	0.08594	1.91277	3.65461
N=	27.	$x =$	−3.92714	−1.88214	0.0812762	1.96304	3.76497
N=	29.	$x =$	−4.0258	−1.93544	0.0770928	2.01218	3.87197
N=	31.	$x =$	−4.12224	−1.98726	0.0733192	2.06026	3.97592

$b = 1$

N=	1.	$x =$	−3.88897	−0.874754	1.76372	$1.5 - 1.32288i$	$1.5 + 1.32288i$
N=	3.	$x =$	−4.68178	−1.57434	1.	2.	3.25612
N=	5.	$x =$	−5.39489	−2.12642	0.560557	2.52299	4.43776
N=	7.	$x =$	−6.04347	−2.57469	0.399957	2.90122	5.31699
N=	9.	$x =$	−6.63991	−2.95589	0.312755	3.22922	6.05382
N=	11.	$x =$	−7.19373	−3.29137	0.257403	3.52515	6.70255
N=	13.	$x =$	−7.7122	−3.59382	0.218965	3.79751	7.28954
N=	15.	$x =$	−8.20088	−3.87118	0.190647	4.05139	7.83003
N=	17.	$x =$	−8.66411	−4.12876	0.168885	4.29019	8.3338
N=	19.	$x =$	−9.10531	−4.37026	0.151622	4.51638	8.80757
N=	21.	$x =$	−9.52723	−4.59838	0.137585	4.73179	9.25624
N=	23.	$x =$	−9.93211	−4.81513	0.125943	4.93783	9.68347
N=	25.	$x =$	−10.3218	−5.02209	0.116128	5.13564	10.0921
N=	27.	$x =$	−10.6978	−5.22048	0.107739	5.32615	10.4844
N=	29.	$x =$	−11.0615	−5.41131	0.100486	5.5101	10.8623
N=	31.	$x =$	−11.414	−5.59539	0.0941509	5.68813	11.2271

$b = 5$

N=	1.	$x =$	−7.06827	−0.861037	4.92931	$1.5 - 3.1225i$	$1.5 + 3.1225i$
N=	3.	$x =$	−9.12978	−2.89737	2.	2.36969	7.65747
N=	5.	$x =$	−10.9342	−4.40461	0.68878	4.75043	9.89959
N=	7.	$x =$	−12.5362	−5.51294	0.45526	5.84134	11.7526
N=	9.	$x =$	−13.9825	−6.41459	0.344015	6.69765	13.3554
N=	11.	$x =$	−15.3071	−7.19146	0.27765	7.43516	14.7857
N=	13.	$x =$	−16.5344	−7.88402	0.233208	8.09639	16.0888
N=	15.	$x =$	−17.6821	−8.51511	0.201238	8.70266	17.2933
N=	17.	$x =$	−18.7633	−9.09894	0.177082	9.26654	18.4186
N=	19.	$x =$	−19.7881	−9.64497	0.158161	9.7963	19.4786
N=	21.	$x =$	−20.7641	−10.1599	0.142927	10.2977	20.4833
N=	23.	$x =$	−21.6976	−10.6486	0.130391	10.7751	21.4407
N=	25.	$x =$	−22.5936	−11.1147	0.119891	11.2316	22.3569
N=	27.	$x =$	−23.4562	−11.5614	0.110964	11.6699	23.2367
N=	29.	$x =$	−24.2888	−11.9907	0.103281	12.092	24.0842
N=	31.	$x =$	−25.0943	−12.4047	0.0965976	12.4997	24.9027

$b = 10$

N=	1.	$x =$	-9.49571	-0.859116	7.35482	$1.5 - 4.4441i$	$1.5 + 4.4441i$
N=	3.	$x =$	-12.5237	-3.93691	2.	3.41554	11.045
N=	5.	$x =$	-15.1536	-6.18624	0.709576	6.52082	14.1094
N=	7.	$x =$	-17.4718	-7.78036	0.463418	8.108	16.6808
N=	9.	$x =$	-19.5539	-9.06292	0.348415	9.34695	18.9215
N=	11.	$x =$	-21.454	-10.1636	0.280425	10.4085	20.9287
N=	13.	$x =$	-23.2101	-11.1433	0.235128	11.3568	22.7615
N=	15.	$x =$	-24.8492	-12.0355	0.202649	12.224	24.4581
N=	17.	$x =$	-26.3911	-12.8607	0.178165	13.0291	26.0446
N=	19.	$x =$	-27.8509	-13.6324	0.15902	13.7844	27.5399
N=	21.	$x =$	-29.2401	-14.3602	0.143625	14.4986	28.9581
N=	23.	$x =$	-30.5679	-15.0509	0.13097	15.1778	30.3099
N=	25.	$x =$	-31.8416	-15.7098	0.120378	15.8271	31.6039
N=	27.	$x =$	-33.0671	-16.3411	0.111381	16.45	32.8469
N=	29.	$x =$	-34.2497	-16.9481	0.103642	17.0497	34.0444
N=	31.	$x =$	-35.3933	-17.5334	0.0969125	17.6287	35.2011

$b = 20$

N=	1.	$x =$	-12.9446	-0.858136	10.8028	$1.5 - 6.30476i$	$1.5 + 6.30476i$
N=	3.	$x =$	-17.3456	-5.42331	2.	4.90515	15.8637
N=	5.	$x =$	-21.1458	-8.73507	0.720469	9.06352	20.0969
N=	7.	$x =$	-24.4773	-11.0107	0.467621	11.3378	23.6825
N=	9.	$x =$	-27.4579	-12.8278	0.350662	13.1123	26.8227
N=	11.	$x =$	-30.1708	-14.3837	0.281836	14.6292	29.6435
N=	13.	$x =$	-32.6736	-15.7676	0.236101	15.9816	32.2235
N=	15.	$x =$	-35.0066	-17.0276	0.203363	17.2166	34.6143
N=	17.	$x =$	-37.1992	-18.1931	0.178712	18.3619	36.8517
N=	19.	$x =$	-39.2734	-19.2831	0.159452	19.4354	38.9616
N=	21.	$x =$	-41.2461	-20.3111	0.143977	20.4498	40.9634
N=	23.	$x =$	-43.1307	-21.2869	0.131261	21.4141	42.8722
N=	25.	$x =$	-44.9378	-22.218	0.120624	22.3354	44.6997
N=	27.	$x =$	-46.6762	-23.1101	0.111591	23.2192	46.4555
N=	29.	$x =$	-48.353	-23.9678	0.103823	24.0696	48.1474
N=	31.	$x =$	-49.9743	-24.795	0.0970707	24.8904	49.7818

13.8 Conclusions

The explicit form of the helicity operator for a symmetric 2nd-rank tensor describing the spin 2 particle is specified in cylindrical coordinates. After separating the variables, the system of ten differential 1st-order equations is derived. It is split into two independent subsystems of four and six equations. The system of four equations is solved straightforwardly in terms of confluent hypergeometric functions, there are corresponding eigenvalues and eigenfunctions. A subsystem of six equations can be reduced to one ordinary differential equation of the 4th-order. Corresponding 4th-order operators are factorised into permutable 2nd-order operators, so the problem reduces to solving two differential equations of the 2nd-order. Their solutions are constructed in terms of Bessel functions. This analysis is extended to the presence of an external uniform magnetic field, when solutions are constructed in terms of confluent hypergeometric functions.

Bibliography

[1] W. Pauli and M. Fierz. Über relativistische Feldleichungen von Teilchen mit beliebigem Spin im elektromagnetishen Feld. *Helvetica Physica Acta*, 12: 297–300, 1939.

[2] M. Fierz and W. Pauli. On relativistic wave equations for particles of arbitrary spin in an electromagnetic field. *Proceedings of the Royal Society of London A*, 173: 211–232, 1939.

[3] L. De Broglie. Sur l'interprétation de certaines équations dans la thórie des particules de spin 2. *Comptes Rendus de l'Académie des Sciences Paris*, 212: 657–659, 1941.

[4] I.M. Gel'fand and A.M. Yaglom. General relativistically invariant equations and infinite-dimensional representations of the Lorentz group. *Journal of Experimental and Theoretical Physics*, 18(8): 703–733, 1948.

[5] E.S. Fradkin. To the theory of particles with higher spins. *Journal of Experimental and Theoretical Physics*, 20(1): 27–38, 1950.

[6] F.I. Fedorov. To the theory of particles with spin 2. *Proceedings of Belarus State University. Ser. Phys.-Math.*, 12: 156–173, 1951.

[7] T. Regge. On properties of the particle with spin 2. *Nuovo Cimento*, 5(2): 325–326, 1957.

[8] V. Ya. Fainberg. On the theory of interaction of particles with higher spins with electromagnetic and meson fields. *Proceedings of the Lebedev Physics Institute of the Academy of Sciences of the USSR*, 6: 269–332, 1955.

[9] H.A. Buchdahl. On the compatibility of relativistic wave equations for particles of higher spin in the presence of a gravitational field. *Nuovo Cimento*, 10: 96–103, 1958.

[10] K. Johnson and E.C.G. Sudarshan. Inconsistency of the local field theory of charged spin 3/2 particles. *Annals of Physics*, 13(1): 121–145, 1961.

[11] K. Johnson and E.C.G. Sudarshan. The impossibility of a consistent theory of a charged higher spin Fermi fields. *Annals of Physics*, 13(1): 126–145, 1961.

[12] H.A. Buchdahl. On the compatibility of relativistic wave equations in Riemann spaces. *Nuovo Cimento*, 25: 486–496, 1962.

[13] B.V. Krylov and F.I. Fedorov. Equations of the first order for graviton. *Doklady of the National Academy of Sciences of BSSR*, 11(8): 681–684, 1967.

[14] F.I. Fedorov. Equations of the first order for gravitational field. *Doklady of the Academy of Sciences of USSR*, 179(4): 802–805, 1968.

[15] A.A. Bogush, B.V. Krylov and F.I. Fedorov. On matrices of the equations for spin 2 particles. *Proceedings of the National Academy of Sciences of BSSR, Physics and Mathematics Series*, 1: 74–81, 1968.

[16] G. Velo and D. Zwanziger. Noncausality and other defects of interaction Lagrangians for particles with spin one and higher. *Physical Review Journals*, 188(5): 2218–2222, 1969.

[17] B.V. Krylov. On the systems of the first order for graviton. *Proceedings of the National Academy of Sciences of BSSR, Physics and Mathematics Series*, 6: 82–89, 1972.

[18] F.I. Fedorov and A.A. Kirilov. The first order equations for graviational field in vacuum. *Acta Physica Polonica B*, 7(3): 161–167, 1976.

[19] W. Cox, First-order formulation of massive spin-2 field theories. *Journal of Physics A*, 15: 253–268, 1982.

[20] V.V. Kisel. On relativistic wave equations for a spin 2 particle. *Proceedings of the National Academy of Sciences of BSSR, Physics and Mathematics Series*, 5: 94–99, 1986.

[21] R.K. Loide. On conformally covariant spin-3/2 and spin-2 equations. *Journal of Physics A*, 19(5): 827–829, 1986.

[22] A.A. Bogush, V.V. Kisel, N.G. Tokarevskaya and V.M. Red'kov. On equations for a spin 2 particle in external electromagnetic and gravitational fields. *Proceedings of the National Academy of Sciences of Belarus, Physics and Mathematics Series*, 1: 62–67, 2003.

[23] V.M. Red'kov, N.G. Tokarevskaya and V.V. Kisel. Graviton in a curved space-time background and gauge symmetry. *Nonlinear Phenom. Complex Syst.*, 6(3): 772–778, 2003.

[24] E.M. Ovsiyuk, V.V. Kisel and V.M. Red'kov. *Maxwell Electrodynamics and Boson Fields in Spaces of Constant Curvature*. Nova Science Publishers Inc., New York, 2014.

[25] V.V. Kisel, E.V. Ovsiyuk, O.V. Veko and V.M. Red'kov. Contribution of gauge degrees of freedom in the energy-momentum tensor of the massless spin 2 field. *Proceedings of the National Academy of Sciences of Belarus, Physics and Mathematics Series*, 2: 58–63, 2015.

[26] V.V. Kisel, E.V. Ovsiyuk, O.V. Veko and V.M. Red'kov. Nonrelartivistic approximation in the theory of a spin two particle. *Doklady of the National Academy of Sciences of Belarus*, 59(3): 21–27, 2015.

[27] V.V. Kisel, E.V. Ovsiyuk, A.V. Ivashkevich and V.M. Red'kov. Fradkin Equation for a Spin 3/2 Particle in Presence of External Electromagnetic and Gravitational Fields. *Ukraine Journal of Physics*, 64(12): 1112–1117, 2019.

[28] A. V. Ivashkevich, E. M. Ovsiyuk, V. V. Kisel and V. M. Red'kov. Spherical solutions of the wave equation for a spin 3/2 particle. *Doklady of the National Academy of Sciences of Belarus*, 63(3): 282–290, 2019.

[29] A.V. Ivashkevich. Solutions with spherical symmetry of the equation for a spin 3/2 Particle. In *Understanding Quaternions*, pages 67–104. Nova Science Publishers Inc., New York, 2020.

[30] A.V. Ivashkevich, E.M. Ovsiyuk, V.V. Kisel and V.M. Red'kov. Massless spin 3/2 field, spherical solutions, eliminating the gauge degrees of freedom. *Doklady of the National Academy of Sciences of Belarus*, 65(6): 668–679, 2021.

[31] A. Ivashkevich, A. Buryy, E. Ovsiyuk, V. Balan, V. Kisel and V. Red'kov. On the matrix equation for a spin 2 particle in pseudo-Riemannian space-time, tetrad method. *Proceedings of Balkan Society of Geometers*, 28: 1–23, 2021.

Index